Mathematical Modeling

Mathematical Modeling

Models, Analysis and Applications
Second Edition

Sandip Banerjee
Indian Institute of Technology Roorkee, India

CRC Press is an imprint of the
Taylor & Francis Group, an **informa** business
A CHAPMAN & HALL BOOK

MATLAB® is a trademark of The MathWorks, Inc. and is used with permission. The MathWorks does not warrant the accuracy of the text or exercises in this book. This book's use or discussion of MATLAB® software or related products does not constitute endorsement or sponsorship by The MathWorks of a particular pedagogical approach or particular use of the MATLAB® software.

Second edition published 2022
by CRC Press
6000 Broken Sound Parkway NW, Suite 300, Boca Raton, FL 33487-2742

and by CRC Press
2 Park Square, Milton Park, Abingdon, Oxon, OX14 4RN

First edition published by CRC Press 2014

CRC Press is an imprint of Taylor & Francis Group, LLC

Library of Congress Cataloging-in-Publication Data

Names: Banerjee, Sandip, 1969- author.
Title: Mathematical modeling : models, analysis and applications / Sandip Banerjee, Indian Institute of Technology Roorkee, India.
Description: Second edition. | Boca Raton : Chapman & Hall/CRC Press, 2022. | Includes bibliographical references and index.
Identifiers: LCCN 2021028880 (print) | LCCN 2021028881 (ebook) | ISBN 9781138495944 (hardback) | ISBN 9781032124353 (paperback) | ISBN 9781351022941 (ebook)
Subjects: LCSH: Mathematical models--Textbooks. | Mathematics--Textbooks.
Classification: LCC QA401 .B365 2022 (print) | LCC QA401 (ebook) | DDC 511.8--dc23
LC record available at https://lccn.loc.gov/2021028880
LC ebook record available at https://lccn.loc.gov/2021028881

ISBN: 9781138495944 (hbk)
ISBN: 9781032124353 (pbk)
ISBN: 9781351022941 (ebk)

DOI: 10.1201/9781351022941

Typeset in CMR 10 font
by KnowledgeWorks Global Ltd.

Access the Support Material: www.Routledge.com/9781138495944

Dedicated to my

MOTHER, SONALI BANERJEE

and to

OTSO OVASKAINEN

who taught me how to do research.

Contents

Foreword

Mathematical modeling is playing an increasingly important role across the breadth of the sciences. While most people are aware of the great impact mathematics has had in physics, it is only during the pandemic of the past 18 months that society at large has realized that mathematical modeling has broader application. This book is therefore very timely as it presents examples of mathematical modeling from areas as diverse as physics, biology, medicine, ecology, epidemiology, and even sociology.

After an in-depth introduction to the art of modeling – why we do it, how we do it, and how we interpret the results – the book goes on to describe a number of the important fundamental mathematical building blocks required in this endeavor, presenting discrete, continuous (ordinary and partial), delayed, and stochastic models. There is also a discussion on how to determine parameter values, a hot research topic at the moment. Each technical section is very well explained, with several practice worked examples to illustrate the typical behaviors exhibited by each model, before going into a more general study. There are many exercises (with solutions and hints at the end of the book) as well as research problems, and some Mathematica and MATLAB codes are also provided.

The extensive range of material covered in this book makes it a great resource for a very broad spectrum of readers, from those moving into the field to those who teach the subject. Some of the examples and methodologies would be accessible to high school students, while other parts of the book are aimed at undergraduate and beginning graduate students.

In summary, not only does this book present many of the techniques that are essential for an applied mathematician who wishes to become a modeler, it also illustrates how the abstract way of thinking, essential for mathematical modeling, unites many problems from across the sciences enabling us to transfer knowledge between seemingly totally unconnected application areas.

Philip K. Maini
Director, Wolfson Centre for Mathematical Biology,
Mathematical Institute, University of Oxford,
October 2021

Preface to Second Edition

Since **Mathematical Modeling: Models, Analysis and Applications** was first published in 2014, the popular reception of the book and the rapid expansion in the field of mathematical modeling calls for a new edition. I have also received several emails from readers commenting on the book, some with suggestions to enhance the already existing value of the book. I realize that there are areas in the book that could be further worked on and improved and hence decided to undertake the challenge of a new edition.

This book is aimed at newcomers who desires to learn mathematical modeling, especially students taking a first course in the subject. Beginning with a step-by-step guidance of model formulation, this book equips the reader with modeling with difference equations (discrete models), ODEs, PDEs, delay and stochastic differential equations (continuous models). This book provides interdisciplinary and integrative overview of mathematical modeling, making it a complete textbook for a wide audience. The book is ideal for an introductory course for undergraduate and graduate students, engineers, applied mathematicians and researchers working in various areas of natural and applied sciences.

Among the new material in existing chapters, **Introduction to Modeling** (Chapter 1) is thoroughly revised. It now offers step-by-step instructions on how to build mathematical models from scratch. Along with that, various mathematical functions and their appropriateness to fit the mathematical models are also highlighted. The renamed chapter **Discrete Models Using Difference Equations** (Chapter 2) contains several new sections, namely, cycles and their stability, bifurcations, and chaos, for a better understanding of richer dynamics of the models. **Continuous Models Using Ordinary Differential Equations** (Chapter 3) has also undergone revision with some interesting models like combat models, zombie models and love affairs models. Bifurcations (1D and 2D) and the chaotic behaviours of models are discussed with substantial examples. The chapter also highlights the technique of estimation of system parameters of the models from data, which is an integral part of mathematical modeling. **Spatial Models Using Partial Differential Equations** (Chapter 4) contains a new section, *Reaction Diffusion Systems*, where stability analysis and numerical simulation of such models are discussed in detail. Slight modifications have been done in

Modeling with Delay Differential Equations (Chapter 5) and **Modeling with Stochastic Differential Equations** (Chapter 6), where preliminaries in both chapters have been rewritten for a better understanding of the said topics. I have also included project problems at the end of each chapter. The Mathematica codes used to plot graphs and figures are also given in separate sections.

A book is always the product not only of its author, but also various people who have contributed toward it, namely, colleagues, research scholars, the work environment, the discussions, engagements, discourses that I have had during years of teaching and research, the many conferences, seminars, university lectures, summer and winter schools that I have attended, referee reports, e-mail correspondence that I have gone through and, most importantly, the patient and constant support of my family. While it would not be possible to name all those who have helped me on this journey, I take this opportunity to thank my colleagues, Prof. Premananda Bera, Prof. Maheshanand and Prof. Manil T. Mohan for fruitful discussions and clarifications of many of my doubts during my book writing process. Prof. Manil T. Mohan also helped me in working on the preliminaries of Chapter 6. My sincere thanks to my research scholars, Arpita Mondal and Rajat Kaushik, who helped me with typing the manuscript. Special mention is due to Dr. Nishi Pulugurtha, who helped me in the composition of the history of mathematical modeling. Finally, a word of thanks to my publisher for their active effort in publishing this book. Working with the CRC Press staff was once again a pleasure. I thank, in particular, Aastha Sharma and Shikha Garg.

Years of teaching the subject have honed my skills in the area of mathematical modeling; however, perfection is something that one ought to strive for. I am sure I have not yet achieved it, and hence would welcome any suggestion for improvements, comments, and events that have criticisms for later editions of the book (write directly to sandipbanerjea@gmail.com).

Sandip Banerjee
Professor
Department of Mathematics
Indian Institute of Technology Roorkee
Roorkee-247667, Uttarakhand, India

Additional material is available from the CRC website
http://www.Routledge.com/9781138495944

Author

Sandip Banerjee is a Professor in the Department of Mathematics, Indian Institute of Technology Roorkee (IITR), India. His area of research is Mathematical Biology. Mathematical modeling is his passion. Prof. Banerjee was the recipient of the Indo-US Fellowship in 2009 and was awarded the IUSSTF Research Fellow medal by the Indo-US Technology Forum. In addition to several national and international projects, Prof. Banerjee is involved in the Virtual Network in Mathematical Biology project, which promotes Mathematical Biology in India. He has also developed several courses like Differential Equations and Numerical Analysis for e-Pathshala and the National Programme on Technology Enhanced Learning (NPTEL) projects, initiated by the Ministry of Human Resource Development (MHDR) India.

Chapter 1

About Mathematical Modeling

1.1 What Is Mathematical Modeling?

Mathematical modeling is the application of mathematics to describe real-world problems and investigate important questions that arise from them. Using mathematical tools, the real-world problem is translated to a mathematical problem that mimics the real-world problem. A solution to the mathematical problem is obtained, which is interpreted in the language of real-world problems to make predictions about the real world.

By real-world problems, I mean problems from biology, chemistry, ecology, environment, engineering, physics, social sciences, statistics, wildlife management, and so on. Mathematical modeling can be described as an activity that allows a mathematician to be a biologist, chemist, ecologist, or economist depending on the problem that he/she is tackling. The primary aim of a modeler is to undertake experiments on the mathematical representation of a real-world problem, instead of undertaking experiments in the real world.

> Challenges in mathematical modeling
> ... not to produce the most comprehensive descriptive model but to produce the simplest possible model that incorporates the major features of the phenomenon of interest.
>
> *Howard Emmons*

1.2 History of Mathematical Modeling

Modeling (from Latin "modellus") is a way of handling reality. It is the ability to create models that distinguish human beings from animals. Models of real objects and things have been in use by human beings since the Stone Age

DOI: 10.1201/9781351022941-1

as is evident in cave paintings. Modeling became important in the Ancient Near East and Ancient Greek civilizations. It is the writing and counting numbers that were the first models. Two other areas where modeling was used in its preliminary forms are astronomy and architecture. By the year 2000 BC, the three ancient civilizations of Babylon, Egypt, and India had a good knowledge of mathematics and used mathematical models in various spheres of life [147].

In the Ancient Greek civilization, the development of philosophy and its close relation to mathematics contributed to a deductive method, which led to probably the first instance of mathematical theory. From about 600 BC, geometry became a useful tool in analyzing reality. Thales predicted the solar eclipse of 585 BC and devised a method for measuring heights by measuring the lengths of shadows using geometry. Pythagoras developed the theory of numbers and, most importantly, initiated the use of proofs to gain new results from already-known theorems. Greek philosophers Aristotle, Eudoxus, and others contributed further, and in the next 300 years after Thales, geometry and other branches of mathematics developed further. The zenith was reached by Euclid of Alexandria who, in circa 300 BC wrote, *The Elements*, a veritable collection of almost all branches of mathematics known at the time. This work included, among others, the first precise description of geometry and a treatise on number theory. It is for this that Euclid's books became important for the teaching of mathematics for many hundreds of years, and around 250 BC, Eratosthenes of Cyrene used this knowledge to calculate the distances between the Earth and the Sun and the Earth and the Moon, and the circumference of the Earth using a geometric model.

A further step in the development of mathematical models was taken up by Diophantus of Alexandria in about 250 AD, who, in his book *Arithmetica*, developed the beginnings of Algebra based on symbolism and the idea of a variable. In the field of astronomy, Ptolemy, influenced by Pythagoras' idea of describing celestial mechanics by circles, developed a mathematical model of the solar system using circles to predict the movement of the Sun, the Moon, and the planets. The model was so accurate that it was used until the early seventeenth century when Johannes Kepler discovered a much more simple and superior model for planetary motion in 1619. This model, with later refinements done by Newton and Einstein, is in use even today.

Mathematical models are used for real-world problems and are hence important for human development. Mathematical models were developed in China, India, and Persia as well as in the Western world. One of the most famous Arabian mathematicians was Muhammad Ibn Musa al-Khwarizmi, who lived in the late eighth century [147]. Interestingly, his name still survives in the word algorithm. His well-known books are *Algoritmo de Numero Indorum* (about the Indian numbers–today called Arabic numbers) and *Al-kitab al-muhtasar fi hisāb al-ǧ abr wa'lmuqābala* (a book about the procedures of calculation balancing) [147]. Both these books contain mathematical models and problem-solving algorithms for use in commerce, survey, and irrigation. The term *algebra* was derived from the title of his second book.

In the Western world, it was only in the sixteenth century that mathematics and mathematical models developed. The greatest mathematician in the Western world after the decline of the Greek civilization was Fibonacci, Leonardo da Pisa. The son of a merchant, Fibonacci made many journeys to the Orient, and familiarized himself with mathematics as it had been practiced in the Eastern world. He used algebraic methods recorded in Al-Hwārizmī's books to improve his trade as a merchant. He first realized the great practical advantage of using the Indian numbers over the Roman numbers which were still in use in Europe at that time. His book *Liber Abaci*, first published in 1202, began with a reference to the ten 'Indian figures' (0, 1, 2,..., 9), as he called them. Also, 1202 is an important year since it saw the number 0 being introduced to Europe. The book itself was meant to be a manual of algebra for commercial use. It dealt in detail with arithmetical rules using numerical examples which were derived from practical use, such as their applications in measure and currency conversion.

The Italian painter Giotto (1267–1336) and the Renaissance architect and sculptor Filippo Brunelleschi (1377–1446) are responsible for the development of geometric principles. In the later centuries, many more and varied mathematical principles were discovered and the intricacy and complexity of the models increased. It is important to note that despite the achievements of Diophant and Al-Hwārizmī, the systematic use of variables was invented by Vieta (1540–1603) [147]. In spite of all these developments, it took many years to realize the true role of variables in the formulation of mathematical theory. It also took time for the importance of mathematical modeling to be completely understood. Physics and its application to nature and natural phenomena is a major force in mathematical modeling and its further development. Later economics became another area of study where mathematical modeling began to play a major role.

1.3 Importance of Mathematical Modeling

A mathematical model, as stated, is a mathematical description of a real life situation. So, if a mathematical model can reflect or mimic the behavior of a real-life situation, then we can get a better understanding of the system through proper analysis of the model using appropriate mathematical tools.

Moreover, in the process of building the model, we discover various factors that govern the system, that are most important to the system, and that reveal how different aspects of the system are related.

The importance of mathematical modeling in physics, chemistry, biology, economics, and even industry cannot be ignored. Mathematical modeling in basic sciences is gaining popularity, mainly in biological sciences, economics, and industrial problems. For example, if we consider mathematical modeling in the steel industry, many aspects of steel manufacture, from mining to distribution, are susceptible to mathematical modeling. In fact, steel companies have participated in several mathematics-industry workshops, where they discussed various problems and obtained solution through mathematical modeling – problems involving control of ingot cooling, heat and mass transfer in blast furnaces, hot rolling mechanics, friction welding, spray cooling and shrinkage in ingot solidification, to mention a few [115]. Similarly, mathematical modeling can be used

(i) to study the growth of plant crops in a stressed environment,
(ii) to study mRNA transport and its role in learning and memory,
(iii) to model and predict climate change,
(iv) to study the interface dynamics for two liquid films in the context of organic solar cells,
(v) to develop multi-scale modeling in liquid crystal science and many more.

For gaining physical insight, analytical techniques are used. However, to deal with more complex problems, numerical approaches are quite handy. It is always advisable and useful to formulate a complex system with a simple model whose equation yields an analytical solution. Then, the model can be modified to a more realistic one that can be solved numerically. Together with the analytical results for simpler models and the numerical solution from more realistic models, one can gain maximum insight into the problem.

1.4 Latest Developments in Mathematical Modeling

Mathematical modeling is an area of great development and research. In recent years, mathematical models have been used to validate hypotheses made from experimental data, and at the same time, the designing and testing of these models have led to testable experimental predictions. There are impressive cases in which mathematical models have provided fresh insight into biological systems, physical systems, decision-making problems, space models, industrial problems, economical problems, and so forth. The development of mathematical modeling is closely related to significant achievements in the field of computational mathematics.

Consider a new product being launched by a company. In the development process, there are critical decisions involved in its launch such as timing, determining price, launch sequence. Experts use and develop mathematical models to facilitate such decision making. Similarly, in order to survive market competition, cost reduction is one of the main strategies for a manufacturing plant, where a large amount of production operation costs are involved. Proper layout of equipment can result in a huge reduction in such costs. This leads to a dynamic facility layout problem for finding equipment sites in manufacturing environments, which is one of the developing areas in the field of mathematical modeling [152].

Mathematical modeling also intensifies the study of potentially deadly flu viruses or Coronavirus (COVID-19) from mother nature and bio-terrorists. Mathematical models are also being developed in optical sciences [8], namely, diffractive optics, photonic band gap structures and waveguides, nutrient modeling, studying the dynamics of blast furnaces, studying erosion, and prediction of surface subsidence.

In geosciences, mathematical models have been developed for talus. Talus is defined by Rapp and Fairbridge [128] as an accumulation of rock debris formed close to mountain walls, mainly through many small rockfalls. Hiroyuki and Yukinori [114] have constructed a new mathematical model for talus development and retreat of cliffs behind the talus, which was later applied to the result of a field experiment for talus development at a cliff composed of chalk. They developed the model which was in agreement with the field observations.

There was tremendous development in the interdisciplinary field of applied mathematics in human physiology in the last decade, and this development still continues. One of the main reasons for this development is the researcher's improved ability to gather data, whose visualization has much better resolution in time and space than just a few years ago. At the same time, this development also constitutes a giant collection of data as obtained from advanced measurement techniques. Through statistical analysis, it is possible to find correlations, but such analysis fails to provide insight into the mechanisms responsible for these correlations. However, when it is combined with mathematical modeling, new insights into the physiological mechanisms are revealed.

Mathematical models are being developed in the field of cloud computing to facilitate the infrastructure of computing resources in which large pools of systems (or clouds) are linked together via the internet to provide IT services

(for example, providing secure management of billions of online transactions) [29]. Development of mathematical models are also noticed
(i) in the study of variation of shielding gas in GTA welding,
(ii) for prediction of aging behavior for Al-Cu-Mg/Bagasse Ash particular composites,
(iii) for public health decision making and estimations,
(iv) for the development of cerebral cortical folding patterns that have fascinated scientists with their beauty and complexity for centuries,
(v) to predict sunflower oil expression,
(vi) in the development of a new three-dimensional mathematical ionosphere model at the European Space Agency/European Space Operators Centre,
(vii) in battery modeling or mathematical description of batteries, which plays an important role in the design and use of batteries, estimation of battery processes and battery design. These are some areas where mathematical modeling plays an important role. However, there are many more areas of application.

1.5 Limitations of Mathematical Modeling

Sometimes although the mathematical model used is well adapted to the situation at hand, it may give unexpected results or simply fail. This may be an indication that we have reached the limit of the present mathematical model and must look for a new refinement of the real-world or a new theoretical breakthrough [129]. A similar type of problem was addressed in [8], which deals with Moire theory, involving the mathematical modeling of the phenomena that occur in the superposition of two or more structures (line gratings, dot screens, etc.), either periodic or not.

In mathematical modeling, more assumptions must be made, as information about real-world systems become less precise or harder to measure. Modeling becomes a less precise endeavor as it moves away from physical systems towards social systems. For example, modeling an electrical circuit is much more straightforward than modeling human decision making or the environment. Since physical systems usually do not change, reasonable past information about a physical system is quite valuable in modeling future performance. However, both social systems and environments often change in ways that are not of the past, and even correct information may be of less value in forming assumptions. Thus, to understand a model's limitations, it is important to understand the basic assumptions that were used to create it.

Real-world systems are complex and a number of interrelated components are involved. Since models are abstractions of reality, a good model must try to incorporate all critical elements and interrelated components of the real-world system. This is not always possible. Thus, an important inherent limitation of a model is created by what is left out. Problems arise when key aspects of the real-world system are inadequately treated in a model or are ignored to avoid complications, which may lead to incomplete models. Other limitations of a mathematical model are that they may assume that the future will be like the past, input data may be uncertain, or the usefulness of a model may be limited by its original purpose.

However, despite all these limitations and pitfalls, a good model can be formulated, if a modeler asks himself/herself the following questions about the model:
(i) Does the structure of the model resemble the system being modeled?
(ii) Why is the selected model appropriate to use in a given application?
(iii) How well does the model perform?
(iv) Has the model been analyzed by someone other than the model authors?
(v) Is adequate documentation of the model available for all who wish to study it?
(vi) What assumptions and data were used in producing model output for the specific application?
(vii) What is the accuracy of the model output?

One should not extrapolate the model beyond the region of fit. A model should not be applied unless one understands the simplifying assumptions on which it is based and can test their applicability. It is also important to understand that the model is not the reality and one should not distort reality to fit the model. A discredited model should not be retained, and one should not limit himself to a single model, as more than one model may be useful for understanding different aspects of the same phenomenon. It is imperative to be aware of the limitations inherent in models. There is no best model, only better models.

1.6 Units

Some standard reference is required to express the result of measurement of a quantity, and this standard is called the unit of measurement for that quantity. In other words, a unit is any standard used for comparison in measurements. For example, if we say that there is a huge stock of wheat in the storage (godown), then this statement is vague and does not give any clear idea. However, if we say that there is 3000 kg of wheat in the storage (godown), then this gives a clear idea of the amount of the wheat stores. Here, weight is the quantity and its unit of measurement is taken as kilogram.

The commonly used system of units are as follows:

(i) **CGS** system: This is a French system developed in 1873 for science, in which **C** stands for centimeter, **G** for gram, and **S** for seconds. This system consists of only three mechanical quantities, namely, mass, length and time.

(ii) **MKS** system: Since the centimeter and gram were felt to be too small for the needs of technology, the **MKS** system (again originated in France) evolved in 1900. This system also deals with only three mechanical quantities mass, length, and time. Here, **M** stands for meter, **K** stands for kilogram, and **S** stands for seconds.

(iii) **SI** system (International System of Units or Le Systeme International d'units): The **SI** system, which came into existence in 1960, is a more rationalized version of the **MKS** system and represents the final variant in the evolution of Units. This system has been prepared by adding some standard units with those of **MKS** system. The fundamental units in SI system are meter (length), kilogram (mass), second (time), ampere (electric current), kelvin (temperature), candela (luminous intensity), and mole (amount of substance).

1.7 Dimensions

A physical quantity can be expressed in terms of the fundamental quantities, namely, length, mass and time or a combination of these three. The particular combination is referred to as the dimension of that physical quantity. Thus, by the dimension of a physical quantity, we mean the power to which the fundamental unit or units need to be raised to obtain units of that quantity. For example, the dimension of volume, which is the product of three units of length, is represented by $L \times L \times L$ or L^3, and the dimension of speed, which is distance/time, is $L/T = LT^{-1}$.

The dimensions are independent of the units used; for example, speed has the dimension LT^{-1} but can be measured in km per hour or meter per second, square brackets [.] are used to denote "the dimension of...", so that

$$\begin{aligned}
[\text{acceleration}] &= \left[\frac{\text{velocity}}{\text{time}}\right] = \frac{[\text{velocity}]}{[\text{time}]} = \frac{LT^{-1}}{T} = LT^{-2}, \\
[\text{force}] &= [\text{mass} \times \text{acceleration}] = [\text{mass}] \times [\text{acceleration}] \\
&= M \times LT^{-2} = MLT^{-2}, \\
[\text{work}] &= [\text{force} \times \text{distance}] = [\text{force}] \times [\text{distance}] \\
&= MLT^{-2} \times L = ML^2T^{-2}.
\end{aligned}$$

However, some quantities are dimensionless or pure number. For example,

$$[\text{angle}] = \left[\frac{\text{arc}}{\text{radius}}\right] = \frac{[\text{arc}]}{[\text{radius}]} = \frac{L}{L} = LL^{-1} = L^0 = \text{a pure number},$$

$$[\text{Specific gravity}] = \left[\frac{\text{mass of a body}}{\text{mass of equal volume of water}}\right] = \frac{M}{M} = \text{a pure number}.$$

In mathematical modeling, models are expressed in the form of an equation, which involves variables, parameters, and constants, all of which represent quantities and, in principle, can be measured. For example, a time of $19s$, where s stands for seconds, is a physical quantity, measured by writing a number followed by a unit of measurement. A measurement without units is totally meaningless, and in model building process, one must be careful to keep track of the units of measurement for all quantities involved at all stages. To start with, one must check on all the equations appearing in a model, if they are dimensionally consistent or not, that is,

$$[\text{left hand side}] = [\text{right hand side}].$$

In other words, they must be dimensionally homogeneous.

For example, suppose a body is moving in a straight line with a force proportional to the cube of its velocity. If we are modeling the force on the moving object, then we will have the model of the form

$$F = kv^3,$$

where k is the constant of proportionality. We now check the dimension of both these equations:

$$[F] = [kV^3] \Rightarrow MLT^{-2} = [k](LT^{-1})^3 \Rightarrow MLT^{-2} = [k]L^3T^{-3}.$$

For consistency, it is required $[k] = ML^{-2}T$ and, k will be measured in $kgm^{-2}s$ in **MKS** or **SI** units.

Notes:

(i) If, in the model $\cos(\alpha t)$ or $e^{\alpha t}$ appears, where t stands for time, then the parameter α must have a dimension T^{-1}, so that αt is dimensionless. Thus, the argument of functions having Taylor series expansions must be dimensionless.

(ii) If an equation involves a derivative, the dimensions of the derivative are given by the ratio of the dimensions, because a derivative is a limiting ratio of two quantities. For example, if v be the velocity of a particle at any time t, then the acceleration is $\frac{dv}{dt}$ and

$$\left[\frac{dv}{dt}\right] = \frac{[v]}{[t]} = \frac{LT^{-1}}{T} = LT^{-2}.$$

Similarly, for the partial derivatives,

$$\left[\frac{\partial v}{\partial x}\right] = \frac{[v]}{[x]} = \frac{LT^{-1}}{L} = T^{-1},$$

and

$$\left[\frac{\partial^2 v}{\partial x^2}\right] = \frac{[v]}{[x^2]} = \frac{LT^{-1}}{L^2} = L^{-1}T^{-1}.$$

(iii) If u be the temperature and x measures distance, then

$$\left[\frac{\partial u}{\partial x}\right] = \frac{[u]}{[x]} = \frac{Q}{L} = QL^{-1}$$

and

$$\left[\frac{\partial^2 u}{\partial x^2}\right] = \frac{[u]}{[x^2]} = \frac{Q}{L^2} = QL^{-2}.$$

In general,

$$\left[\frac{\partial^{m+n} u}{\partial x^m t^n}\right] = QL^{-m}T^{-n}.$$

(iv) Two quantities may be added if they have the same dimensions. Quantities of different dimensions may be multiplied or divided.

(v) Index law: If $[A] = M^{a_1}L^{b_1}T^{c_1}$ and $[B] = M^{a_2}L^{b_2}T^{c_2}$, then

$$[AB] = M^{a_1+a_2}L^{b_1+b_2}T^{c_1+c_2}.$$

(vi) Pure numbers are dimensionless and multiplying a physical quantity by a pure number does not change its dimensions.

(vii) The dimension of an integral

$$\int_a^b p dq$$

is given by $[p][q]$.

Example 1.7.1 *Fourier's law, relating heat flux to temperature gradient is given by*

$$J = -\alpha \frac{\partial u}{\partial x},$$

where J is the heat flux, u the temperature, x denotes distance and k is the conductivity. Determine the dimension of α, that is, $[\alpha]$.

Solution: Heat flux J is the heat energy per unit area per unit time. So,

$$[J] = \frac{[\text{Energy}]}{[\text{area}][\text{time}]}.$$

Now, energy has the dimensions of work done, which is force $\times$ distance, so [energy] $= MLT^{-2} \times L$ and [area] $= L^2$. Therefore,

$$[J] = \frac{ML^2T^{-2}}{L^2T} = MT^{-3}.$$

Now, $[u] = Q$ and $[x] = L$, so

$$\left[\frac{\partial u}{\partial x}\right] = \frac{[u]}{[x]} = \frac{Q}{L} = QL^{-1}.$$

Since $[\alpha] = [J] \times \left[\frac{\partial u}{\partial x}\right]^{-1}$, then $[\alpha] = MT^{-3} \times (QL^{-1})^{-1} = MLT^{-3}Q^{-1}$.

(In SI units, α is measured in $kg\ m\ s^{-3}\ K^{-1}$.)

1.8 Dimensional Analysis

When the dimension of a quantity is found and expressed in the form of an equation, the equation is called the dimensional equation . Dimensional analysis is a method of using the fact that the dimension of each term in an equation must be the same, to suggest a relationship between the physical quantities involved.

Suppose, we are trying to develop a model which will predict the frequency (n) of vibration of a stretched string. We assume that the frequency depends on the tension (t_s) of the stretched string, the length (l), and mass (m) per unit length of the string. We assume that

$$n = k\ t_s^a\ l^b\ m^c,$$

where a, b, c are real numbers and k is the proportionality constant (dimensionless quantity).

Considering dimensions, we get,

$$[n] = [k\ t_s^a\ l^b\ m^c] = k\ [t_s^a]\ [l^b]\ [m^c], \text{ where}$$

t_s = tension= force $\Rightarrow$ [force] $= MLT^{-2}$,

l = length $\Rightarrow$ [length] = L,

m = mass per unit length = $\frac{\text{mass}}{\text{length}} \Rightarrow [m] = ML^{-1}$, and

n = frequency of stretched string = vibration per unit time $\Rightarrow [n] = T^{-1}$.
Therefore,

$$\begin{aligned} T^{-1} &= k\,(MLT^{-2})^a\, L^b\,(ML^{-1})^c \\ \Rightarrow M^0L^0T^{-1} &= kM^{a+c}L^{a+b-c}T^{-2a}. \end{aligned}$$

Equating the powers of M, L, and T on both sides, we obtain

$$a+c=0,\ a+b-c=0,\ \ -2a=-1$$
$$\Rightarrow\ a=1/2,\ c=-1/2,\ b=-1.$$

This gives

$$n = k\, t_s^{1/2}\, l^{-1}\, m^{-1/2}$$
$$\Rightarrow n = \frac{k}{l}\sqrt{\frac{t_s}{m}}.$$

Example 1.8.1 *Given* $\frac{dp}{dr} - \frac{\rho v}{r^2} = 0$*, where* p *is the pressure,* ρ *is the density, and* v *is the velocity. Check if the equation is dimensionally homogeneous (dimensionally correct).*

Solution: We have

$$\begin{aligned} \text{Pressure} &= \frac{\text{Force}}{\text{Area}} \Rightarrow [\text{pressure}] = \frac{[\text{Force}]}{[\text{Area}]} = \frac{MLT^{-2}}{L^2} = ML^{-1}T^{-2}, \\ \text{Density} &= \frac{\text{Mass}}{\text{Volume}} \Rightarrow [\text{Density}] = \frac{[\text{Mass}]}{[\text{Volume}]} = \frac{M}{L^3} = ML^{-3}, \\ \text{Velocity} &= \frac{\text{Distance}}{\text{Time}} \Rightarrow [\text{Velocity}] = \frac{[\text{Distance}]}{[\text{Time}]} = \frac{L}{T} = LT^{-1}. \end{aligned}$$

Now, the given equation is

$$\frac{dp}{dr} - \frac{\rho v}{r^2} = 0,$$

where r is the radial distance. Considering dimensions, we get,

$$\left[\frac{dp}{dr}\right] = \left[\frac{\rho v}{r^2}\right].$$

$$\text{L.H.S:}\ \left[\frac{dp}{dr}\right] = \frac{[p]}{[r]} = \frac{ML^{-1}T^{-2}}{L} = ML^{-2}T^{-2},$$

$$\text{R.H.S:}\ \left[\frac{\rho v}{r^2}\right] = \frac{[\rho][v]}{[r^2]} = \frac{ML^{-3}LT^{-1}}{L^2} = ML^{-4}T^{-1}.$$

Thus, **L.H.S** $\neq$ **R.H.S** and hence, the given equation is dimensionally incorrect.

1.9 Scaling

The fundamental idea of scaling is to make the independent and dependent variables dimensionless. It is an extremely useful technique in mathematical modeling and numerical simulation. It reduces the number of independent physical parameters in the model and makes the size of the independent and dependent variables about unity.

Mathematically, for any variable z, we will introduce a corresponding dimensionless variable

$$\bar{z} = \frac{z - z_0}{z_c},$$

where z_0 is the reference value of z (generally taken as zero) and z_c is a characteristic size of $\mid z \mid$, often referred to a scale. $\bar{z}$ is a dimensionless number as the numerator and denominator have the same dimensions. There is no hard and fast rule for finding z_c and sometimes it becomes quite challenging. Let me illustrate the process with the help of an example.

Consider the spruce budworm outbreak model [100]

$$\frac{dB}{dt} = rB\Big(1 - \frac{B}{k}\Big) - \frac{\beta B^2}{\alpha^2 + B^2}, \qquad B(0) = B_0,$$

where $B(t)$ is the density of the budworm at any time t, r is the intrinsic growth rate, and k is the carrying capacity (maximum size of the population that can be reached with the available resources) of the budworm density. The predation term $\frac{\beta B^2}{\alpha^2+B^2}$ was chosen by Ludwig et al. [100], to take into account the observed facts about spruce budworms and their predators (birds), where β is the rate of predation and α is the saturation factor of spruce budworms.

Basically, it is a first-order differential equation, with initial condition $B(0) = B_0$. We now give a step-by-step approach to non-dimensionalize this initial value problem.

Step I: We first identify the variables, parameters and their dimensions and put them in tabular form (Table 1.1).

Step II: We now distinctly separate the variables into two parts: a new variable, which has no dimension and a quantity, which represents the units of measurements and carries physical dimensions. Thus, we get,

$$B = u\, B^* \text{ and } t = z\, t^*.$$

(Please note that we have chosen B^* and t^* in such a manner that $u = \frac{B}{B^*}$

Variables	Dimensions
t	T (time)
B	$ML^{-3} = \rho$ (density)
Parameters	**Dimensions**
r (intrinsic growth rate)	T^{-1}
k (carring capacity)	$\rho = ML^{-3}$
α (saturation factor)	$\rho = ML^{-3}$
β (predation factor)	$ML^{-3}T^{-1} = \rho T^{-1}$
B_0 (initial density)	$ML^{-3} = \rho$

TABLE 1.1: *Variables, parameters and their dimensions.*

and $z = \frac{t}{t^*}$ are dimensionless variables.)

Step III: We now use the chain rule to form a new differential equation:

$$\frac{dB}{dt} = \frac{dB}{du}\frac{du}{dz}\frac{dz}{dt} = B^*\frac{du}{dz}\frac{1}{t^*}.$$

Substituting $u = \frac{B}{B^*}$ and $z = \frac{t}{t^*}$, we obtain,

$$\begin{aligned} \frac{B^*}{t^*}\frac{du}{dz} &= ruB^*\left(1 - \frac{uB^*}{k}\right) - \frac{\beta u^2 B^{*2}}{\alpha^2 + u^2 B^{*2}} \\ \frac{du}{dz} &= rt^* u\left(1 - \frac{u}{k/B^*}\right) - \frac{\beta B^* t^* u^2}{\alpha^2 + B^{*2}u^2}. \end{aligned}$$

If we choose $B^* = \alpha$ and $t^* = \frac{\alpha}{\beta}$, then

$$[u] = \frac{[B]}{[B^*]} = \frac{ML^{-3}}{[\alpha]} = \frac{ML^{-3}}{ML^{-3}} = 1 \text{ (a pure number)},$$

and

$$[z] = \frac{[t]}{[t^*]} = \frac{T}{[\alpha/\beta]} = \frac{T[\beta]}{[\alpha]} = \frac{T.ML^{-3}T^{-1}}{ML^{-3}} = 1 \text{ (a pure number)}.$$

Thus, both u and z are dimensionless. Substituting $B^* = \alpha$ and $t^* = \frac{\alpha}{\beta}$, we obtain

$$\begin{aligned} \frac{du}{dz} &= r\frac{\alpha}{\beta}u\left(1 - \frac{u}{k/\alpha}\right) - \frac{\beta\alpha\frac{\alpha}{\beta}u^2}{\alpha^2 + \alpha^2 u^2}, \\ \frac{du}{dz} &= \left(r\frac{\alpha}{\beta}\right)u\left(1 - \frac{u}{k/\alpha}\right) - \frac{u^2}{1 + u^2}. \end{aligned}$$

We now introduce two dimensionless parameters,

$$a = \frac{r\alpha}{\beta} \text{ and } b = \frac{k}{\alpha},$$

and the equation becomes

$$\frac{du}{dz} = au\left(1 - \frac{u}{b}\right) - \frac{u^2}{1+u^2}.$$

Clearly,

$$[a] = \left[\frac{r\alpha}{\beta}\right] = \frac{[r][\alpha]}{[\beta]} = \frac{T^{-1}ML^{-3}}{ML^{-3}T^{-1}} = 1 \text{ (a pure number)},$$

and

$$[b] = \left[\frac{k}{\alpha}\right] = \frac{[k]}{[\alpha]} = \frac{ML^{-3}}{ML^{-3}} = 1 \text{ (a pure number)}.$$

Thus, a and b are dimensionless parameters.

Step IV: We now look into the initial condition: $B(0) = B_0$. Since, $B = uB^*$, we get,

$$u(0)B^* = B_0 \qquad (B = uB^* \Rightarrow B(0) = u(0)B^*)$$
$$\Rightarrow u(0) = \frac{B_0}{B^*} = \frac{B_0}{\alpha} = \gamma \text{ (say)}.$$

Note that,

$$[\gamma] = \left[\frac{B_0}{\alpha}\right] = \frac{[B_0]}{[\alpha]} = \frac{ML^{-3}}{ML^{-3}} = 1 \text{ (a pure number)}.$$

The initial condition becomes $u(0) = \gamma$ (a dimensionless quantity).

After non-dimensionalization, the simplified form of the equation is

$$\frac{du}{dz} = au\left(1 - \frac{u}{b}\right) - \frac{u^2}{1+u^2}, \; u(0) = \gamma,$$

where the number of parameters has been reduced from 5 to 3 (including initial condition), thereby making the analysis of the equation simpler.

Note: There is no unique way of choosing B^* and t^*, some experimentation may be necessary (trial and error method).

1.10 How to Built Mathematical Models

1.10.1 Step I (The Start)

Before we start formulating a model, we should be clear about our objectives. The objectives decide the further course of the task in two different ways:

(i) Firstly, the amount of detail incorporated into the model depends on the purpose for which the model will be utilized. For example, in modeling the growth of skin cancer (melanoma) to act as an aid for oncologists, an empirical model containing terms for the most significant determinants of skin cancer growth may be quite adequate and the model can be viewed as a summary of the current understanding of melanoma (skin cancer). However, such a model is clearly of limited use as a research tool to investigate the growth of exocrine tumors (pancreatic cancer).

(ii) Secondly, there should be a clear-cut division between the system to be modelled and its environment. Care should be taken that the division is well-made, which means that the environment may affect the dynamics of the system but the system does not affect the environment.

1.10.2 Step II (The Assumption)

Once the system which needs to be modelled is decided, a basic framework of the model needs to be constructed. This will depend on the underlying assumption that we make in the construction of the model. It is to be noted that whenever assumptions are made, they will be considered as true while doing analysis of the system and results of such an analysis are only as valid as the assumptions. Thus, one of the fundamental differences between classical mechanics and relativity theory is that Newton assumed that mass is the universal constant, whereas Einstein considered mass as a variable.

Let us consider the growth of a population. It is a common assumption that a population will grow at a rate that is proportional to its size, in absence of limiting factors. In continuous time, such a population growth will be represented by the differential equation

$$\frac{dx}{dt} = rx,$$

where $x(t)$ is the population size at time t and r is the constant rate of growth. Solving the differential equation we get

$$x(t) = x(0)\, e^{rt},$$

where $x(0)$ is the initial size of the population at time $t = 0$. From the solution, we conclude that the population grow in size at an exponential rate. Clearly, that is not true, not all population grow exponentially. So, **where is the flaw?**

Please note that the basic framework of the model is based on the assumption that the growth of the population takes place in absence of limiting factor. However, in reality, population are subjected to constraints such as food supply or space or habitat, which restrict the range of sustainable population sizes. Thus, the appropriateness of the model depends on all the assumptions that have been made to construct the basic framework of the model. The growth model is represented by a differential equation, which implies that we have assumed that growth takes place continuously. Suppose, the population consists of discrete generations, then it would be proper to represent the growth model by the difference equation

$$x_{n+1}(t) = r x_n(t),$$

where $x_n(t)$ is the size of the nth generation at any time t. The solution of the proposed difference equation is

$$x_n(t) = x_0 \; r^n,$$

where x_0 is the initial population.

In both the cases (continuous and discrete), we assume that the population behaves according to the deterministic law. However, if we assume randomness in the model, we can intuitively expect stochastic variability to be important. In that case, the model would be represented by a stochastic differential equation.

Another assumption we made in formulating the simple population growth model is that the population is uniformly distributed in space and has no contact with other populations. To obtain the basic framework of such a model, we need to introduce a spatially explicit model with rates describing the movement of individuals from one area to another, in other words, if we are interested in the spatial distribution of our population, we need to represent the model by a partial differential equation with diffusion term.

Thus, we notice that to obtain a simple population growth model, how many different assumptions have to be made. Therefore, we must proceed with extreme caution while formulating models of complicated systems. To get the result close to the real world, it is often advised to examine different versions (for example, deterministic, stochastic, spatial) based on the same basic model and compare the results derived from the model.

1.10.3 Step III (Schematic or Flow Diagrams)

Whether the system is simple or complex, it is always advisable not to simply jump from an assumption to an equation. A much more methodical step in the form of schematic or flow diagrams is desirable, which provides a visual aid to the problem. A flow diagram or schematic diagram is a pictorial representation depicting the flow of steps in a model. The elements of the schematic diagram are represented by circles, rectangles, diamonds, or other shapes, with lines and arrows representing connections between events and the direction or order in which they occurs. The schematic diagram shows a starting point, an ending or stopping point, along with the sequence and decision points. Since, it is easier to understand a process in a visual form than in a verbal description, a schematic or flow diagram is always useful to start a modeling process.

Consider a susceptible-infected-recovered-susceptible (SIRS) model, where a person, susceptible to a particular disease, gets infected at a rate α when he/she comes in contact with an infected one. From the infected class, a person gets cured at a rate β and move to the recovered class. The recovered person is also susceptible to the disease at a rate δ. The schematic diagram for this scenario is given in fig. 1.1.

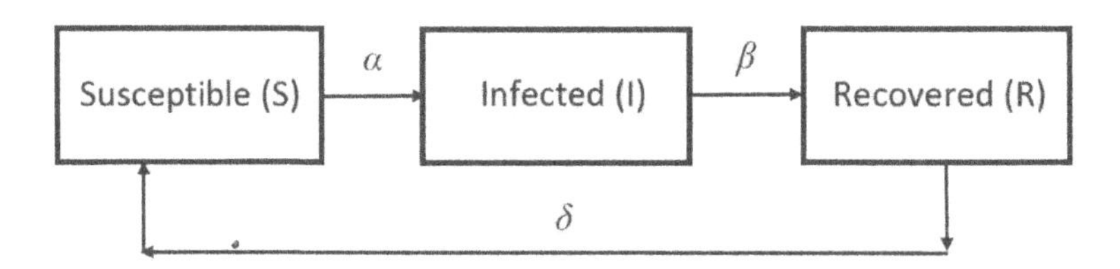

FIGURE 1.1: *The schematic diagram for susceptible-infected-recovered-susceptible (SIRS) model.*

1.10.4 Step IV (Choosing Mathematical Equations)

One can now choose the appropriate mathematical equations, once the structure of the model has been decided. A good starting point would be to look equations from literatures, where somebody else may have published an equation relating to the quantities you are interested in. But, it is necessary to proceed with caution as equations in the literatures may not be expressed exactly in the form required for the model.

1.10.5 Step V (Solving Equations)

In general, obtaining an analytical solution of the mathematical model of a real-life problem is rare. If the model consists of linear system of differential equations, or if the model consists of just one differential equations (even non-linear), there is a good chance of obtaining analytical solutions. However, when models contain a system of non-linear differential equations, it is difficult to solve them analytically. Then, we use numerical methods to obtain approximate solutions, which can never have the same generality as analytical solutions, but numerical solutions of the model generally mimic the process described in the model.

1.10.6 Step VI (Interpretation of the Result)

In the beginning, before the formulation of the model, you have asked a question, on the basis of which the model has been constructed. By looking into the solution of the variables of the model, the behavior of the model can be described. Basically, you have to rephrase the results of step V in non-technical terms, namely, in terms of question posed. Anyone, who can understand the statement of the question as it was presented to you originally, should be able to understand your answer. A schematic diagram for mathematical modeling is shown in fig. 1.2.

1.11 Mathematical Models and Functions

The objective of this section is to examine different kinds of functions, their behavior and how they can be used to represent the realistic situation. When we write a function to represent a situation and use a function to determine key aspects of a situation (for example, how to minimize the surface of a box or minimize its volume), we are engaging in mathematical modeling.

1.11.1 Linear Models

Consider a situation where a sales representative has to sell mini vacuum cleaner for \$50 per unit and how much money can they brings in, if he/she sells x units. One can model this situation with linear functions as some quantity in the situation has a constant rate of change. This means, in this example of selling mini vacuum cleaner, each unit costs \$50, implying that the profits increase at a constant rate. Therefore, the mini vacuum cleaner selling model can be represented by the linear equation

$$y = 50\,x.$$

The function necessarily contains the point $(0, 0)$, which means if the sales representative does not sell any unit of mini vacuum cleaner, he/she does not bring in any money.

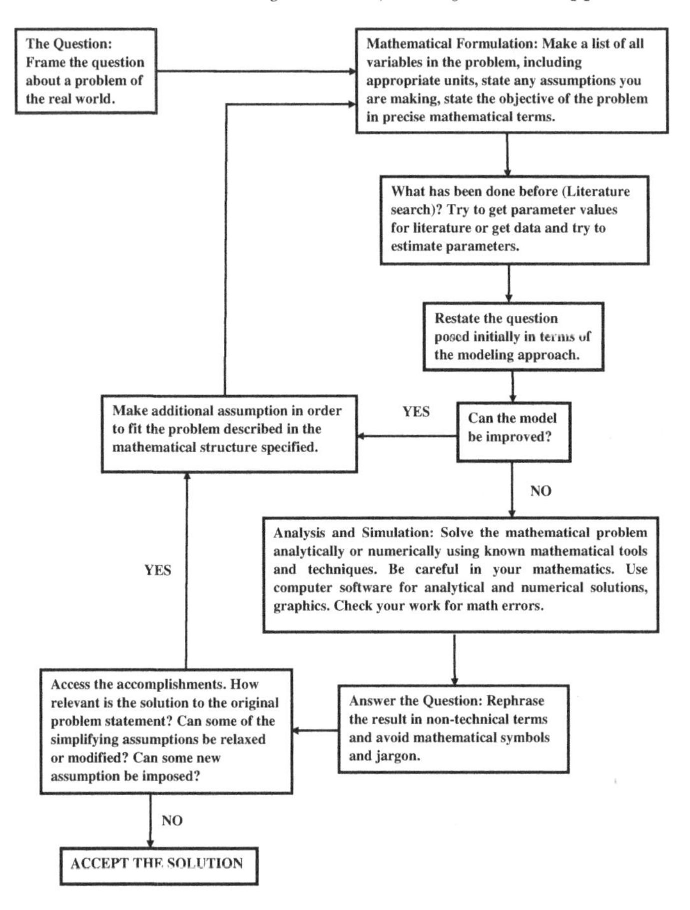

FIGURE 1.2: *Schematic diagram for mathematical modeling.*

Consider another situation, where a person has borrowed \$200 from a friend and agrees to return at a constant rate of \$25 per month. We can consider this as a linear model as follows:

$$y(t) = -25\, t + 200,$$

where t represents the number of months required to pay-off the debt and $y(t)$ is the amount remaining after paying off each month. The function is linear because of the constant rate of change, that is, constant return of \$25 per month. Clearly, the whole debt is paid off in 8 months ($y(t) = 0$).

Thus, linear functions are used to model a situation of constant change, either increase or decrease.

1.11.2 Quadratic Models

Consider a situation where a person is standing on a platform 10 m off the ground. The person throws a ball into air with an initial velocity 2 m/s, so that it will land on the ground (not on the platform). The obvious questions are (i) how high will the ball go? (ii) when will it reaches ground? and (iii) when will it reaches its maximum height? We can model this situation using the function of the form

$$h(t) = -g\, t^2 + v_0\, t + h_0, \tag{1.1}$$

where v_0 is the initial velocity of the ball, h_0 is the initial height of the ball, g is the gravitational constant, which pulls the ball down and hence it is negative, the variable t represents the time and $h(t)$ is the height above the ground.

Please note that (1.1) is the famous distance-time relation $s = ut - 1/2gt^2$, which we are familiar with in physics, only with change of notations and initial condition $s(0) = h_0$ (instead of $s(0) = 0$). Here, $h_0 = 10$ m, $v_0 = 2$ m/s and we assume $g = 10$ m/s^2, then (1.1) becomes

$$\begin{aligned} h(t) &= -10\, t^2 + 2\, t + 10 \\ (t - 0.1)^2 &= -0.1(h(t) - 10.1). \end{aligned} \tag{1.2}$$

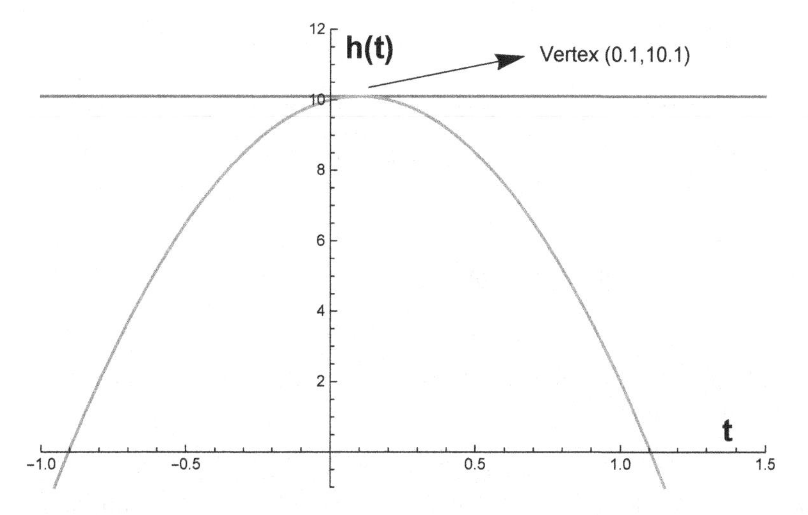

FIGURE 1.3: *The figure shows the parabola $h(t) = -10\, t^2 + 2\, t + 10$, whose vertex is at (0.1,10.1), which is also its point of maxima.*

Equation (1.2) represents a parabola whose vertex is at the point $(0.1, 10.1)$, which is the point of maxima (fig. 1.3). This means that the ball will reach the maximum height of 10.1 m, 0.1 seconds after the ball is thrown into the air.

Thus, a quadratic function, whose vertex is a global maximum, can be used to model a falling object. Similarly, in a situation where some quantity decreases and then increases, a quadratic function whose vertex is a global minimum, can be used.

1.11.3 Cubic Models

Consider a situation where a rectangular box is made by cutting squares out of the four corners of a rectangular piece of cardboard of dimensional 20 inches by 12 inches. Let x be the length of the squares cut out of the corners (fig. 1.4). Then, the volume of the box will be represented by

$V(x) = (20 - 2x)\ (12 - x)\ x = 4x^3 - 64x^2 + 240x = 4(x^3 - 16x^2 + 60x).$

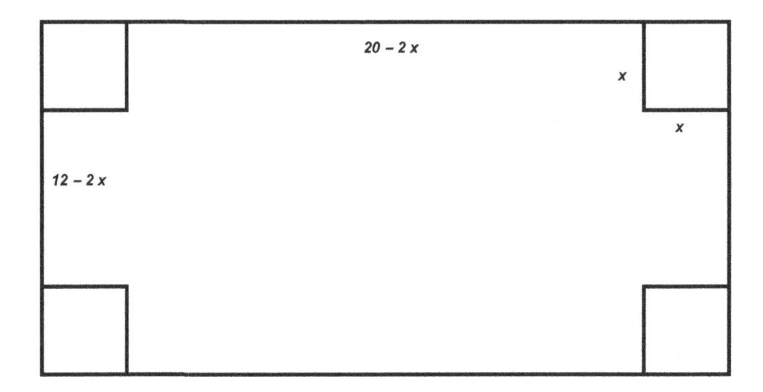

FIGURE 1.4: *The figure shows a rectangular box of dimension 20 inches by 12 inches, where x inch of the length of the square has been cut out from the corner.*

If you want to cut out a square of length 2 inches, then the volume of the box will be

$$V(2) = 256\ (inch)^3.$$

If we want to find the volume of the box to be 100 $(inch)^3$, we put $V(x) = 100$, to calculate x, implying

$$4x^3 - 64x^2 + 240x = 100 \Rightarrow\ x = 0.4751, 5, 10.5249.$$

If we take $x = 0.4751$ inch or 5 inches, the volume will be approximately 100 $(inch)^3$.

Thus, we can use cubic functions to model situations that involve volume. They can also be used to model situations that follow particular growth patterns.

1.11.4 Logistic Function and Logistic Growth Model

A logistic function or a logistic curve is given by

$$f(t) = \frac{k}{1 + e^{-r(t-t_0)}},$$

where, k is the maximum value of the curve, r is the logistic growth rate, and t_0 is some initial value. The logistic curve is S-shaped and is also known as sigmoid curve.

Consider a population with plenty of resources and no threat from predators. In such cases, the growth rate of the population is proportional to the population, that is,

$$\frac{dx}{dt} = r\ x,$$

where x is the population density at time t and r is the proportionality constant (constant growth rate). The solution of the differential equation is

$$x = x_0 e^{rt},$$

where x_0 is the population density at time $t = 0$. The solution represents exponential (unlimited) growth.

Now, in reality, most populations have limited resources. To put a constraint or unlimited (exponential) growth, the growth term $(r\ x)$ needs to be multiplied by a factor $(1 - \frac{x}{k})$. This factor is close to 1 (no effect), when x is much smaller than k and close to zero, when x is close to k. The modified model is

$$\frac{dx}{dt} = rx\left(1 - \frac{x}{k}\right), \tag{1.3}$$

which is known as the logistic growth model or Verhulst model. The solution $x(t)$, is given by,

$$x(t) = \frac{kx_0}{x_0 + (k - x_0)e^{-rt}}, \tag{1.4}$$

where x_0 is the population density at time $t = 0$, r is the intrinsic growth rate of the population and k is called carrying capacity (maximum size of the population that can be reached with the available resources) of the population density (fig. 1.5(a)).

1.11.5 Gompertz Function and Gompertz Growth Model

Gompertz function or Gompertz curve, named after Benjamen Gompertz, is a sigmoid function. The Gompertz function is represented by the following

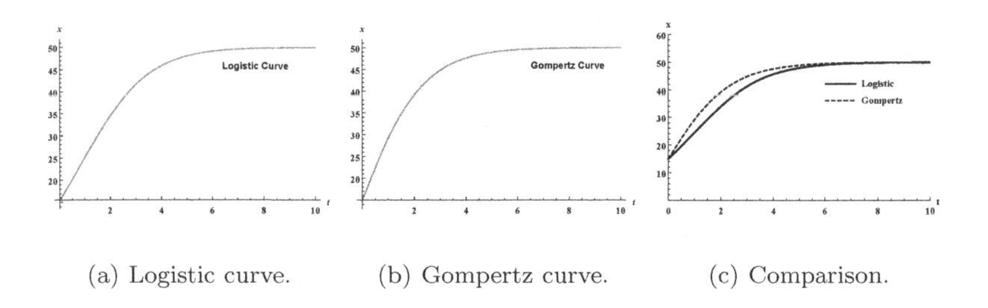

(a) Logistic curve. (b) Gompertz curve. (c) Comparison.

FIGURE 1.5: *The figure shows the plot of the Logistic function and the Gompertz function for* $k = 50, r = 0.8, x_0 = 15$ *and its comparison.*

differential equation of first order,

$$\frac{dx}{dt} = rx \ln\left(\frac{k}{x}\right), \; x(0) = x_0, \tag{1.5}$$

where $x(t)$ represents the number of individuals at any time, x_0 is the initial number of individuals at time $t = 0$, r is the intrinsic growth rate of the individual and k is the carrying capacity. The solution $x(t)$, is given by,

$$x(t) = k \; exp\left[\ln\left(\frac{x_0}{k}\right) e^{-rt}\right], \tag{1.6}$$

which is the Gompertz function or Gompertz curve. It was originally derived to estimate human mortality [24]. This function represents a modified model for a time series, where growth is the slowest at the start and end of a time period (fig. 1.5(b)).

1.12 Functional Responses in Population Dynamics

Crawford Stanley Holling, a Canadian ecologist, introduced the term "functional response", to describe the change in the rate of consumption of prey by a predator when the density of prey is varied. A simple way of finding this is to plot the number of preys consumed per unit time as a function of prey density.

1.12.1 Holling Type I Functional Response

If there is a linear relationship between the number of preys consumed and prey density, Holling proposed Type I functional response. Simply, it is assumed that the probability of a given predator encountering prey in a fixed time interval T, within a fixed spectral region, depends linearly on the prey density [24]. Let X be the number of preys consumed by one predator, then

$$Y = aT_sX,$$

where X is the prey density, T_s is the searching time, and a is the constant of proportionality or the discovery rate as termed by Holling.

If we neglect the handling time (time spend to handle the prey), then all the time has been used in searching, implying $T_s = T$. Let P be the density of the predator acting independently, the total number of prey will be reduced by a quantity

$$Y = (aTX)P$$

and we have the type I functional response.

Consider the predatory nature of the spider, which is passive in nature. The number of insects caught in the web is proportional to the fly density and hence Holling Type I functional response is found in them (fig. 1.6(a)). Prey mortality due to predator is constant here (fig. 1.6(b)).

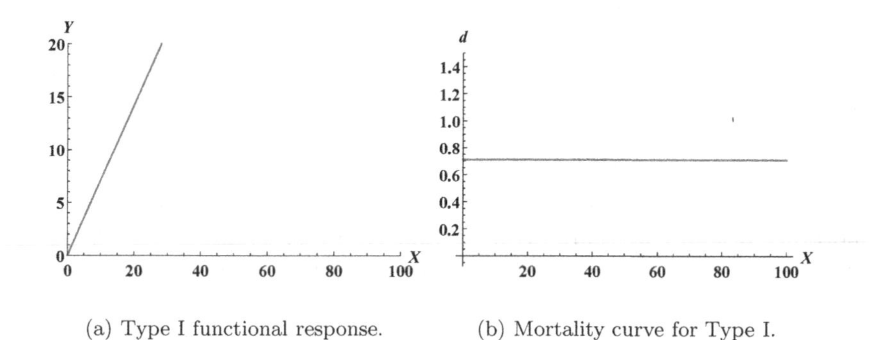

(a) Type I functional response. (b) Mortality curve for Type I.

FIGURE 1.6: *The figure shows the plot of (a) Holling Type I functional response and (b) prey mortality* $d = \frac{Y}{X}$. *Parameter values are* $a = 0.71$ *and* $T_s = T = 1$ *[59].*

1.12.2 Holling Type II Functional Response

We now assume that each predator requires a handling time b (beginning from the time the predator finds its prey to the time the prey is consumed by the predator, which includes chasing, killing, eating, and digesting) for each individual prey that is consumed, then the searching time for the predator is reduced, which is given by,

$$T_s = T - bY.$$

Using Type I functional response, we get,

$$\begin{aligned} Y &= aT_sX = a(T-bY)X \Rightarrow Y = aTX - abXY, \\ \Rightarrow Y &= \frac{aTX}{1+abX}. \end{aligned}$$

Therefore, the total number of prey reduced by the quantity $\frac{aTX}{1+abX}P$ and we have Type II functional response (fig. 1.7(a)).

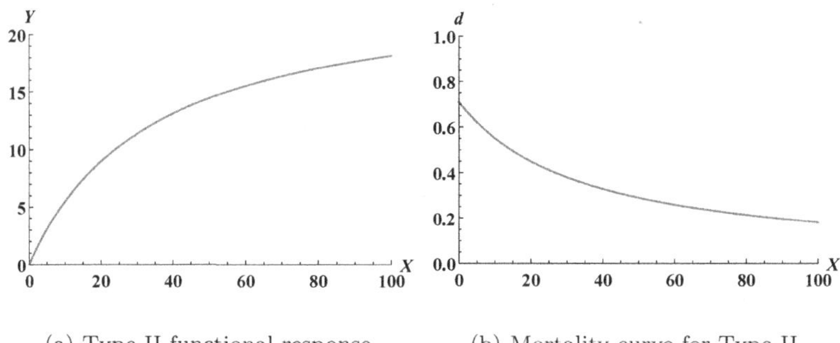

(a) Type II functional response.

(b) Mortality curve for Type II.

FIGURE 1.7: *The figure shows the plot of (a) Holling Type II functional response and (b) prey mortality* $d = \frac{Y}{X}$. *Parameter values are* $a = 0.71$, $b = 0.041$, *and* $T = 1$ *[59].*

In Holling Type II functional response, search rate is constant and prey mortality declines with prey density (fig. 1.7(b)). Maximum damage of this kind of predation happens at low prey density. For example, when the population of gypsy moths is sparse, birds and small mammals destroy most of the moth pupae, whereas a negligible proportion of pupae is killed by them, when the population of gypsy moth is dense.

1.12.3 Holling Type III Functional Response

Type III functional response is the generalization of Type II functional response, given by

$$Y = \frac{aTX^k}{1+abX^k}.$$

This functional form is generally used by theoretical ecologists to study the effect of k on community and population dynamics, by varying its magnitude. In other contexts, $k = 2$ is the general assumption (fig. 1.8(a)).

Many predators respond to chemicals emitted by prey and increase their searching activity. Type III functional response is used in the case of such

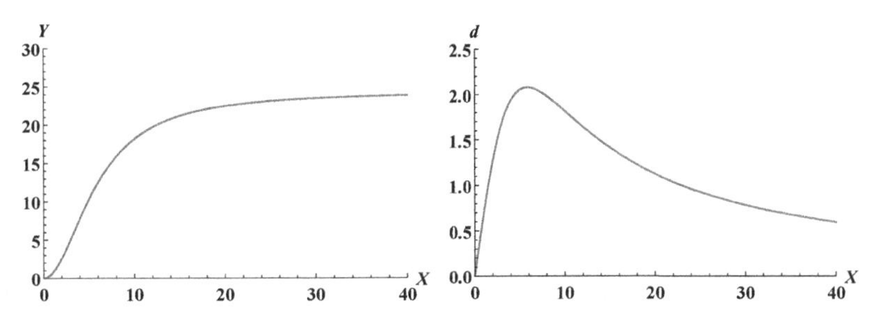

(a) Type III functional response. (b) Mortality curve for Type III.

FIGURE 1.8: *The figure shows the plot of (a) Holling Type III functional response and (b) prey mortality* $d = \frac{Y}{X}$ *for* $k = 2$. *Parameter values are* $a = 0.71$, $b = 0.041$ *and* $T = 1$ *[59].*

predators, who increase their searching activity with increasing prey density. In this type of functional response, the mortality first increases with increasing prey density and then decreases (fig. 1.8(b)).

1.13 Miscellaneous Examples

Problem 1.13.1 *The following equations are meant to predict the volume flow rate of fluid through a small hole in the side of a large tank filled with fluid to a height H above the hole. Which one is dimensionally correct?*

$$(i)\ Q = \frac{AH^2}{g},\ (ii)\ Q = A\sqrt{2gH},\ (iii)\ Q = H\sqrt{2Ag},$$

where A is cross-sectional area of the hole and g is the acceleration due to gravity.

Solution: $[Q] = [\text{Volume of fluid flowing per unit time}] = L^3T^{-1}$.

(i) $\left[\frac{AH^2}{g}\right] = \frac{[A][H^2]}{[g]} = \frac{L^2L^2}{LT^{-2}} = L^3T^2$.

(ii) $[A\sqrt{2gH}] = [A]\sqrt{2}[\sqrt{g}][\sqrt{H}] = \sqrt{2}L^2(\sqrt{LT^{-2}})^{1/2}L^{1/2} = \sqrt{2}L^3T^{-1}$.

(iii) $[H\sqrt{2Ag}] = \sqrt{2}[H][\sqrt{g}][\sqrt{A}] = \sqrt{2}L(L^2)^{1/2}(LT^{-2})^{1/2} = \sqrt{2}L^{5/2}T^{-1}$.

(ii) is dimensionally correct.

Problem 1.13.2 *The distance travelled by a particle in time t is given by $s = a + bt + ct^2 + dt^3$. Find the dimensions of a, b, c and d.*

Solution: $s = a + bt + ct^2 + dt^3$. If the equation is dimensionally correct, we must have the same dimension on both sides. Considering dimensions we get,

$$[s] = [a] + [bt] + [ct^2] + [dt^3] \Rightarrow \; L = [a] + [b]T + [C]T^2 + [d]T^3.$$

L.H.S. dimension is L.
R.H.S. $[a] = L$ and $[b] = LT^{-1}$, so that when multiplied by T, it becomes L. Similarly, $[c] = LT^{-2}$, $[d] = LT^{-3}$.

Problem 1.13.3 *A toy manufacturing company uses the function*

$$f(x) = 0.02x^3 - 0.1x^2 + x + 5$$

to model the cost of running the company, where x is the number of toys (in thousands) and f is the total cost (in thousands).

(i) Draw the graph of this function. What will be the interval of domain?

(ii) What is the cost of manufacturing 2500 toys and what is the cost per toy?

(iii) If the total cost is \$30,000, how many toys have been produced?

Solution: (i) The graph of $f(x) = 0.02x^3 - 0.1x^2 + x + 5$ is shown in (fig. 1.9). Domain: $x \geq 0$.

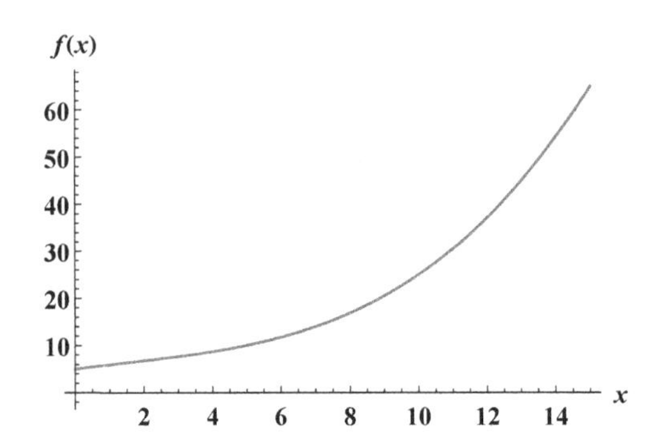

FIGURE 1.9: *Graph of f(x)=* $0.02x^3 - 0.1x^2 + x + 5$, *which represents a cubic function.*

(ii) Cost of manufacturing 2500 toys is $f[2.5] = 0.02(2.5)^3 - 0.1(2.5)^2 + (2.5) + 5 = 7.1875$ and hence $\$7.1875 \times 1000 = \$7187.50$. Cost per toy: $\dfrac{\$7187.50}{2500} = \$2.875$

(iii) If the total cost is \$30000, then the number of toys produced is obtained by solving $0.02x^3 - 0.1x^2 + x + 5 = 30 \Rightarrow x = 10.9135$ thousands $\approx 10,914$.

Problem 1.13.4 *The temperature distribution of a certain place is approximated by the function $T(h)$, representing the temperature (in Celsius) in terms of hours after midnight.*

$$\begin{aligned} T(h) &= 20, \quad 0 \leq h < 6, \\ &= 20 + 3(h-6), \quad 6 \leq h < 14, \\ &= 44 - (h-14)^2, \quad 14 \leq h < 18, \\ &= 28, \quad 18 \leq h < 20, \\ &= 68 - 2h, \quad 20 \leq h \leq 24. \end{aligned}$$

(i) Draw the graph of $T(h)$ and explain, what is happening during each interval.

(ii) There are two times (in terms of hours) when the temperature has reached $20^0 C$, find them.

Solution: (a) The graph starts from midnight (0:00 hrs). In the first 6 hours (0:00–06:00 hrs), the temperature remains constant at 20^0C. In the next 8 hours (6:00–14:00 hrs), there is a steep rise in the temperature at a constant rate, which reaches the maximum value of 44^0C. Then, the temperature starts to fall in the next 4 hours (14:00–18:00 hrs), till it reaches the value 28^0C and remains constant for the next 2 hrs (18:00-20:00 hrs). In the last 4 hours (20:00–24:00 hrs), there is a decline in the temperature at a constant rate till it reaches 20^0C.

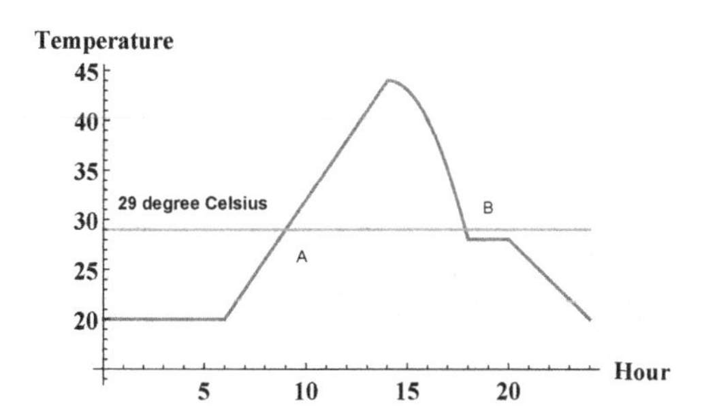

FIGURE 1.10: *Graph of the function $T(h)$, which represents a piecewise function.*

(b) From fig. 1.10 or from the function, it is clear that the temperature of 29^0C is achieved in 9 hrs (y-coordinate of point A). Second time is obtained from the y-coordinate of point B, which is happening between 14 hrs and 18 hrs. Therefore, $44 - (h-14)^2 = 29 \Rightarrow (h-14)^2 = 15 \Rightarrow (h-14) = \pm 3.87 \Rightarrow h = (14 \pm 3.87)$. Clearly, $(14 - 3.87) = 10.13$ hrs lie outside the range $(16, 20)$, hence, $(14+3.87) = 17.87$ hrs is the time when the temperature reaches 29^0C for the second time.

Problem 1.13.5 *Obtain the solution of the Logistic growth model by solving the differential equation*

$$\frac{dx}{dt} = rx\left(1 - \frac{x}{k}\right), \; x(0) = x_0,$$

analytically. Hence, obtain the limiting value of $x(t)$ as $t \to \infty$.

Solution: Separation of variables gives

$$\left[\frac{1}{x-k} - \frac{1}{x}\right] dx = -rdt \Rightarrow \ln\left(\frac{x-k}{x}\right) = -rt + \text{Constant}.$$

$$\text{At } t = 0, x = x_0 \Rightarrow \text{Constant} = \ln\left(\frac{x_0 - k}{x_0}\right).$$

$$\Rightarrow \ln\left(\frac{x-k}{x}\right) = -rt + \ln\left(\frac{x_0 - k}{x_0}\right) \Rightarrow \frac{\dfrac{x-k}{x}}{\dfrac{x_0 - k}{x_0}} = e^{-rt}.$$

$$\text{On simplification, } x(t) = \frac{kx_0}{x_0 + (k - x_0)e^{-rt}}.$$

$$\text{Hence, } \lim_{t\to\infty} x(t) = \lim_{t\to\infty}\left[\frac{kx_0}{x_0 + (k - x_0)e^{-rt}}\right] = \frac{kx_0}{x_0} = k.$$

1.14 Exercises

1. The velocity u of propagation of deep ocean waves is a function of the wavelength λ, acceleration g due to gravity and density ρ of the liquid. By assuming that $u = k\lambda^a g^b \rho^c$, where k is a dimensionless constant, a, b, c are real numbers, find the relation between u,λ, g, and ρ.

2. Which of the following equations contain an error and which are dimensionally correct?

   (i) $v^2 = u^2 + 2gz$,

   (ii) $p = \dfrac{\rho l u}{z}$,

   (iii) $F = -pA$,

   (iv) $p + \dfrac{1}{2}\rho u^2 = -mgz$,

where p is the pressure, ρ is the density, u and v are velocities, g is the acceleration due to gravity, F is a force, m is a mass, A is an area, z and l are lengths.

3. A man plans to build a garden, triangular in shape, such that the height of the triangle is 4 ft shorter than the length of the base and also plans to put a fence along the hypotenuse of the triangle. Write an equation to model the length of the hypotenuse and obtain the value (in ft) that will minimize the length of the hypotenuse.

4. The time of oscillation T_1 of a small drop of liquid under surface tension depends on the density ρ, the radius r and the surface tension s and is given by $T_1 = \sqrt{\frac{\rho r^3}{s}}$. Check whether the equation is dimensionally correct or not.

5. Due to the increase in global population, many environmentalists believe that the carrying capacity of the Earth (the maximum population Earth can sustain) depends not only on the finite natural resources of the planet-water, land, air, and materials, but also on how the people use and preserve the resources. The graphs show four different ways that a growing population can approach its carrying capacity over time. Explain each of the scenarios by analyzing the graphs.

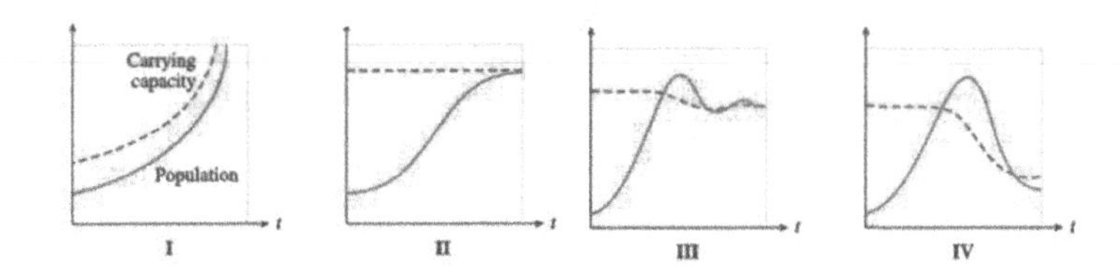

6. A person running a business, makes an initial investment of \$2000 and the cost of manufacturing each item is \$15. Write a function to model the total costs. Also, write a function to model the average cost per item. Graphically, find the minimum of this function and interpret the result.

7. Suppose a company decides to make plastic container, which can hold 0.001 m^3 (1 litre) of some powder. The container should have a circular base, with double thickness at the top and the bottom. Let the cost of the plastic sheet be \$0.20 per square meter. Using the concept of mathematical modeling, obtain the most economical size of the plastic

container for the powder.

8. In an environment, a new species has been introduced, which can affect the growth of an existing species in various ways. The following graphs show four hypothetical scenarios after Species A is introduced into an environment where Species B is already present and living. Explain each of the scenarios by analyzing the graphs.

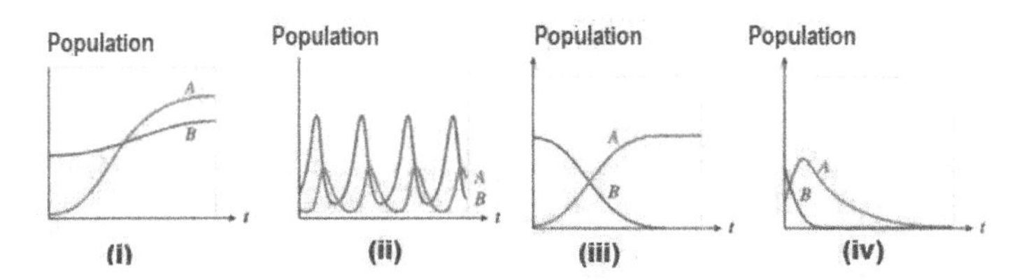

9. Obtain the solution of the Gompertz growth model by solving the differential equation

$$\frac{dx}{dt} = rx \ln\left(\frac{k}{x}\right), \quad x(0) = x_0,$$

analytically. Hence, obtain the limiting value of $x(t)$ as $t \to \infty$.

10. Four different functions are described in figs. (A), (B), (C), and (D). Match each description with the appropriate graph.

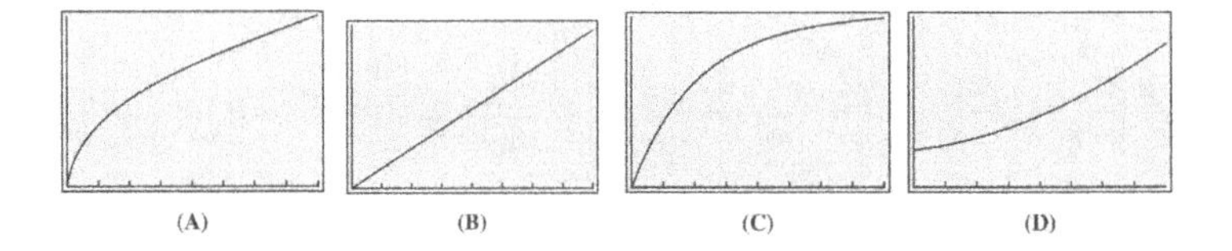

(i) The concentration of the pollutant that pours into a lake is a function of time and initially increases quite rapidly, but due to the natural mixing and self-cleansing action of the lake, the concentration levels off and stabilizes at some saturation level.

(ii) A train travels at a constant speed and its distance from the point of origin is a function of time.

(iii) The population of a city, which is a function of time, starts increasing rather slowly, but continues to grow at a faster and faster rate.

(iv) The production level at a manufacturing plant is a function of the amount of money invested in the plant (capital outlay). At first, small increases in capital outlay result in large increases in production, but eventually, the investors begin to experience diminishing returns on their money, so that although production continues to increase, it is at a disappointingly slow rate.

11. The reaction of lead nitrate and potassium iodide in solution is lead iodide, which precipitates as a yellow solid at the bottom of the container. More lead iodine is produced if you add more lead nitrate to the solution and continues till all the potassium iodide is used up. The table shows the height of the precipitate in the container as a function of the amount of lead nitrate added [61].

Lead nitrate solution (cc)	0.5	1.0	1.5	2.0	2.5	3.0	3.5	4.0
Height of precipitate (mm)	2.8	4.8	6.2	7.4	9.5	9.6	9.6	9.6

(i) Plot the data and sketch a piecewise linear function with two parts to fit the data points.

(ii) Write a mathematical formula for the piecewise function.

(iii) Interpret your graph in the context of the problem.

12. A person goes into skydiving and let $f(t)$ approximates his altitude as a function time t (in seconds), starting from ground, till he reaches the ground.

$$\begin{aligned} f(t) &= 30t, \quad 0 \leq t < 300, \\ &= 9000, \quad 300 \leq t < 400, \\ &= 9000 - 16(t-400)^2, \quad 400 \leq t < 420, \\ &= 2600 - 130(t-420), \quad 420 \leq t \leq 440. \end{aligned}$$

(i) Draw the graph of $f(t)$ and explain, what is happening during each interval.

(ii) There are two times that the person is at an altitude of 3600 ft, find them.

13. A person runs a business of wooden jewelery boxes. The cost of manufacturing the wooden box is \$6 per box, plus an initial investment of \$250 is needed for other materials. Another \$3 is needed to decorate the boxes. Write a function to model the total cost of the boxes.

14. Consider the damped Lotka Volterra predator–prey model

$$\begin{aligned}\frac{dx}{dt} &= rx(1-\frac{x}{k}) - \alpha xy,\\ \frac{dy}{dt} &= -\beta y + \gamma xy, \ x(0) = x_0, \ y(0) = y_0,\end{aligned}$$

where x and y are densities of prey and predator, respectively, r is the intrinsic growth rate of the prey and k is its carrying capacity (maximum resources the environment can sustain indefinitely), α is the predation rate per individual per unit time, γ is the reproduction rate of predators per individual per unit time. Here, $r, k, \alpha, \beta, \gamma$ are positive parameters and the initial conditions are $x(0) = x_0$, $y(0) = y_0$. Non-dimensionalize the model.

15. A chocolate-making company markets its products to individuals, to retailers, and to wholesale distributors. Three different prices are offered by the company for its chocolates. If you order 15 or fewer boxes, the price is $4.50 each. If you order more than 15 but no more than 60 boxes, the price is $3.75 each. If you order more than 60 boxes, the price is $2.75 each. Write a function $f(x)$ to model the ordering x number of boxes of chocolates and plot the graph of $f(x)$.

Chapter 2

Discrete Models Using Difference Equations

2.1 Difference Equations

The term *difference equation* sometimes refers to a specific type of recurrence relation. However, difference equation is frequently used to refer to any recurrence relation. It is a discrete analog of a differential equation.

So, how do you define a difference equation?

A difference equation is an equation involving differences. One can define a difference equation as a sequence of numbers that are generated recursively using a rule to the previous numbers in the sequence [74]. For example, the sequence of triangular numbers

$$0,\ 1,\ 3,\ 6,\ 10,\ 15,\ 21,\ 28,\ 36,\ 45, \ldots,$$

is obtained from the difference equation

$$T_{n+1} = T_n + (n+1)$$

with $T_0 = 0$.

Clearly, $T_1 = T_0 + (0+1) = 0 + (0+1) = 1\ (n=0);$
$T_2 = T_1 + (1+1) = 1 + (1+1) = 3\ (n=1);$
$T_3 = T_2 + (2+1) = 3 + (2+1) = 6\ (n=2);$
....
and so on.

One can also define difference equation as an iterated map

$$x_{n+1} = f(x_n),$$

if the sequence is observed as an iterated function: $x_0,\ f(x_0),\ f(f(x_0)), \ldots,$ where $f(x_0)$ is the first iterate of x_0 under f.

DOI: 10.1201/9781351022941-2

Let us now see how a difference equation is formulated. Consider the relation

$$u_n = cn - 3,$$

where c is an arbitrary constant. Now,

$$u_{n+1} = c(n+1) - 3.$$

The required difference equation is obtained by eliminating c from u_n and u_{n+1}, which gives

$$u_{n+1} = \frac{u_n + 3}{n}(n+1) - 3 \Rightarrow nu_{n+1} = (n+1)u_n + 3.$$

Difference equations can be linear or non-linear. A linear difference equation is an equation that relates a term in a sequence to previous terms using recursion. The word linear refers to the fact that previous terms are arranged as a first-degree polynomial in the recurrence relation [126]. If the previous terms are polynomials of degree more than one, the difference equation is non-linear. For example, $x_{n+1} = 6x_n + 5$ is a linear difference equation, whereas $x_{n+1} = 6x_n^2 + 5$ is non-linear.

A linear difference equation is of the form

$$x_{n+1} = \alpha_n x_n + \beta_n,$$

where, α_n and β_n are known real constants. If $\alpha_n = \alpha$ and $\beta_n = \beta$, for all n, then the equation is said to be autonomous, otherwise, it is said to be non-autonomous. For example, $x_{n+1} = 5x_n + 2$ is an autonomous equation and $x_{n+1} = \frac{5}{n+1}x_n + \frac{2}{n+2}$ is non-autonomous .

2.1.1 Linear Difference Equation with Constant Coefficients

Consider the linear difference equation of the form

$$c_0 u_n + c_1 u_{n-1} + c_2 u_{n-2} = f(n). \tag{2.1}$$

The difference equation is homogeneous if $f(n) = 0$, otherwise it is non-homogeneous. The order of the difference equation is the difference between the largest (n) and smallest ($n-2$) arguments appearing in the difference equation with unit interval. Thus, the order of equation (2.1) is 2.

Equation (2.1) is a linear difference equation with constant coefficients as the coefficients of the successive differences are constants and the differences of successive orders are of the first degree.

2.1.2 Solution of Homogeneous Equations

(a) Consider the first-order homogeneous linear difference equation

$$u_n - k(u_{n-1}) = 0. \tag{2.2}$$

Putting n = 1,2,3,... we get,

$$\begin{aligned} u_1 &= ku_0, \\ u_2 = ku_1 &= k^2u_0, \\ u_3 = ku_2 &= k^3u_0, \\ &- - - - - - - - - - - - \\ u_n = ku_{n-1} &= k^n u_0. \end{aligned}$$

Therefore, $u_n = c\,k^n$ (c is an arbitrary constant) is a general solution of (2.2).

(b) Consider the first-order linear difference equation of the form

$$u_n = au_{n-1} + b = 0 \quad (\text{n} = 1,2,\ldots),$$

where a and b are constants. Now,

$$\begin{aligned} u_n &= a(au_{n-2} + b) + b = a^2u_{n-2} + b(a+1) \\ &= a^2(au_{n-3} + b) + b(a+1) = a^3u_{n-3} + b(a^2 + a + 1) \\ &= \ldots\ldots\ldots\ldots\ldots\ldots\ldots\ldots \\ &= a^n u_0 + b(a^{n-1} + a^{n-2} + \ldots\ldots + a^2 + a + 1) \\ &= a^n u_0 + b\left(\frac{1-a^n}{1-a}\right) \quad (\text{if a} < 1), \\ &= a^n u_0 + b\left(\frac{a^n - 1}{a-1}\right) \quad (\text{if a} > 1), \\ &= a^n u_0 + nb \quad (\text{if a} = 1), \end{aligned}$$

which is the required solution of the first-order linear difference equation $u_n = au_{n-1} + b$.

(c) Consider the second-order homogeneous linear difference equation

$$a_0u_n + a_1u_{n-1} + a_2u_{n-2} = 0. \tag{2.3}$$

Let the solution of (2.3) be of the form $u_n = c\,k^n$ $(c \neq 0)$. Substituting it in (2.3), we obtain

$$\begin{aligned} a_0ck^n + a_1ck^{n-1} + a_2ck^{n-2} &= 0 \\ \Rightarrow \quad a_0k^2 + a_1k + a_2 &= 0, \end{aligned}$$

which is called the auxiliary equation.

(i) If the auxiliary equation has two distinct real roots, m_1 and m_2 (say), then,

$$c_1m_1^n + c_2m_2^n$$

is the general solution of (2.3), c_1 and c_2 are arbitrary constants.

(ii) If the roots of the auxiliary equation are real and equal, $m_1 = m_2 = m$ (say), then,

$$(c_1 + c_2 n)m^n$$

is the general solution of (2.3), c_1 and c_2 are arbitrary constants.

(iii) If the auxiliary equation has imaginary roots (which occur in conjugate pairs), $\alpha + i\beta$ and $\alpha - i\beta$ (say), then,

$$r^n(c_1 \cos n\theta + c_2 \sin n\theta)$$

is the general solution of (2.3), $r = \sqrt{\alpha^2 + \beta^2}$ and $\theta = \tan^{-1}(\frac{\beta}{\alpha})$, c_1 and c_2 are arbitrary constants.

Note: Solutions for non-homogeneous equations can be obtained by particular integral methods, undetermined coefficients, Z-Transform, Laplace Transform, etc. Interested readers can look into [31, 47, 106] for detailed information.

Example 2.1.1 Obtain the difference equation by eliminating the arbitrary constants from $u_n = A2^n + B(-3)^n$.

Solution: Given,

$$\begin{aligned} u_n &= A2^n + B(-3)^n, \\ \Rightarrow u_{n+1} &= A2^{n+1} + B(-3)^{n+1}, \\ \Rightarrow u_{n+2} &= A2^{n+2} + B(-3)^{n+2}. \end{aligned}$$

Therefore,

$$\begin{aligned} u_{n+1} &= 2A2^n - 3B(-3)^n, \\ u_{n+2} &= 4A2^n + 9B(-3)^n. \end{aligned}$$

Solving, we get

$$A = \frac{3u_{n+1} + u_{n+2}}{10\ 2^n} \text{ and } B = \frac{u_{n+2} - 2u_{n+1}}{15\ (-3)^n}.$$

Hence, the required difference equation is

$$u_n = \frac{3u_{n+1} + u_{n+2}}{10} + \frac{u_{n+2} - 2u_{n+1}}{15},$$

$$\Rightarrow u_{n+2} + u_{n+1} - 6u_n = 0.$$

Example 2.1.2 Find u_n if $u_0 = 0, u_1 = 1$ and $u_{n+2} + 16u_n = 0$.

Solution: Let $u_n = c\,k^n (c \neq 0)$ be a solution of $u_{n+2} + 16u_n = 0$, then the required auxiliary equation is

$$k^2 + 16 = 0 \Rightarrow k = \pm 4i.$$

The general solution is

$$\begin{aligned} u_n &= c_1(4i)^n + c_2(-4i)^n, \\ &= 4^n[c_1 e^{in\pi/2} + c_2 e^{-in\pi/2}], \\ &= 4^n[A_1 \cos(n\pi/2) + A_2 \sin(n\pi/2)], \end{aligned}$$

where A_1 and A_2 are arbitrary constants. Now, $u_0 = 0$ and $u_1 = 1$ implies $A_1 = 0$ and $A_2 = \frac{1}{4}$. Therefore,

$$u_n = 4^{n-1} \sin\left(\frac{n\pi}{2}\right)$$

is the required solution.

Note: The solutions of homogeneous linear difference equations with constant coefficients are composed of linear combinations of the basic expressions of the form $u_n = ck^n$. The qualitative behavior of the basic solution will depend on the real values of k, namely, on the four possible ranges [30]:

$$k \geq 1, \quad k \leq -1, \quad 0 < k < 1, \quad -1 < k < 0.$$

For $k > 1$, the solution $u_n = ck^n$ becomes unbounded as n increases (fig. 2.1(a)); for $0 < k < 1$, k^n goes to zero as n increases, hence u_n decreases (fig. 2.1(b)); for $-1 < k < 0$, k^n oscillates between positive and negative values, with diminishing magnitude to zero (fig. 2.1(c)) and for $k < -1$, k^n oscillates between positive and negative values with increasing magnitude (fig. 2.1(d)).

The marginal points $k = 1, k = 0$ and $k = -1$ correspond to constant solution ($u_n = c$), zero solution ($u_n = 0$) and an oscillatory solution between $-c$ and $+c$, respectively. Fig. 2.1 illustrates different behaviors of the solution for different ranges of k.

Example 2.1.3 Solve: $x_{n+1} = \frac{x_n}{4} + y_n,\ y_{n+1} = 3\frac{x_n}{16} - \frac{y_n}{4}$.

Solution:

$$x_{n+1} = \frac{x_n}{4} + y_n \Rightarrow \quad y_n = x_{n+1} - \frac{x_n}{4} \Rightarrow \quad y_{n+1} = x_{n+2} - \frac{x_{n+1}}{4}.$$

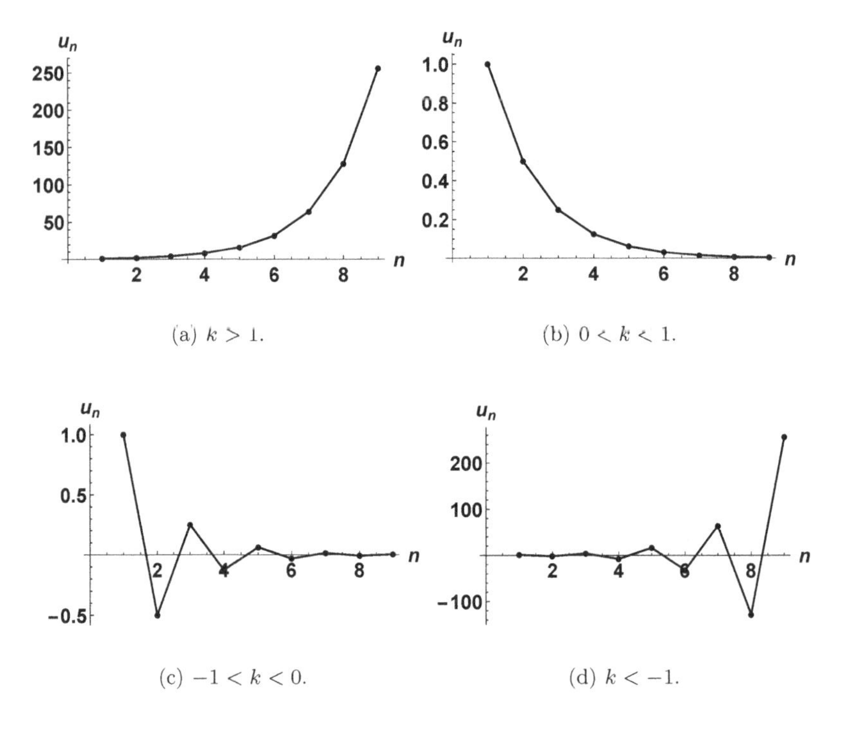

FIGURE 2.1: *The figures show the qualitative behavior of $u_n = ck^n$ for different ranges of k.*

Eliminating y_{n+1} from both the equations, we get,

$$4x_{n+2} - x_n = 0.$$

Let $x_n = ck^n (c \neq 0)$ be a solution of $4x_{n+2} - x_n = 0$. The required auxiliary equation is

$$4k^2 - 1 = 0 \Rightarrow k = \pm\frac{1}{2}.$$

The general solution is

$$x_n = c_1 \left(\frac{1}{2}\right)^n + c_2 \left(\frac{-1}{2}\right)^n,$$

where c_1 and c_2 are arbitrary constants. Similarly, it can be shown,

$$y_n = d_1 \left(\frac{1}{2}\right)^n + d_2 \left(\frac{-1}{2}\right)^n,$$

where d_1 and d_2 are arbitrary constants.

2.1.3 Difference Equations: Equilibria and Stability

2.1.3.1 Linear Difference Equations

We consider an autonomous linear discrete equation of the form

$$u_n = a\, u_{n-1} + b \ (a \neq 1).$$

By equilibrium point (or fixed points or steady-state solutions), it is meant that there is no change from generation $(n-1)$ to generation n. If u^* be the equilibrium solution of the model, then

$$u_n = u_{n-1} = u^* \ \Rightarrow au^* + b = u^* \ \Rightarrow u^* = \frac{b}{1-a}.$$

The equilibrium point u^* is said to be stable if all the solutions of $u_n = a\ u_{n-1} + b$ approach $u^* = \dfrac{b}{1-a}$ as $n \to \infty$ (as n becomes large). The equilibrium point u^* is unstable if all solutions (if exists) diverge from u^* to $\pm\infty$.

The stability of the equilibrium point u^* depends on a. The fixed point (equilibrium point) u^* of the autonomous discrete equation $u_n = a\ u_{n-1} + b$ is
(i) stable if $|a| < 1$,
(ii) unstable if $|a| > 1$ and
(iii) if $a = \pm 1$, the case is ambiguous.

Note: If $a > 0$, the solutions converge monotonically to u^* and if $a < 0$, the solutions converge to u^* with oscillations.

Example 2.1.4 Find the equilibrium point of

$$x_{n+1} = a(x_n - 1)$$

for $a = \frac{4}{5}$ and determine its stability. Explain the dynamics when $a = -\frac{4}{5}, \frac{5}{4}, -\frac{5}{4}$.

Solution: (i) If x^* be the equilibrium point of $x_{n+1} = \frac{4}{5}(x_n - 1)$, then

$$\begin{aligned} x_{n+1} &= x_n = x^*, \\ \Rightarrow x^* &= \frac{4}{5}(x^* - 1), \\ \Rightarrow x^* &= -4. \end{aligned}$$

The given equation is of the form

$$x_{n+1} = ax_n + b,$$

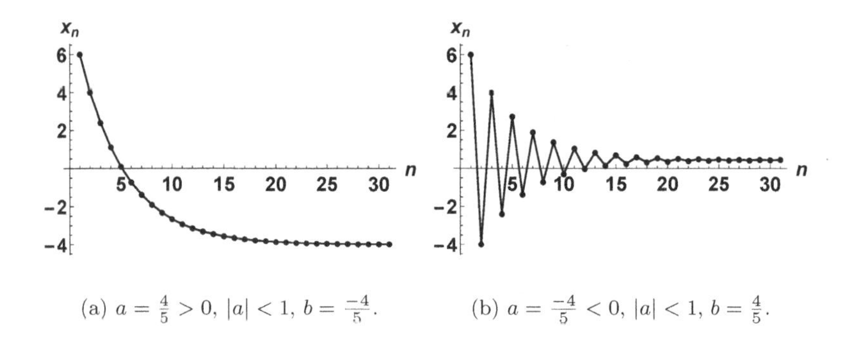

(a) $a = \frac{4}{5} > 0$, $|a| < 1$, $b = \frac{-4}{5}$.

(b) $a = \frac{-4}{5} < 0$, $|a| < 1$, $b = \frac{4}{5}$.

FIGURE 2.2: *The figures show the plot of* $x_{n+1} = ax_n + b$ *for different values of* a *and* b *with* $a_0 = 6$. *(a) The solution converges monotonically to* $x^* = -4$. *(b) The solution converges with oscillations to* $x^* = 0.444$.

where $|a| = |\frac{4}{5}| = \frac{4}{5} < 1$. Therefore, the equilibrium point is stable.

Since, $a > 0$, the solution with converge monotonically (fig. 2.2(a)).

(ii) For $a = \frac{-4}{5}$, the equilibrium point is given by

$$\begin{aligned} x^* &= \frac{-4}{5}(x^* - 1), \\ \Rightarrow x^* &= \frac{4}{9} = 0.444. \end{aligned}$$

The given equation is of the form

$$x_{n+1} = ax_n + b,$$

where $|a| = |\frac{-4}{5}| = \frac{4}{5} < 1$. Therefore, the equilibrium point is stable.

Since, $a < 0$, the solution converges to x^* with oscillations (fig. 2.2(b)).

(iii) For $a = \frac{5}{4}$, the equilibrium point is given by

$$\begin{aligned} x^* &= \frac{5}{4}(x^* - 1), \\ \Rightarrow x^* &= 5. \end{aligned}$$

The given equation is of the form

$$x_{n+1} = ax_n + b,$$

where $|a| = |\frac{5}{4}| = \frac{5}{4} > 1$. Therefore, the equilibrium point is unstable.

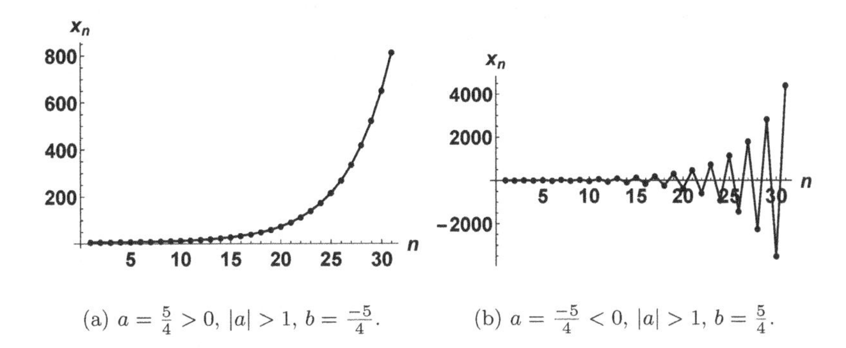

(a) $a = \frac{5}{4} > 0$, $|a| > 1$, $b = \frac{-5}{4}$. (b) $a = \frac{-5}{4} < 0$, $|a| > 1$, $b = \frac{5}{4}$.

FIGURE 2.3: *The figures show the plot of $x_{n+1} = ax_n + b$ for different values of a and b with $a_0 = 6$. (a) The solution diverges monotonically from $x^* = 5$. (b) The solution diverges with oscillations from $x^* = 0.556$.*

Since, $a > 0$, the solution diverges monotonically from x^* (fig. 2.3(a)).

(iv) For $a = \frac{-5}{4}$, the equilibrium point is given by

$$\begin{aligned} x^* &= \frac{-5}{4}(x^* - 1), \\ \Rightarrow x^* &= \frac{5}{9} = 0.556. \end{aligned}$$

The given equation is of the form

$$x_{n+1} = ax_n + b,$$

where $|a| = |\frac{-5}{4}| = \frac{5}{4} > 1$. Therefore, the equilibrium point is unstable.

Since, $a < 0$, the solution diverges from x^* with oscillation (fig. 2.3(b)).

2.1.3.2 System of Linear Difference Equations

For a system of difference equations, it is possible to determine the stability of the system using eigenvalues. We consider the linear homogeneous system,

$$\begin{aligned} u_{n+1} &= \alpha u_n + \beta v_n, \\ v_{n+1} &= \gamma u_n + \delta v_n, \end{aligned} \tag{2.4}$$

which can be expressed in the matrix form as

$$\begin{pmatrix} u_{n+1} \\ v_{n+1} \end{pmatrix} = \begin{pmatrix} \alpha & \beta \\ \gamma & \delta \end{pmatrix} \begin{pmatrix} u_n \\ v_n \end{pmatrix},$$

$$\Rightarrow w_{n+1} = Aw_n,$$

$$\text{where } w_n = \begin{pmatrix} u_n \\ v_n \end{pmatrix} \text{ and } A = \begin{pmatrix} \alpha & \beta \\ \gamma & \delta \end{pmatrix}.$$

Clearly $(0, 0)$ is the equilibrium point of the homogeneous system.

Theorem 2.1.3.1 *Let λ_1 and λ_2 be two real distinct eigenvalues of the coefficient matrix A of the homogeneous linear system (2.5). Then, the equilibrium point $(0, 0)$ is*
(i) stable if both $|\lambda_1| < 1$ and $|\lambda_2| < 1$,
(ii) unstable if both $|\lambda_1| > 1$ and $|\lambda_2| > 1$,
(iii) saddle if $|\lambda_1| < 1$ and $|\lambda_2| > 1$ or if $|\lambda_1| > 1$ and $|\lambda_2| < 1$.

Note: *$\lambda_1 = \lambda_2 = 1$ is a rare borderline case and will not be considered here.*

Theorem 2.1.3.2 *Let $\lambda_1 = \lambda_2 = \lambda^*$ be real and equal eigenvalues of the coefficient matrix A of the homogeneous linear system (2.5), then the equilibrium point $(0, 0)$ is*
(i) stable if $|\lambda^*| < 1$,
(ii) unstable if $|\lambda^*| > 1$.

Theorem 2.1.3.3 *Let $a+ib$ and $a-ib$ be the complex conjugate eigenvalues of the coefficient matrix A of a homogeneous linear system, then the equilibrium point $(0, 0)$ is*
(i) stable focus or spiral if $|a \pm ib| < 1$,
(ii) unstable focus or spiral if $|a \pm ib| > 1$.

Consider a non-homogeneous linear system of the form

$$w_{n+1} = Aw_n + b,$$

$$\text{where } w_n = \begin{pmatrix} u_{n+1} \\ v_{n+1} \end{pmatrix} \text{ and } A = \begin{pmatrix} \alpha & \beta \\ \gamma & \delta \end{pmatrix} \text{ and } b = \begin{pmatrix} k_1 \\ k_2 \end{pmatrix}.$$

The equilibrium solution $w^* = (u^*, v^*)^T$ of the system is obtained by solving

$$u^* = \alpha u^* + \beta v^* + k_1, \text{ and } v^* = \gamma u^* + \delta v^* + k_2. \tag{2.5}$$

$$\Rightarrow\ u^* = \frac{\beta k_2 - (\delta - 1)k_1}{(\alpha - 1)(\delta - 1) - \beta\gamma},\quad v^* = \frac{\gamma k_1 - (\alpha - 1)k_2}{(\alpha - 1)(\delta - 1) - \beta\gamma}.$$

For stability, the same results hold as for the homogeneous system. This is due to the fact that the non-homogeneous system, with a unique equilibrium point, can be converted to a homogeneous system. The system (2.5) can be written as

$$w^* = Aw^* + b \Rightarrow A\, w^* - w^* + b = \mathbf{0} \text{ (null matrix)},$$

Substituting $w_n = z_n + w^*$ in $w_{n+1} = Aw_n + b$ we get,

$$z_{n+1} + w^* = A(z_n + w^*) + b \Rightarrow z_{n+1} = Az_n + Aw^* - w^* + b$$
$$\Rightarrow z_{n+1} = Az_n \quad (\text{since } Aw^* - w^* + b = \mathbf{0}),$$

which is a linear homogeneous system, whose stability has already been discussed.

Example 2.1.5 Find the equilibrium point of the linear homogeneous system

$$x_{n+1} = -x_n - 4y_n, \ y_{n+1} = x_n - y_n.$$

and check its stability.

Solution: If (x^*, y^*) be the equilibrium solution, it is obtained by solving

$$x^* = -x^* - 4y^* \Rightarrow x^* + 2y^* = 0,$$
$$y^* = x^* - y^* \Rightarrow x^* - 2y^* = 0.$$

Clearly, $(0, 0)$ is the only solution of the system as the coefficient matrix

$\begin{pmatrix} 1 & 2 \\ 1 & -2 \end{pmatrix}$ is non-singular, that is, $\begin{vmatrix} 1 & 2 \\ 1 & -2 \end{vmatrix} = -4 \neq 0.$

The coefficient matrix of the linear homogeneous system is $\begin{pmatrix} 1 & 2 \\ 1 & -2 \end{pmatrix}$, whose eigenvalues are obtained by solving $\begin{vmatrix} 1-\lambda & 2 \\ 1 & -2-\lambda \end{vmatrix} = 0,$
$\Rightarrow \lambda = -1 \pm 2\imath$, and $|-1 \pm 2\imath| = \sqrt{1+4} = \sqrt{5} > 1$. Hence, the equilibrium point $(0, 0)$ is unstable.

Example 2.1.6 Find the equilibrium point of the linear homogeneous system

$$x_{n+1} = \alpha x_n + 0.12y_n, \ y_{n+1} = 3x_n + \alpha y_n.$$

Find all the real values of α for which the equilibrium point is stable.

Solution: Clearly, $(0, 0)$ is the equilibrium point. The coefficient matrix of the linear homogeneous system is $\begin{pmatrix} \alpha & 0.12 \\ 3 & \alpha \end{pmatrix}$, whose eigenvalues are obtained by solving

$$\begin{vmatrix} \alpha-\lambda & 0.12 \\ 3 & \alpha-\lambda \end{vmatrix} = 0 \Rightarrow (\alpha - \lambda)^2 = 0.36 \Rightarrow \lambda = \alpha \pm 0.6.$$

The equilibrium point $(0, 0)$ will be stable if $|\alpha + 0.6| < 1$ and $|\alpha - 0.6| < 1$.

$|\alpha + 0.6| < 1 \Rightarrow -1 < \alpha + 0.6 < 1 \Rightarrow -1.6 < \alpha < 0.4$. Similarly,

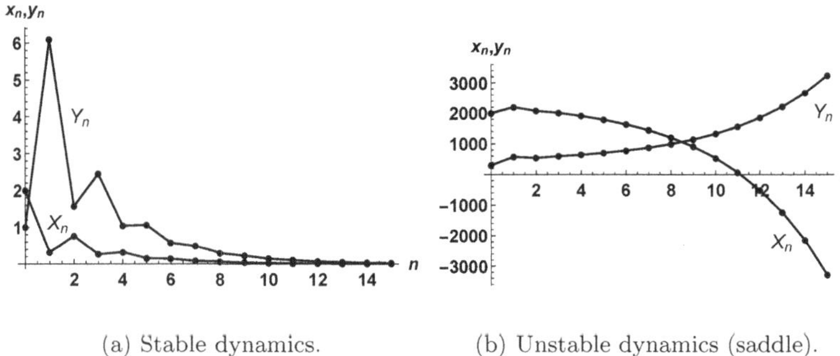

(a) Stable dynamics. (b) Unstable dynamics (saddle).

FIGURE 2.4: *The figures show the dynamics of* x_n *and* y_n *(a) for* $\alpha = 0.1$ *and initial value* $x[0] = 2, y[0] = 1$, *(b) initial value* $x[0] = 2000, y[0] = 300$.

$$|\alpha - 0.6| < 1 \;\Rightarrow\; -1 < \alpha - 0.6 < 1 \;\Rightarrow\; -0.4 < \alpha < 1.6.$$

Combining, we obtain $-0.4 < \alpha < 0.4$ (common region of the two inequalities), which gives all the real values of α for which the equilibrium point is stable (fig. 2.4(a)).

Example 2.1.7 Find the equilibrium point of the linear non-homogeneous system

$$x_{n+1} = 0.75x_n - y_n + 1000, \; y_{n+1} = -0.5x_n + 0.25y_n + 1500$$

and check its stability.

Solution: If (x^*, y^*) be the equilibrium solution of the given non-homogeneous system, it is obtained by solving

$$x^* = 0.75x^* - y^* + 1000 \;\Rightarrow\; 0.25x^* + y^* = 1000,$$
$$y^* = -0.5x^* + 0.25y^* + 1500 \;\Rightarrow\; 0.5x^* + 0.75y^* = 1500.$$

Solving, we get $(x^*, y^*) = (2400, 400)$ as the unique solution of the system. The coefficient matrix of the linear non-homogeneous system is

$$\begin{pmatrix} 0.75 & -1 \\ -0.5 & 0.25 \end{pmatrix},$$

whose eigenvalues are obtained by solving

$$\begin{vmatrix} 0.75 - \lambda & -1 \\ -0.5 & 0.25 - \lambda \end{vmatrix} = 0,$$

$\Rightarrow \; \lambda_1 = 1.25$ and $\lambda_2 = -0.25$. Now, $|\lambda_1| = 1.25 > 1$ and $|\lambda_2| = 0.25 < 1$. Hence, the equilibrium point $(2400, 400)$ is a saddle (fig. 2.4(b)).

2.1.3.3 Non-Linear Difference Equations

Non-linear difference equations are to be handled with special techniques and cannot be solved by simply setting $u_n = c\ k^n$. Here, we shall not discuss about the solutions of non-linear difference equations but focus on the qualitative behaviors, namely, equilibrium solution (fixed point or steady state), stability, cycles, bifurcations and chaos.

In the context of difference equations, x^* is the steady-state solution (equilibrium solution) of the non-linear difference equation

$$x_{n+1} = f(x_n), \text{ if } x_{n+1} = x_n = x^*,$$

that is, there is no change from generation n to generation (n+1).

By definition, the steady-state solution is stable if for $\epsilon > 0, \exists$ a $\delta > 0$ such that $|x_0 - x^*| < \delta$ implies that for all $n > 0$, $|f^n(x_0) - x^*| < \epsilon$. The steady-state solution is asymptotically stable if, in addition, $\lim_{x \to \infty} x_n = x^*$ holds.

After obtaining the equilibrium solution, we look into its stability, that is, given some value x_n close to x^*, does x_n tends towards x^* or move away from it? To address this issue, we give a small perturbation to the system about the steady state x^*. Mathematically, this means replacing x_n by $x^* + \epsilon_n$, where ϵ_n is small. Then,

$$\begin{aligned} x_{n+1} &= f(x_n), \\ \Rightarrow \quad x^* + \epsilon_{n+1} &= f(x^* + \epsilon_n), \\ &\approx f(x^*) + \epsilon_n f'(x^*) \quad \text{(by Taylor series expansion)}, \\ &= x^* + \epsilon_n f'(x^*). \end{aligned}$$

Since x^* is the equilibrium solution, $x^* = f(x^*)$, which implies

$$\epsilon_{n+1} \approx \epsilon_n f'(x^*). \tag{2.6}$$

The solution of (2.6) will decrease if $|f'(x^*)| < 1$, and increase if $|f'(x^*)| > 1$ [106].

Theorem 2.1.3.4 *The equilibrium solution x^* of $x_{n+1} = f(x_n)$ is stable if $|f'(x^*)| < 1$ and unstable if $|f'(x^*)| > 1$. No definite conclusion if $|f'(x^*)| = 1$. Also, if $f'(x^*) < 0$, then the solution oscillates locally around x^*, but if $f'(x^*) > 0$, they do not oscillate (f(x) must be differentiable at $x = x^*$).*

The equilibrium solution (u^*, v^*) of a non-linear discrete system of the form $u_{n+1} = f(u_n, v_n)$ and $v_{n+1} = g(u_n, v_n)$ is obtained by solving

$$u^* = f(u^*, v^*) \text{ and } v^* = g(u^*, v^*).$$

Its stability analysis near the equilibrium point (u^*, v^*) can be determined by linearizing the system about the equilibrium point.

Theorem 2.1.3.5 *Let (u^*, v^*) be an equilibrium solution of non-linear systems $u_{n+1} = f(u_n, v_n)$ and $v_{n+1} = g(u_n, v_n)$ and A be the corresponding matrix of partial derivatives (also known as the Jacobian matrix) given by,*

$$A = \begin{pmatrix} f_u(u^*, v^*) & f_v(u^*, v^*) \\ g_u(u^*, v^*) & g_v(u^*, v^*) \end{pmatrix},$$

then (u^*, v^*) is stable if each eigenvalue of A has modulus less than 1 and unstable if one of the eigenvalues of A has modulus greater than 1 [106].

Example 2.1.8 Find the positive equilibrium point of the non-linear discrete equation

$$x_{n+1} = x_n\sqrt{2 - x_n}$$

and check its stability. Also, find its interval of existence and stability.

Solution: If x^* be the equilibrium solution of the given non-linear discrete equation, it is obtained by solving

$$x^* = x^*\sqrt{2 - x^*} \;\Rightarrow\; x^*(\sqrt{2 - x^*} - 1) = 0 \;\Rightarrow\; x^* = 0,\ 1.$$

Therefore, the positive equilibrium point is $x^* = 1$. Let

$$f(x) = x\sqrt{2 - x} \;\Rightarrow\; f'(x^*) = \frac{2 - 2x}{\sqrt{2 - x}}.$$

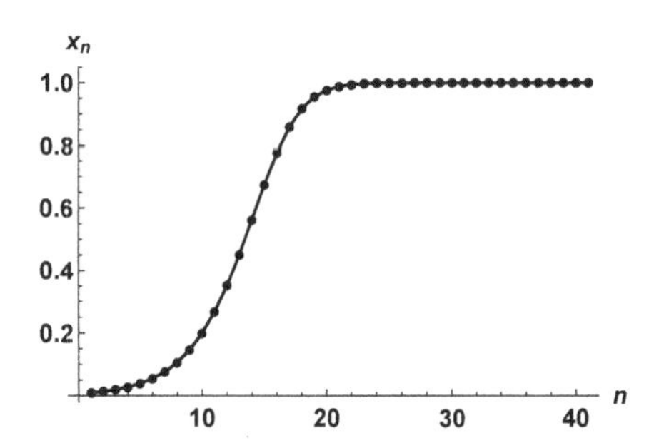

FIGURE 2.5: *The figure shows that the system approaches the stable equilibrium solution $x^* = 1$.*

Clearly, $|f'(x^* = 1)| = 0 < 1$, implying that the equilibrium point $x^* = 1$ is stable (fig. 2.5). The interval of the existence of the equilibrium point is $2 - x > 0 \Rightarrow\; x < 2$ and the interval of its stability is given by

$$\begin{aligned} |f'(x^*)| = \left|\frac{2 - 2x}{\sqrt{2 - x}}\right| < 1 \Rightarrow \quad -1 < \frac{2 - 2x}{\sqrt{2 - x}} < 1 \\ \Rightarrow \quad -\sqrt{2 - x} < 2 - 2x \;\text{ and }\; 2 - 2x < \sqrt{2 - x} \\ \Rightarrow \; \sqrt{2 - x} > 2x - 2 \;\text{ and }\; 2 - 2x < \sqrt{2 - x} \end{aligned}$$

Both the inequalities give $4x^2 - 7x + 2 < 0$, which implies

$$\frac{7-\sqrt{17}}{8} < x < \frac{7+\sqrt{17}}{8} \quad \Rightarrow \quad 0.36 < x < 1.39.$$

Example 2.1.9 Find the equilibrium point of the non-linear discrete system

$$x_{n+1} = x_n + 2.5y_n - 0.1(x_n)^2 - 1, \; y_{n+1} = y_n + \frac{5}{x_n} - 1$$

and check its stability.

Solution: If (x^*, y^*) be the equilibrium solution of the given non-linear discrete system, it is obtained by solving

$$\begin{aligned} x^* &= x^* + 2.5y^* - 0.1(x^*)^2 - 1, \\ y^* &= y^* + \frac{5}{x^*} - 1. \end{aligned}$$

Solving, we get $(x^*, y^*) = (5, 1.4)$ as the unique equilibrium solution of the system. The Jacobian matrix of the system is

$$\begin{pmatrix} 1 - 0.2x^* & 2.5 \\ -\dfrac{5}{(x^*)^2} & 1 \end{pmatrix} = \begin{pmatrix} 0 & 2.5 \\ -0.2 & 1 \end{pmatrix},$$

whose eigenvalues are obtained by solving

$$\begin{vmatrix} 0-\lambda & 2.5 \\ -0.2 & 1-\lambda \end{vmatrix} = 0 \Rightarrow \lambda = \frac{1 \pm i}{2}, \text{ where } |\lambda| = \frac{1}{\sqrt{2}} < 1.$$

Hence, the equilibrium point $(5, 1.4)$ is stable (fig. 2.6).

2.2 Introduction to Discrete Models

In discrete models, the state variables change only at a countable number of points in time. These points in time are the ones at which the event occurs/changes in state. Thus, in discrete-time modeling, there is a state transition function which computes the state at the next time instant given the current state and input. In many situations, the changes are really discrete which occur at well-defined time intervals. Moreover, in many cases, the data are usually discrete rather than continuous. Hence, due to the limitations of the available data, we may be compelled to work with the discrete model, even though the underlying model is continuous.

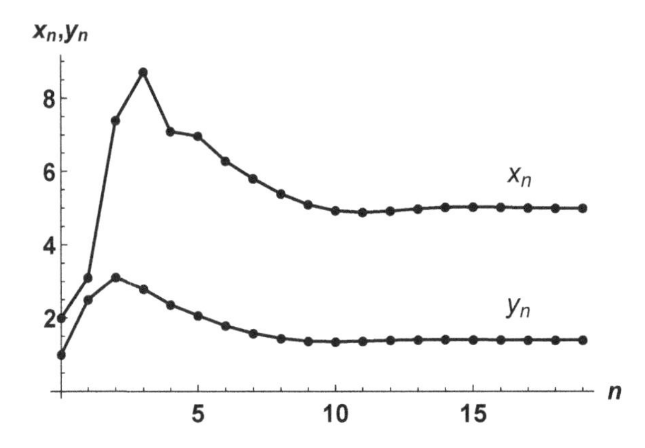

FIGURE 2.6: *The figure shows that the system approaches the stable equilibrium solution* $(x^*, y^*) = (5, 1.4)$.

2.3 Linear Models

2.3.1 Population Model Involving Growth

Suppose a population of Royal Bengal tigers in the Sunderban area, West Bengal, India, is decreasing at the rate of 3% per year. Let p_0 be the size of the initial population of tigers and p_n, the number of tigers after n years, then

$$p_{n+1} = p_n - 0.03\, p_n = 0.97\, p_n \quad \text{for} \quad n = 0, 1, 2, 3, \ldots \tag{2.7}$$

Equation 2.7 represents the number of tigers in a given year, with the number of tigers in the previous years, and is called a first-order linear difference equation. It gives the value of a specific p_{n+1}, provided we know the value of p_n.

If, in 2010, the population of tigers is 150 and we want to know the population after 4 years, then,

$$\begin{aligned}
p_1 &= 0.97\, p_0 = 0.97 \times 150 = 145.5 \simeq 146. \\
p_2 &= 0.97\, p_1 = 0.97 \times 145.5 = 141.135 \simeq 141. \\
p_3 &= 0.97\, p_2 = 0.97 \times 141.135 = 136.9 \simeq 137. \\
p_4 &= 0.97\, p_3 = 0.97 \times 136.9 = 132.79 \simeq 133.
\end{aligned}$$

Thus, after 4 years (in 2014), we can expect 133 tigers in the Sunderban area.

We can also find p_4 explicitly in terms of p_0, by working backwards as follows:

$$\begin{aligned} p_4 &= (0.97)\, p_3 = (0.97)\,(0.97)\, p_2 = (0.97)^2\, p_2 \\ &= (0.97)^2\,(0.97)\, p_1 = (0.97)^3\, p_1 \\ &= (0.97)^3\,(0.97)\, p_0 = (0.97)^4\, p_0 \\ &= (0.97)^4 \times 150 = 0.88529 \times 150 = 132.7935 \simeq 133. \end{aligned}$$

Thus, we conclude that p_4 can be computed without reference to any of the values p_3, p_2 and p_1, provided we know p_0. The general solution of the difference equation of the form $x_{n+1} = kx^n \ \ (n = 0, 1, 2, 3, ...)$ is $x_{n+1} = k^n x_0$, where $k = 0.97$ in this case. Fig. 2.7(a) shows that for this value of k, the tiger population will become extinct in approximately 165 years.

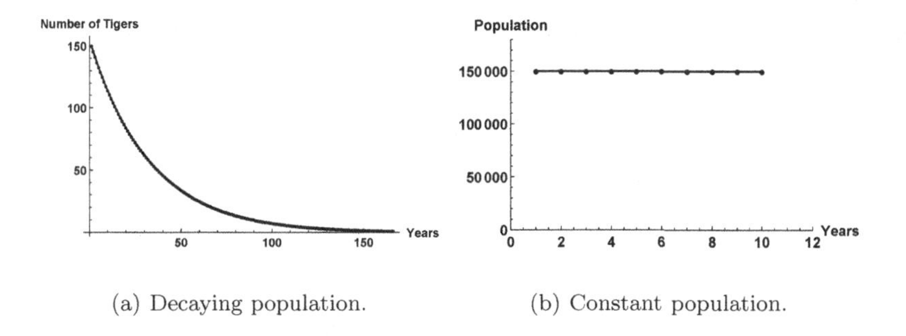

(a) Decaying population. (b) Constant population.

FIGURE 2.7: *The figures show (a) the dynamics of the tiger population, (b) the effect of immigration on the population.*

Example 2.3.1 Consider a population growing each year by 2%. Immigration occurs at a constant rate of 3000 per year. Find the strength of the population after 10 years if the current strength is 150000.

Solution: If P_n be the strength of the population at the end of year n, then

$$P_{n+1} = P_n + 0.02P_n - 3000 \Rightarrow P_{n+1} = 1.02P_n - 3000,$$

gives the required model, which is represented by a non-homogeneous discrete linear equation of the form $x_{n+1} = a\, x_n + b$, whose solution is

$$x_{n+1} = a^{n+1} x_0 + b\, \frac{a^{n+1} - 1}{a - 1} \ \ (a > 1).$$

Here, $a = 1.02$, $b = -3000$ and $x_0 = 150000$. The strength of the population after 10 years is given by (putting $n = 9$)

$$x_{10} = (1.02)^{10} \times 150000 + (-3000)\, \frac{(1.02)^{10} - 1}{1.02 - 1} = 182849 - 32849 = 150000.$$

This implies that the population will remain constant throughout (fig. 2.7(b)).

2.3.2 Newton's Law of Cooling

Suppose a cup of coffee, initially at a temperature of 190 °F, is placed in a room, which is held at a constant temperature of 70 °F. After 1 minute, the coffee has cooled to 180 °F. If we need to find the temperature of the coffee after 15 minutes, we will use Newton's law of cooling, which states that the rate of change of the temperature of an object is proportional to the difference between its own temperature and the ambient temperature (that is, the temperature of its surroundings). Mathematically, this means

$$t_{n+1} - t_n = k(S - t_n),$$

where t_n is the temperature of the coffee after n minutes, S is the temperature of the room, and k is the constant of proportionality.
We first make use of the information given about the change in the temperature of the coffee during the first minute to determine the value of the constant of proportionality k. Thus,

$$\begin{aligned} t_1 - t_0 &= k(S - t_0), \\ \Rightarrow 180 - 190 &= k(70 - 190) \Rightarrow k = \frac{1}{12}, \\ \Rightarrow t_{n+1} - t_n &= \frac{1}{12}(70 - t_n) \Rightarrow t_{n+1} = \frac{11}{12}t_n + \frac{70}{12}. \end{aligned}$$

This is of the form $u_n = au_{n-1} + b$ (see 2.1.2 b), whose solution is given by

$$\begin{aligned} t_{n+1} &= \left(\frac{11}{12}\right)^{n+1} 190 + 70\left[1 - \left(\frac{11}{12}\right)^{n+1}\right], \\ &= 70 + 120\left(\frac{11}{12}\right)^{n+1}, \quad \text{for n = 0,1,2,... .} \end{aligned}$$

For $n = 14$, $t_{15} = 70 + 120\left(\frac{11}{12}\right)^{15} = 102.54$. Hence, after 15 minutes, the coffee has cooled to 102.54 °F.

Since, $\lim_{x\to\infty}\left(\frac{11}{12}\right)^{n} = 0$, the temperature of the coffee will approach the equilibrium temperature of 70 °F (the room temperature) as n increases (fig. 2.8(a)).

Example 2.3.2 A soda–can is taken out from the refrigerator, and its temperature is recorded after 1/2 an hour. After a wait of another 1/2 hour, the temperature is recorded again. If the two readings are 45 °F and 55 °F respectively, what is the temperature inside the refrigerator (assume the room temperature to be 70 °F)?

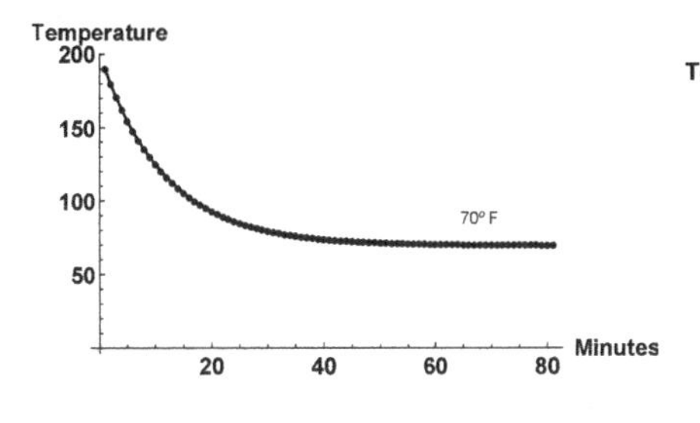

(a) Coffee reaches room temperature.

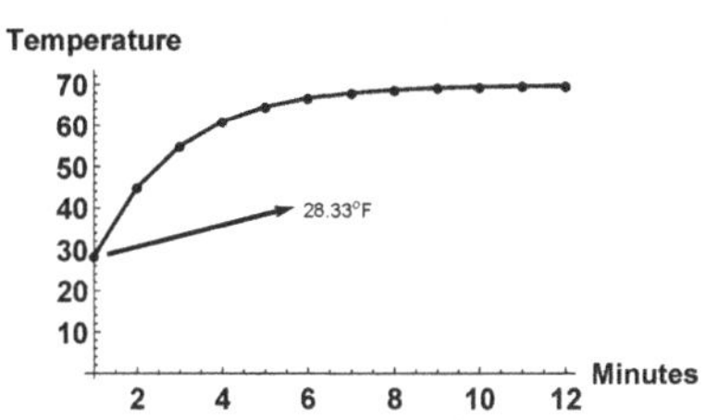

(b) Temperature inside refrigerator.

FIGURE 2.8: *The figures show (a) that the cup of coffee initially at* 190 °F *reaches the room temperature of* 70 °F *as n increases (b) the temperature inside the refrigerator is* 28.33 °F.

Solution: Let t_0 be the temperature of the soda–can 1/2 hour after it was removed from the refrigerator (zero-time) and t_1 be the temperature after waiting 1/2 hour more. Then, t_{-1} will give the temperature when the soda–can was inside the refrigerator. Using Newton's law of cooling, we get

$$\begin{aligned} t_1 - t_0 &= k(S - t_0), \\ \Rightarrow 55 - 45 &= k(70 - 45) \Rightarrow k = \frac{2}{5}, \\ \Rightarrow t_{n+1} - t_n &= \frac{2}{5}(70 - t_n) \Rightarrow t_{n+1} = \frac{3}{5}t_n + 28. \end{aligned}$$

This is of the form $u_n = au_{n-1} + b$ (see 2.1.2 b), whose solution is given by

$$t_{n+1} = \left(\frac{3}{5}\right)^{n+1} 45 + 28\left[\frac{1 - \left(\frac{3}{5}\right)^{n+1}}{1 - \left(\frac{3}{5}\right)}\right] = 70 - 25\left(\frac{3}{5}\right)^{n+1}.$$

Putting $n = -1, 0$, it can be easily checked

$$t_0 = 70 - 25\left(\frac{3}{5}\right)^0 = 45 \text{ and } t_1 = 70 - 25\left(\frac{3}{5}\right)^1 = 55.$$

Therefore, temperature inside the refrigerator is (putting $n = -2$)

$$t_{-1} = 70 - 25\left(\frac{3}{5}\right)^{-1} \simeq 28.3\,°\text{F. (fig. 2.8}(b)\text{.)}$$

2.3.3 Bank Account Problem

Suppose a savings account is opened that pays 4% interest compounded yearly with an initial deposit of \$10000, and a deposit of \$5000 is made at the end of each year. For a savings account that is compounded yearly, the interest is added to the principal at the end of each year. If a_n be the amount at the end of year n ($n = 0, 1, 2, 3, ...$), then

$$a_1 = a_0 + ra_0 = (1+r)a_0,$$
$$a_2 = a_1 + ra_1 = (1+r)a_1,$$
$$\cdots\cdots\cdots\cdots\cdots\cdots\cdots\cdots$$
$$\cdots\cdots\cdots\cdots\cdots\cdots\cdots\cdots$$
$$a_{n+1} = a_n + ra_n = (1+r)a_n,$$

where r is the rate of interest. Now, if a deposit of $5000 is made at the end of each year, then the dynamic model which describes this scenario is given by

$$a_{n+1} = (1+r)a_n + 5000 = (1+0.04)a_n + 5000 = 1.04a_n + 5000.$$

Thus, the amount for three consecutive years will be

$$\begin{aligned} a_1 &= 1.04a_0 + 5000 = 1.04 \times 10000 + 5000 = 10400 + 5000 = 15400, \\ a_2 &= 1.04 \times 15400 + 5000 = 16016 + 5000 = 21016, \\ a_3 &= 1.04 \times 21016 + 5000 = 21856.64 + 5000 = 26856.64, \end{aligned}$$

and so on. Let us now consider a different scenario, where no deposits are made, but $2000 is withdrawn at the end of each year. We want to find out how much money be deposited initially, so that we never run out of cash. The model for this scenario is

$$a_{n+1} = 1.04a_n - 2000,$$

where we assume that the money is withdrawn after the interest from previous years has been added, and we are not penalized for withdrawing money each year. The equilibrium value is given by

$$a^* = 1.04\, a^* - 2000 \Rightarrow a^* = \frac{2000}{0.04} = 50000.$$

Therefore, if the initial deposit (a_0) in the account is $50000 and we withdraw $2000 each year, then the account will always have the same amount at the end of each year (fig. 2.9(a)).

An obvious question is what happen if $a_0 < 50000$ or $a_0 > 50000$. Fig. 2.9(a) shows that if a_0 is less than 50000, the amount in the account decreases to 0, and the amount grows without bound if a_0 is greater than 50000. Thus, the system approaches zero or increases without bound if $a_0 \neq 50000$, and therefore, this equilibrium value is unstable.

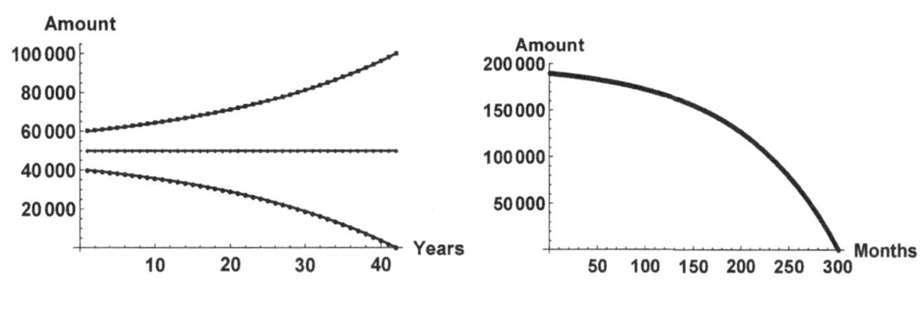

(a) $a_0 > 50000, a_0 = 50000, a_0 < 50000$.

(b) House loan paid in 300 months.

FIGURE 2.9: *The figures show (a) different dynamics with different initial deposits a_0, (b) initial house loan amount P_0 = \$189894 is repaid in 300 months.*

Example 2.3.3 A couple like to purchase their first house with their combined annual income of \$84000. They also have a savings of \$40000, which they want to use as a down payment. For the rest, they can take a loan from the bank with 12% annual interest paid monthly for 25 years. However, the bank will not allow their monthly installment to exceed 1/4 of their monthly income. (a) What is the maximum budget, the couple can afford for the house under these conditions? (b) To afford a house costs \$290000; what would be their annual income?

Solution: Let P_n be the amount of loan to be paid after n months, then

$$P_{n+1} = P_n + rP_n - d,$$

where r is the monthly rate of interest and d is the monthly installment to be paid. This is of the form (2.1.2 b), whose solution is given by

$$P_{n+1} = P_0(1+r)^{n+1} - d\frac{(1+r)^{n+1} - 1}{r},$$

where P_0 is the initial amount of loan.

(a) The monthly income of the couple is $\dfrac{96000}{12}$ = \$8000, so their maximum repayment of the loan is $\dfrac{8000}{4}$ = \$2000 = d and $r = \frac{0.12}{12} = 0.01$. After 25 years, that is, 300 months, the loan amount will be zero, which implies $P_{300} = 0$ (fig. 2.9(b)). Therefore,

$$P_{300} = P_0(1+0.01)^{299+1} - 2000\,\frac{(1+0.01)^{299+1} - 1}{0.01} = 0,$$

$$\Rightarrow\; P_0 \times 19.79 = \frac{2000}{0.01}(19.79 - 1) \;\Rightarrow\; P_0 = 189893.886 \simeq \$189894.$$

Therefore, the maximum budget that the couple can afford for the house is \$189894 + \$40000 = \$229894.

(b) To afford a house which costs \$290000, the couple has to take a load of (290000 – 40000) = \$250000 and their monthly installment will be

$$250000\,(1+0.01)^{299+1} - d\,\frac{(1+0.01)^{299+1}-1}{0.01} = 0,$$

$$\Rightarrow d = \frac{250000 \times 19.79 \times 0.01}{18.79} \simeq \$2633.$$

Hence, the annual income of the couple is \$2633 × 4 × 12 = \$126384.

2.3.4 Drug Delivery Problem

Suppose a patient is given a drug to treat some infection. He/she is given the same dose of the medicine at equally spaced time intervals. The body metabolizes some of the drugs so that, after some time, only a portion r of the original amount of the drug remains. After each dose, the amount of the drug in the body is equal to the amount of the given dose b, plus the amount of the drug remnant from the previous dose. The dynamic model which describes this scenario is given by

$$x_{n+1} = rx_n + b.$$

The equilibrium point is given by $a^* = ra^* + b \;\Rightarrow\; a^* = \dfrac{b}{1-r}$.

$$\begin{aligned}
\text{Let } x_0 &= 0 \text{ (no drug), then } x_1 = b \text{ (first dose).}\\
x_2 &= b \text{ (second dose)} + r\; b \text{ (amount remaining from previous dose).}\\
x_3 &= b \text{ (third dose)} + r\; (b + rb) \text{ (amount remaining from previous dose).}\\
&\cdots\cdots\cdots\\
x_{n+1} &= b + r\;(b + rb + ... + r^{n-1}b) = b(1 + r + r^2 + ... + r^n) = b\frac{1-r^{n+1}}{1-r}.
\end{aligned}$$

Since $r < 1, x_{n+1} \longrightarrow \dfrac{b}{1-r}$, the stable equilibrium point.

Suppose, the amount of drug in the patient's bloodstream decreases at the rate of 80% per hour (this means 20% of the drug remains in the body). To sustain the drug to a certain level, an injection is given at the end of each hour that increases the amount of drug in the bloodstream by 0.2 unit. The dynamic model which describes this scenario is given by

$$a_{n+1} = a_n - 0.8a_n + 0.2 \;\Rightarrow\; a_{n+1} = 0.2a_n + 0.2,$$

where a_n is the amount of drug in the blood at the end of n hours. The equilibrium solution of this model is given by

$$a^* = 0.2a^* + 0.2 \;\Rightarrow\; a^* = 0.25.$$

The long-term behavior of the system will depend on the initial value a_0. The figure shows that no matter what is the value of a_0, the system always approaches the value of $\frac{b}{1-r} = \frac{0.2}{1-0.2} = 0.25$, implying that 0.25 is a stable equilibrium point (fig. 2.10(a)).

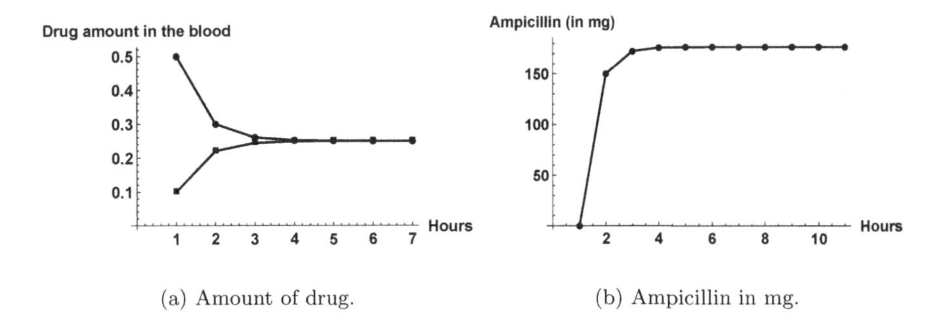

(a) Amount of drug. (b) Ampicillin in mg.

FIGURE 2.10: *The figures show (a) the amount of drug in a patient's bloodstream always reaches the steady-state value 0.25, independent of the initial value a_0, implying a stable equilibrium, (b) ampicillin reaches steady-state value* 176.47 *mg.*

Example 2.3.4 A person with an ear infection takes 150 mg ampicillin tablet once every 4 hours. About 15% of the drug in the body at the start of a four-hour period is still there at the end of that period. What quantity of ampicillin is in the body (a) right after taking the third tablet? (b) at the steady-state level right after taking a tablet? (c) at the steady-state level right before taking a tablet?

Solution: (a) The quantity of ampicillin right after taking the third tablet is

$$x_3 = b\,\frac{1-r^3}{1-r} = 150\,\frac{1-(0.15)^3}{1-0.15} = 175.87 \text{ mg}.$$

(b) The quantity of ampicillin at the steady-state level right after taking a tablet is (fig. 2.10(b))

$$\frac{b}{1-r} = \frac{150}{1-0.15} = 176.47 \text{ mg}.$$

(c) The quantity of ampicillin at the steady-state level right before taking a tablet is

$$\frac{b}{1-r} - b = \frac{150}{1-0.15} - 150 = 26.47 \text{ mg}.$$

2.3.5 Harrod Model (Economic Model)

The Harrod model [111], which was developed in the 1930s, gives some insight into the dynamics of economic growth. The model aims to determine an equilibrium growth rate for the economy. Let G_n be the Gross Domestic Product (GDP) on national income, which is one of the primary indicators to determine a country's economy, and $S(n)$ and $I(n)$ be the savings and investment of the people. The Harrod model assumed that in a country people's savings depend on GDP or national income; that is, savings is a constant proportion of current income, which implies

$$S_n = a\ G_n,\ \text{where } a\ (> 0) \text{ is a constant of proportionality.} \tag{2.8}$$

Harrod further assumed that the investment made by the people depends on the difference between the GDP of the current year and the last year, that is,

$$I_n = b\,(G_n - G_{n-1})\,,\ b > a. \tag{2.9}$$

Finally, the Harrod model assumed that all the savings made by the people are invested, that is,

$$S_n = I_n. \tag{2.10}$$

From (2.8), (2.9) and (2.10), we obtain

$$b\,(G_n - G_{n-1}) = S_n = a\ G_n \Rightarrow G_n = \frac{b\ G_{n-1}}{b-a},\ \text{whose solution is}$$

$$G_n = G(0)\left(\frac{b}{b-a}\right)^n.$$

Thus, Harrod's model concludes that GDP or national income increases geometrically with time.

2.3.6 Arms Race Model

We consider two countries engaged in an arms race. We assume that the two countries have similar economic strengths and the same level of distrust for each other. Let T_n be the total amount of money spent by the two countries on arms after n years. Let g (>0) measures the restraint of growth due to economic strength (or weakness) of the countries and d (>0), the level of distrust between the two countries. Both the countries also spent a constant amount (say, k) of money for buying arms irrespective of involving in an arms race. Then, the dynamic discrete model for the total amount of money T_n

spent on arms by each country after n years is given by

$$\begin{aligned} T_n &= (1-g)\, T_{n-1} + d\, T_{n-1} + k, \\ \Rightarrow T_n &= (1-g+d)\, T_{n-1} + k, \\ T_n &= (1-g+d)^n T_0 + k\left(\frac{1-(1-g+d)^n}{1-(1-g+d)}\right), \\ T_n &= (1-g+d)^n T_0 + k\left(\frac{1-(1-g+d)^n}{g-d}\right). \end{aligned}$$

The equilibrium solution is

$$\begin{aligned} T_n &= T_{n-1} = T^*, \\ \Rightarrow (1-g+d)T_{n-1} + k &= T_{n-1}, \\ \Rightarrow T^* &= \frac{k}{g-d}, \ (g > d). \end{aligned}$$

Thus, as time increases, the total amount of money spent on arms reaches a steady state, and both the countries have a "stable" arms race (fig. 2.11).

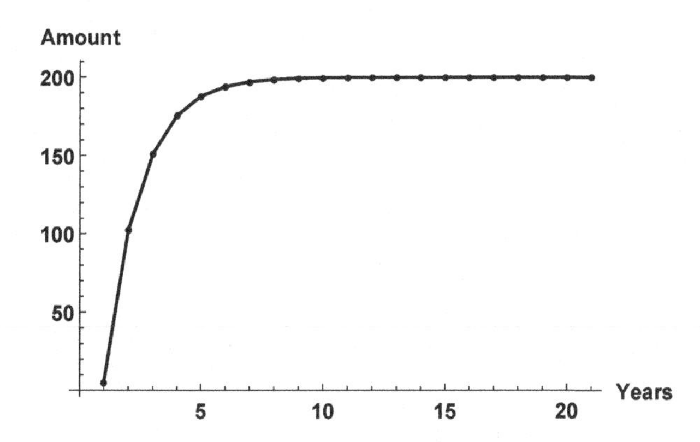

FIGURE 2.11. *The amount of money spent by both the countries on arms reaches a steady-state value with increasing time. Parameter values g = 0.6, d = 0.1 and k = 100.*

2.3.7 Lanchester's Combat Model

F.W. Lanchester, a British Engineer, developed one of the first mathematical models for analyzing combats [86], whose greatest strength lies in its simplicity. The models helped in better planning, prediction of battles and their possible outcomes. Consider two adversaries, namely, A-team and B-team. Let A_n and B_n be the number of units of A-team and B-team, respectively, remaining in the battle after time n. By units, we will mean fighter planes, ships, tanks, soldiers, etc., depending on the context of the battle. It is assumed that the combat loss rate of both the teams is proportional to the size of their respective enemies. Under this assumption, the discrete dynamical system is

$$A_{n+1} = A_n - \beta B_n + r_1,$$
$$B_{n+1} = B_n - \alpha A_n + r_2,$$

where α (> 0) is the fighting effectiveness of A-team, β (> 0) is the fighting effectiveness of B-team, r_1 and r_2 are the respective numbers of reinforcements for A-team and B-team, respectively for each time step. The equilibrium solution is given by $(A^*, B^*) = \left(\frac{r_2}{\alpha}, \frac{r_1}{\beta}\right)$. The Jacobian matrix of the system is $\begin{pmatrix} 1 & -\beta \\ -\alpha & 1 \end{pmatrix}$, whose eigenvalues are obtained by solving $\begin{vmatrix} 1-\lambda & -\beta \\ -\alpha & 1-\lambda \end{vmatrix} = 0 \Rightarrow \lambda_{1,2} = 1 \pm \sqrt{\alpha\beta}$. Clearly, $|\lambda_1| = 1 + \sqrt{\alpha\beta} > 1$, and $|\lambda_2| = 1 - \sqrt{\alpha\beta} < 1$. This implies that the system is a saddle about the equilibrium point. Figs. 2.12 show the dynamics of the combat model.

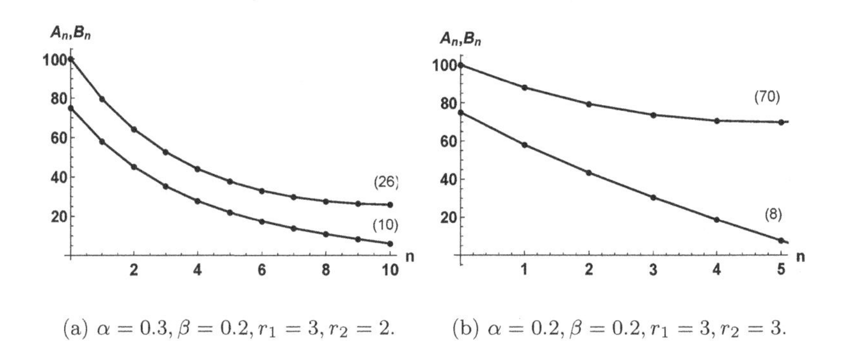

(a) $\alpha = 0.3, \beta = 0.2, r_1 = 3, r_2 = 2$.   (b) $\alpha = 0.2, \beta = 0.2, r_1 = 3, r_2 = 3$.

FIGURE 2.12: *The figures show (a) team-B is in winning position in* 10 *time units with more initial units and less fighting effectiveness and reinforcement, (b) team-B wins in* 5 *time units with equal fighting effectiveness and reinforcement. Initial units:* $A(0) = 75, B(0) = 100$. *The numbers in parenthesis indicate the number of units at* 10 *and* 5 *time units, respectively.*

2.3.8 Linear Predator–Prey Model

We consider a forest containing tigers (predator) and deer (prey). The tigers kill their prey, that is, deer, for food. Let T_n and D_n be the respective population of tigers and deer at the end of year n. We try to formulate the model to examine the long-term behavior of the two species under a few assumptions. We assume that

(i) Deer are the only source of food for the tigers, and tigers are the only predators for deer.

(ii) The deer population will grow exponentially if there are no tigers and without the deer population, the tigers will die out due to starvation.

(iii) The rate at which the tiger population grows increases with the presence of the deer population, and the rate at which the deer population grows decreases with the presence of the tiger population.

Under these assumptions, the dynamic model for this scenario is given by two-dimensional linear discrete dynamical system as [4]

$$\begin{aligned} T_{n+1} &= (1-\alpha)T_n + \beta D_n \equiv f_1(T_n, D_n), \\ D_{n+1} &= -\delta T_n + (1+\gamma)D_n \equiv f_2(T_n, D_n), \end{aligned}$$

where α, β, γ and δ are positive constants, $0 < \alpha, \gamma < 1$. Here, α is the rate at which the tigers die if no deer are available for food and β is the rate at which the tiger population grows when the food (deer) is available. Similarly, the deer population grows at a rate γ when no tigers are around and decreases at a rate δ in the presence of a tiger population.

Clearly, $(0,0)$ is the only equilibrium point. For stability, we calculate the Jacobian matrix at $(0,0)$, given by

$$\begin{pmatrix} 1-\alpha & \beta \\ -\delta & 1+\gamma \end{pmatrix}.$$

The eigenvalues of the Jacobian matrix are

$$2 - \alpha + \gamma - \sqrt{(\alpha+\gamma)^2 - 4\alpha\beta}$$

and

$$2 - \alpha + \gamma + \sqrt{(\alpha+\gamma)^2 - 4\alpha\beta}.$$

For stability, both the eigenvalues must be numerically less than 1, that is,

$$|2 - \alpha + \gamma - \sqrt{(\alpha+\gamma)^2 - 4\alpha\beta}| < 1 \;\text{ and }\; |2 - \alpha + \gamma + \sqrt{(\alpha+\gamma)^2 - 4\alpha\beta}| < 1$$

Let $\alpha = 0.5$, $\beta = 0.4$, $\gamma = 0.1$ and $\delta = 0.17$, then the eigenvalues of the Jacobian matrix

$$\begin{pmatrix} 1-\alpha & \beta \\ -\delta & 1+\gamma \end{pmatrix} = \begin{pmatrix} 0.5 & 0.4 \\ -0.17 & 1.1 \end{pmatrix}$$

are 0.948 and 0.652, whose modulus is less than 1. Hence, the system is stable, showing that both the species go extinct (fig. 2.13(a)). Now, we change δ (death rate of deer in the presence of tigers) to 0.05 and observe the change in dynamics. The eigenvalues now are 1.06 and 0.535. Clearly, one of the eigenvalues is 1.06, whose modulus is greater than 1, and therefore, the system is unstable (fig. 2.13(b)).

We now fix the initial population of deer at 200 and would like to see how the dynamics change with the varying tiger population. With $T_0 < 500$, both the populations grow unboundedly (fig. 2.13(c)) but for sufficient large T_0 (say, 2300), both the species goes to extinction (fig. 2.13(d)). Readers can check for similar dynamics with a fixed population of tigers (say, 600) and varying the initial population of deer.

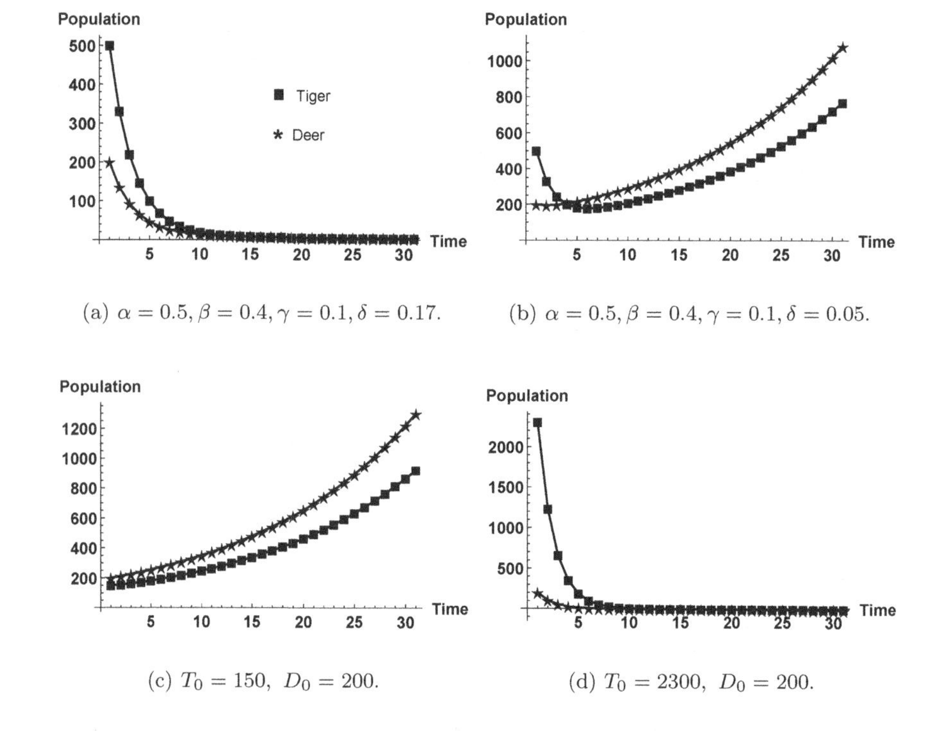

(a) $\alpha = 0.5, \beta = 0.4, \gamma = 0.1, \delta = 0.17$.

(b) $\alpha = 0.5, \beta = 0.4, \gamma = 0.1, \delta = 0.05$.

(c) $T_0 = 150,\ D_0 = 200$.

(d) $T_0 = 2300,\ D_0 = 200$.

FIGURE 2.13: *The figures show different behaviors of tiger and deer populations with changing parameter values and initial conditions: (a) stable system, (b) unstable system, (c) unstable system (d) stable system (parameter values of (b) are used in (c) and (d) with different initial conditions).*

2.4 Non-Linear Models

Density dependence is not considered by linear models, which assume that the same growth characteristics are applied to the population regardless of their sizes. In the natural world, linear growths are seldom seen (except for bacteria and viruses). Non-linear models or density-dependent models are quite successful in this regard. The non-linear models successfully capture the density dependence and their varying effects, which is reflected in the qualitative behavior of the solutions of the models.

2.4.1 Density-Dependent Growth Models

2.4.1.1 Logistic Model

We now consider the growth of a population x_n of a species after n generations, which is density dependent. Let the population grow linearly at a rate r and its growth is inhibited due to overpopulation. This results in a non-linear discrete equation as

$$x_{n+1} = rx_n\left(1 - \frac{x_n}{k}\right) \quad (r, k > 0) \quad \text{or} \quad x_{n+1} = ax_n - bx_n^2 \quad (a > 0, b > 0).$$

Both the equations are called discrete logistic equations (or logistic maps) and are perhaps the most commonly used in models of density dependence because of their simplicity.

If x^* be the equilibrium solution of the logistic equation, it is obtained by solving

$$rx^*\left(1 - \frac{x^*}{k}\right) = x^* \Rightarrow x^* = 0,\ k\frac{r-1}{r}.$$

Now, $f'(x) = r - \frac{2rx}{k}$. Clearly, $|f'(x^* = 0)| = |r| < 1 \Rightarrow 0 < r < 1$ (since, $r > 0$), implying that the equilibrium point $x^* = 0$ is stable if $r \in (0, 1)$ (fig. 2.14(a), bottom curve).

The equilibrium point $x^* = k\frac{r-1}{r}$ exists if $r > 1$ (since it is a population model, the equilibrium solution has to be non-negative). For stability, we have

$$\left|f'(x^* = k\frac{r-1}{r})\right| < 1 \Rightarrow |2 - r| < 1 \Rightarrow 1 < r < 3 \text{ (fig. 2.14(a), top curve)},$$

implying that the equilibrium point $x^* = k\frac{r-1}{r}$ is stable if $r \in (1, 3)$.

Note: When $x^* = 0$ is stable, $x^* = k\frac{r-1}{r}$ does not exist and when $x^* = k\frac{r-1}{r}$ is stable, $x^* = 0$ is unstable.

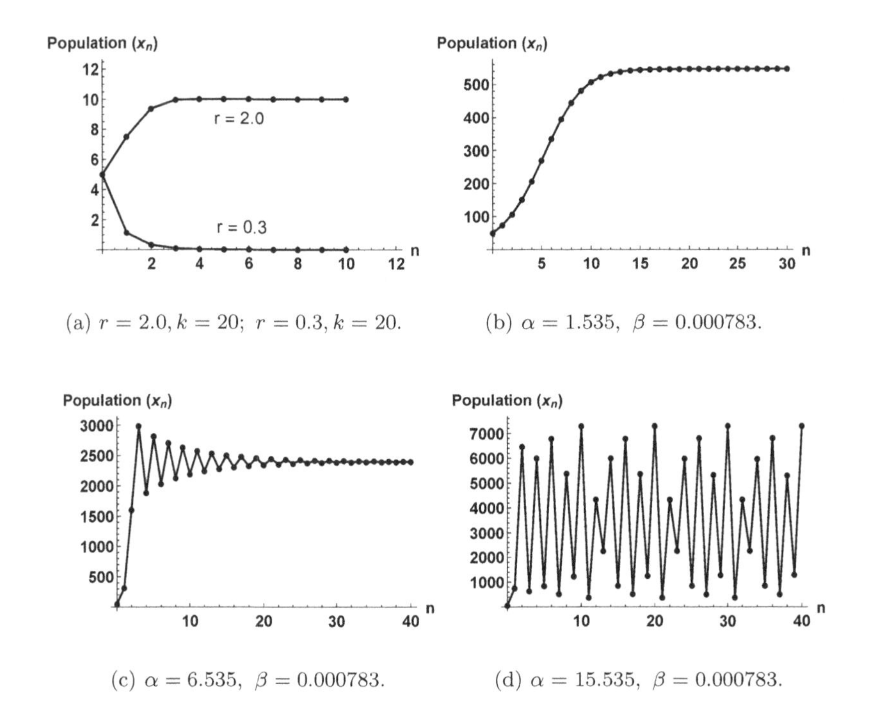

FIGURE 2.14: *The figure shows the dynamics of growth models. (a) Stability of logistic model with $r > 1$ and $r < 1$, (b) Stability of Richer's model for $1 < \alpha < e$, (c) Stability of Richer's model for $e < \alpha < e^2$, (d) Richer's model is unstable for $\alpha > e^2$.*

2.4.1.2 Richer's Model

Richer's model is another example for a density-dependent model for the population x_n of a species after n generations and is given by

$$x_{n+1} = \alpha x_n e^{-\beta x_n} \quad (\alpha > 0, \beta > 0),$$

where α represents the maximal growth rate of the organism and β is the inhibition of growth caused by overpopulation.

If x^* be the equilibrium solution of Richer's model, then

$$\alpha x_n e^{-\beta x^*} = x^* \Rightarrow x^*_{1,2} = 0, \ \frac{\ln(\alpha)}{\beta}.$$

Now, $f'(x) = \alpha e^{-\beta x} - \alpha\beta x e^{-\beta x}$. Clearly, $|f'(x_1^* = 0)| = |\alpha| < 1 \Rightarrow 0 < \alpha < 1$ (since, $\alpha > 0$), implying that the equilibrium point $x_1^* = 0$ is stable if $\alpha \in (0, 1)$.

The equilibrium point $x_2^* = \dfrac{\ln(\alpha)}{\beta}$ is stable if

$$|f'(x^* = x_2^*)| < 1 \Rightarrow |1 - \ln(\alpha)| < 1 \Rightarrow 1 < \alpha < e^2,$$

implying that the equilibrium point $x^* = x_2^*$ is stable if $\alpha \in (1, e^2)$.

For $1 < \alpha < e$ (≈ 2.7183), $f'(x^* = x_2^*) > 0$, hence, the solution of Richer's model is stable and monotonically approaches the equilibrium point x_2^* (fig. 2.14(b)).
For $e < \alpha < e^2$ (≈ 7.389), $f'(x^* = x_2^*) < 0$, hence, the solution of Richer's model is stable and oscillates as it approaches the equilibrium x_2^* (fig. 2.14(c)).
For $\alpha > e^2$ (≈ 7.389), the solution of Richer's model is unstable and oscillates as it grows away from the equilibrium point x_2^* (fig. 2.14(d)).

2.4.2 The Learning Model

When we learn a new topic, the following principle may apply. If the present amount learned is L_n, then L_{n+1} equals L_n, minus the fraction r of L_n forgotten, plus the new amount learned, which we assume is inversely proportional to the amount already learned ($\propto 1/L_n$). We also assume that the person learning the new topic cannot forget the whole part of the topic learned ($r \neq 1$). Under the following assumptions, the model is given by

$$L_{n+1} = L_n - rL_n + \frac{k}{L_n} \quad (0 \leq r < 1, k > 0),$$

where k is the constant of proportionality. The steady-state solution is given by

$$L^* = L^* - rL^* + \frac{k}{L^*} \Rightarrow L_n^* = \sqrt{\frac{k}{r}}.$$

Let $f(L) = L - rL + \dfrac{k}{L}$, then $f'(L) = 1 - r - \dfrac{k}{L^2}$. For stability, we must have

$$|f'(L^*)| < 1 \Rightarrow |1 - 2r| < 1 \Rightarrow 0 < r < 1.$$

Therefore, the system is stable if $0 < r < 1$ and unstable if $r > 1$.

2.4.3 Dynamics of Alcohol: A Mathematical Model

Some teenagers and young adults have made drinking as an integral part of their social life. They believe that the quality of their social life is enhanced by moderate drinking without understanding the risks involved in the consumption of alcohol. About 45-55 percent of road traffic deaths in India occur under the influence of alcohol.

BAC or Blood alcohol content of 0.1 means 1 gram of alcohol is present in 100 ml of blood. In India, the BAC legal limit is 0.03 percent per 100 ml of blood or 30 mg of alcohol in 100 ml of blood. In the United States, BAC legal limit is 0.08 and in most European countries, it is 0.05. A standard alcoholic drink, which is equivalent to 350 ml of beer, will raise the BAC of a 68 kg adult male to approximately 20-22 mg of alcohol in 100 ml of blood (count is different in a female). The alcohol consumed is absorbed into the body primarily through the lining of the stomach and that is the reason why a blood alcohol content (BAC) peak is obtained within 20 minutes of the last drink. The main content of alcohol is ethanol (a chemical compound) and the body deals with this chemical in the blood stream in two ways. The first is the metabolism of ethanol by liver enzymes and the second is the filtration by the kidneys. The average rate alcohol leaves the body is 15 mg per 100 ml per hour (a standard alcoholic drink for men). Thus, if a person 75 mg of ethanol in his body at the beginning of an hour, then the body will eliminate 15 mg in the next hour.

Let a_n be the amount of alcohol in a person's body at the beginning of hour n and let the person average eliminates about 15 percent of the alcohol from his/her body each hour, then the amount of alcohol eliminated during $(n-1)^{\text{th}}$ hour is 0.15 a_{n-1}, where a_{n-1} is the amount of alcohol in a person's body at the beginning of hour $(n-1)$. Therefore, the amount of alcohol in a person's body is modelled by the dynamical system [143]

$$a_n = a_{n-1} - r\ a_{n-1} + A,$$

where r is the fraction of alcohol filtrated by the kidneys during each time period and A is the amount of alcohol consumed in each time period (for alcohol $r = 0.15$).

Though we have assumed that alcohol leaves the body at a constant rate of 15 percent, this does not happen in reality. Metabolism of ethanol does not happen at a constant rate, the speed is different for different people. Therefore, we consider r as a function of a, the amount of alcohol in the body. For alcohol, we assume that $r(a)$ goes to zero as the amount of alcohol (a) in the body gets larger. To model this situation, we assume the functional form of $r(a)$ as:

(i) $r(a) = \dfrac{1}{1+a}$ (a rational function), or

(ii) $r(a) = \dfrac{1}{e^a}$ (an exponential function),

as both the functions are decreasing in nature (see fig. 2.15(a)).

If $f(a)$ be the filtration rate of alcohol by the kidneys per hour, then the amount of alcohol filtrated (or eliminated) is the product of the rate r and

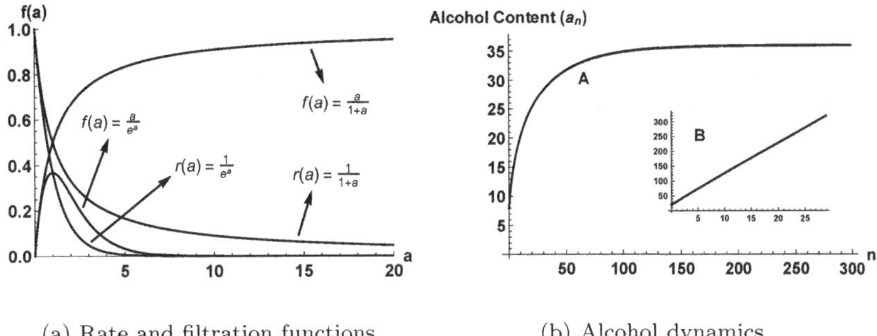

(a) Rate and filtration functions.

(b) Alcohol dynamics.

FIGURE 2.15: *Figure (a) shows different rate functions $r(a)$ and filtration functions $f(a)$. Figure (b) shows the dynamics of alcohol content when the system is stable (A) and unstable (B).*

the amount of alcohol a in the body, that is,

$$f(a) = r(a)\ a = \left(\frac{1}{1+a}\right) a \text{ or } \left(\frac{1}{e^a}\right) a.$$

Fig. 2.15(a) gives the plot of both the functions. The rational function suggests that as the amount of alcohol in the body increases, the amount filtrated (eliminated) also increases (it reaches the maximum value of 1 gram per hour, the horizontal asymptote of the function $f(a) = \frac{a}{1+a}$). On the other hand, the exponential function implies that the amount of alcohol decreases with the increased intake of alcohol. Therefore, the rational function is more reasonable to accept as there is no reason to believe that the amount of alcohol filtrated decreases with the consumption of more alcohol. Using any loss of generality, we assume the form of rational function [144]

$$r = \frac{\beta}{\gamma + a_{n-1}}, \tag{2.11}$$

where β and γ are constants depending on a particular person and a_{n-1} is the number of grams of alcohol in the body at the beginning of $(n-1)^{\text{th}}$ hour. The next step is to estimate the values of β and γ.

Consider a case where a blood test is performed on a person after he has consumed 21 grams of alcohol, shows 40 percent filtration. The same person consuming 36 grams of alcohol shows 25 percent filtration after 1 hour. This means when $a = 21, \gamma = 0.4$ and when $a = 36, \gamma = 0.25$. Substituting the values of a and γ in (2.11), we obtain

$$0.4 = \frac{\beta}{\gamma + 21}, \quad 0.25 = \frac{\beta}{\gamma + 36} \quad \Rightarrow \beta = 10, \gamma = 4.$$

Therefore, for this particular person, the function r=$\frac{10}{4+a_{n-1}}$ approximates the fraction of alcohol as a function of alcohol in the body in $(n-1)^{\text{th}}$ hour. Thus, the mathematical model for the amount of alcohol (in grams) in a person's body at the beginning of the nth hour is given by

$$a_n = a_{n-1} - \frac{10}{4+a_{n-1}} a_{n-1} + A,$$

where A is the constant amount of alcohol consumed in each time period.
Note: The numbers $\beta = 10, \gamma = 4$ are actually used for the elimination rate for males. For women, β is generally less.

The positive equilibrium point is given by

$$a^* = a^* - \frac{10}{4+a^*} a^* + A \Rightarrow a^* = \frac{4A}{10-A}, \quad (A < 10).$$

For stability, we must have,

$$|f'(a^*)| < 1 \Rightarrow \left|1 - \frac{40}{(4+a^*)^2}\right| < 1 \Rightarrow \left|1 - \frac{(10-A)^2}{40}\right| < 1$$

$$\Rightarrow 10 - 4\sqrt{5} < A < 10 + 4\sqrt{5} \Rightarrow 1.06 < A < 18.94.$$

Therefore, the system is stable if $1.06 < A < 18.94$ (fig. 2.15(b)A) and unstable if $A < 1.06$ and $A > 18.94$ (fig. 2.15(b)B).

2.4.4 Two Species Competition Model

In a certain forest, black bears and grizzly bears compete with each other for food. Suppose that in the absence of any competition or hunting, the black bear population will grow by α_1 per year, while the grizzly bear population will grow by α_2. Each year the competition between the two types of bears leads to the death of a certain number of each type of bear (due to fighting and food shortages). The number of black bears that die is equal to the product of the black and grizzly bear populations at a rate β_1. The number of grizzly bears that die is equal to the product of the black and grizzly populations at a rate β_2. Let B_n and G_n be the population of black bears and grizzly bears at year n, respectively. Under the following assumptions, the model is given by [156]

$$\begin{aligned} B_{n+1} &= B_n + \alpha_1 B_n - \beta_1 B_n G_n, \\ G_{n+1} &= G_n + \alpha_2 G_n - \beta_2 B_n G_n. \end{aligned}$$

The equilibrium points are obtained by solving

$$\begin{aligned} &B^* = B^* + \alpha_1 B^* - \beta_1 B^* G^*, \\ &G^* = G^* + \alpha_2 G^* - \beta_2 B^* G^*, \\ &\Rightarrow (B^*, G^*) = (0,0) \text{ and } \left(\frac{\alpha_2}{\beta_2}, \frac{\alpha_1}{\beta_1}\right). \end{aligned}$$

For stability, we calculate

$$\begin{pmatrix} \dfrac{\partial f}{\partial B} & \dfrac{\partial f}{\partial G} \\ \dfrac{\partial g}{\partial B} & \dfrac{\partial g}{\partial G} \end{pmatrix}_{(B^*,G^*)} = \begin{pmatrix} 1+\alpha_1-\beta_1\ G^* & -\beta_1 B^* \\ -\beta_2 G^* & 1+\alpha_2-\beta_2\ B^* \end{pmatrix},$$

where $f(B,G) = B + \alpha_1\ B - \beta_1\ B\ G,\ g(B,G) = G + \alpha_2\ G - \beta_2\ B\ G.$

At $(0,0)$, the matrix is $\begin{pmatrix} 1+\alpha_1 & 0 \\ 0 & 1+\alpha \end{pmatrix}$, whose eigenvalues are $1+\alpha_1$ and $1+\alpha_2$. Clearly, $|1+\alpha_1| > 1,\ |1+\alpha_2| > 1$, implying that the system is unstable at the origin. This means both the species will not become extinct simultaneously.

At $\left(\frac{\alpha_2}{\beta_2}, \frac{\alpha_1}{\beta_1}\right)$, the matrix is $\begin{pmatrix} 1 & \frac{\beta_1\alpha_2}{\beta_2} \\ \frac{\beta_2\alpha_1}{\beta_1} & 1 \end{pmatrix}$, whose eigenvalues are $1 \pm \sqrt{\alpha_1\alpha_2}$. Clearly, $|1+\sqrt{\alpha_1\alpha_2}| > 1$, and $|1-\sqrt{\alpha_1\alpha_2}| < 1$, implying that the system is saddle at $\left(\frac{\alpha_2}{\beta_2}, \frac{\alpha_1}{\beta_1}\right)$.

Since the model is unstable at both the equilibrium points, and we modify it by adding a self-competition term in both the bear species. We assume that each bear species also compete for food among each other. This will modify the model as

$$\begin{aligned} B_{n+1} &= B_n + \alpha_1\ B_n - \beta_1\ B_n\ G_n - \gamma_1\ B_n^2, \\ G_{n+1} &= G_n + \alpha_2\ G_n - \beta_2\ B_n\ G_n - \gamma_2\ G_n^2. \end{aligned}$$

The equilibrium points are obtained by solving

$B^* = B^* + \alpha_1\ B^* - \beta_1\ B^*\ G^* - \gamma_1\ (B^*)^2,$

$G^* = G^* + \alpha_2\ G^* - \beta_2\ B^*\ G^* - \gamma_2\ (G^*)^2$, and we get

$(B^*, G^*) = (0,0),\ \left(\dfrac{\alpha_1}{\gamma_1}, 0\right),\ \left(0, \dfrac{\alpha_2}{\gamma_2}\right)$ and $\left(\dfrac{\alpha_1\gamma_2-\alpha_2\beta_1}{\gamma_1\gamma_2-\beta_1\beta_2}, \dfrac{\alpha_2\gamma_1-\alpha_1\beta_2}{\gamma_1\gamma_2-\beta_1\beta_2}\right).$

For stability, we obtain the matrix

$$A = \begin{pmatrix} 1+\alpha_1-\beta_1\ G^*-2\gamma_1\ B^* & -\beta_1 B^* \\ -\beta_2 G^* & 1+\alpha_2-\beta_2\ B^*-2\gamma_2\ G^* \end{pmatrix},\ \text{where}$$

$f(B,G) = B+\alpha_1\ B-\beta_1\ B\ G-\gamma_1\ B^2,\ g(B,G) = G+\alpha_2\ G-\beta_2\ B\ G-\gamma_2\ G^2.$

At $(0,0)$, the system is unstable (just like before).

At $\left(\dfrac{\alpha_1}{\gamma_1}, 0\right)$, $A = \begin{pmatrix} 1+\alpha_1 & -\frac{\alpha_1\beta_1}{\gamma_1} \\ 0 & 1+\alpha_2-\frac{\alpha_1\beta_1}{\gamma_1} \end{pmatrix}$, whose eigen values are

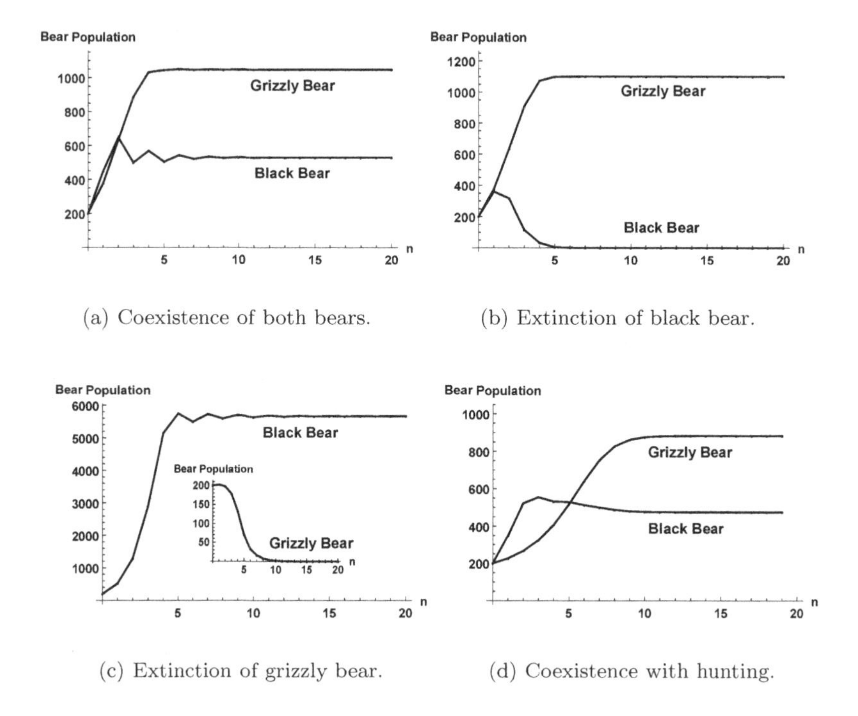

(a) Coexistence of both bears.

(b) Extinction of black bear.

(c) Extinction of grizzly bear.

(d) Coexistence with hunting.

FIGURE 2.16: *The figure shows the dynamics of a competition model of black bear and grizzly bear over time. (a) Both bears coexist* ($\alpha_1 = 1.9, \beta_1 = 0.0003, \gamma_1 = 0.003, \alpha_2 = 1.1, \beta_2 = 0.0001, \gamma_2 = 0.001$)*. (b) Black bear dies off* ($\beta_1 = 0.0025$*, rest same as (a)). (c) Grizzly bear goes to extinction* ($\alpha_1 = 1.7, \beta_1 = 0.0003, \gamma_1 = 0.0003, \alpha_2 = 0.05, \beta_2 = 0.0001, \gamma_2 = 0.0001$)*. (d) Survival and coexistence of both the bears despite of hunting (same as (a)).*

$\lambda_1 = 1 - \alpha_1,\ \lambda_2 = 1 + \alpha_2 - \dfrac{\alpha_1\beta_2}{\gamma_1}$. The system will be stable if $|1 - \alpha_1| < 1$ and $\left|1 + \alpha_2 - \dfrac{\alpha_1\beta_2}{\gamma_1}\right| < 1$.

At $\left(0, \dfrac{\alpha_2}{\gamma_2}\right)$, $A = \begin{pmatrix} 1 + \alpha_1 - \frac{\alpha_2\beta_1}{\gamma_2} & 0 \\ -\frac{\alpha_2\beta_2}{\gamma_2} & 1 - \alpha_2 \end{pmatrix}$, whose eigen values are

$\lambda_1 = 1 - \alpha_2,\ \lambda_2 = 1 + \alpha_1 - \dfrac{\alpha_2\beta_1}{\gamma_2}$. The system will be stable if $|1 - \alpha_2| < 1$ and $\left|1 + \alpha_1 - \dfrac{\alpha_2\beta_1}{\gamma_2}\right| < 1$.

At $\left(\frac{\alpha_1\gamma_2 - \alpha_2\beta_1}{\gamma_1\gamma_2 - \beta_1\beta_2}, \frac{\alpha_2\gamma_1 - \alpha_1\beta_2}{\gamma_1\gamma_2 - \beta_1\beta_2}\right)$, the Jacobian matrix is given by

$$A = \begin{pmatrix} \dfrac{\beta_1(\beta_2 - \alpha_2\gamma_1) - (1-\alpha_1)\gamma_1\gamma_2}{\beta_1\beta_2 - \gamma_1\gamma_2} & \dfrac{\beta_1(\alpha_1\gamma_2 - \alpha_2\beta_1)}{\beta_1\beta_2 - \gamma_1\gamma_2} \\ \dfrac{\beta_2(\alpha_2\gamma_1 - \alpha_1\beta_2)}{\beta_1\beta_2 - \gamma_1\gamma_2} & \dfrac{\beta_1\beta_2 - (\alpha_1\beta_2 + (1-\alpha_2)\gamma_1)\gamma_2}{\beta_1\beta_2 - \gamma_1\gamma_2} \end{pmatrix},$$

whose eigenvalues are $\lambda_{1,2}$ (say). The system will be stable if $|\lambda_{1,2}| < 1$.

The model is now tested numerically. Fig. 2.16(a) shows that both the bears co-exist even though there is intra-specific and inter-specific competition among them. Reducing the inter-specific competition parameter β_1 to 0.0025, it is observed that the black bear goes to extinction in approximately 5 years (fig. 2.16(b)). With some proper choice of parameters, the extinction of grizzly bears is seen in fig. 2.16(c).

Suppose the hunters kill 100 black bears and 150 grizzly bears every year. Fig. 2.16(d) shows that both the bears survive and coexist over time, but their numbers becomes less compared to the case when there are no hunters (fig. 2.16(a)).

2.4.5 2-cycles

Consider a non-linear difference equation $x_{n+1} = f(x_n)$. A pair of distinct points x_1 and x_2 is called a 2-cycle of $x_{n+1} = f(x_n)$ if

$$f(x_1) = x_2 \quad \text{and} \quad f(x_2) = x_1.$$

Each point is called a point of period 2 for $f(x)$.

Note: (i) $f(x_2) = x_1 \Rightarrow f(f(x_1)) = x_1$ (since, $x_2 = f(x_1)$. If $h(x) = f(f(x))$, then $h(x_1) = x_1 \Rightarrow x_1$ is the fixed point of $h(x)$. Therefore, any point of period 2 for $f(x)$ is actually a fixed point of the composition $f(f(x))$.

(ii) If x^* is a fixed point of $f(x)$, then $f(x^*) = x^*$, which implies $f(f(x^*)) = f(x^*) = x^*$, that is x^* is also a fixed point of $f(f(x))$. Therefore, solving $f(f(x)) = x$, not only gives point of period 2 but also all the fixed points.

2.4.6 Stability of 2-cycles

Local stability of a 2-cycles of $x_{n+1} = f(x_n)$ means each point of the cycle is a stable fixed point of f(f(x)), otherwise the 2-cycle is unstable. The condition for stability is given by

$$|f'(x_1)f'(x_2)| < 1 \text{ (locally stable)}, \ |f'(x_1)f'(x_2)| > 1 \text{ (unstable)},$$

where x_1 and x_2 are two distinct points of 2-cycles of $x_{n+1} = f(x_n)$.

Example 2.4.1 Find a 2-cycle of (i) $x_{n+1} = \dfrac{-3x_n}{x_n^2+1}$ (ii) $x_{n+1} = {x_n}^2 + x_n - 4$, and determine its stability.

Solution: (i) Let $f(x) = \dfrac{-3x}{x^2+1}$. The equilibrium points are obtained by solving

$$x^* = \frac{-3x^*}{x^{*2}+1},$$
$$\Rightarrow\ x^*\ (x^*+4) = 0\ \Rightarrow\ x^* = 0, \pm 2i.$$

We next compute

$$f(f(x)) = \frac{3(x^2-1)}{(x^2+1)^2}.$$

Now, any point of period 2 for $f(x)$ is actually a fixed point of the composition $f(f(x))$ and we solve

$$f(f(x)) = x\ \Rightarrow \frac{9x(x^2+1)}{1+11x^2+x^4} = x,$$
$$\Rightarrow\ x(x^4+2x^2-8) = 0.$$

This equation also gives the fixed points of f(x). Hence, dividing by $x(x^2+4)$ we get,

$$x^2 - 2 = 0\ \Rightarrow x = \pm\sqrt{2}.$$
$$\text{Now, } f(\sqrt{2}) = -\sqrt{2} \text{ and } f(-\sqrt{2}) = \sqrt{2}.$$

Therefore $\sqrt{2}$ and $-\sqrt{2}$ constitute a 2-cycle. Now,

$$f'(x) = 2x + 1\frac{3(x^2-1)}{(x^2+1)^2},$$
$$\Rightarrow\ f'(\sqrt{2}) = \frac{1}{3},\ f'(-\sqrt{2}) = \frac{1}{3},$$
$$\text{and } |f'(\sqrt{2})f'(-\sqrt{2})| = \left|\frac{1}{9}\right| < 1,$$
$$\Rightarrow \text{the 2-cycle is stable (fig. 2.17(a)).}$$

(ii) Let $f(x) = x^2 + x - 4$. We first obtain the fixed points of

$$x_{n+1} = {x_n}^2 + x_n - 4$$

by solving

$$x^* = {x^*}^2 + x^* - 4\ \Rightarrow x^* = -2, 2.$$

We next compute

$$f(f(x)) = (x^2 + x - 4)^2 + (x^2 + x - 4) - 4 = x^4 + 2x^3 - 6x^2 - 7x + 8.$$

Now, any point of period 2 for $f(x)$ is actually a fixed point of the composition $f(f(x))$ and we solve

$$f(f(x)) = x \Rightarrow x^4 + 2x^3 - 6x^2 - 7x + 8 = x \Rightarrow x^4 + 2x^3 - 6x^2 - 8x + 8 = 0.$$

This equation also gives the fixed points of $f(x)$. Hence, dividing by $(x^2 - 4)$ we get,

$$x^2 + 2x - 2 = 0 \Rightarrow x = (-1 + \sqrt{3}) \text{ and } (-1 - \sqrt{3}). \text{ Now,}$$

$$f(-1+\sqrt{3}) = (-1+\sqrt{3})^2 + (-1+\sqrt{3}) - 4 = (-1-\sqrt{3}) \text{ and } f(-1-\sqrt{3}) = (-1+\sqrt{3}).$$

Therefore $(-1 + \sqrt{3})$ and $(-1 - \sqrt{3})$ constitute a 2-cycle. Now,

$$
\begin{aligned}
&f'(x) = 2x + 1,\\
&f'(-1+\sqrt{3}) = 2(-1+\sqrt{3}) + 1 = -1 + 2\sqrt{3},\\
&f'(-1-\sqrt{3}) = 2(-1-\sqrt{3}) + 1 = -1 - 2\sqrt{3},\\
&f'(-1+\sqrt{3})f'(-1-\sqrt{3}) = (-1+2\sqrt{3})(-1-2\sqrt{3}) = 1 - 12 = -11,\\
&|f'(-1+\sqrt{3})f'(-1-\sqrt{3})| = |-11| = 11 > 1 \Rightarrow \text{the 2-cycle is unstable}
\end{aligned}
$$

(fig. 2.17(b)).

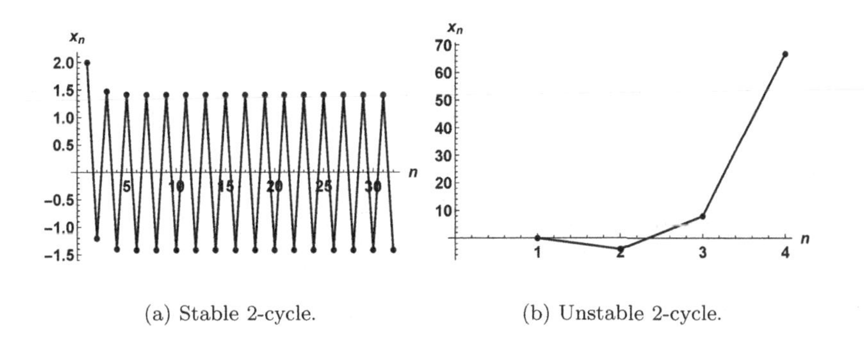

(a) Stable 2-cycle. (b) Unstable 2-cycle.

FIGURE 2.17: *The figures show the dynamics of 2-cycles, (a) stable and (b) unstable.*

2.4.7 3-cycles

Consider a non-linear difference equation

$$x_{n+1} = f(x_n).$$

A set of three distinct points x_1, x_2, x_3 is called a 3-cycle of $x_{n+1} = f(x_n)$ if $f(x_1) = x_2, f(x_2) = x_3, f(x_3) = x_1$. Each point of a 3-cycle is called a point of period 3. By taking compositions, it is possible to generate all other points, starting with any of these points (say, x_1):

$$f(x_1) = x_2,\ f^2(x_1) = f(f(x_1)) = f(x_2) = x_3,$$
$$f^3(x_1) = f(f(f(x_1))) = f(f^2(x_1)) = f(x_2) = x_3$$

Local stability of 3-cycle of $x_{n+1} = f(x_n)$ means each point of the cycle is a stable fixed point of $f(f^2(x))$, otherwise the 3-cycle is unstable. The condition for stability is given by

$$\begin{aligned} |f'(x_1)f'(x_2)f'(x_3)| &< 1 \text{ (locally stable)},\\ |f'(x_1)f'(x_2)f'(x_3)| &> 1 \text{ (unstable)}, \end{aligned}$$

where x_1, x_2, x_3 are three distinct points of 3-cycles of $x_{n+1} = f(x_n)$.

Example 2.4.2 For the system $x_{n+1} = 1-2|x_n|$, find the points in its 3-cycle, where $\frac{7}{9}$ is one such point. Also, determine its stability.

Solution: Let $f(x) = 1 - 2|x|$. Now,

$$\begin{aligned} f\left(\frac{7}{9}\right) &= 1 - 2\left|\frac{7}{9}\right| = -\frac{5}{9},\\ f^2\left(\frac{7}{9}\right) &= f\left(f\left(\frac{7}{9}\right)\right) = f\left(-\frac{5}{9}\right) = 1 - 2\left|-\frac{5}{9}\right| = -\frac{1}{9},\\ f^3\left(\frac{7}{9}\right) &= 1 - f\left(f^2\left(\frac{7}{9}\right)\right) = 1 - f\left(-\frac{1}{9}\right) = 1 - 2\left|-\frac{1}{9}\right| = -\frac{7}{9}. \end{aligned}$$

This implies $\frac{7}{9}, -\frac{5}{9}, -\frac{1}{9}$ is a 3-cycle of $x_{n+1} = 1-2|x_n|$. Since, $f(x) = 1-2|x|$,

$$\begin{aligned} f'(x) &= -2,\ x > 0,\\ &= 2,\ x < 0. \end{aligned}$$

Therefore,

$$f'(\frac{7}{9}) = -2, f'(-\frac{5}{9}) = 2, f'(-\frac{1}{9}) = 2,$$

and

$$\left|f'\left(\frac{7}{9}\right)f'\left(-\frac{5}{9}\right)f'\left(-\frac{1}{9}\right)\right| = |-8| = 8 > 1,$$

implying that the 3-cycle is unstable.

2.5 Bifurcations in Discrete Models

Consider a one-dimensional discrete dynamical system

$$x_{n+1} = f(x_n, r), \tag{2.12}$$

where $r \in \mathbf{R}$ is a parameter.

A fixed point x^* of (2.12) undergoes bifurcation at $r = r^*$ if it changes its number, location, form and stability. Discrete=time bifurcation systems may experience saddle node, transcritical and pitchfork bifurcations, which are entirely analogous to bifurcations of fixed points (equilibrium points) in continuous dynamical systems. However, the flip bifurcation (period-doubling) is unique to discrete-time dynamical systems [36]. Thus, at $r = r^*$, the fixed point x^* of (2.12) may change from stable to unstable, and either another stable point of equilibrium (fixed point) or a stable 2-cycle emerges from it.

Theorem 2.5.0.1 *Consider one-dimensional discrete dimensional system*

$$x_{n+1} = f(x_n, r), \tag{2.13}$$

where f is a smooth function from $\mathbf{R} \times \mathbf{R} \to \mathbf{R}$ and $r \in \mathbf{R}$ (Set of real numbers). Let x^* be the fixed point and r^* be the bifurcation point for (2.13), satisfying the necessary condition for the bifurcation of fixed points, namely,

$$f(x^*, r^*) = x^*, \quad \frac{\partial f}{\partial x}(x^*, r^*) = 1.$$

(a) If (i) $\dfrac{\partial f(x^*, r^*)}{\partial r} \neq 0$, *(ii)* $\dfrac{\partial^2 f(x^*, r^*)}{\partial x^2} \neq 0$,

then the system undergoes a saddle-node bifurcation at (x^*, r^*).

(b) If (i) $\dfrac{\partial f(x^*, r^*)}{\partial r} = 0$, *(ii)* $\dfrac{\partial^2 f(x^*, r^*)}{\partial x \partial r} \neq 0$, *(iii)* $\dfrac{\partial^2 f(x^*, r^*)}{\partial x^2} \neq 0$, *then the system undergoes a transcritical bifurcation at (x^*, r^*).*

(c) If (i) $\dfrac{\partial f(x^*, r^*)}{\partial r} = 0$, *(ii)* $\dfrac{\partial^2 f(x^*, r^*)}{\partial x \partial r} \neq 0$, *(iii)* $\dfrac{\partial^2 f(x^*, r^*)}{\partial x^2} = 0$, *(iv)* $\dfrac{\partial^3 f(x^*, r^*)}{\partial x^3} \neq 0$, *then the system undergoes a pitchfork bifurcation at (x^*, r^*).*

Theorem 2.5.0.2 *Consider a one-dimensional discrete dimensional system*

$$x_{n+1} = f(x_n, r), \tag{2.14}$$

where f is a smooth function from $\mathbf{R} \times \mathbf{R} \to \mathbf{R}$ and $r \in \mathbf{R}$ (Set of real numbers). Let x^* be the fixed point and r^* be the bifurcation point for (2.14), satisfying the necessary condition for the bifurcation of fixed points, namely,

$$f(x^*, r^*) = x^*, \quad \frac{\partial f(x^*, r^*)}{\partial x} = -1.$$

If $(i)\ \dfrac{\partial^2 f(x^*, r^*)}{\partial x \partial r} \neq 0, \quad (ii)\ \dfrac{1}{2}\left[\dfrac{\partial^2 f(x^*, r^*)}{\partial x^2}\right]^2 + \dfrac{1}{3}\dfrac{\partial^3 f(x^*, r^*)}{\partial x^3} \neq 0,$

then a period-doubling bifurcation (flip bifurcation) occurs at (x^*, r^*) for (2.14).

Example 2.5.1 Classify the bifurcation for one-dimensional systems with the single parameter r:

$$\begin{aligned}
(i)\ x_{n+1} &= r + x_n - {x_n}^2, \\
(ii)\ x_{n+1} &= (1 + r)x_n - {x_n}^2, \\
(iii)\ x_{n+1} &= (1 + r)x_n - {x_n}^3, \\
(iv)\ x_{n+1} &= -(1 + r)x_n + {x_n}^3.
\end{aligned}$$

Solution: (i) Let $f(x, r) = r + x - x^2$. Fixed points (equilibrium points) are given by

$$r + x^* - {x^*}^2 = x^* \ \Rightarrow x^* = \pm\sqrt{r}.$$

Clearly, $f(x, r)$ has two fixed points at $x = \pm\sqrt{r}$ and there are no fixed points for $r < 0$. As the parameter varies, the two equilibria existing on one side of the bifurcation disappear on the other side of the bifurcation, that is, two equilibria move towards each other, coincide and are destroyed. Also, $f'(x) = 1 - 2x = 1$ at $x = 0$ for $r = 0$, implying there is a neutral[1] point. Now,

$$f(0, 0) = 0, \frac{\partial f(0, 0)}{\partial x} = 1 - 2x|_{(0,0)} = 1, \frac{\partial f(0, 0)}{\partial r} = 1 \neq 0, \frac{d\partial^2 f(0, 0)}{\partial x^2} = -2 \neq 0.$$

Hence, a saddle-node bifurcation occurs at $(x^*, r^*) = (0, 0)$ (fig. 2.18(a)).

(ii) Let $f(x, r) = (1 + r)x - x^2$. Fixed points are given by

$$(1 + r)x^* - {x^*}^2 = x^* \ \Rightarrow x^*(r - x^*) = 0 \ \Rightarrow x^* = 0, \ r.$$

[1] A fixed point x^* is called neutral if $|f'(x^*)|$=1.

Also, $f'(x) = 1 + r - 2x = 1$ at $x = 0$ for $r = 0$, implying there is a neutral point . Now,

$$f(0,0) = 0, \frac{\partial f(0,0)}{\partial x} = 1 + r - 2x|_{(0,0)} = 1,$$
$$\frac{\partial f(0,0)}{\partial r} = 0, \frac{d\partial^2 f(0,0)}{\partial x \partial r} = 1 \neq 0, \frac{\partial^2 f}{\partial x^2}(0,0) = -2 \neq 0.$$

Hence, a transcritical bifurcation occurs at $(x^*, r^*) = 0$ (fig. 2.18(b)).

(iii) Let $f(x,r) = (1+r)x - x^3$. Equilibrium points are given by

$$(1+r)x^* - {x^*}^3 = x^* \Rightarrow x^*(r - {x^*}^2) = 0 \Rightarrow x^* = 0, \pm\sqrt{r}.$$

f(x,r) has three fixed points at $x^* = 0, \pm\sqrt{r}$. For $\sqrt{r} < 0$, $f(x)$ has only one fixed point. Also, $f'(x) = 1 + r - 3x^2 = 1$ at $x = 0$ for $r = 0$, implying there is a neutral fixed point. Now,

$$f(0,0) = 0, \frac{\partial f}{\partial x}(0,0) = 1, \frac{\partial f}{\partial r}(0,0) = x|_{(0,0)} = 0,$$
$$\frac{\partial f^2}{\partial x^2} = -6x|_{(0,0)} = 0, \frac{\partial f^2(0,0)}{\partial x \partial r} = 1 \neq 0, \frac{\partial f^3(0,0)}{\partial x^3} = -6 \neq 0.$$

Hence, a pitchfork bifurcation occurs at $(x^*, r^*) = (0,0)$ (fig. 2.18(c)).

(iv) Let $f(x,r) = -(1+r)x + x^3$. Fixed points are given by

$$-(1+r)x^* + {x^*}^3 = x^* \Rightarrow x^* = 0, x^* = \pm\sqrt{(2+r)}.$$

For $r < -2$, $f(x)$ has one fixed point, namely, $x^* = 0$. Now, $f'(x) = -(1 + r) + 3x^2$. Also,

$$f(0,0) = 0, \frac{\partial f(0,0)}{\partial x} = -1, \frac{\partial f}{\partial r} = -x, \frac{\partial^2 f(0,0)}{\partial x \partial r} = -1 \neq 0,$$
$$\frac{\partial f^2(0,0)}{\partial x^2} = 6x|_{(0,0)} = 0,$$
$$\frac{\partial f^3(0,0)}{\partial x^3} = 6 \neq 0, \frac{1}{2}\left[\frac{\partial f^2(0,0)}{\partial x^2}\right]^2 + \frac{1}{3}\frac{\partial f^3(0,0)}{\partial x^3} = 0 + \frac{1}{3} \times 6 = 2 \neq 0.$$

Hence, the system undergoes a period doubling bifurcation at $(x^*, r^*) = (0,0)$ (fig. 2.18(d)).

Theorem 2.5.0.3 *Consider a two-dimensional discrete dynamical system with one parameter r*

$$x_{n+1} = f(x_n, y_n, r),$$
$$y_{n+1} = g(x_n, y_n, r).$$

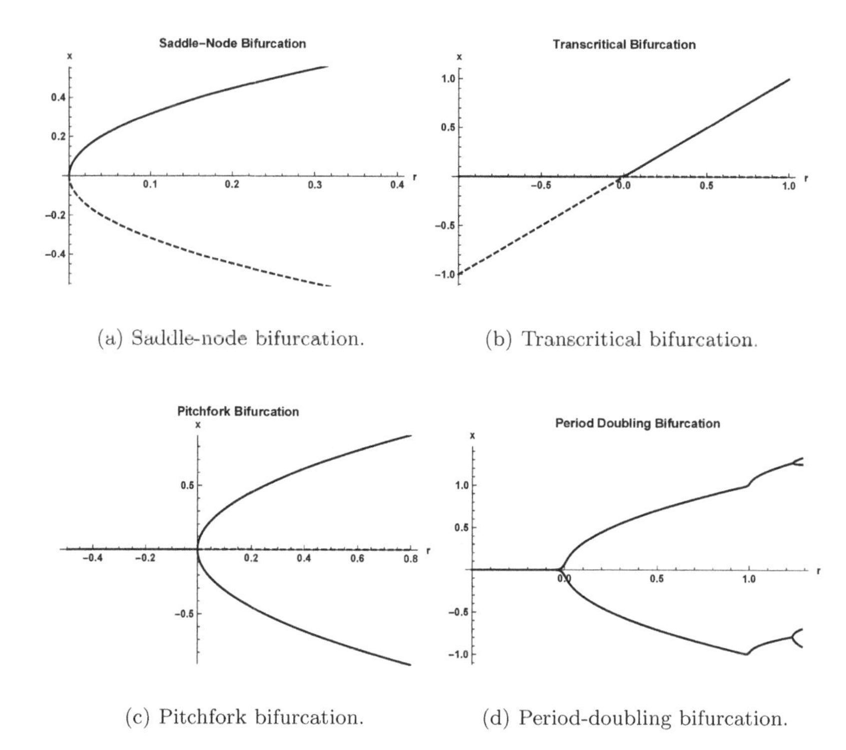

(a) Saddle-node bifurcation. (b) Transcritical bifurcation.

(c) Pitchfork bifurcation. (d) Period-doubling bifurcation.

FIGURE 2.18: *The figures show different bifurcations for one-dimensional discrete systems with the single parameter r.*

Suppose that for each r, it has a smooth family of equilibrium points $(x^*(r), y^*(r))$, at which the eigenvalues are complex conjugates, namely, $\lambda_{1,2} = \alpha(r) \pm i\beta(r)$. If there exists a critical value r^* of the parameter r such that

$$(i)\ |\lambda_{1,2}(r^*)| = \sqrt{\alpha^2(r^*) + \beta^2(r^*)} = 1,$$

$$(ii)\ \frac{d}{dr}|\lambda_{1,2}(r^*)|_{r=r^*} \neq 0,$$

then, the system exhibits Neimark-Sacker bifurcation (also called Hopf bifurcation for map) at $r = r^*$ [53].

Example 2.5.2 Consider the two dimensional discrete system

$$\begin{aligned} x_{n+1} &= r\, x_n(1 - x_n) + e^r y_n \\ y_{n+1} &= r\, x_n(1 - x_n) + e^r y_n - x_n,\ r \in \mathbf{R}. \end{aligned}$$

Show that Neimark-Sacker bifurcation (Hopf bifurcation for map) occurs at origin when $r = 0$.

Solution: The points of equilibria are given by,

$$rx^*(1-x^*) + e^r y^* = x^*, \tag{2.15}$$
$$rx^*(1-x^*) + e^r y^* - x^* = y^*. \tag{2.16}$$

Subtracting (2.15) from (2.16), we get $y^* = 0$ and $x^* = 0, \dfrac{r-1}{r}$. Therefore,

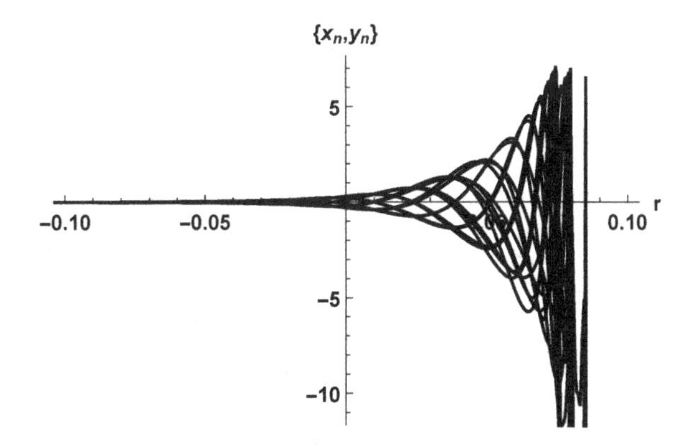

FIGURE 2.19: *The figure shows Neimark-Sacker bifurcation (Hopf bifurcation for map) at the origin* $(0,0)$ *when* $r = 0$.

the steady states are $(0,0)$, $\left(\dfrac{r-1}{r}, 0\right)$. The Jacobian matrix is given by $J(r) = \begin{bmatrix} r-2rx^* & e^r \\ r-1-2rx^* & e^r \end{bmatrix}$, whose eigenvalues are given by the characteristic equation

$$\begin{vmatrix} r-2rx^*-\lambda & e^r \\ r-1-2rx^* & e^r-\lambda \end{vmatrix} = 0 \Rightarrow \lambda^2 - (r+e^r)\lambda + e^r = 0 \text{ (since, } x^* = 0),$$
$$\Rightarrow \lambda_{1,2} = \frac{(r+e^r) \pm \sqrt{((r+e^r)^2 - 4e^r)}}{2} = \frac{(r+e^r) \pm i\sqrt{(4e^r - (r+e^r)^2)}}{2}.$$

Clearly, $\lambda_1(r) = \overline{\lambda_2(r)}$; $\lambda_{1,2}(r^* = 0) = \left|\dfrac{1 \pm i\sqrt{3}}{2}\right| = 1$ and

$$\begin{aligned}\frac{d}{dr}(|\lambda_{1,2}(r^*=0)|) =& \frac{d}{dr}\sqrt{(r+e^r)^2 + 4e^r - (r+e^r)^2}|_{r=0} \\ =& \frac{d}{dr}(\sqrt{4e^r})|_{r=0} = 2\frac{d}{dr}(e^{r/2})|_{r=0} = 1 \neq 0\end{aligned}$$

Therefore, the system exhibits Neimark-Sacker bifurcation at the origin $(0,0)$ when $r = 0$ (fig. 2.19).

2.6 Chaos in Discrete Models

A dynamical system exhibits chaos if it has solutions that appear to be quite random and the solutions exhibit sensitive dependence on initial conditions [113]. According to Tabor [158], by a chaotic solution to a deterministic equation, we mean a solution whose outcome is very sensitive to initial conditions and whose evolution through phase space appears to be quite random.

Henri Poincare, who is also known as the father of chaos theory, was studying orbits in the solar system. Earlier work by Sir Isaac Newton gave a well-developed theory on a solar system in which the orbits worked with the help of well-ordered equations, which is subtle. Poincare extended his study by adding more elements in those equations, already developed by Sir Newton. In his study, Poincare obtained different results when he changed the initial starting point of some of the orbits in the solar system. He observed that it is very difficult, in fact, it is almost impossible to predict how the orbit might work, if there is a small change in the initial condition of the equations. This is the history behind the chaos theory, which was discovered by Henri Poincare.

In the 1960s, Ed Lorenz faced a similar problem while working on a computer program based on complex mathematical formulas used to generate weather patterns. His research generated a number of variables, which could be used to predict weather patterns. After few months, Lorenz re-entered the data and ran the program to see the weather pattern again. To his surprise he got a completely different result. The reason, he later found out is making an error in re-entering the data in the decimal place, which had a drastic impact on the pattern generated by the computer. This again supported what Poincare had discovered before, that is, a small change in initial conditions lead to great differences in outcome [23]. This phenomenon is known as the butterfly effect. This phrase refers to the idea that the flapping of a butterfly's wing might cause tiny changes in the atmosphere that ultimately cause a tornado to appear (or prevent a tornado from appearing). The flapping of the butterfly's wing represents a small change in the initial condition of the system, which causes a chain of events leading to large-scale phenomena [64].

2.6.1 Criteria of Chaos for Discrete Dynamical System

Consider the one-dimensional system $x_{n+1} = f(x_n), (n \geq 0)$, where $f : X \to X$ is a map and (X, d) is a metric space. A continuous map f is said to be chaotic on X (in the sense of Devaney [28]) if
(i) f is transitive,

(ii) the periodic points of f are dense in X,
(iii) f has sensitive dependence on initial conditions.

2.6.2 Quantification of Chaos: Lyapunov Exponent

One of the quantitative measures for a discrete map to be chaotic is to calculate the Lyapunov exponent of the map. If all the points in a neighbourhood of a trajectory converge toward the same orbit, the attractor is a fixed point or a limit cycle and the rate of attraction is exponentially fast. If the map processes the property of sensitive dependence on initial conditions, the neighbouring orbits should move apart exponentially, which gives the mean rate of separation of trajectories of the system. This means rate or average exponential rate, also known as Lyapunov exponent, needs to be positive for the neighbouring orbits to be diverging. Hence, a positive value of the Lyapunov exponent is a signature of chaos [90].

Consider a one-dimensional discrete map

$$x_{n+1} = f(x_n).$$

The question which we ask here is how the point sequence $x_0, x_1, x_2, ...$ differs from the point sequence $x'_0, x'_1, x'_2, ...$, which is obtained from a slightly modified initial condition $x'_0 = x_0 + \delta x_0$. In general, we get

$$\begin{aligned} & x'_n = x_n + \delta x_n, \\ & \Rightarrow x'_n = f(x_{n-1} + \delta x_{n-1}), \\ & \Rightarrow x_n + \delta x_n = f(x_{n-1}) + \delta x_{n-1} f'(x_{n-1}) + ---, \\ & \Rightarrow \delta x_n \approx f'(x_{n-1})\delta x_{n-1}, \end{aligned}$$

which gives the deviation in n^{th} step in linear approximation.

The mathematical definition of the Lyapunov exponent for a one-dimensional discrete map is given by [154]

$$\lambda(x_0) = \lim_{n \to \infty} \frac{1}{n} \sum_{i=0}^{n-1} ln|f'(x_i)|,$$

which depends on the initial condition x_0. Thus, the Lyapunov exponent is a logarithmic measure for the mean expansion rate per iteration (per unit time) of the distance between two infinitesimally close trajectories.

A dynamical system with a positive Lyapunov exponent ($\lambda > 0$) is called chaotic, and for stable fixed points and cycles, λ is negative.

Note: Does positive Lyapunov exponent always mean Chaos? The short answer is NO (but why?).

Consider two systems, $x_{n+1} = 4x_n(1-x_n)$ and $x_{n+1} = 4{x_n}^2$. Both the systems have a positive Lyapunov exponent, but the first one exhibits chaos and not the second one. This shows that positive Lyapunov exponent is not enough, a bounded phase space is also needed to exclude the trivial case of an unstable system ($x_{n+1} = 4{x_n}^2$), where trajectories diverge exponentially for all times. For the logistic system, any finite error after a few iterations grows to the size of the whole system $[0,1]$. In the second case ($x_{n+1} = 4{x_n}^2$), the system is infinite. If the system is actually bounded to be finite by wrapping everything around on a torus, it will exhibit chaos.

Example 2.6.1 Find the Lyapunov exponent for the tent map

$$f(x) = \begin{cases} 2rx, & 0 \le x \le \frac{1}{2}, \\ 2r(1-x), & \frac{1}{2} < x \le 1. \end{cases}$$

Solution: Differentiating $f(x)$ with respect to x, we get

$$f'(x) = \begin{cases} 2r, & 0 \le x < \frac{1}{2}, \\ -2r, & \frac{1}{2} < x \le 1. \end{cases}$$

Therefore, $|f'(x)| = 2r$, for all $x \in [0,1]$, except at $x = \frac{1}{2}$, the point of non-differentiability. The Lyapunov exponent of the tent map is given by,

$$\begin{aligned} \lambda &= \lim_{n\to\infty} \frac{1}{n} \sum_{i=0}^{n-1} ln|f'(x_i)| = ln(2r) \lim_{n\to\infty} \frac{1}{n} \sum_{i=0}^{n-1} 1, \\ &= ln(2r) \lim_{n\to\infty} \frac{1}{n} \times n = ln(2r). \end{aligned}$$

Since $\lambda > 0$ for $2r > 1$, that is, for $r > \frac{1}{2}$, the tent map is chaotic and non-chaotic for $r < \frac{1}{2}$ (figs. 2.20(a) and (b)).

Theorem 2.6.2.1 *Suppose $f : [a,b] \to \mathbf{R}$ is a continuous map. If f has an orbit of period 3, then f is chaotic [90].*

Example 2.6.2 Show that the Tent map $f : [0,1] \to [0,1]$ defined by, $f(x) = \begin{cases} 2x, & 0 \le x < \frac{1}{2}, \\ 2(1-x), & \frac{1}{2} \le x \le 1 \end{cases}$ is chaotic.

Solution: The Tent map is a one-dimensional map which is continuous on $[0,1]$, but not differentiable at $x = \frac{1}{2}$. If the map has 3-cycle (an orbit of period 3), then it is chaotic (see theorem 2.6.2.1). The second composition of $f(x) = f(f(x)) = f^2(x)$ is given by

$$f^2(x) = \begin{cases} 4x & 0 \le x \le \frac{1}{4}, \\ 4x-1 & \frac{1}{4} \le x \le \frac{1}{2}, \\ 4x-2 & \frac{1}{2} \le x \le \frac{3}{4}, \\ 4x-3 & \frac{3}{4} \le x \le 1. \end{cases}$$

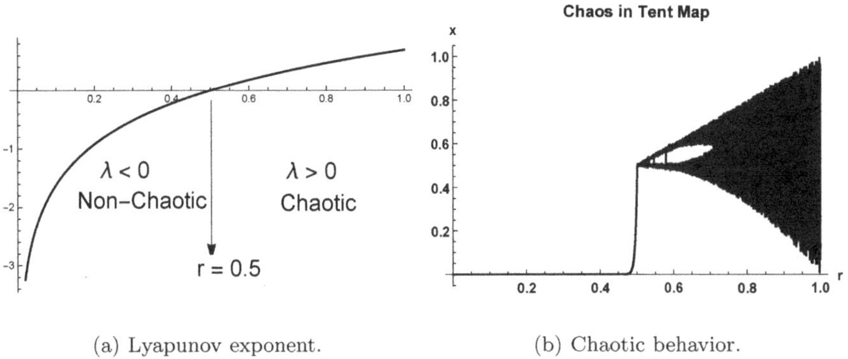

(a) Lyapunov exponent. (b) Chaotic behavior.

FIGURE 2.20: *The figures show (a) positive Lyapunov exponent in* $0.5 < r < 1$, *implying chaotic behavior in that region (b) chaotic behavior in tent map in* $0.5 < r < 1$.

Clearly, $f^2(x) = f(f(x)) = f(2x) = 2.2x,\ 0 \le 2x \le \frac{1}{2} = 4x,\ 0 \le x \le \frac{1}{4}$,
$f^2(x) = f(f(x)) = f(2x) = 2.2x - 1,\ \frac{1}{2} \le 2x \le 1 = 4x - 1,\ \frac{1}{4} \le x \le 1$,
$f^2(x) = f(2x-1) = 2.(2x-1),\ 0 \le 2x - 1 \le \frac{1}{2} = 4x - 2,\ \frac{1}{2} \le x \le \frac{3}{4}$,
$f^2(x) = f(f(x)) = 2.(2x-1) - 1, \frac{1}{2} \le 2x - 1 \le 1 = 4x - 3,\ \frac{3}{4} \le x \le 1$.

In a similar manner, $f^3(x) = f(f^2(x)) = \begin{cases} 8x, & 0 \le x \le \frac{1}{8}, \\ 8x-1, & \frac{1}{8} \le x \le \frac{1}{4}, \\ 8x-2, & \frac{1}{4} \le x \le \frac{3}{8}, \\ 8x-3, & \frac{3}{8} \le x \le \frac{1}{2}, \\ 8x-4, & \frac{1}{2} \le x \le \frac{5}{8}, \\ 8x-5, & \frac{5}{8} \le x \le \frac{3}{4}, \\ 8x-6, & \frac{3}{4} \le x \le \frac{7}{8}, \\ 8x-7, & \frac{7}{8} \le x \le 1, \end{cases}$

Fixed points of $f^3(x)$ are obtained by solving $f^3(x^*) = x^*$ and we get the eight fixed points as $x^* = 0, \frac{1}{7}, \frac{2}{7}, \frac{3}{7}, \frac{4}{7}, \frac{5}{7}, \frac{6}{7}$.
($8x^* = x^* \Rightarrow x^* = 0, 8x^* - 1 = x^* \ \Rightarrow 7x^* = 1 \ \Rightarrow x^* = \frac{1}{7}$ and so on). Also,

$$f(\frac{1}{7}) = \frac{2}{7},\ f^2(\frac{1}{7}) = f(f(\frac{1}{7})) = f(\frac{2}{7}) = \frac{4}{7},\ f^3(\frac{1}{7}) = f(f^2(\frac{1}{7})) = f(\frac{4}{7}) = \frac{1}{7}.$$

This implies that the set $\{\frac{1}{7}, \frac{2}{7}, \frac{4}{7}\}$ forms a 3-cycle of the map $f(x)$ and the given map is chaotic on the interval [0,1] (theorem2.6.2.1).

2.7 Miscellaneous Examples

Problem 2.7.1

Let t_n be the temperature in degrees centigrade and n be the number of meters above the ground. The air cools by about $0.02\,^\circ\mathrm{C}$ for each meter rise above the ground level.
(i) Formulate a discrete dynamical system to model this situation.
(ii) If the current temperature at ground level is $30\,^\circ\mathrm{C}$, find the temperature 500 m above the ground.
(iii) Find the height above the ground level at which the temperature is $0\,^\circ\mathrm{F}$.

Solution: (i) The discrete dynamical model is given by

$$t_n = t_{n-1} - 0.02$$

(negative because temperature decreases (cools) as height increases), where t_n is the temperature in degrees centigrade, n meters above the ground.

(ii) The temperature 0.5 km. (500 m.) above the ground is given by

$$\begin{aligned}
&t_{500} = t_{499} - 0.02,\\
&t_{499} = t_{498} - 0.02,\\
&t_{498} = t_{497} - 0.02,\\
&-\,-\,-\,-\,-\,-\,-\,-\,-\\
&-\,-\,-\,-\,-\,-\,-\,-\,-\\
&t_2 = t_1 - 0.02,\\
&t_1 = t_0 - 0.02.\\
&\text{Adding, we get,}\ \ t_{500} = t_0 - 0.02 \times 500\\
&= 30 - 0.02 \times 500 = 30 - 10 = 20\,^\circ\mathrm{C}.\\
&\text{In general, } t_n = t_0 - 0.02\, n.
\end{aligned}$$

(iii) If n be the height above the ground at which the temperature is about $0\,^\circ\mathrm{C}$, then, $0 = 30 - 0.02 \times n \Rightarrow n = 1500$ m, that is, 1.5 km.

Problem 2.7.2 *The Doppler effect states that if one travels toward a sound, the frequency of the sound seems higher. The frequency of middle C on a piano keyboard is 256 cycles per second. For each mile per hour one increases speed, the apparent frequency of the sound increases by $\frac{256}{760}$ cycles per second, where 760 miles/h is the speed of the sound.*
(i) Formulate a discrete dynamical system to model this situation.
(ii) How fast does one need to travel for the middle C of a keyboard to sound like C#, which is 271 cycles per second?
(iii) How fast does one need to travel for the middle C of the keyboard to sound like the C that is 1 octave higher, that is, C is 512 cycles/second?

Solution: (i) Let f_n be the frequency of the sound that one hears, in cycles per speed, when traveling at a speed of n miles/h towards the sound, then the discrete dynamical model is

$$f_n = f_{n-1} + \frac{256}{760}, \text{ which can also be expressed as } f_n = f_0 + \frac{256n}{760}.$$

(ii) C# sounds like 271 cycles per second, which implies that $f_n = 271$ and $f_0 = 256$, and we get

$$271 = 256 + \frac{256n}{760} \quad \Rightarrow n = 44.53 \text{ miles/h}.$$

Thus, one has to travel approximately at the rate of 44.53 miles/h towards the sound to sound like C#, which is 271 cycles per second.

(iii) Putting $f_n = 512$, $f_0 = 256$, we get

$$512 = 256 + \frac{256n}{760} \quad \Rightarrow n = 760 \text{ miles/hour}.$$

This shows that one needs to travel at the speed of sound for the middle C of a keyboard to sound like the C that is 1 octave higher.

Problem 2.7.3 *Let U_n and V_n be the total amount of pollutants in lakes A and B, respectively, in year n, and that 38% of the pollutant from lake A and 13% of the pollutant from lake B are removed every year. The pollutant that is removed from lake A is added to lake B due to the flow of water from lake A to lake B. Also, it is assumed that 3 tons of pollutants are directly added to lake A and 9 tons of pollutants are added to lake B [143].*
(i) Develop a discrete dynamical system from the above information. Find the equilibrium points and state whether they are stable or not.
(ii) Suppose it is determined that an equilibrium level of a total of 10 tons of pollutants in lake A and a total of 30 tons in lake B would be acceptable. What restrictions should be placed upon the total amounts of pollutants that are added directly, so that these equilibria can be achieved?

Solution: From the schematic diagram (fig. 2.21), the discrete dynamical system is formulated as

$$\begin{aligned} U_n &= U_{n-1} - 0.38U_{n-1} + 3, \\ V_n &= V_{n-1} + 0.38U_{n-1} - 0.13V_{n-1} + 9, \end{aligned}$$

where U_n and V_n are the total amounts of pollutants in lake A and lake B, respectively, in year n. The equilibrium point is obtained by solving

$$\begin{aligned} &U^* = U^* - 0.38U^* + 3 \text{ and } V^* = V^* + 0.38U^* - 0.13V^* + 9, \\ &\Rightarrow U^* = 7.9, \; V^* = 92.3. \end{aligned}$$

Therefore, the equilibrium point of the system is $(7.9, 92.3)$. The system can be rewritten as

$$U_n = 0.62U_{n-1} + 3,$$
$$V_n = 0.38U_{n-1} + 0.87V_{n-1} + 9.$$

The coefficient matrix of the system is given by $A = \begin{pmatrix} 0.62 & 0 \\ 0.38 & 0.87 \end{pmatrix}$, whose eigenvalues are obtained from

$$| A - \lambda I | = 0 \Rightarrow \begin{vmatrix} 0.62 - \lambda & 0 \\ 0.38 & 0.87 - \lambda \end{vmatrix} = 0 \Rightarrow \lambda_{1,2} - 0.62, 0.87.$$

Clearly, $|\lambda_{1,2}| < 1$, and hence the system is stable.

(ii) Suppose the equilibrium values are set to be $U^* = 10$ tons and $V^* = 30$ tons, and let x and y be the amounts of pollutants that are added directly to the lakes, respectively. Then,

$$U^* = U^* - 0.38U^* + x \text{ and } V^* = V^* + 0.38U^* - 0.13V^* + y,$$

where $U^* = 10$ and $V^* = 30$. Solving, we get $x = 3.8$ tons, $y = 0.1$ tons.

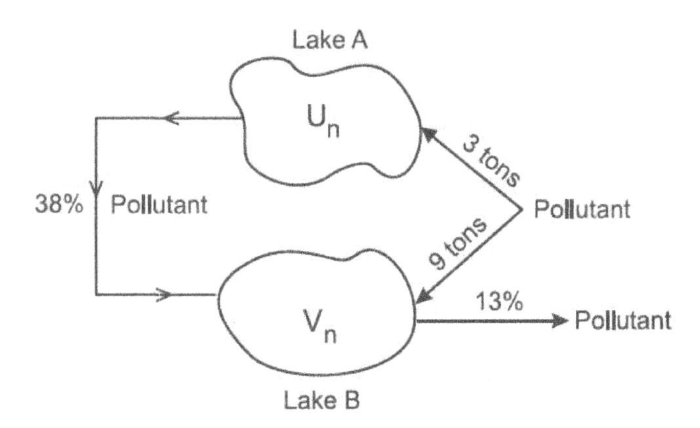

FIGURE 2.21: *The schematic diagram of Problem 2.7.3, where U_n and V_n are the total amount of pollutants in lakes A and B, respectively, in year n.*

Problem 2.7.4 *You buy a house for Rs. 900000 (Indian Rupees). You put Rs. 100000 as down payment and mortgage the rest for 30 years at 9.75% annual interest.*
(i) Write a discrete dynamical system that models how the amount owed on this mortgage changes from month to month and solve it analytically.
(ii) Find the equilibrium value of this model.

Solution: (i) According to the problem, cost of the house = Rs. 900000, down payment = Rs. 100,000, mortgage term = 30 years = 360 months, loan amount = Rs. 800000. Let L_n be the loan amount left at the end of n payment, i be the monthly rate of interest and M be the monthly payment. Then, the discrete dynamical model is given by

$$\begin{aligned} L_n &= L_{n-1} + iL_{n-1} - M, \\ \Rightarrow\ L_n &= (1+i)L_{n-1} - M. \end{aligned}$$

Now,

$$\begin{aligned}
L_1 &= (1+i)L_0 - M, \\
L_2 &= (1+i)L_1 - M = (1+i)^2 L_0 - M[1+(1+i)], \\
L_3 &= (1+i)L_2 - M = (1+i)^3 L_0 - M[1+(1+i)+(1+i)^2], \\
\ldots &\ \ldots \ldots\ldots\ldots\ldots\ldots\ldots\ldots \\
L_n &= (1+i)L_{n-1} - M = (1+i)^n L_0 - M[1+(1+i)+(1+i)^2 \\
&+ \ldots + (1+i)^{n-1}], \\
\therefore L_n &= (1+i)^n L_0 - \frac{M[(1+i)^n - 1]}{i}.
\end{aligned}$$

(ii) We observe that $L_0 = 800,000, L_{360} = 0, i = 9.75\%/12 = (\frac{9.75}{12})/100 = 0.008125$ (as the rate of interest is annual). Therefore,

$$0 = (1.008125)^{360} \times 8 \times 10^5 = \frac{M[(1.008125)^{360} - 1]}{0.008125} \quad \Rightarrow \quad M = \text{Rs. } 6873.00.$$

(iii) The equilibrium value is given by

$$L^* = (1+i)L^* - M \ \Rightarrow M = iL^* 6873/0.008125 \ \Rightarrow\ L^* = \text{Rs. } 845937.00.$$

Problem 2.7.5 *The dynamical system that models the amount of alcohol in a person's body is given by* $U_{n+1} = U_n - \frac{9U_n}{4.2+U_n} + d$*, where* U_n *is the number of grams of alcohol in the body at the beginning of hour* n *and* d *is the constant amount consumed per hour. Find the equilibrium value, given that this person consumes 7 gms of alcohol per hour. Is the system stable?*

Solution: The equilibrium point is given by

$$U_{n+1} = U_n = U^* \ \Rightarrow\ U^* - \frac{9U^*}{4.2+U^*} + 7 = U^* \ \Rightarrow\ U^* = 14.7.$$

Let $f(x) = x - \frac{9x}{4.2+x} + 7$. For stability, we calculate

$$f'(x) = 1 - \frac{4.2 \times 9}{(4.2+x)^2} \ \Rightarrow\ f'(14.2) = 0.894 \ \Rightarrow\ \mid f'(14.2) \mid < 1.$$

Therefore, the system is stable.

Problem 2.7.6 *Consider the price model $P_{n+1} = \frac{1}{P_n} + \frac{P_n}{2} - 1$. Find the two equilibrium points and determine their stability.*

Solution: The equilibrium points are given by

$$\frac{1}{P^*} + \frac{P^*}{2} - 1 = P^* \Rightarrow P^* = 0.732, -2.732.$$

$$\text{Let } f(P) = \frac{1}{P} + \frac{P}{2} - 1 \Rightarrow f'(P) = -\frac{1}{P^2} + \frac{1}{2}.$$

Now, $f'(0.732) = -1.36 \Rightarrow \mid f'(0.732) \mid > 1 \Rightarrow$ the system is unstable about $P^* = 0.732$, and $f'(-2.732) = 0.366 \Rightarrow \mid f'(-2.732) \mid < 1 \Rightarrow$ the system is stable about $P^* = -2.732$.

Problem 2.7.7 *Suppose that each day, 3% of material A decays into material B and 9% of material B decays into lead. Suppose that initially, there are 50 grams of A and 7 grams of B.*
(i) Formulate a discrete dynamical system to model this situation. How much of each material will be left after 5 days?
(ii) Make a graph of $A(n)$ and $B(n)$ for n going from 0 to 50, and observe how they behave.
(iii) Suppose that after 30 days, there are 20 grams of material B left, but there were only 10 grams of B to start with. How many grams of material A was there, to begin with, to the nearest gram?

Solution: (i) Material $A \rightarrow$ Material $B \rightarrow$ Lead.
Let A_n, B_n be the amounts of materials A and B, respectively after n days, A_0, B_0 be the initial amounts of materials A and B respectively, and a, b are the rates of decay of A and B. Then, according to the problem, the linear discrete model is given by

$$\begin{aligned} A_n &= A_{n-1} - aA_{n-1}, \\ B_n &= B_{n-1} - bB_{n-1} + aA_{n-1}. \end{aligned}$$

$$\begin{aligned} &\text{Now, } A_n = (1-a)A_{n-1} \Rightarrow A_1 = (1-a)A_0, \\ &A_2 = (1-a)A_1 = (1-a)^2 A_0, \\ &A_3 = (1-a)A_2 = (1-a)^3 A_0, ..., \\ &\Rightarrow A_n = (1-a)^n A_0 \end{aligned}$$

and

$$\begin{aligned} &B_1 = (1-b)B_0 + aA_0, \\ &B_2 = (1-b)B_1 + aA_1 = [(1-b)^2 B_0 + a(1-b)A_0] + a(1-b)A_0, \\ &B_3 = (1-b)B_2 + aA_2 = (1-b)^3 B_0 + a(1-b)^2 A_0 + a(1-a)(1-b)A_0 \\ &\quad + a(1-a)^2 A_0, \\ &B_n = (1-b)^2 B_0 + aA_0[(1-b)^{n-1} + (1-b)^{n-2}(1-a) + ... + (1-a)^{n-1}], \end{aligned}$$

$$\Rightarrow B_n = (1-b)^n B_0 + \frac{a[(1-b)^n - (1-a)^n]}{a-b}.$$

Given $a = 3\%$, $b = 9\%$, $A_0 = 50$ g, $B_0 = 7$ g. Therefore, after 5 days the amounts of material that will be left are

$$A_5 = (.97)^5 \times 50 = 42.949 \text{ g}, \; B_5 = (.91)^5 \times 7 + \frac{.03 \times 50(.97^5 - .91^5)}{.06} = 10.24 \text{ g}.$$

(ii) From fig. 2.22, it is clear that material A slowly decreases, whereas material B increases slightly due to the addition from material A and then slowly decreases as the decay rate of material B is high.

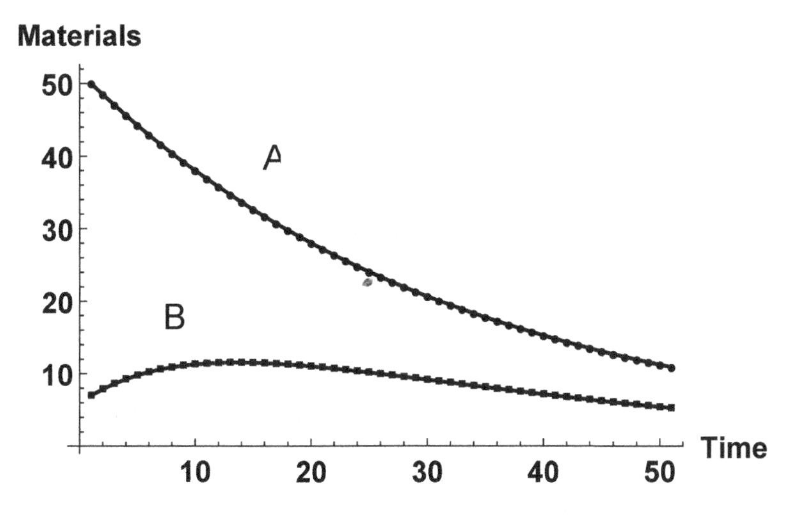

FIGURE 2.22: *The behavior of material $A(n)$ and material $B(n)$ with respect to time in days.*

(iii) Here, a = 0.03, b = 0.09, $B_{30} = 20$ and $B_0 = 10$,

$$\begin{aligned} \Rightarrow B_{30} &= (1-0.09)^{30} \times 10 + \frac{0.03 \times A_0(0.97^{30} - 0.91^{30})}{0.06} = 20, \\ \Rightarrow A_0 &= 113.51 \approx 114 \text{ g}. \end{aligned}$$

Problem 2.7.8 *Suppose you have a roll of paper, such as paper towels. Let the radius of the cardboard core be 0.5 inches. Suppose the paper is 0.002 inches thick. Let $R(n)$ represent the radius, in inches, of the roll when the paper has been wrapped around the core n times. Let $L(n)$ be the total length of the paper when it is wrapped about the core n times. Note that $R(0)=0.5$ and $L(0)=0$.*
(i) Develop a dynamical system for $R(n)$ and $L(n)$.
(ii) What is the length of paper on the roll when it has a radius of 2 inches?

Solution: (i) Let R_n be the radius when the paper is wrapped n times and L_n be the length of paper used for wrapping till n times.

$$\begin{aligned}
R_n &= R_{n-1} + t \text{ (thickness of paper)},\\
R_1 &= R_0 + t,\\
R_2 &= R_1 + t = R_0 + 2t,\\
R_3 &= R_2 + t = R_0 + 3t,\\
\cdots\cdots &\ \cdot\cdot \ \cdots\cdots\cdots\cdots\\
\Rightarrow R_n &= R_0 + nt.
\end{aligned}$$

Again,

$$\begin{aligned}
L_n &= L_{n-1} + 2\pi R_{n-1},\\
L_1 &= L_0 + 2\pi R_0 = 2\pi R_0 \ (\because \ L_0 = 0),\\
L_2 &= L_1 + 2\pi R_1 = 2\pi R_0 + 2\pi R_0 + 2\pi t = 4\pi R_0 + 2\pi t,\\
L_3 &= L_2 + 2\pi R_2 = 4\pi R_0 + 2\pi t + 2\pi R_0 + 4\pi t,\\
L_3 &= 6\pi R_0 + 6\pi t = 3\ 2\pi + 3\ (3-1)\ \pi t,\\
L_4 &= L_3 + 2\pi R_3 = 6\pi R_0 + 6\pi t + 2\pi R_0 + 6\pi t\\
&= 8\pi R_0 + 12\pi t = 4\ 2\pi + 4\ (4-1)\ \pi t,\\
\cdots\cdots &\ \cdot\cdot \ \cdots\cdots\cdots\cdots\\
\therefore L_n &= 2\pi n R_0 + n(n-1)\pi t.
\end{aligned}$$

(ii) Now, given $R_0 = 0.5$ inches, $t = 0.002$ inches and $L_0 = 0$, which implies,

$$R_n = 0.5 + 0.002n, \quad \text{and} \quad L_n = \pi n + 0.002\pi n(n+1).$$

We have to find the length of the paper when the radius of the roll is 2 inches, that is, $R_n = 2$. Therefore,

$$\begin{aligned}
2 &= 0.5 + 0.002n,\\
\Rightarrow n &= \frac{1.5}{0.002} = 750. \text{ Therefore,}\\
L_{750} &= \pi \times 750 + 750 \times 749\pi \times 0.002\\
&= 5888.14 \text{ inches.}
\end{aligned}$$

Problem 2.7.9 *Presently, you weigh* 169 *pounds. You consume* x *pounds worth of calories each week. Assume your body burns off the equivalent of* 3% *of its weight each week through normal metabolism. In addition, you burn off* $\frac{1}{4}$ *pound of weight through daily exercise each week. Find* x *to one decimal place if you want to weigh between* 144 *and* 146 *pounds in* 1 *year (*52 *weeks).*

Solution: Let W_n be the weight after n weeks. The calories consumed each week are x pounds and W_0 = initial weight = 169 pounds.

$$\begin{aligned} W_n &= W_{n-1} - 0.03W_{n-1} - 0.25 + x, \\ W_n &= 0.97W_{n-1} + x - 0.25, \\ W_1 &= 0.97W_0 + x - 0.25, \\ W_2 &= 0.97W_1 + x - 0.25 = 0.97^2W_0 + (x - 0.25)[1 + 0.97], \\ W_3 &= 0.97W_2 + x - 0.25 = 0.97^3W_0 + (x - 0.25)[1 + 0.97 + 0.97^2], \\ W_n &= 0.97^nW_0 + (x - 0.25)\frac{[1 - 0.97^n]}{1 - 0.97}. \end{aligned}$$

Now, according to the problem,

$$144 < W_{52} < 146,$$
$$\Rightarrow\ 144 < (0.97)^{52} \times 169 + \tfrac{(x-0.25)}{0.03}[1 - (0.97)^{52}] < 146,$$
$$\Rightarrow\ 4.375 < x < 4.45\ \Rightarrow\ x = 4.4 \text{ pounds worth of calories.}$$

Problem 2.7.10 *The air pressure at sea level is* 1000 g/cm^2. *The pressure increases by* 2.08 g/cm^2 *for each cm. of depth in salt water. Write a discrete mathematical model for this discrete dynamical system.*

Solution: Let P_0 = Pressure at sea level, P_n = Pressure of air below n cm of sea level, r = rate of pressure increase per cm of depth, then

$$P_n = P_{n-1} + r,\ P_1 = P_0 + r,\ P_2 = P_1 + r = P_0 + 2r,$$
$$P_3 = P_2 + r = P_0 + 3r, ...,\ P_n = P_0 + nr.$$
$$\text{Given } P_0 = 1000,\ r = 2.08,\ \Rightarrow\ P_n = 1000 + 2.08\ n.$$

Problem 2.7.11 *A certain drug is effective in treating a disease if the concentration remains above* 100 mg/L. *The initial concentration is* 640 mg/L. *It is known from laboratory experiments that the drug decays at the rate of* 20% *of the amount present each hour.*

(i) Formulate a linear discrete system that models the concentration after each hour.

(ii) Find graphically at what hour the concentration reaches 100 mg/L.

(iii) Modify your model to include a maintenance dose administered every hour.

(iv) Check graphically or otherwise to determine the maintenance doses that will keep the concentration above the minimum effective level of 100 mg/L *and below the maximum safe level of* 800 mg/L.

(v) Working with the maintenance doses you found in (iv), try varying the initial concentration. What do you observe about the tendency to stay within the necessary bounds, as well as the long-term tendency?

Solution: (i) Let C_n be the concentration of drug at hour n. Since the drug decays at the rate of 20% of the amount present each hour, the linear discrete model is given by

$$\begin{aligned} C_n &= C_{n-1} - \frac{20}{100}C_{n-1} \Rightarrow C_n = 0.8C_{n-1}, \\ \Rightarrow C_1 &= 0.8C_0 \text{ } (C_0 \text{ being the initial concentration of the drug}), \\ \Rightarrow C_2 &= 0.8C_1 = (0.8)^2C_0 \Rightarrow C_3 = 0.8C_2 = (0.8)^3C_0, ..., \\ \Rightarrow C_n &= (0.8)^nC_0 = (0.8)^n \times 640. \end{aligned}$$

(ii) From the graph (fig. 2.23(a)), we can see that after 9 hours, the concentration reaches 100 mg/L. Thus, doses must be provided before this time for recovery.

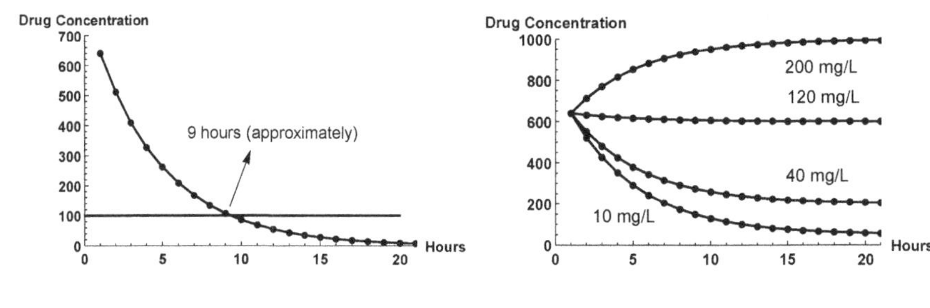

(a) Effective level reached in 9 hours. (b) Different maintenance doses.

FIGURE 2.23: *The figures show the dynamics of the concentration of drugs. (a) Concentration of drug reaches* 100 *mg/L in* 9 *hours. (b) The effect of different maintenance doses, namely,* $x = 10, 40, 120, 200$ *on the concentration of drugs* C_n.

(iii) Let x ml/L be the hourly maintenance dose. Then, $C_n = 0.8C_{n-1} + x$.

(iv) Equilibrium solution is given by $C^* = 0.8C^* + x \Rightarrow C^* = 5x$. Now, according to the problem, $100 < 5x < 800 \Rightarrow 20 < x < 160$. Hence, the value of the maintenance dose must lie between 20 and 160, which is clearly understood from the graph (fig. 2.23(b)).

(v) The final concentration does not depend on the value of the initial dose. Hence, our bounds remain the same even with the change in initial dose (this can be easily verified from the graph and is left for the readers).

Problem 2.7.12 *Consider the disease model*

$$x_{n+1} = 0.80\ x_n + rx_n\left(1 - \frac{x_n}{10^6}\right),$$

where r (>0) is the infection rate. Find the value of the infection rate at which a fixed point bifurcates and classify the bifurcation. Discuss its biological significance.

Solution: Let $f(x) = 0.80x + rx\left(1 - \frac{x}{10^6}\right)$. Fixed points are given by

$$0.80x^* + rx^*(1 - \frac{x^*}{10^6}) = x^* \Rightarrow x^*\left(-0.20 + r - \frac{rx^*}{10^6}\right) = 0,$$

$$\Rightarrow x^* = 0,\ {x_r}^* = 10^6\frac{(r-0.2)}{r}.$$

For the existence of the positive equilibrium, we must have $r > 0.2$. Now,

$$\frac{\partial f}{\partial x} = 0.80 + r - \frac{2xr}{10^6}.\text{ For stability, } \left|\frac{\partial f'}{\partial x}(x^*, r)\right|_{x^*=0} < 1 \Rightarrow |0.80 + r| < 1,$$

$$\Rightarrow -1 < 0.80 + r < 1 \quad -1.80 < r < 0.20 \Rightarrow \ 0 < r < 0.20 \text{ (since } r > 0).$$

Therefore, $x^* = 0$ is stable for $0 < r < 0.2$ and unstable for $r > 0.20$. There is a change in stability, from stable to unstable, at r=0.20 and this occurs simultaneously with ${x_r}^*$ coming into existence at the same location, since ${x_r}^* = 0$ when r=0.20. Clearly,

$$\left|\frac{\partial f}{\partial x}({x_r}^*, r)\right| = 0.80 + r - \frac{2r}{10^6}\left(\frac{r - 0.20}{r}\right) \times 10^6 = 1.20 - r.$$

Therefore, ${x_r}^*$ is stable for all values of r slightly greater than 0.20. Also,

$$f({x_r}^*, r^*) = f(0, 0.20) = 0,\ \frac{\partial f}{\partial x}(0, 0.20) = 0.80 + 0.20 = 1,$$

$$\frac{\partial f}{\partial r}(0, 0.20) = x^*(1 - \frac{x^*}{10^6})|_{x^*=0} = 0,$$

$$\frac{\partial^2 f}{\partial x \partial r}(0, 0.20) = \left(1 - \frac{2x^*}{10^6}\right)|_{x^*=0} = 1 \neq 0,\ \frac{\partial^2 f}{\partial x^2}(0, 0.20) = -\frac{2}{10^6} \neq 0.$$

This implies that the system undergoes transcritical bifurcation at $({x_r}^*, r^*)$=(0,0.20). The point ${x_r}^* = 0$ bifurcates at $r = 0.20$ into the positive stable fixed point $x = {x_r}^*$.

Biological Significance:
There is a change in dynamics in the disease model at the infection rate $r = 0.20$. If the infection rate is less than 0.20, the disease will eventually die out on its own (since $x^* = 0$ is stable). However, if $r > 0.20$, the disease will remain in the population and never be eliminated. A proportion of the population will always remain infected since ${x_r}^*$ is stable [144].

Problem 2.7.13 *The Gross National Product (GNP) is the total value (capital) of all finished goods and services produced by a country's citizens* *(labor) in a given financial year, irrespective of their location. In 1982, Day [25] investigated a highly simplified model for the GNP of a country, given by*

$$G_{n+1} = \sigma \frac{BG_n^{\beta}(m - G_n)^{\gamma}}{1+\lambda}, \quad \beta, \gamma, m > 0,$$

where G_n is the capital–labor ratio, σ is the saving ratio, $BG_n^{\beta}(m - G_n)^{\gamma}$ is the per capital production function, and λ is the natural rate of population growth.
(i) For $\sigma = 0.5, \beta = 0.3, \gamma = 0.2, \lambda = 0.2, m = 1, B = 1$, find the non-zero equilibrium point and check its stability.
(ii) Changing B to 3.3 and keeping other parameter values the same, obtain the non-zero equilibrium point and check its stability. What is the difference, if compared with (ii)?
(iii) Obtain the time series graph with the parameter values mentioned in (i) and (ii) and initial condition $G(0) = 0.01$ and comment on the dynamics of the model.

Solution: (i) The equilibrium points are obtained by solving

$$G^* = \sigma \frac{B(G^*)^{\beta}(m - G^*)^{\gamma}}{1+\lambda}.$$

For $\sigma = 0.5, \beta = 0.3, \gamma = 0.2, \lambda = 0.2, m = 1, B = 1$, the non-zero equilibrium point is 0.26. Let

$$f(G) = 0.4167(1-G)^{0.2}G^{0.3} \Rightarrow f'(G) = \frac{0.125(1-G)^{0.2}}{G^{0.7}} - \frac{0.0833G^{0.3}}{(1-G)^{0.8}}.$$

Now, $f'(0.26) = 0.23 < 1 \Rightarrow$ the system is stable about $G^* = 0.26$.

(ii) Putting $B = 3.3$ (other parameter values remain the same), the real non-zero point of equilibrium is 0.87. For this change of B,

$$f(G) = 1.375(1-G)^{0.2}G^{0.3}, \Rightarrow f'(G) = \frac{0.4125(1-G)^{0.2}}{G^{0.7}} - \frac{0.275G^{0.3}}{(1-G)^{0.8}}.$$

Now, $f'(0.87) = -1.05 \Rightarrow |f'(0.87)| > 1 \Rightarrow$ the system is unstable about $G^* = 0.87$. Since $f'(0.87) < 0$, the solution will oscillate about $G^* = 0.87$.

In (i), the system is stable, hence the solution approaches the steady-state solution (equilibrium solution) $G^* = 0.26$; in (ii), the system is unstable and the solution oscillates about $G^* = 0.87$ as $f'(0.87) < 0$.

(iii) Figs. 2.24(a) and (b) show the dynamics of the models for $B = 1$ and $B = 3.3$, respectively, which supports the analytic results stated in (i) and (ii).

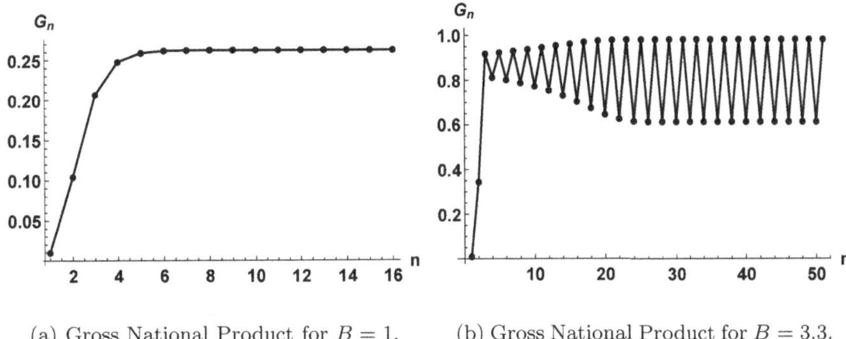

(a) Gross National Product for $B = 1$. (b) Gross National Product for $B = 3.3$.

FIGURE 2.24: *The figures show the dynamics of the Gross National Product (GNP) for changing values of the parameter B. (a) Stable monotonic behavior, (b) unstable oscillatory behavior.*

Problem 2.7.14 *Consider the non-linear discrete model*

$$x_{n+1} = x_n + a\sin(b\ x_n).$$

Check the stability of the model at the equilibrium point $x = 0$ $(a > 0, b > 0)$.

Solution: To check the stability of the model, we define the function

$$f(x) = x + a\ \sin(b\ x)\ \Rightarrow\ f'(x) = 1 + a\ b\cos(bx),$$
$$\Rightarrow\ |f'(0)| = 1 + a\ b > 1,\ \text{as } a, b > 0.$$

Therefore, the model is unstable at the equilibrium point $x = 0$ (fig. 2.25).

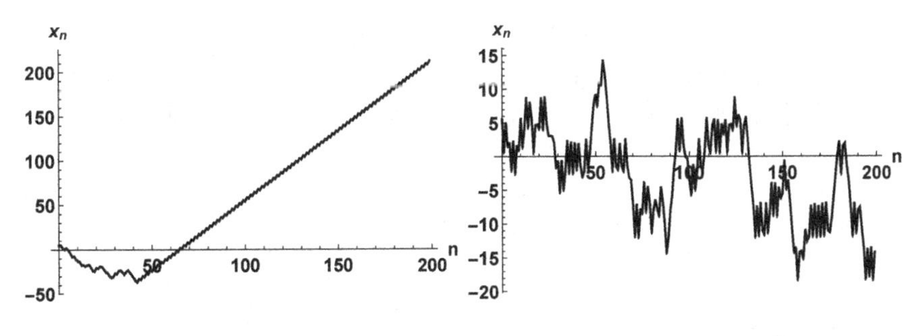

(a) Dynamics with $a = 4,\ b = 2$. (b) Dynamics with $a = 5,\ b = 2$.

FIGURE 2.25: *The figures show that $x_{n+1} = x_n + a\sin(b\ x_n)$ is unstable about the equilibrium point $x = 0$.*

Problem 2.7.15 *The activity of a recurrent two-neuron module at time n is given by [121]*

$$x_{n+1} = \theta_1 + \frac{w_{11}}{1+e^{-x_n}} + \frac{w_{12}}{1+e^{-y_n}},$$
$$y_{n+1} = \theta_2 + \frac{w_{21}}{1+e^{-x_n}},$$

where θ_1, θ_2 are neuron biases; w_{11}, w_{12}, w_{21} are weights.
(i) For $\theta_1 = -2, \theta_2 = 3, w_{11} = -20, w_{12} = 6, w_{21} = -6$, find the equilibrium point and check its stability.
(ii) Obtain the time series graph with the parameter values mentioned in (i) with initial condition $(x_0, y_0) = (5, 5)$ for $n = 50$ and comment on the dynamics of the model.

Solution:(i) The equilibrium points (x^*, y^*) are obtained by solving

$$x^* = \theta_1 + \frac{w_{11}}{1+e^{-x^*}} + \frac{w_{12}}{1+e^{-y^*}}, \qquad y^* = \theta_2 + \frac{w_{21}}{1+e^{-x^*}}.$$

For $\theta_1 = -2, \theta_2 = 3, w_{11} = -20, w_{12} = 6, w_{21} = -6$, the equilibrium points are $(1.28038, 1.69508)$. The Jacobian matrix is given by

$$J = \begin{bmatrix} -3.40371 & 0.786301 \\ -1.02111 & 0 \end{bmatrix},$$

whose eigenvalues are $\lambda_1 = -3.14872$ and $\lambda_2 = -0.254993$. Since $|\lambda_1| > 1$ and $|\lambda_2| < 1$ the system is a saddle and hence unstable.

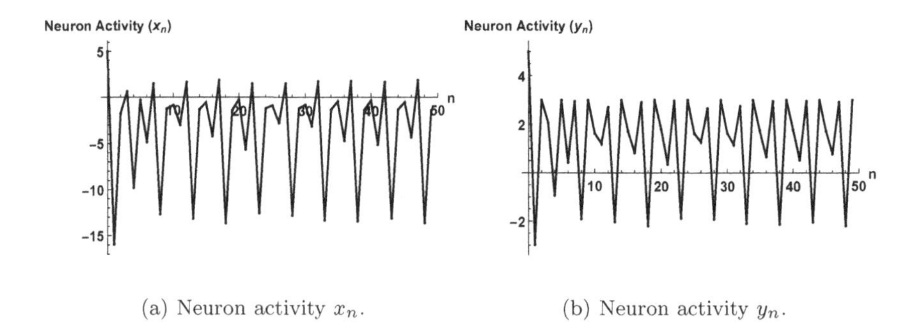

(a) Neuron activity x_n. (b) Neuron activity y_n.

FIGURE 2.26: *The figures show a recurrent two neuron module with an excitatory neuron (x_n) and an inhibitory neuron with self-connection (y_n).*

(ii) The solutions exhibit a global chaotic attractor with the given set of parameters (fig. 2.26). The Lyapunov exponents ($\Lambda_1 = 0.22$, $\Lambda_2 = -3.3$) have been calculated in [121], confirming that the attractor is chaotic.

Problem 2.7.16 *Consider the following two-dimensional discrete systems:*

$$x_{n+1} = -rx_n - (1+r){x_n}^2 + ry_n,$$
$$y_{n+1} = -rx_n - (1+r){x_n}^2 + ry_n - x_n.$$

Show that a Neimark-Sacker bifurcation (Hopf bifurcation map) occurs at the origin when $r = 1$.

Solution: The points of equilibria are given by,

$$-r\ x^* - (1+r)(x^*)^2 + r\ y^* = x^*, \tag{2.17}$$
$$-r\ x^* - (1+r)(x^*)^2 + r\ y^* - x^* = y^*. \tag{2.18}$$

Subtracting (2.17) from (2.18), we get $y^* = 0$ and $x^* = 0, -1$. Therefore, the

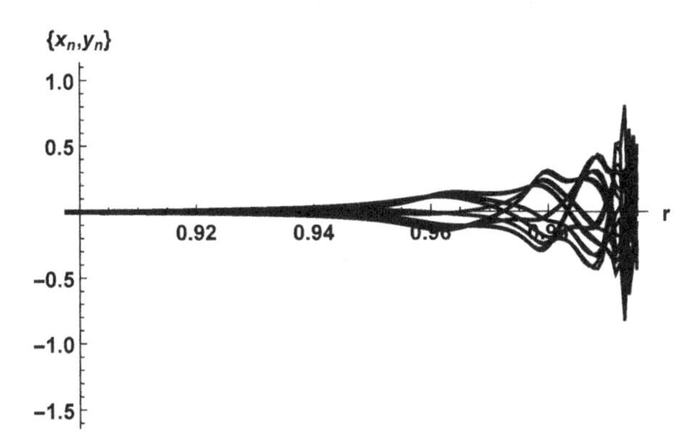

FIGURE 2.27: *The figure shows Neimark-Sacker bifurcation (Hopf bifurcation for map) at the origin* $(0,0)$ *when* $r = 1$.

steady states are $(0,0), (-1,0)$. The Jacobian matrix is given by

$$J(r) = \begin{bmatrix} -r - 2(1+r)x^* & r \\ -r - 2(1+r)x^* - 1 & r \end{bmatrix},$$

whose eigenvalues are given by the characteristic equation

$$\begin{vmatrix} -r - 2(1+r)x^* - \lambda & r \\ -r - 2(1+r)x^* - 1 & r - \lambda \end{vmatrix} = 0,$$
$$\Rightarrow (\lambda + r)(\lambda - r) + r(r+1) = 0 \text{ (since, } x^* = 0),$$
$$\Rightarrow \lambda^2 + r = 0, \ \Rightarrow \lambda_{1,2} = \pm i\sqrt{r}.$$

Clearly, $\lambda_1(r) = \overline{\lambda_2(r)}$; $|\lambda_{1,2}|_{r=1} = |\pm i| = 1$; $\dfrac{d}{dr}(|\lambda_{1,2}(r)|)_{r=1} = \dfrac{d}{dr}(r)|_{r=1}$ $= 1 \neq 0$.

Therefore, the system exhibits Neimark-Sacker bifurcation at the origin $(0,0)$ when $r = 1$ (fig. 2.27).

Problem 2.7.17 *Let U_n and I_n be the measures of unemployment and inflation at time n. An unemployment-inflation model [3] is given by*

$$U_{n+1} = U_n - b(m - I_n),$$
$$I_{n+1} = I_n - (1-c)(\beta_1 + \beta_2 e^{-U_n}) + (\beta_1 + \beta_2 e^{-[U_n - b(m-I_n)]}),$$

where $b, m, c, \beta_1, \beta_2$ are constants.
(i) Obtain the equilibrium point.
(ii) For $m = 2, \beta_1 = -2.5, \beta_2 = 20, c = 0.2$, find the eigenvalues of the system in terms of b. Putting $b = 0.1$, obtain the eigenvalues and comment on the stability of the system.
(iii) Obtain the time series graph with the parameter values mentioned in (ii) and initial condition $(U_0, I_0) = (1, 6)$ for $n = 80$ and comment on the dynamics of the model. Also, obtain the phase portrait.

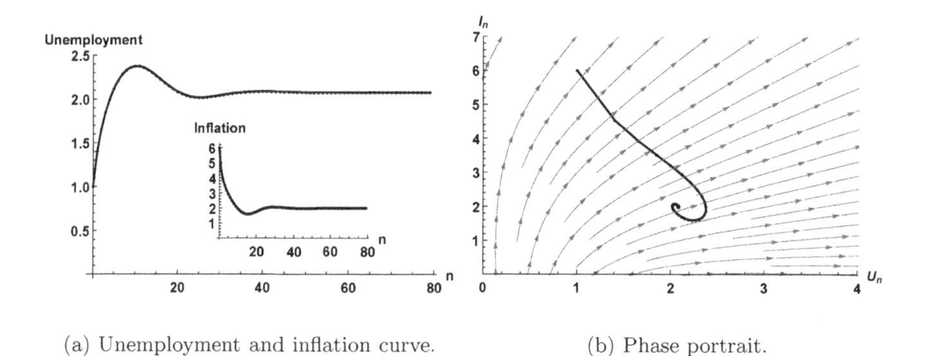

(a) Unemployment and inflation curve. (b) Phase portrait.

FIGURE 2.28: *The figures show the inverse trade-off between unemployment and inflation.*

Solution: (i) The equilibrium points (U^*, I^*) are obtained by solving

$$U^* = U^* - b(m - I^*) \Rightarrow I^* = m,$$
$$I^* = I^* - (1-c)\left(\beta_1 + \beta_2 e^{-U^*}\right) + \left(\beta_1 + \beta_2 e^{-[U^* - b(m-I^*)]}\right),$$
$$\Rightarrow c\left(\beta_1 + \beta_2 e^{-U^*}\right) = 0 \Rightarrow U^* = \ln\left(\frac{-\beta_2}{\beta 1}\right).$$

(ii) The Jacobian matrix is given by

$$J(r) = \begin{bmatrix} 1 & b \\ c\beta_1 & 1 + b\beta_1 \end{bmatrix},$$

whose eigenvalues are given by the characteristic equation

$$\begin{vmatrix} 1-\lambda & b \\ c\beta_1 & 1+b\beta_1-\lambda \end{vmatrix} = 0 \Rightarrow (\lambda-1)^2 - b\beta_1(\lambda-1) - bc\beta_1 = 0,$$

$$\Rightarrow (\lambda-1)^2 + b\frac{5}{2}(\lambda-1) + \frac{b}{2} = 0 \Rightarrow \lambda = 1 - \frac{5b}{4} \pm \frac{\sqrt{25b^2-8b}}{4}.$$

For $b = 0.1$, the eigenvalues are $0.875 \pm 0.185i$, whose modulus is $0.894 < 1$. Therefore, the system is stable.

(iii) As unemployment is high, inflation is low, which is evident from fig. 2.28(a). The relationship between unemployment and inflation has traditionally been an inverse correlation, which is clear from the phase portrait (fig. 2.28(b)).

Problem 2.7.18 *Consider the non-linear discrete system describing the interaction between the price P_n and the demand D_n:*

$$P_{n+1} = P_n + a_1 D_n - b_1 {P_n}^2 + h_1,$$
$$D_{n+1} = D_n + \frac{c_1}{P_n} - k_1,$$

where $a_1(>0), b_1(>0), c_1(>0), k_1(>0)$ are parameters, h_1 can take any sign.
(i) Explain the formation of the model in details, by describing each of the parameters and what they signify in the context of the model.
(ii) For $a_1 = 2.5, b_1 = 0.1, h_1 = -1, c_1 = 5, k_1 = 1$, find all the equilibrium points and check its stability. Obtain the time series graph (P_n, D_n) with the given parameter values and initial condition $(P_1, D_1) = (4, 2)$ for $n = 30$ and comment on the dynamics of the model. Also obtain the phase portrait.
(iii) Repeat the process as stated in (ii) with parameter set $a_1 = 2, b_1 = 0.1, h_1 = -2, c_1 = 4, k_1 = 2$.
(iv) For the given parameter values $a_1 = 2.5, b_1 = 0.06, h_1 = -1, c_1 = 3.6, k_1 = 1$, show that the model has a 2-cycle and check its stability.
(v) For $a_1 = 2.5, h_1 = -1, k_1 = 1, b_1 = 0.1$, numerically estimate the value of the parameter $c_1(>0)$ at which the equilibrium point bifurcates into a stable cycle.

Solution: (i) Here, P_n and D_n are the price and demand for a commodity at time n, respectively. As the demand for a commodity increases, its price also increases from its present level, which explains the term $a_1 D_n$. The price falls $(-b_1 P_n^2)$ when its supply increases ($b_1 P_n^2$ is the supply function, which depends on the price). Again, the demand for a commodity falls from its present level if its price increases and vice-versa, implying that demand is inversely proportional to the price of a commodity. This explains the term $\frac{c_1}{P_n}$ in the demand equation.

(ii) The equilibrium points (P^*, D^*) are obtained by solving

$$P^* = P^* + a_1 D^* - b_1 {P^*}^2 + h_1, D^* = D^* + \frac{c_1}{P^*} - k_1.$$

For $a_1 = 2.5, b_1 = 0.1, h_1 = -1, c_1 = 5, k_1 = 1$, the equilibrium points is $(5.0, 1.4)$. The Jacobian matrix is given by

$$J = \begin{bmatrix} 0 & 2.5 \\ -0.2 & 1 \end{bmatrix},$$

whose eigenvalues are $\lambda_{1,2} = 0.5 \pm 0.5i$ and $|0.5 \pm 0.5i| = 0.7071 < 1 \Rightarrow$ the system is stable (fig. 2.29(a)). The phase portrait is shown in (fig. 2.29(b)).

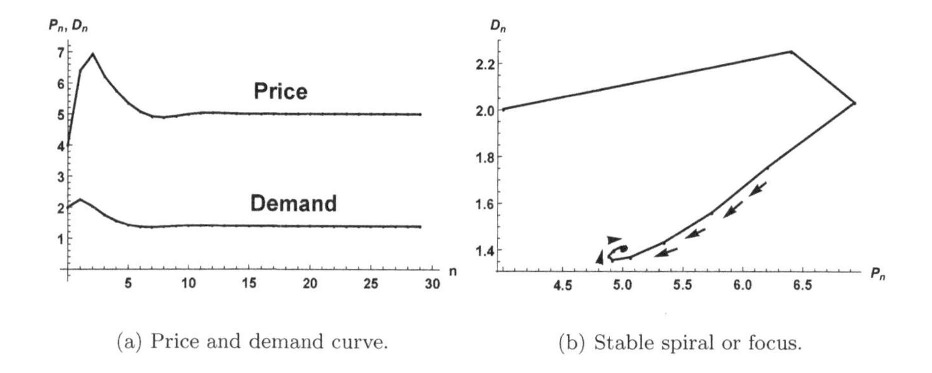

(a) Price and demand curve. (b) Stable spiral or focus.

FIGURE 2.29: *The figures show the dynamics of price and demand for a commodity. (a) The time series graph of price and demand, showing the demand for a commodity decreases as its price increases, (b) phase portrait of the system showing stable spiral or focus.*

(iii) For $a_1 = 2, b_1 = 0.1, h_1 = -2, c_1 = 4, k_1 = 2$, the equilibrium point is $(2.0, 1.2)$. The Jacobian matrix is given by

$$J = \begin{bmatrix} 0.6 & 2.0 \\ -1 & 1 \end{bmatrix},$$

whose eigenvalues are $\lambda_{1,2} = 0.8 \pm 1.4i$ and $|0.8 \pm 1.4i| = 1.61245 > 1 \Rightarrow$ the system is unstable (fig. 2.30(a)). The phase portrait is shown in (fig. 2.30(b)).

(iv) For $a_1 = 2.5, b_1 = 0.06, h_1 = -1, c_1 = 3.6, k_1 = 1$, the equilibrium point is $(3.6, 0.71)$. Let

$$F(P, D) = \left(P + 2.5D - 0.06P^2 - 1, D + \frac{3.6}{P} - 1\right).$$

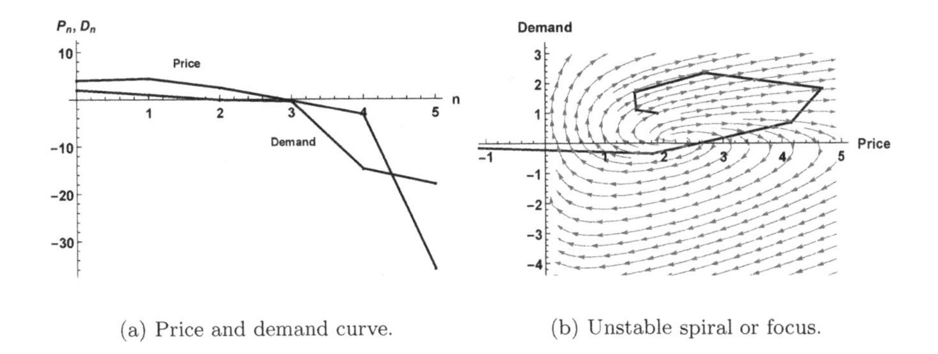

(a) Price and demand curve. (b) Unstable spiral or focus.

FIGURE 2.30: *The figures show the dynamics of price and demand for a commodity. (a) The time series graph of price and demand shows that they both reach negative values as the system is unstable, (b) phase portrait of the system showing unstable spiral or focus.*

We next compute

$$F(F(P,D)) = F\left(-1 + P - 0.06P^2 + 2.5D, -1 + \frac{3.6}{P} + D\right),$$
$$= \Bigg(-2 + 2.5D + 2.5(-1 + D + \frac{3.6}{P}) + P$$
$$- 0.06P^2 - 0.06(-1 + 2.5D + P - 0.06P^2)^2,$$
$$- 2 + D + \frac{3.6}{P} + \frac{3.6}{(-1 + 2.5D + P - 0.06P^2)}\Bigg).$$

Now, any point of period 2 for $F(P,D)$ is actually a fixed point of the composition $F(F(P,D))$ and we solve numerically $F(F(P,D)) = (P,D)$, that is, solving

$$- 2 + 2.5D + 2.5(-1 + D + \frac{3.6}{P}) + P$$
$$- 0.06P^2 - 0.06(-1 + 2.5D + P - 0.06P^2)^2$$
$$= P \text{ and } - 2 + D + \frac{3.6}{P} + \frac{3.6}{(-1 + 2.5D + P - 0.06P^2)} = D.$$

The solutions are $(32.63, 13.67)$,$(1.91, 12.78)$,$(-2.20, 1.79)$, $(0.99, -0.85)$ and $(3.6, 0.71)$. Now,

$$F(32.63, 13.67) = (1.91, 12.78) \text{ and } F(1.91, 12.78) = (32.63, 13.67).$$

Therefore, $(32.63, 13.67)$ and $(1.91, 12.78)$ constitute a 2-cycle.

To check the stability of the 2-cycle, we calculate

$$D_1 = \begin{pmatrix} 1-0.12P & 2.5 \\ -3.6/P^2 & 1 \end{pmatrix}_{(P,D)\to(32.63,13.67)} = \begin{pmatrix} -5.526 & 2.5 \\ -0.0047/P^2 & 1 \end{pmatrix}, \text{ and}$$

$$D_2 = \begin{pmatrix} 1-0.12P & 2.5 \\ -3.6/P^2 & 1 \end{pmatrix}_{(P,D)\to(1.91,12.78)} = \begin{pmatrix} 0.62 & 2.5 \\ -1.37 & 1 \end{pmatrix}.$$

Now, $D_2D_1 = \begin{pmatrix} -3.43 & 4.05 \\ 7.57 & -2.43 \end{pmatrix}$, whose eigenvalues are $\lambda_{1,2} = -8.48, 2.63$. Since $|\lambda_{1,2}| > 1$, the 2-cycle is unstable.

(v) The bifurcation occurs at $c_1 = 2.6$ approximately (trial and error method) and is shown in fig. 2.31.

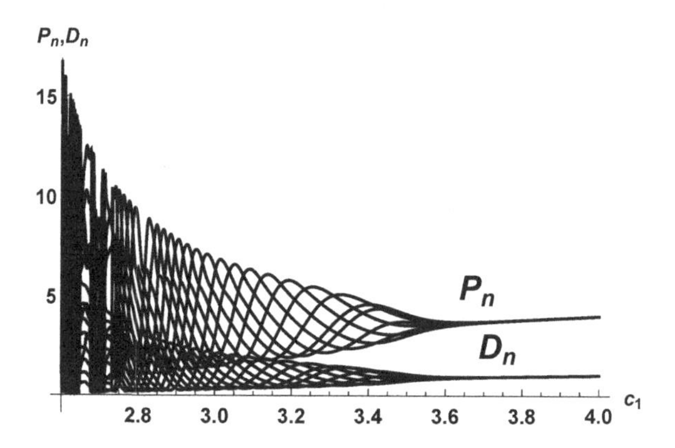

FIGURE 2.31: *The figure shows that the model bifurcates to steady cycles once the bifurcation parameter c_1 crosses the value 2.6.*

Problem 2.7.19 *In 1977, Lasota [89] investigated a red blood cell population model. Let B_n be the red blood cell count per unit volume in the nth time interval, then*

$$B_{n+1} = (1-\alpha)B_n + \beta\ B_n^r e^{-sB_n},\ \beta, r, s > 0, 0 < \alpha \leq 1,$$

where α is the fraction of blood cells which are not destroyed (by natural ageing process or by some accidental factor).

(i) For $\beta = 1.1\times10^6, r = 8, s = 16, \alpha = 0.2$, find the equilibrium points and check their stability. Obtain the time series graph with the given parameter values and initial condition $(B_0) = 1.2$ for $n = 30$ and comment on the dynamics of the model.

(ii) Changing α to 0.3 and keeping other parameter values the same, obtain the equilibrium points and check their stability. Obtain the time series graph with the new set of parameter values and initial condition $(B_0) = 1.2$ for $n = 30$ and comment on the dynamics of the model. What is the difference, if compared with (i)? How does the dynamics of the model changes if $\alpha = 0.39$?

(iii) Obtain the graph of B_n by changing the parameter α from 0.1 to 0.4 and comment.

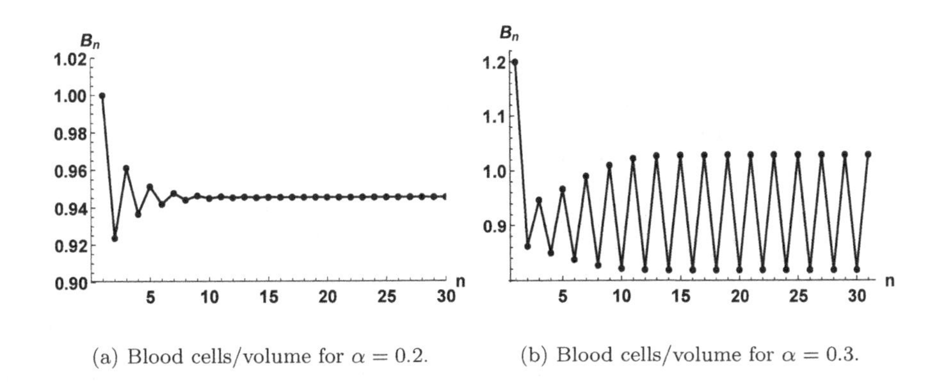

(a) Blood cells/volume for $\alpha = 0.2$.

(b) Blood cells/volume for $\alpha = 0.3$.

FIGURE 2.32: *The figures show the behavior of blood cell population in many clinical cases: (a) The trajectory converges to a steady-state solution, implying that in normal conditions, the number of blood cells is stable, (b) a periodic solution of period 1, implying the existence of stable oscillation in the blood cell population.*

Solution:(i) The equilibrium points are obtained by solving

$$B^* = (1-\alpha)B^* + \beta\ (B^*)^r e^{-sB^*}.$$

For $\beta = 1.1 \times 10^6, r = 8, s = 16, \alpha = 0.2$, the real points of equilibria are 0.16 and 0.95. Let

$f(B) = 0.8B + 1.1 \times 10^6 B^8 e^{-16B}$, therefore,
$f'(B) = 0.8 + 8.8 \times 10^6 B^7 e^{-16B} - 1.76 \times 10^7 B^8 e^{-16B}$. Now,
$f'(0.16) = 2.04 > 1 \Rightarrow$ the system is unstable about $B^* = 0.16$;
$f'(0.95) = -0.59 \Rightarrow |f'(0.95)| < 1 \Rightarrow$ the system is stable about $B^* = 0.95$.
Since $f'(0.95) < 0$, the solution will oscillate with diminishing magnitude and converge to $B^* = 0.95$ (fig. 2.32(a)).

(ii) Putting $\alpha = 0.3$ (other parameter values remain same), the real points of equilibria are 0.17 and 0.90. For this change of α,

$f(B) = 0.7B + 1.1 \times 10^6 B^8 e^{-16B}$, therefore,
$f'(B) = 0.7 + 8.8 \times 10^6 B^7 e^{-16B} - 1.76 \times 10^7 B^8 e^{-16B}$. Now,
$f'(0.17) = 2.27 > 1 \Rightarrow$ the system is unstable about $B^* = 0.17$;
$f'(0.90) = -1.18 \Rightarrow |f'(0.90)| > 1 \Rightarrow$ the system is unstable about $B^* = 0.95$.
Since $f'(0.90) < 0$, the solution will oscillate about $B^* = 0.90$ (fig. 2.32(b)).

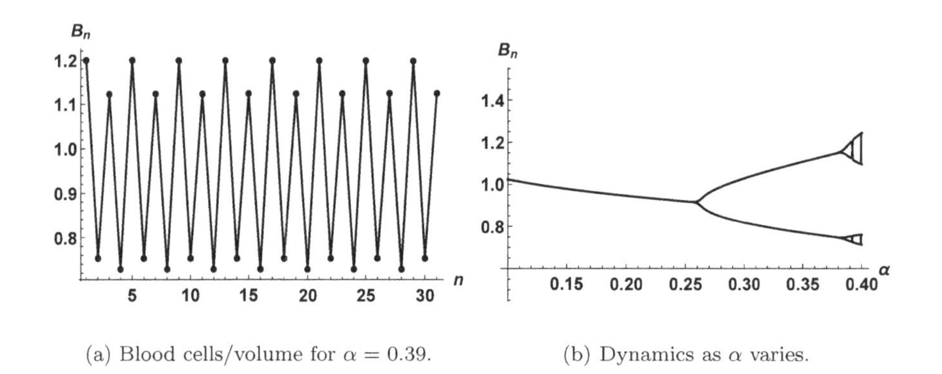

(a) Blood cells/volume for $\alpha = 0.39$.

(b) Dynamics as α varies.

FIGURE 2.33: *The figures show the behavior of the blood cell population in many clinical cases: (a) A periodic solution of period 2, (b) bifurcations in the blood cell population.*

The difference between (i) and (ii) is, in the second case, the system is unstable for both the equilibrium points, and hence the solution will never converge to the points of equilibria.

For $\alpha = 0.39$, the system will be unstable like the previous one, and the solution will demonstrate period oscillations (fig. 2.33(a)). In this case, three different values may be discerned that keep repeating, whereas it was two in the previous case.

(iii) A bifurcation is obtained at $\alpha = 0.26$ (fig. 2.33(b)).

2.8 Mathematica Codes

2.8.1 Lanchester's Combat Model (Figure 2.12(a))

```
α = 0.3; β = 0.2; r1 = 3.0; r2 = 2.0;
rr2 = RSolve [{a[n + 1] = a[n] − βb[n] + r1, b[n + 1] = b[n] − αa(n) + r2,
 a[0] = 75, b[0] = 100} , {a[n], b[n]}, n]
cc11 = Table[Evaluate[a[n]/. rr2], {n, 0, 10}]
aa11 = Table[{n, cc11[[n + 1]][[1]]}, {n, 0, 10}]
ll1 = ListPlot[aa11, Joined → True, PlotMarkers → {Automatic, Small},
PlotRange → All, AxesOrigin → {0, 0}, PlotStyle → Black,
AxesLabel → {n, An, Bn}, BaseStyle → {FontWeight → Bold, FontSize → 15}]
cc22 = Table[Evaluate[b[n]/. rr2], {n, 0, 10}]
bb11 = Table[{n, cc22[[n + 1]][[1]]}, {n, 0, 10}]
ll2 = ListPlot[bb11, Joined → True, PlotMarkers → {Automatic, Small},
PlotRange → All, AxesOrigin → {0, 0}, PlotStyle → Black,
AxesLabel → {n, An, Bn}, BaseStyle → {FontWeight → Bold, FontSize → 15}]
Show [ll1, ll2, PlotRange → {{0, 10}, {0, 100}}]
```

Note: Figs. (2.2–2.13, 2.22, 2.23) can be obtained by modifying the code in 2.8.1.

2.8.2 Lyapunov Exponent (Figure 2.20(a))

```
f(r_):=Function[x, Piecewise[{{2rx, 0 ≤ x ≤ 1/2}, {2r(1-x), 1/2 ≤ x ≤ 1}}]];
Orbit[map_, x0_, n_]:=NestList[map, x0, n];
IterativeProcess[map_, x0_, {min_, max_}]:=
Module[{fr, orb}, orb = Orbit(map, x0, 50);
fr = MapThread[Line[{{#1, #1}, {#1, #2}, {#2, #2}}]&,
{Drop[orb, −1], Drop[orb, 1]}];
Show[Plot[{map[x], x}, {x, min, max}], Graphics[{fr}]]]
LyapunovExp(map_, x0_, n_):=
Plus@@(Function[x,
Evaluate[Log[Abs[D[map[x], x]]]]]/@Drop[Orbit[map, x0, n + 500], 500])/n
Plot[LyapunovExp[f[r], 0.5, 500], {r, 0, 1}, PlotStyle → Black]
```

2.8.3 Two Species Competition Model (Figure 2.16(a))

```
α1 = 1.9; β1 = 0.0003; γ1 = 0.003; α2 = 1.1; β2 = 0.0001; γ2 = 0.001;
rr1 = RecurrenceTable [{B[n + 1] = B[n] + α1 B[n] − β1 B[n]G[n] − γ1 B[n]^2,
 G[n + 1] = G[n] + α2 G[n] − β2 B[n]G[n] − γ2 G[n]^2, B[0] = 200, G[0] = 200.},
{B, G}, {n, 0, 30}]
l1 = ListPlot[Table[{n − 1, rr1[[n]][[1]]}, {n, 1, 21}], Joined → True,
PlotMarkers → {Automatic, 4}, PlotRange → All, AxesOrigin → {0, 0},
PlotStyle → Black, AxesLabel → {n, Bear Population},
BaseStyle → {FontWeight → Bold, FontSize → 12}]
l2 = ListPlot[Table[{n − 1, rr2[[n]][[1]]}, {n, 1, 21}], Joined → True,
PlotMarkers → {Automatic, 4}, PlotRange → All, AxesOrigin → {0, 0},
PlotStyle → Black, AxesLabel → {n, Bear Population},
BaseStyle → {FontWeight → Bold, FontSize → 12}]
Show [l1, l2, PlotRange → {{0, 20}, {0, 1100}}]
```

Note: Figs. (2.14–2.17, 2.24–2.26, 2.28–2.30, 2.32, 2.33) can be obtained by modifying the code in 2.8.2.

2.8.4 Saddle-Node Bifurcation (Figure 2.18(a))

```
Bif1D[fn_ , var_ , r_ ]:=Module[{TT, Iter, h},
h[xx_ ]:=fn/.var → xx;
Iter[k_ ]:=Nest[h, 0.4, k];
TT:=Table[Iter[n], {n, 100, 105}];
Plot[{TT}, r, AxesOrigin → {r[[2]], 0}, PlotStyle → Black]]
p1 = Bif1D [r + x − x^2, x, {r, 0, 0.8}]
p2 = Plot [−√r, {r, 0, 0.8}, PlotStyle → {Dashed, Black}]
Show [p1, p2, PlotRange → {{0, 0.4}, {−0.5, 0.5}}, AxesLabel → {r, x},
 BaseStyle → {FontWeight → Bold, FontSize → 10}]
```

Note: Figs. (2.18(b), 2.18(c), 2.18(d), 2.20(b)) can be obtained by modifying the code in 2.8.3.

2.8.5 Neimark-Sacker Bifurcation (Figure 2.19)

```
Bif2D(fg_, varx_, vary_, x0_, y0_, param_, interval_):=
Module[{F, G, X, Y, p0, lll}, F(x_, y_):=fg[[1]]/.{varx → x, vary → y};
G(x_, y_):=fg[[2]]/. {varx → x, vary → y};
X[a_]:=F[a[[1]], a[[2]]]; Y[b_]:=G[b[[1]], b[[2]]];
p0 = {x0, y0};
Iterasyon(d_):=Nest[{X({#1[[1]], #1[[2]]}), Y({#1[[1]], #1[[2]]})}&, p0, d];
Tx:=Table[Iterasyon[n][[1]], {n, 80, 87}];
Ty:=Table[Iterasyon[n][[2]], {n, 80, 87}];
Plot[{Tx, Ty}, {param, interval[[1]], interval[[2]]}, PlotRange → All,
PlotStyle → Black, AxesLabel → {param, {x_n, y_n}}]]
pp1 = Bif2D[{r x (1-x) + Exp[r] y, r x (1-x) + Exp[r] y − x},
x, y, 0.2, 0.3, r, {−0.1, 0.1}]
Show[pp1, PlotRange → {{−0.1, 0.1}, {−10.7, 6.7}},
AxesLabel → {"r", "x_n,y_n"},
BaseStyle → {FontWeight → Bold, FontSize → 12}]
```

Note: Figs. (2.19, 2.27, 2.31) can be obtained by modifying the code in 2.8.4.

2.9 Matlab Codes

2.9.1 Lanchester's Combat Model (Figure 2.12(a))

```
n1 = [1 : 10]; alpha1 = 0.3; beta1 = 0.2; r11 = 3; r22 = 2;
x1(1) = 75.; y1(1) = 100.;
for m1 = 1 : 9;
x1(m1+1) = x1(m1) − beta1 * y1(m1) + r11;
y1(m1+1) = y1(m1) − alpha1 * x1(m1) + r22;
end
plot(n1, x1,'o−', n1, y1,'o−')
xlabel ('n')
ylabel ('A_n, B_n')
```

2.9.2 Saddle-Node Bifurcation (Figure 2.18(a))

```
kk1 = 50; x = zeros(1, kk1); kk2 = 30;
kk3 = kk1 - (kk2 - 1);
for r = 0 : 0.01 : 0.8
x = 0.4; xo = x;
for n = 2 : kk1
xn = r + xo - xo^2;
x(n) = xn;
xo = xn;
end
plot(r * ones(kk2), x(kk3 : kk1), 'black.', 'MarkerSize', 4)
holdon
end
xlabel('r')
ylabel('x_n')
x = linspace(0, 0.8);
y1 = -sqrt(x);
plot(x, y1)
```

2.9.3 Chaotic Behavior in Tent Map (Figure 2.20(b))

```
r = linspace(0, 1, 1000); x = 0.23;
for i = 1 : 100
y = 2 * r. * (0.5 - abs(x - 0.5));
x = y;
end
A = zeros(800, 1000);
for i = 1 : 800
y = 2 * r. * (0.5 - abs(x - 0.5));
A(i, :) = y;
x = y;
end
plot(A', 'k.', 'Markersize', 2);
xlabel('r') ylabel('x')
```

2.9.4 2D Bifurcation (Figure 2.31)

```
kk1 = 100; x = zeros(1, kk1); y = zeros(1, kk1);
kk2 = 30; kk3 = kk1 − (kk2 − 1);
a1 = 2.5; b1 = 0.1; h1 = −1.; k1 = 1.;
for c1 = 2.6 : 0.005 : 4.
x = 4.0; y = 1.0;
xo = x;  yo = y;
for n = 2 : kk1
xn = xo + a1 * yo − b1 * xo^2 + h1;  yn = yo + c1/xo − k1;
x(n) = xn;
xo = xn;
y(n) = yn;
yo = yn;
end
plot (c1kk2ones, x(kk3 : kk1),'r.','MarkerSize', 2)
hold on
plot (c1kk2ones, y(kk3 : kk1),'b.','MarkerSize', 2)
end
xlabel ('c1')
ylabel ('xn, yn')
```

2.10 Exercises

1. Let A_n be the total value of an investment at the end of n-terms, earning a compound interest at a rate of r per term, with an additional deposit of d at the end of each term.
(i) Construct an Annuity Saving Model, assuming A_0 to be the original investment.
(ii) Find the total value after 3 years if the initial investment is \$10000, each month the account gets 1% interest and \$1000 is deposited each month.
(iii) Find the total value after 7 years if the initial investment is \$20000, the account gets 6% annual interest compounded monthly and a deposit of \$3000 is made each month.

2. Let L_n be the unpaid balance on a loan with an interest rate of r per term. A payment of P is made at the end of each year.
(i) Assuming L_0 to be the original amount borrowed, construct a loan payment model for the unpaid balance at the end of each n years.
(ii) Find the balance after 3 years when payments of \$500 are made quarterly on an original loan of \$9,600 with an interest rate of 10.5% per quarter.
(iii) Find the balance after 5 years when payments of \$400 are made monthly on an original loan of \$20,000 with an annual interest of 8%.

3. Let P_n be the population of the nth generation, growing linearly at a rate r and undergoes either immigration or migration at a constant level k.
(i) Formulate a linear discrete immigration/migration model.
(ii) Find the population of the 4th generation when the initial population is 3900, the growth rate is 7% per generation and immigration is occurring at a constant rate of 190 per generation.
(iii) How the population of the 4th generation changes if there is a migration of 190 per generation, instead of immigration.

4. Let T_n be the amount of pollutant present in a contaminated lake, which is cleaned by filtering out a certain fraction α of all pollutants present at that time, but another β tons of pollutants seep in $(0 < \alpha < 1, \beta > 0)$.
(i) Formulate a mathematical model, assuming T_0 tons of pollutants are initially present in the lake.
(ii) If each week 10% of all pollutants present can be removed, but another 2 tons seep in, find the values of α and β and write the discrete equation for this process.
(iii) How many tons of pollutants will be there in the lake after 2 years, if initially it is contaminated with 70 tons of pollutants?
(iv) How long will it take before the pollutant level falls below 20 tons?

5. Let a pendulum swings in such a manner that the greatest negative (or positive) angle it makes on one side of the vertical is always a certain negative fraction of the greatest positive (or negative) angle it previously made on the other side of the vertical.
(i) Formulate a mathematical model for this process if θ_n is the nth greatest angle (positive or negative) and θ_0 is the initial angle of the pendulum when released.
(ii) Suppose, the greatest negative (or positive) angle it makes on one side of the vertical is always 90% of the greatest positive (or negative) angle it previously made on the other side of the vertical, which always occurs 4 seconds earlier, how many times will the pendulum cross the vertical before the magnitude of the angle it achieves is 1° or less?
(iii) Approximately, how long will it take for this to happen?

6. Suppose a person is saving for retirement by depositing equal amounts

every quarter into a retirement plan to earn 9% annual interest compounded quarterly. If the account initially had zero balance,
(i) how much will be available at retirement time in 25 years if each deposit is \$10,000?
(ii) how much should each deposit be, to have \$2,000,000 at retirement in 25 years?

7. In a certain forest, 3% of the trees are destroyed naturally each year. Also, 4000 trees are harvested for timber (cut down for industrial use) but 7000 new trees are either planted or sprout up on their own.
(i) Formulate a discrete model for the yearly number of trees T_n in the forest.
(ii) What is the maximum number of trees the forest will have if the currently estimated number of trees in that forest is 70,000?

8. A credit card company charges 3% interest on any previous unpaid balance from a person and the person pays off 20% of that previous balance. Also, another \$100 is charged on that credit card from the person for not paying the full amount.
(i) Formulate a discrete model for the monthly unpaid balance.
(ii) What amount will the unpaid balance gradually approach?

9. Let a town be affected by viral fever. Each day 10% of those have the viral fever in the town recover from it, while another 500 people are affected by the viral fever. If there are currently 2000 cases of viral fever, formulate a discrete mathematical model and find out how many cases will be there, two weeks from now. At that level, will the number of cases eventually stabilize?

10. It is known that as the demand for a product increases, its prices increases, and the price of a product decreases when its supply increases. Let D_n be the demand, S_n be the supply, and P_n be the prices of a certain product. We assume that the demand force increases continuously from negative to positive as the current demand D_n increases and the same is the behavior for the supply S_n. Also, as the price of a product increases, its supply increases, but demand from consumers decreases, that is, supply is directly proportional to the price, and demand is inversely proportional to the price.
(i) Formulate a non-linear discrete price model for the price of a product at time n from the above information.
(ii) What happens to the above model when the price is close to zero ? How can that be resolved?

11. Let I_n be the number of infected population at any given time n and r be the fraction of the infected population who have recovered in time $(n+1)$. It is assumed that the numbers of the newly infected population are directly proportional to the size of the infected population I_n and to

the size of the susceptible population $N - I_n$, where N is the population size.
(i) Formulate a discrete contagious disease model for the infected population I_{n+1}.
(ii) In a population of size 10^6, each week 80% of those infected recover. If 1000 are infected in one week, then the following week, 1500 are infected. Formulate a mathematical model.
(iii) In a population of size 4×10^6, 40% infected are still sick in a week, and the maximum number of new infected cases possible in any week is 50000, formulate a model with this information.
(iv) In a population of size 10000, if 100 are infected, the number doubled in the following week. Formulate a model assuming recovery is not possible.
(v) Write down the models when the number of cases is proportional to (a) I_n^2 and $(1 - \frac{I_n}{N})^2$, (b) I_n^2 and $(1 - \frac{I_n^2}{N^2})$, (c) I_n and $\exp(\frac{-I_n}{N})$ and repeat as given in (ii), (iii) and (iv).

12. A radioactive element is known to decay at the rate of 2% every 20 years.
(i) If initially, you had 165 grams of this element, how much would you have in 60 years?
(ii) What is the half-life of this element?
(iii) Suppose that the bones of a certain animal maintains a constant level of this element while the animal is living, but the element begins to decay as soon as the animal dies. If a bone of this animal is found and is determined to have only 10% of its original level of this element, how old is the bone?

13. The population of weasels is growing at the rate of 3% per year. Let w_n be the number of weasels, n years from now and suppose that there are currently 350 weasels.
(i) Formulate a linear discrete model which describes how the population changes from year to year.
(ii) Solve the difference equation of part (i). If the population growth continues at the rate of 3%, how many weasels will there be 15 years from now?
(iii) Plot w_n versus n for $n = 0, 1, 2, ..., 100$.
(iv) How many years will it take for the population to double?
(v) Find $\lim_{n \to \infty} w_n$. What does this say about the long-term size of the population? Will this really happen?
(vi) If the rate of growth of the weasel population was 5% instead of 3%, how many years would it take for the population to double?

14. Assume the temperature of a roast in the oven increase at a rate proportional to the difference between the oven (set to 368 °F) and the roast. If the roast enters the oven at 48 °F and is measured one hour later

to be at 88 °F, when should the table be set if the eating temperature is 168 °F?

15. The forensics team from a local police department is called to the scene of an apparent murder. The body of the victim is found in a room where the thermostat was set at 72 °F. The police investigators believe that the victim died in the room and that the body has been there ever since.
(i) Formulate a linear discrete system that describes how the temperature of the dead body changes from hour to hour and solve it.
(ii) At 5 A.M., when the investigators first arrived, they took the temperature of the body and found that it was 87.5 °F. Just before the body was to be removed from the room an hour later, they measured the temperature to be 80.4 °F. Find the time when the victim was murdered (normal body temperature of the victim = 98.6 °F).

16. An industry dumps a metal-based pollutant into a local lake. Currently, there are 350 pounds of pollutants in the lake, and 20 pounds are added each year, all at once at the end of the year. It is known that the amount of pollutants lost through evaporation and decomposed chemically in a year is 10% of the amount in the lake.
(i) Formulate a linear discrete system that models the amount of the pollutant in the lake after each year.
(ii) Show graphically, what will happen in the long run if the situation described above persists? How does the outcome depend on the initial amount of the pollutant in the lake?
(iii) The maximum safe amount of pollutants in this lake is 125 pounds of pollutants. If we set a goal to achieve this safe level in 15 years, determine how much pollutants can be added to the lake each year.
(iv) What if the amount of pollutants that is added each year is divided into equal parcels, are added each month throughout the year? How does this change the model in part (i) and the results obtained in parts (ii) and (iii)?

17. A person has signed a contract to write a book, where he has already written the first 90 pages. But, the publisher wants the book to be 300 pages long. The author has started writing k pages of the book, each day, starting from today.
(i) Formulate a linear discrete system that models w_n, the total number of pages written after n days.
(ii) The publisher has given the author 21 days to finish the book. Find the approximate number of pages the author needs to write each day to complete the book.

18. You are counseling a recent graduate on retirement planning. You estimate that he will earn 9% per year (or 0.75% per month) on a retirement account. With his high-paying job, he will be able to invest

$5000 per month initially. To allow for inflation and pay raises, you suggest increasing this amount by 0.5% per month. Assume that he makes his first deposit at the end of his first month of employment.
(i) Formulate a discrete dynamical system that models the amount in this person's retirement account and solve it.
(ii) Graphically, determine the balance in the account after 20 years and use the analytic solution to verify it.
(iii) How does the analytic solution change if the monthly investment increases by 0.75% per month instead of 0.5% per month?

19. Let the kidney's filter out 45% of the vitamin A in the plasma each day and the liver absorbs 35% of the vitamin A in the plasma each day. Also, 2% of the vitamin A in the liver is absorbed back into the plasma each day. Let 15% of the vitamin A in the plasma be converted to a chemical B and 5% of the chemical B is converted back to vitamin A. Suppose 10% of chemical B is filtered out by the kidneys each day. Assume that the daily intake is 2 mg of vitamin A and 1 mg of the chemical B each day.
(i) Formulate a discrete dynamical model for vitamin A in the plasma, vitamin A in the liver and chemical B.
(ii) Find the equilibrium solution and check for its stability.
(iii) Suppose a person ingests 1.4 mg of vitamin A each day and wants the equilibrium for vitamin A in the plasma to be 3 mg. How much of the chemical B should you ingest each day to accomplish this? What will be the equilibrium amount of the chemical B and vitamin A in the liver with this consumption?

20. Consider a linear predator–prey model. Let the growth rate of the prey be 1.2 and that of the predator be 1.3; the prey's population diminishes by 0.3 times that of the predator; the predator's population is increased by 0.4 times that of the prey. Construct a discrete model from the given information and study its stability.

21. Let the two species X_n and Y_n compete for food, water, habitat, etc., at any time n. Both the species grow at the rate r_1 and r_2, respectively, and they diminish by an amount directly proportional to the size of the other, say at a rate s_1 and s_2, respectively.
(i) Formulate a linear competition model with the above information.
(ii) Suppose there is a constant migration or constant immigration, k_1 and k_2, respectively, how will the model change?
(iii) Let the growth rate r_1 of the first species be 1.5 and that of the second species be 1.25. Also, each is diminished by 0.4 times the population of the other. Construct a discrete model and comment on its stability.
(iv) In addition to the information in (iii), if the first species undergo immigration of 2500 at each time step and the second one migrates at

1200 per time step, formulate a new model, find the equilibrium point and comment on its stability.

22. A soccer player gets a contract for \$50000 per game and a \$3,00,000 signing bonus.
(i) Formulate a discrete dynamical system to model this player's earning over the season.
(ii) How much this player will earn if he plays 120 games and how many games he needs to play to earn at least \$1,000,000.

23. A car holds 30 liters of petrol (full tank) and gets 10 km to a liter of petrol.
(i) Formulate a discrete dynamical system to model the amount of petrol left in the car's tank after driving for n km.
(ii) How much petrol will be left in the tank after you have driven 120 km and how many km can you drive before running out of petrol?

24. Research evidence has shown that the number of chirps (high-pitched sound) some crickets make depends on the temperature. At 70 °F), one species of crickets makes about 124 chirps per minute. The number of chirps increases by 4 per minute for each degree the temperature rises. Formulate a discrete dynamical system to model the situation that relates the number of chirps per minute in terms of the temperature and use this expression to find approximate temperature if you count about 24 chirps in 10 seconds?

25. The pavement on a bridge is 1000 m long at 70 °F). The length of the pavement grows by 0.012 m for each degree rise in temperature. Formulate a discrete dynamical system to model the length of the bridge in terms of the temperature. What will be the length of the bridge if the temperature reaches 105 °F)?

26. Consider the Gaussian map

$$x_{n+1} = e^{-\alpha x_n} + \beta.$$

(i) Plot the bifurcation diagram for $\beta = -0.5$, with bifurcation parameter α, where $0 \leq \alpha \leq 20$.
(ii) Plot the bifurcation diagram for $\alpha = 20$, with bifurcation parameter β, where $-1 \leq \beta \leq 1$.

27. Most of us keeps the water tap open while shaving. Let W_n be the amount of water that is wasted per day by n number of people who let the water tap running while shaving.
(i) Formulate a discrete dynamical system to model this situation.
(ii) Assuming you shave five times a week, estimate how much water would be wasted per day if you leave the water tap running while shaving.

28. Each year, a person who smokes a single pack of cigarettes a day, will absorb about 2.7 mg of cadmium (an extremely dangerous heavy-metal pollutant and is most toxic when inhaled). For simplicity, it is assumed that cadmium is absorbed at the end of the year. Some people eliminate about 8% of the cadmium from their bodies each year. Formulate a dynamical system for C_n, the amount of cadmium in such a person's body after n years as a result of smoking and hence find the amount of cadmium in such a person's body as a result of smoking for 20 years.

29. Suppose a person's body burns 130 kcal per week for each pound it weighs. Suppose this person presently weighs 165 pounds and consumes about 21,000 kcal per week. Suppose this person decides to eat 200 kcal less each week than the week before. Let C_n represents the number of kilocalories this person consumes during the n_{th} week of this diet and W_n be the weight of this person after n weeks of this diet. Develop a dynamical system of two equations, one for C_n and one for W_n, assuming $C_1 = 20,800$ and $W_0 = 165$ (To work this problem, you need to know that burning 3600 kcal would reduce a person's weight by 1 pound and consuming 3600 kcal would result in the gaining 1 pound).

30. A boy absorbs 0.4 mg of lead into his plasma each day, starting from today. Each day, 35% of the lead in his plasma is absorbed into his bones and 9% of the lead in his plasma is eliminated in the urine (approximately). Also, 0.118% of the lead in his bones is absorbed back into his plasma each day. Formulate a discrete mathematical model for this situation, assuming P_n to be the amount of lead in this boy's plasma and B_n to be the amount of lead in this boy's bones, at the beginning of day n. Hence, find P_4 and B_4, where $P_0 = 0$ and $B_0 = 0$.

31. Find the positive fixed points and their intervals of existence and stability of the non-linear discrete system $x_{n+1} = x_n\sqrt{(a - x_n)}$.

32. Find all the equilibrium points of the model

    (i) $P_{n+1} = \dfrac{rP_n^2}{1000}\left(1 - \dfrac{P_n}{1000}\right)$, and determine their stability for $r = 5$.

    (ii) $P_{n+1} = rP_n e^{\frac{-P_n}{1000}}$, and determine their stability for $r = 0.5, 5.0$.

    (iii) $P_{n+1} = \dfrac{2}{P_n} + r$, and determine their stability for $r = -1, 1$.

33. Consider the non-linear discrete disease model

$$I_{n+1} = I_n - rI_n + sI_n\left(1 - \frac{I_n}{10^6}\right)^2.$$

(i) Obtain all the fixed points. Obtain the interval of stability for which disease will eventually die out of its own (disease-free equilibrium). If

$r = 0.25$, obtain the largest infection rates for disease-free equilibrium.
(ii) Find the condition of stability for non-zero fixed points.

34. Consider a density-dependent population model with constant migration:

$$x_{n+1} = rx_n\left(1 - \frac{x_n}{K}\right) - A.$$

Find the largest value of A that could be migrated each generation, so that the population has a stable positive fixed point, given $r = 3$, $K = 6000$.

35. Consider the price model

$$P_{n+1} = \frac{r}{P_n} + P_n - A, (r, n > 0).$$

Find the largest value of A, for which the price will approach a stable positive equilibrium for any initial value $P_0(> 0)$ when $r = 2$.

36. For the population model

$$P_{n+1} = r\frac{{P_n}^2}{K}\left(1 - \frac{P_n}{K}\right),$$

where r (> 0) is the growth rate and K (> 0) is the carrying capacity, find the condition that a positive equilibrium point exists. Also, find the conditions that the equilibrium point is stable.

37. Consider a price model

$$x_{n+1} = \frac{1}{x_n} + \frac{x_n}{2} - 1.$$

(i) Find the equilibrium points of the model and determine their stability.
(ii) Find a 2-cycle of the model and determine its stability.

38. Consider a density-dependent population model with constant migration:

$$x_{n+1} = rx_n\left(1 - \frac{x_n}{k}\right) - A,$$

with parameter values $r = 4$, $k = 800$ and $A = 100$.
(i) Find the equilibrium points of the model and determine their stability.
(ii) Find a 2-cycle of the model and determine its stability.

39. For the one-dimensional discrete dynamical system

$$x_{n+1} = \frac{rx_n}{1 + {x_n}^2},$$

find all the fixed points (equilibrium points) and classify them. Find the value of r for which the system bifurcates, classify the bifurcation and obtain the bifurcation diagram.

40. For the one-dimensional discrete dynamical system

$$x_{n+1} = {x_n}^2 + r,$$

find all the equilibrium points and obtain the stability condition. Also, obtain the values of r at which the equilibrium points bifurcate and classify those bifurcations. Find the values of r for which there is a stable 2-cycle. When is it super stable?

41. Consider the two-dimensional discrete system

$$\begin{aligned} x_{n+1} &= -rx_n - (1+r){x_n}^2 + (2rx_n + 2r + r^2)y_n, \\ y_{n+1} &= -rx_n - (1+r){x_n}^2 + (2rx_n + 2r + r^2)y_n - x_n \; r \in \mathbf{R}. \end{aligned}$$

Show that a Neimark-Sacker bifurcation occurs at origin when $r = \frac{\sqrt{5}-2}{2}$.

42. Consider the Henon map

$$\begin{aligned} x_{n+1} &= 1 - \alpha x_n^2 + y_n, \\ y_{n+1} &= \beta x_n, \; (\alpha > 0, \; -1 < \beta < 1). \end{aligned}$$

(i) Find the equilibrium points and obtain the condition of their existence.
(ii) For $\alpha = 1.4$ and $\beta = 0.3$, check the stability of the equilibrium points. Show that the system exhibits chaotic behavior with these parameter values.
(iii) Show that the system undergoes a period doubling bifurcation $\alpha = \frac{3}{4}(1-\beta)^2$ for fixed β. Verify it numerically for $\beta = 0.4, 0.5$.

43. For the one-dimensional discrete dynamical system

$$x_{n+1} = rx_n - {x_n}^3,$$

find all the equilibrium points and obtain the condition for their existence. Also, find the condition of their stability. Obtain the values of r at which the equilibrium points bifurcate and classify those bifurcations. Find the values of r for which there is a stable 2-cycle.

44. Find the Lyapunov exponent for the Bernoulli's shift map

$$f(x) = \begin{cases} 2x, & 0 \le x \le \frac{1}{2}, \\ 2x - 1, & \frac{1}{2} \le x \le 1. \end{cases}$$

What conclusion you draw from the value of the Lyapunov exponent? Graphically justify the conclusion.

45. Find the Lyapunov exponent for the system $x_{n+1} = 1 - r|x_n|$, $(0 \le r \le 2)$ for any non-zero solution.

46. Hassel's model, which is an alternative to the logistic growth model and Richer's model, is often used to study the dynamics of the insect population. The model is given by the equation

$$P_{n+1} = \frac{aP_n}{(1+bP_n)^c}.$$

Obtain the equilibrium points and comment on their stability. Numerically, solve the equation for $a = 3.269, b = 0.00745, c = 0.8126$ [103], $P_0 = 100$ and comment on the dynamics of the model.

47. Consider the non-linear discrete system describing the interaction between the price P_n and the demand D_n:

$$P_{n+1} = P_n + a_1 D_n - b_1 f(P_n) + h_1,$$
$$D_{n+1} = D_n + \frac{c_1}{P_n} - k_1, \; f(P_n) = \sqrt{P_n}\,,$$

where $a_1(>0), b_1(>0), h_1, c_1(>0), k_1$ are parameters.
(i) Explain the formation of the model in details, by describing each of the parameters and what they signify in the context of the model.
(ii) For $a_1 = 2.5, h_1 = -1, k_1 = 1, b_1 = 0.11, c_1 = 2.6$, find all the equilibrium points and check its stability.
(iii) Obtain the time series graph (P_n, D_n) with the parameter values mentioned in (ii) and initial condition $(P_0, D_0) = (900, 110)$ for $n = 100$ and comment on the dynamics of the model. What changes are observed for (a) $b_1 = 0.125, c_1 = 2.34$ (b) $b_1 = 0.06, c_1 = 3.6$ (c) $b_1 = 0.05, c_1 = 3.82$?
(iv) For $a_1 = 2.5, h_1 = -1, k_1 = 1, b_1 = 0.1$, numerically estimate the value of the parameter $c_1(>0)$ at which the equilibrium point bifurcates into a stable cycle. Identify the bifurcation.
(v) Put $f(P_n) = e^{P_n}$ and repeat the process from (i)–(iv).

48. Consider the following two-dimensional discrete systems:

$$x_{n+1} = -rx_n - (1+r){x_n}^2 + r^2 y_n,$$
$$y_{n+1} = -rx_n - (1+r){x_n}^2 + r^2 y_n - x_n.$$

Show that a Neimark-Sacker bifurcation occurs at the origin when $r = 1$.

49. Consider the following two-dimensional discrete model:

$$\begin{aligned} x_{n+1} &= x_n + 2x_n\,(1 - x_n) - x_n\,y_n, \\ y_{n+1} &= y_n + 2y_n\,(1 - y_n) - x_n\,y_n. \end{aligned}$$

Find all the positive equilibrium points. Calculate the Jacobian matrix at the equilibrium points and obtain the eigenvalues. Based on the result, classify the equilibrium points into one of the following: stable point, unstable point, saddle point, stable spiral focus or unstable spiral focus.

50. Let L_n be the population of Larva, P_n be the population of Pupa and A_n be the adult population of flour beetle (with different stages of life) [22] at the nth time step. At $(n+1)^{th}$ time step, the population model is given by,

$$\begin{aligned} L_{n+1} &= bA_n, \\ P_{n+1} &= (1-\mu_L)L_n, \\ A_{n+1} &= (1-\mu_P)P_n + (1-\mu_A)A_n. \end{aligned}$$

(i) Explain the formation of the model in details, by describing each of the parameters b, μ_L, μ_P, μ_A and what they signify in the context of the model.
(ii) For $b = 4.88, \mu_L = 0.2, \mu_P = 0.001, \mu_A = 0.01$, find all the equilibrium points and check their stability. Also, obtain the time series graph with initial condition $(L_0, P_0, A_0) = (300, 110, 200)$ for $n = 100$ and comment on the dynamics of the model.

2.11 Projects

1. Consider the discrete dynamical system which shows a predator-prey relationship (say, fox and rabbit) between two interacting species:

$$\begin{aligned} R_{n+1} &= R_n + a\ R_n - b\ F_n\ R_n, \\ F_{n+1} &= F_n - c\ F_n + d\ R_n\ F_n. \end{aligned}$$

(i) Explain the formation of the model in details, by describing each of the parameters a, b, c, d and what they signify in the context of the model.

(ii) Find all the equilibrium points. Obtain the interval of existence and interval of stability. Discuss the implications of these results in the context of the model.

(iii) For $a = 0.039, b = 0.0003, c = 0.12$ and $d = 0.0001$, obtain the time series graph of the two interacting species with $(R_0, F_0) = (900, 110)$ for $n = 500$ and comment on the behaviour of the species. Vary the parameters b and d to change the dynamics of the system. Which of the species is the dominating one? Is any of the species go extinct?

(iv) Suppose a club organizes a fox hunt in one of its consecutive two meetings, where the first meeting will be held in 36 months from now and the second will be 36 months after the first one. If the members of the club restrict themselves to killing 45 foxes, how will it affect the long-term behaviour of the species?

(v) The model is now modified with density-dependent growth rates as follows:

$$R_{n+1} = r_1 R_n \left(1 - \frac{R_n}{k_1}\right) - b\ R_n\ F_n,$$
$$F_{n+1} = r_2 F_n \left(1 - \frac{F_n}{k_2}\right) + d\ R_n\ F_n.$$

Find all the equilibrium points. Obtain the interval of existence and interval of stability. Discuss the implications of these results in the context of the model.

(vi) For $r_1 = 1.039, k_1 = 1000, b = 0.0003, r_2 = 0.88, k_2 = 150$ and $d = 0.0001$, obtain the time series graph of the two interacting species with $(R_0, F_0) = (900, 110)$ for $n = 500$ and comment on the behaviour of the species. What happens if $r_1 = 0.88$?

(vii) Consider another density-dependent growth rate predator-prey model:

$$R_{n+1} = \frac{r_1 R_n}{1 + \left(\frac{R_n}{k_1}\right)^2} - b\ R_n\ F_n,$$
$$F_{n+1} = \frac{r_2 F_n}{1 + \left(\frac{F_n}{k_2}\right)^2} + d\ R_n\ F_n.$$

Obtain the time series graph for $r_1 = 1.039, k_1 = 1000, b = 0.0003, r_2 = 0.88, k_2 = 150$ and $d = 0.0001$, with initial condition $(R_0, F_0) = (900, 110)$ for $n = 500$ and comment on the behaviour of the species. What happens if $r_1 = 0.88$?

2. Consider the discrete dynamical system that models the spread of infectious disease in a population of size N:

$$I_{n+1} = I_n - rI_n + s\ f(I_n, R_n),$$
$$R_{n+1} = R_n - tR_n + rI_n.$$

Here, I_n represents the number infected and R_n ,who has recovered,

r, s, t are parameters, where $0 < r, t \leq 1, s > 0$.

$$(a)\ f(I_n, R_n) = I_n\left(1 - \frac{I_n + R_n}{N}\right),$$
$$(b)\ f(I_n, R_n) = {I_n}^2\left(1 - \frac{I_n + R_n}{N}\right),$$
$$(c)\ f(I_n, R_n) = {I_n}^2\left(1 - \frac{I_n + R_n}{N}\right)^2,$$
$$(d)\ f(I_n, R_n) = I_n\left(1 - \frac{I_n + R_n}{N}\right)^2.$$

(i) Explain the formation of the model in detail, by describing each of the parameters and what they signify in the context of the model for (a),(b)and (c).

(ii) Find the points of equilibria and obtain the condition(s) (if any) for their existence and stability in all of (a),(b)and (c).

(iii) For $r = 0.75, s = 2, t = 0.5, N = 1000$, find all the equilibrium points and check their stability for (a),(b)and (c). Solve the system numerically and comment on the dynamics of the system from the graph obtained. Vary the values of r and t between 0 and 1 and comment on the change in dynamics (if any). Repeat the same for $r = 0.295, s = 5, t = 0.5, N = 1000$.

(iv) For $r = 0.5, t = 0.5, N = 10000$, in (a), find all the positive equilibrium points for all $s \geq 0$ and determine their stability. Obtain the interval of existence and interval of stability. Discuss the implications of these results in the context of the model. Solve the model numerically for $s = 2, 4, 6, 7, 7.3$, and comment on the graphs obtained.

(v) For $r = 1.0, s = 9.0, t = 1.0, N = 6000$, find all the fixed points and check the stability numerically. Show that for some small initial values of I_0 and R_0, eradication of the disease is possible, but for a larger initial population, this may not occur.

(vi) Show that with $r = 0.1, s = 4.5, N = 1000$, the model with (a) has a period-doubling cascade as t (>0) increases, which ultimately leads to chaos.

Chapter 3

Continuous Models Using Ordinary Differential Equations

3.1 Introduction to Continuous Models

Continuous models are systems whose inputs and outputs are capable of changing at any instant of time. A continuous model consists of a dependent continuous variable, varying with some other independent continuous variables. We use a first-order ordinary differential equation (or a system of first-order ordinary differential equations) to model a continuous system if we have some information or assumption about the rate of change of the dependent variable(s) with respect to the independent variable.

Let us model a situation of rumor propagation or spreading of rumors. Suppose, a company has 2000 employees. On Monday, a rumor begins to spread among them that the CEO of the company has been hospitalized after being infected with COVID-19, the infectious disease caused by the newly discovered coronavirus.

We assume that the spread of rumors is proportional to the number of possible meetings between employees who have heard the rumor and those who have not. Let $x(t)$ be the number of employees who have heard the rumor after t days; then the scenario is modeled with a first-order ordinary differential equation as

$$\frac{dx}{dt} = kx(2000 - x), \tag{3.1}$$

where k is a constant of proportionality.

We also assume that initially (when $t = 0$), only 40 employees heard the rumor, as they all attended the same meeting. After 2 days, 200 employees heard the rumor. Mathematically, this means $y(0) = 40$ and $y(2) = 200$.

DOI: 10.1201/9781351022941-3

Equation(3.1) can be easily solved using the separation of variables as follows:

$$\frac{dx}{x(2000-x)} = k\ dt \Rightarrow \frac{1}{2000}\int\left[\frac{1}{x}+\frac{1}{2000-x}\right]dx = k\int dt,$$
$$\Rightarrow \frac{1}{2000}\ln\left(\frac{x}{2000-x}\right) = kt + C,$$
$$\Rightarrow \frac{x}{2000-x} = A\ e^{2000kt},\ \text{where } A = e^{2000C}.$$
$$\Rightarrow x(t) = \frac{2000\ A\ e^{2000kt}}{1 + A\ e^{2000kt}}.\ \text{Given } x(0) = 40 \Rightarrow 40 = \frac{2000A}{1+A} \Rightarrow A = \frac{1}{49}.$$
$$\Rightarrow x(t) = \frac{2000\ e^{2000kt}}{49 + e^{2000kt}}.\ \text{After 2 days, 200 employees heard the rumor.}$$
$$\Rightarrow 200 = \frac{2000\ e^{4000k}}{49 + e^{4000k}} \Rightarrow k = \frac{1}{4000}\ \ln\left(\frac{49}{9}\right).$$

Therefore, the number of employees who have heard the rumor is given by

$$x(t) = \frac{2000\ e^{\frac{t}{2}\ln\left(\frac{49}{9}\right)}}{49 + e^{\frac{t}{2}\ln\left(\frac{49}{9}\right)}} \Rightarrow x(t) = \frac{2000}{49\ e^{-\frac{t}{2}\ln\left(\frac{49}{9}\right)} + 1}. \tag{3.2}$$

We now pose a question that how much time will the rumor take to reach 1000 employees? Substituting $x(t) = 1000$, we obtain

$$1000 = \frac{2000}{49\ e^{-\frac{t}{2}\ln\left(\frac{49}{9}\right)} + 1} \Rightarrow t = \frac{2\ln 49}{\ln(49/9)} = 4.6\ \text{days}.$$

Therefore, the rumor reaches 1000 employees in 4 days, 14 hours and 24 minutes. So, if the rumor starts on Monday 9:30 am, it will reach 1000 employees by Friday 11:54 pm.

We observe from (3.2) that $x(t) \to 2000$ as $t \to \infty$. The obvious question is why $x(t)$ approaches a constant value 2000 as time t becomes large? Why is it not approaching any other value? What is the significance of the value 2000? Answer to these questions with possible reasons will be discussed in the next section.

3.2 Steady-State Solution

We consider a system of n non-linear, autonomous (does not explicitly depend on time, that is, time-invariant) differential equations

$$\begin{aligned} \frac{d\tilde{x}}{dt} &= f(\tilde{x}), \\ \text{where } \tilde{x} &= (x_1, x_2, \ldots\ldots\ldots\ldots, x_n)^T, \\ \text{and } f(\tilde{x}) &= (f_1(\tilde{x}), f_2(\tilde{x}), \ldots\ldots\ldots, f_n(\tilde{x}))^T. \end{aligned}$$

A steady-state solution or equilibrium solution or critical point is a constant solution, that is, where the value of $\tilde{x}$ does not change over time and is obtained by putting

$$\frac{d\tilde{x}}{dt} = 0.$$

Please note that in order for the values of $\tilde{x}$ to be the same (constant) over time, there must not be any change in $\tilde{x}$, implying $\frac{d\tilde{x}}{dt} = 0$. Therefore, the only value(s) of $\tilde{x}$ for which this can happen is $f(\tilde{x}) = 0$, and so $f(\tilde{x}) = 0$ gives a steady-state solution or an equilibrium solution or a fixed point.

In short, if we consider a nonlinear time-invariant system

$$\frac{d\tilde{x}}{dt} = f(\tilde{x}), \ \tilde{x}(t_0) = \tilde{x_0}, \text{ where } f : \Re \to \Re,$$

a point $\tilde{x_e}$ is an equilibrium point or a steady-state solution of the system if $f(\tilde{x_e}) = 0$.

A steady-state solution or equilibrium point or critical point is said to be isolated if there is a neighborhood to the equilibrium point that does not contain any other equilibrium points. In what follows, we shall assume that every critical point is isolated. Consider a simple growth model of a population:

$$\frac{dx}{dt} = ax - bx^2 \quad (a, b > 0).$$

Here, the population grows linearly and there is a crowding effect or intra-specific competition $(-bx^2)$, which depresses the rate of growth of the population. The steady-state solution is given by

$$\frac{dx}{dt} = 0 \Rightarrow ax - bx^2 = 0 \Rightarrow x = 0 \text{ and } \frac{a}{b}.$$

Therefore, 0 and $\frac{a}{b}$ are two steady states or equilibrium solutions of this population.

Again, consider a two species predator–prey system

$$\frac{dx}{dt} = rx\left(1-\frac{x}{k}\right) - \alpha xy,$$
$$\frac{dx}{dt} = -\beta y + \gamma xy.$$

The steady-state solution is given by $\frac{dx}{dt}=0$, $\frac{dy}{dt}=0$. Solving, we get the steady-state solutions of the system as $(0,0), (k,0)$ and $\left(\frac{\beta}{\gamma}, \frac{r(k\gamma-\beta)}{k\alpha\gamma}\right)$. The last solution is feasible if $k\gamma - \beta > 0$ because populations are positive in real life.

3.3 Stability

In layman's language, we say that an equilibrium point or a steady-state solution $\tilde{x_e}$ is locally stable, if all solutions that start near $\tilde{x_e}$ (that is, the initial conditions are in the neighborhood of $\tilde{x_e}$) remain near $\tilde{x_e}$ for all the time. Furthermore, if all the solutions starting near $\tilde{x_e}$ approach $\tilde{x_e}$ as $t \to \infty$, we say that the equilibrium point or steady-state solution $\tilde{x_e}$ is locally asymptotically stable.

Consider the dynamical system satisfying

$$\frac{d\tilde{x}}{dt} = f(\tilde{x}),\ \tilde{x}(t_0) = \tilde{x_0},\ \text{where } f : \Re \to \Re, \tag{3.3}$$

In the sense of Lyapunov, the equilibrium point or the steady-state solution $\tilde{x_e}$ of (3.3) is stable at $t = t_0$ if for any $\epsilon > 0$, there exists a $\delta(t_0, \epsilon) > 0$ such that

$$\| \tilde{x}(t_0) - \tilde{x_e} \| < \delta \ \Rightarrow\ \| \tilde{x}(t) - \tilde{x_e} \| < \epsilon,\ \forall\ t \geq t_0.$$

The equilibrium point or the steady-state solution $\tilde{x_e}$ of (3.3) is asymptotically stable (in the sense of Lyapunov) at $t = t_0$ if

(i) $\tilde{x_e}$ is stable,

(ii) there exists $\delta(t_0) > 0$ such that

$$\| \tilde{x}(t_0) - \tilde{x_e} \| < \delta \ \Rightarrow\ \lim_{t\to\infty} \tilde{x}(t) = 0.$$

An equilibrium point or a steady-state solution is unstable if it is not stable. Please note that the definitions of stability and asymptotic stability are local definitions as they describe the behavior of a system near an equilibrium point.

3.3.1 Linearization and Local Stability Analysis

This method is known as Lyapunov's first method or reduced method, where the stability analysis of a steady-state or equilibrium point is done by studying the stability of the corresponding linearized system in the neighborhood (vicinity) of the steady state.

We consider the model of the form

$$\frac{dx}{dt} = f(x),$$

whose local stability analysis we want to perform about the equilibrium point x^* (obtained by putting f(x) = 0). We give a small perturbation to the system about the equilibrium point x^*. Mathematically, this means we put $x = X + x^*$ into the above equation and get

$$\frac{dX}{dt} = f(x^* + X) = f(x^*) + Xf'(x^*) + \ldots \text{ (higher-order terms)},$$
$$\frac{dX}{dt} \approx f'(x^*)X, \quad \text{since } f(x^*) = 0 \text{ and neglecting higher-order terms.}$$

Therefore, we conclude that the system is stable if $f'(x^*) < 0$ (decreasing function) and unstable if $f'(x^*) > 0$ (increasing function). If $f'(x^*) = 0$, no definite conclusion can be drawn from linear stability analysis.

Let us now consider the model given by the system of differential equations of the form

$$\frac{dx}{dt} = f(x,y), \quad \frac{dy}{dt} = g(x,y). \tag{3.4}$$

Let (x^*, y^*) be the steady-state solution of (3.4), then $f(x^*, y^*) = 0$ and $g(x^*, y^*) = 0$. We now give a small perturbation to the system about the steady state, and mathematically this means we put $x = X + x^*$ and $y = Y + y^*$. This implies

$$\frac{dX}{dt} = f(x^* + X, y^* + Y)$$
$$= f(x^*, y^*) + Xf_x(x^*, y^*) + Yf_y(x^*, y^*) + \ldots \text{ higher-order terms}$$
(by Taylor series expansion of two variables).

Similarly,

$$\frac{dY}{dt} = g(x^*, y^*) + Xg_x(x^*, y^*) + Yg_y(x^*, y^*) + \ldots \text{ higher-order terms},$$

where $f_x(x^*, y^*)$ is $\frac{\partial f}{\partial x}$ evaluated at the steady state (x^*, y^*). Since by definition, $f(x^*, y^*) = 0$, $g(x^*, y^*) = 0$, by neglecting second and higher-order terms, we obtain

$$\begin{aligned}\frac{dX}{dt} &= f_x(x^*, y^*)X + f_y(x^*, y^*)Y, \\ \frac{dY}{dt} &= g_x(x^*, y^*)X + g_y(x^*, y^*)Y,\end{aligned}$$

which can be put in matrix form as

$$\frac{d\tilde{x}}{dt} = A\tilde{x}, \text{ where } \tilde{x} = \begin{pmatrix} x \\ y \end{pmatrix} \text{ and } A = \begin{pmatrix} f_x & f_y \\ g_x & g_y \end{pmatrix}. \tag{3.5}$$

Let $\tilde{x} = \widehat{v}e^{\lambda t}$ be a trial solution of (3.5), where $\widehat{v}$ ($\neq 0$) is some fixed vector which needs to be determined. Then

$$\frac{d\tilde{x}}{dt} = \widehat{v}\lambda e^{\lambda t} = A\widehat{v}e^{\lambda t} \Rightarrow A\widehat{v} = \lambda\widehat{v}.$$

From linear algebra, it can be easily concluded that λ is the eigenvalue of the matrix A, whose eigenvector is $\widehat{\nu}$. The eigenvalues are obtained by solving

$$\det(A - \lambda I) = 0 \Rightarrow \begin{vmatrix} f_x - \lambda & f_y \\ g_x & g_x - \lambda \end{vmatrix} = 0,$$

$$\Rightarrow \lambda^2 - (f_x + g_y)\lambda + f_x g_y - f_y g_x = 0 \Rightarrow \lambda^2 - \text{trace}(A)\ \lambda + \det(A) = 0.$$

Let λ_1 and λ_2 be the two eigenvalues of the matrix A. The necessary and sufficient condition that λ_1 and λ_2 will be negative (if real) or have negative real parts (if complex) is

$$\begin{aligned}\text{trace}(A) &= f_x + g_y < 0, \\ \det(A) &= f_x g_y - f_y g_x > 0.\end{aligned}$$

Routh-Hurwitz Stability Criteria:

(i) **Quadratic Equation:** $\lambda^2 + a_1\lambda + a_2 = 0$,
Stability Criteria: $a_1 > 0, a_2 > 0$.

(ii) **Cubic Equation:** $\lambda^3 + a_1\lambda^2 + a_2\lambda + a_3 = 0$,
Stability Criteria: $a_1 > 0, a_2 > 0, a_3 > 0, a_1a_2 - a_3 > 0$.

(iii) **Fourth-Order Equation:** $\lambda^4 + a_1\lambda^3 + a_2\lambda^2 + a_3\lambda + a_4 = 0$,
Stability Criteria: $a_1 > 0, a_2 > 0, a_3 > 0, a_4 > 0, a_1a_2 - a_3 > 0$,
$a_1a_2a_3 - a_3^2 - a_1^2a_4 > 0$.

Example 3.3.1 The fish growth model by Von Bertalanffy [42] is given by

$$\frac{dF(t)}{dt} = \alpha F^{3/2}(t) - \beta F(t),$$

where $F(t)$ denotes the weight of the fish, and α and β are positive constants.

Solution: The equilibrium solution of the model is given by

$$\frac{dF(t)}{dt} = 0 \Rightarrow \alpha F^{3/2} - \beta F = 0,$$

$$\Rightarrow F(\alpha F^{1/2} - \beta) = 0 \Rightarrow F^* = 0 \text{ and } F^* = \frac{\beta^2}{\alpha^2}.$$

$$\text{Let } W(F) = \alpha F^{3/2} - \beta F, \text{ then } W'(F^*) = \frac{3}{2}\alpha F^{*1/2} - \beta.$$

Now, $W'(0) = -\beta < 0$ implies that the equilibrium point $F^* = 0$ is stable and $W'(\beta/\alpha) = \frac{1}{2}\beta > 0$ implies that the equilibrium point $F^* = \beta^2/\alpha^2$ is unstable.

Example 3.3.2 *Consider the autonomous non-linear system*

$$\begin{aligned}\frac{dx}{dt} &= \beta(2-x) + x^2y,\\ \frac{dy}{dt} &= x - x^2y \ (\beta > 1).\end{aligned}$$

Find the equilibrium point(s) and comment on the stability (local) of the equilibrium point(s).

Solution: The equilibrium points of the model is given by

$$\beta(2-x) + x^2y = 0, \tag{3.6}$$

$$x - x^2y = 0. \tag{3.7}$$

(3.7) gives $x = 0$ and $xy = 1$. But, $x = 0$ does not satisfy (3.6). Hence, substituting $y = \frac{1}{x}$ in (3.6), we get $x = \frac{2\beta}{\beta-1}$. Therefore, the only equilibrium point is $(x^*, y^*) = \left(\frac{2\beta}{\beta-1}, \frac{\beta-1}{2\beta}\right)$. Linearizing about the steady state (x^*, y^*), we obtain the system as

$$\begin{pmatrix} \frac{dx}{dt} \\ \frac{dy}{dt} \end{pmatrix} = \begin{pmatrix} 2-\beta & \frac{4\beta^2}{(\beta-1)^2} \\ -1 & \frac{-4\beta^2}{(\beta-1)^2} \end{pmatrix} \begin{pmatrix} x \\ y \end{pmatrix}.$$

The characteristic equation is given by

$$\begin{vmatrix} 2-\beta-\lambda & \frac{4\beta^2}{(\beta-1)^2} \\ -1 & \frac{-4\beta^2}{(\beta-1)^2} - \lambda \end{vmatrix} = 0,$$

$$\Rightarrow \lambda^2 + \frac{\beta^3 + 5\beta - 2}{(\beta-1)^2} + \frac{4\beta^2}{\beta - 1} = 0.$$

Clearly, $\beta^3 + 5\beta - 2 > 0$, for all $\beta > 1$. Therefore, by Routh Hurwitz's criteria, the system is always stable about the equilibrium point $(x^*, y^*) = \left(\frac{2\beta}{\beta-1}, \frac{\beta-1}{2\beta}\right)$.

3.3.2 Lyapunov's Direct Method

Lyapunov's direct method, also known as Lyapunov's second method, determines the stability of a system without explicitly integrating the differential equation

$$\frac{d\tilde{x}}{dt} = f(\tilde{x}), \ \tilde{x}(t_0) = \tilde{x_0}, \text{ where } f : \Re \to \Re.$$

The method is a generalization of the idea that if the potential energy has a relative minimum at the equilibrium point, then the equilibrium point is stable; otherwise, it is unstable. Russian mathematician Aleksandr Mikhailovich Lyapunov generalized this principle to obtain a method for studying the stability of the general autonomous system.

3.3.2.1 Lyapunov's Condition for Local Stability

Consider the autonomous system

$$\frac{d\tilde{x}}{dt} = f(\tilde{x}), \ \tilde{x}(t_0) = \tilde{x_0}, \text{ where } f : \Re \to \Re, \tag{3.8}$$

having isolated critical point at the origin $\tilde{x} = \mathbf{0}$ and $f(\tilde{x})$ has continuous partial derivatives for all $\tilde{x}$. Let $V(\tilde{x})$ be a positive definite function in a neighborhood **S** of the origin $\tilde{x} = 0$ and the derivative $\dot{V}(\tilde{x})$ of $V(\tilde{x})$ with respect to the system (3.8) is negative semi-definite in the neighborhood **S** of the origin $\tilde{x} = \mathbf{0}$, then $V(\tilde{x})$ is called a Lyapunov function for the system (3.8). Mathematically, $V(\tilde{x})$ is called a Lyapunov function if

(i) $V(\tilde{x}) > 0$ in the nbd. of the origin $\tilde{x} = \mathbf{0}$,

(ii) $V(\mathbf{0}) = 0$ for all $\tilde{x} = \mathbf{0}$,

(iii) $\dot{V}(\tilde{x}) \leq 0$ in the nbd. of the origin $\tilde{x} = \mathbf{0}$,

(iv) $\dot{V}(\mathbf{0}) = 0$ for all $\tilde{x} = \mathbf{0}$.

If there exists a Lyapunov function $V(\tilde{x})$ for the system (3.8) in the nbd. of the origin $\tilde{x} = \mathbf{0}$, then the steady-state solution or the equilibrium point $\tilde{x} = \mathbf{0}$ is stable (locally).

If the derivative $\dot{V}(\tilde{x})$ of $V(\tilde{x})$ is negative semi-definite in the neighborhood **S** of the origin $\tilde{x} = \mathbf{0}$, then the steady state solution or the equilibrium point $\tilde{x} = \mathbf{0}$ is locally asymptotically stable (LAS).

Please note that there is no general method for constructing a Lyapunov function, but if one can construct a Lyapunov function for the system (3.8), then one can directly obtain information about the steady state of the equilibrium point $\tilde{x} = \mathbf{0}$ and hence the name Lyapunov's direct method.

3.3.2.2 Lyapunov's Condition for Global Stability

The equilibrium point $\tilde{x} = \mathbf{0}$ is globally asymptotically stable if it is a basin of attraction in the entire state space or entire domain. Thus, $\tilde{x} = \mathbf{0}$ is globally asymptotically stable (GAS) for the system (3.8) if

(i) $\tilde{x} = \mathbf{0}$ is locally asymptotically stable,

(ii) The Lyapunov function $V(\tilde{x})$ is radially unbounded, that is,

$$| V(\tilde{x}) | \to \infty \text{ as norm}(\tilde{x}) = \| \tilde{x} \| \to \infty.$$

Example 3.3.3 Consider the system

$$\dot{x} = \frac{dx}{dt} = -x - y, \quad \dot{y} = \frac{dy}{dt} = x - y^3.$$

Check the local and global stability of the stable.

Solution: We first show that the system is locally asymptotically stable (LAS).

Lyapunov's Indirect Method

Clearly, $(0, 0)$ is the only steady-state solution (equilibrium point) of the given system. Linearizing about the steady state $(0, 0)$, we obtain the system as

$$\begin{pmatrix} \dot{x} \\ \dot{y} \end{pmatrix} = \begin{pmatrix} \frac{dx}{dt} \\ \frac{dy}{dt} \end{pmatrix} = \begin{pmatrix} -1 & -1 \\ 1 & 0 \end{pmatrix} \begin{pmatrix} x \\ y \end{pmatrix}.$$

The variational matrix $\begin{pmatrix} -1 & -1 \\ 1 & 0 \end{pmatrix}$ has eigen values $\frac{-1}{2} \pm i\frac{\sqrt{3}}{2}$, whose real part is negative. Hence, by Lyapunov's indirect method, the system is locally asymptotically stable (LAS) about $(0, 0)$.

Lyapunov's Direct Method

We consider a function $V(x, y) = ax^2 + by^2$, where a and b are some positive constants to be determined. Clearly, the function $V(x, y)$ is positive definite as $V(x, y) > 0, \ \forall \ (x, y) \neq (0, 0)$ and $V(0, 0) = 0$. Also,

$$\begin{aligned} \dot{V}(x, y) = \frac{dV}{dt} &= \frac{\partial V}{\partial x}\dot{x} + \frac{\partial V}{\partial y}\dot{y} \\ &= 2ax(-x - y) + 2by(x - y^3) \\ &= -2ax^2 - 2by^4 + 2xy(b - a). \end{aligned}$$

If we choose b = a = 1, then

$$\begin{aligned}\dot{V}(x,y) = \frac{dV}{dt} &= -2x^2 - 2y^4 < 0, \ \forall \ (x,y) \neq (0,0), \\ &= 0, \ \forall \ (x,y) = (0,0).\end{aligned}$$

This implies $\dot{V}(x,y)$ is negative definite. Hence, by Lyapunov's direct method, (0,0) is locally asymptotically stable (LAS).

Global Stability

We now use Lyapunov's conditions to prove that the system is globally asymptotically stable (GAS). Already, we have shown that the function is locally asymptotically stable. Now, the function $V(x,y) = x^2 + y^2$ (a=1, b=1) is also radially unbounded as $V(x,y) \to \infty$ as $\textbf{norm}(x,y) = \sqrt{x^2+y^2} \to \infty$. Therefore, by Lyapunov's stability conditions, the system is globally asymptotically stable (GAS).

3.4 Phase Plane Diagrams of Linear Systems

We consider a two-dimensional autonomous linear system of the form

$$\begin{aligned}\frac{dx}{dt} &= \lambda_1 x + \lambda_2 y, \\ \frac{dy}{dt} &= \lambda_3 x + \lambda_4 y, \end{aligned} \tag{3.9}$$

which can be written in matrix form as

$$\frac{d\tilde{x}}{dt} = A\,\tilde{x}, \text{ where } A = \begin{pmatrix} \lambda_1 & \lambda_2 \\ \lambda_3 & \lambda_4 \end{pmatrix} \text{ and } \tilde{x} = \begin{pmatrix} x \\ y \end{pmatrix}. \tag{3.10}$$

Clearly, the linear system has a unique steady state $(0,0)$, provided $det(A) = \lambda_1\lambda_4 - \lambda_2\lambda_3 \neq 0$. The solution of (3.10) can be visualized as trajectories moving in the xy-plane and can be sketched, which are known as phase portraits.

For a better understanding of the system, we consider a much more simplified linear system of the form

$$\frac{dx}{dt} = \lambda_1 x \text{ and } \frac{dy}{dt} = \lambda_4 y, \tag{3.11}$$

where $(0,0)$ is the unique equilibrium solution of (3.11). This can be put in matrix form as

$$\frac{d}{dt}\begin{pmatrix} x \\ y \end{pmatrix} = \begin{pmatrix} \lambda_1 & 0 \\ 0 & \lambda_4 \end{pmatrix}\begin{pmatrix} x \\ y \end{pmatrix}.$$

The characteristic equation of the matrix $B = \begin{pmatrix} \lambda_1 & 0 \\ 0 & \lambda_4 \end{pmatrix}$ is

$$\begin{vmatrix} \lambda_1 - k & 0 \\ 0 & \lambda_4 - k \end{vmatrix} = 0 \Rightarrow (\lambda_1 - k)(\lambda_4 - k) = 0 \Rightarrow k = \lambda_1, \lambda_4,$$

which are the roots of the characteristic equation, also called the eigenvalues of the matrix B.

The solution to (3.11) is

$$x = x(0)e^{\lambda_1 t}, \quad y = y(0)e^{\lambda_4 t},$$

where $x(0)$ and $y(0)$ are the initial values.

Case I: If both the eigenvalues λ_1 and λ_4 are negative, all the trajectories approach $(0, 0)$, that is, all the solutions of the system converge to the equilibrium solution $(0, 0)$, no matter what the initial conditions may be (fig. 3.1(a)). The steady state $(0, 0)$ is called a stable node (fig. 3.1(b)).

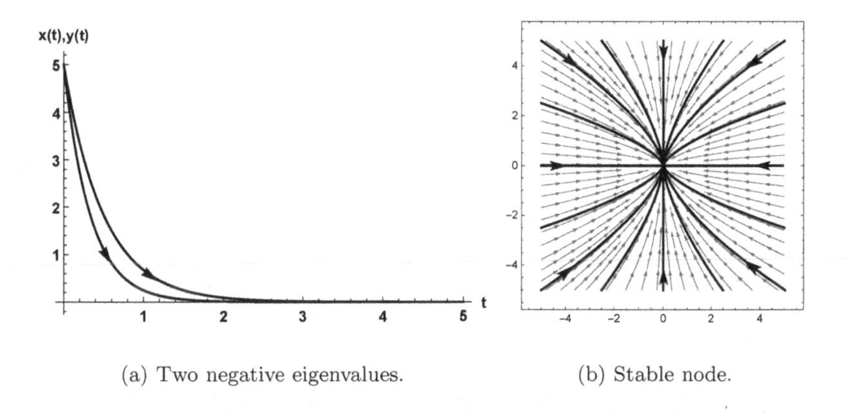

(a) Two negative eigenvalues. (b) Stable node.

FIGURE 3.1: *Phase portrait showing all the trajectories approach* $(0, 0)$*, a stable node.*

Case II: If both the eigenvalues λ_1 and λ_4 are positive, all the trajectories move away from $(0, 0)$, that is, all the solutions of the system diverge from the equilibrium solution $(0, 0)$, irrespective of the initial conditions (fig. 3.2(a)). In this case, the steady state $(0, 0)$ is called an unstable node (fig. 3.2(b)).

Case III: If the eigenvalues are opposite in sign, say, $\lambda_1 < 0$ and $\lambda_4 > 0$, then $x(t)$ decreases whereas $y(t)$ increases exponentially (fig. 3.3(a)). All the solutions, in this case, approach the steady state (0,0) for some time until they come close to it, and then they start moving away from the steady state,

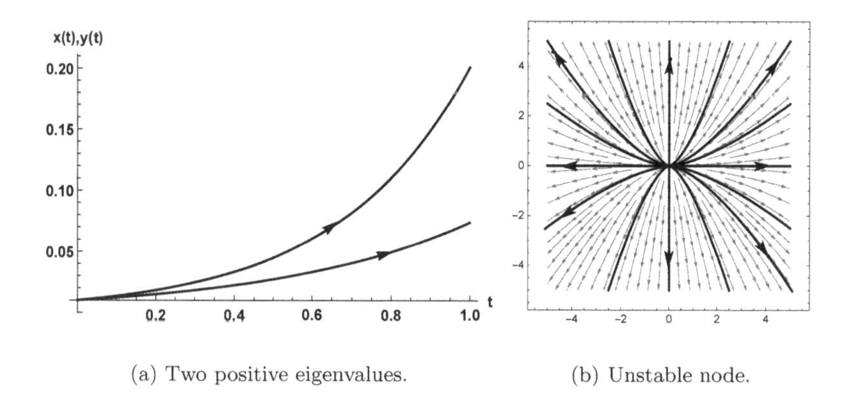

(a) Two positive eigenvalues. (b) Unstable node.

FIGURE 3.2: *Phase portrait showing all the trajectories move away from* $(0,0)$*, an unstable node.*

irrespective of the initial solutions. The steady state $(0,0)$ is called a saddle point (fig. 3.3(b)).

Case IV: If the eigenvalues are complex conjugates, say, $\lambda_1 = a + ib$ and $\lambda_4 = a - ib$, then the solutions are of the form

$$\begin{aligned} x(t) &= x(0)e^{at}(\cos b + i \sin b), \\ y(t) &= y(0)e^{at}(\cos b - i \sin b). \end{aligned}$$

(i) If $a < 0$, then both the eigenvalues have negative real parts, the term e^{at} decays for increasing t, and hence the solution decreases (fig. 3.4(a)). In this case, all the trajectories spiral towards the steady state $(0,0)$, irrespective of the initial conditions and the steady state is known as a stable spiral or stable focus (fig. 3.4(b)).

(ii) If $a > 0$, both the eigenvalues have positive real parts, the term e^{at} grows exponentially and so is the solution, for $t > 0$ (fig. 3.5(a)). In this case, all the trajectories spiral away from the steady state $(0,0)$, irrespective of the initial conditions and the steady state is called an unstable spiral or unstable focus (fig. 3.5(b)).

(iii) If $a = 0$, both the eigenvalues are purely imaginary. In this case, all the trajectories are closed orbits about the steady state $(0,0)$. The solutions are periodic (fig. 3.6(a)) and the steady state is called a center (fig. 3.6(b)).

Please note that for simplicity, we have considered the linear system of the form

$$\frac{dx}{dt} = \lambda_1 x \text{ and } \frac{dy}{dt} = \lambda_4 y. \tag{3.12}$$

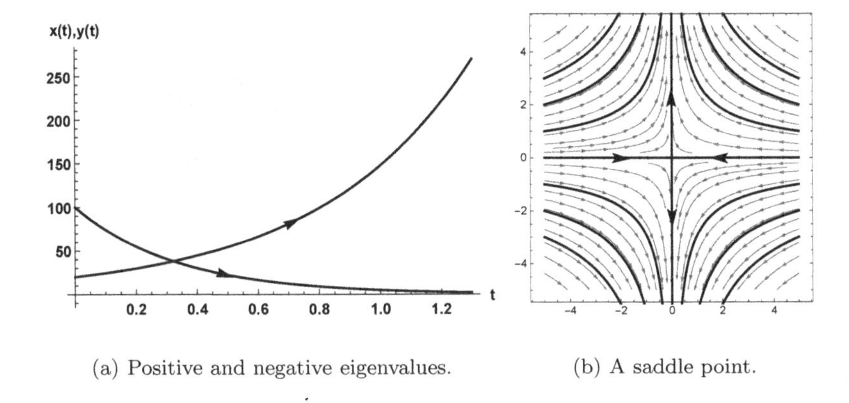

(a) Positive and negative eigenvalues.

(b) A saddle point.

FIGURE 3.3: *Phase portrait shows all the trajectories approach the steady state (0,0) for some time until they come close to it, and then they start moving away from it, a saddle.*

A similar approach with (3.9) would have given the same result. In that case, the characteristic equation of the matrix $A = \begin{pmatrix} \lambda_1 & \lambda_2 \\ \lambda_3 & \lambda_4 \end{pmatrix}$ is

$$\begin{vmatrix} \lambda_1 - k & \lambda_2 \\ \lambda_3 & \lambda_4 - k \end{vmatrix} = 0 \Rightarrow k^2 - (\lambda_1 + \lambda_4)k + (\lambda_1\lambda_4 - \lambda_2\lambda_3) = 0,$$

$$\Rightarrow k^2 - \text{Trace}(A)\ k + \text{Det}(A) = 0 \Rightarrow\Rightarrow k = \frac{1}{2}\left[\text{Trace}(A) \pm \sqrt{\text{Discriminant}}\right],$$

where Discriminant $= (\text{Trace(A)})^2 - 4\text{Det}(A)$.

Case I: Let Det(A) $= -$D (<0, D being a positive quantity), then Discriminant $= (\text{Trace(A)})^2 - 4\text{Det}(A) = (\text{Trace(A)})^2 + 4D >$ Trace(A). Therefore, $\frac{1}{2}\left[\text{Trace}(A) + \sqrt{\text{Discriminant}}\right]$ and $\frac{1}{2}\left[\text{Trace}(A) - \sqrt{\text{Discriminant}}\right]$ will have opposite signs and the phase portrait will be a saddle. The sign of Trace(A) is irrelevant in this case.

Case II: Let Det(A) > 0 and Discriminant $= (\text{Trace(A)})^2 - 4\text{Det}(A) > 0$, that is,

$$\sqrt{\text{Discriminant}} = \text{Trace(A)}\sqrt{1 - \frac{\text{Det(A)}}{(\text{Trace(A)})^2}} < \text{Trace(A)}.$$

Therefore, $\frac{1}{2}\left[\text{Trace}(A) + \sqrt{\text{Discriminant}}\right]$ and $\frac{1}{2}\left[\text{Trace}(A) - \sqrt{\text{Discriminant}}\right]$ will have the same signs. The eigen values will be positive if Trace(A) >0 and negative if Trace(A) < 0.

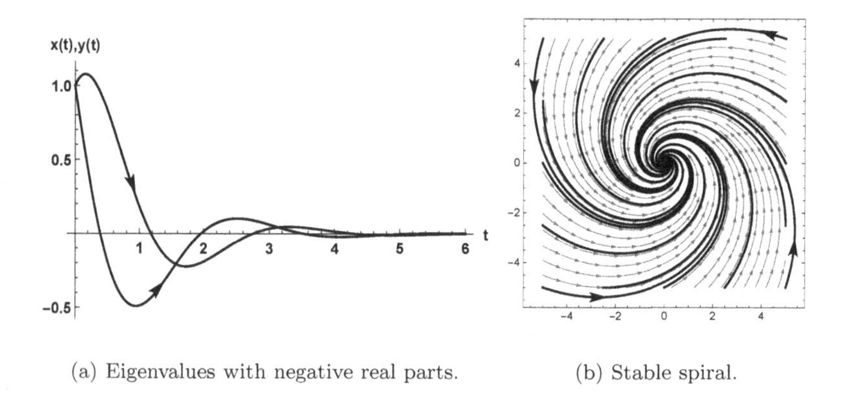

(a) Eigenvalues with negative real parts. (b) Stable spiral.

FIGURE 3.4: *Phase portrait showing all the trajectories spiral towards* $(0, 0)$, *a stable spiral or a stable focus.*

Conclusions:
(i) Discriminant > 0, Trace(A) $< 0 \Rightarrow$ phase portrait is a stable node.
(ii) Discriminant > 0, Trace(A) $> 0 \Rightarrow$ phase portrait is an unstable node.
(iii) If Discriminant $= 0$, then the phase portrait is stable or unstable node according as Trace(A) < 0 or Trace(A) > 0.

Case III: Let Discriminant $= (\text{Trace(A)})^2 - 4\text{Det(A)} < 0$, then the values of k are complex conjugates, namely,

$$\frac{1}{2}\left[\text{Trace}(A) + i\sqrt{\text{Discriminant}}\right] \text{ and } \frac{1}{2}\left[\text{Trace}(A) - i\sqrt{\text{Discriminant}}\right].$$

The phase portrait will be a stable spiral if Trace(A) <0 and an unstable spiral if Trace(A) > 0. If Trace(A) $= 0$, the phase portrait will be center.

Let λ and μ be the roots of the characteristic equation or the eigenvalues of the matrix A. The possible cases and the results are now shown in tabular form.

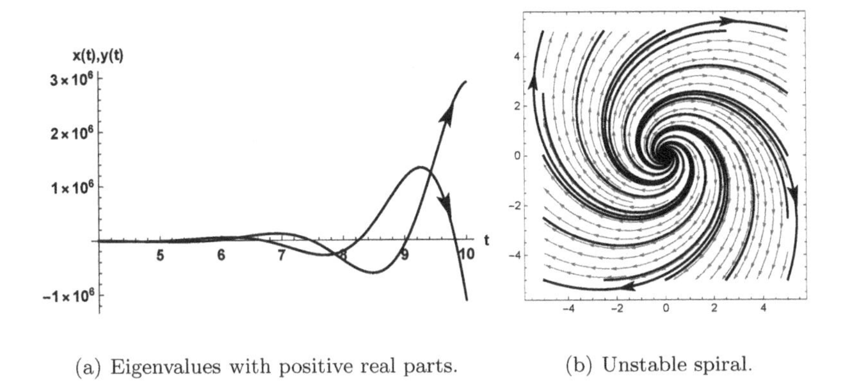

(a) Eigenvalues with positive real parts. (b) Unstable spiral.

FIGURE 3.5: *Phase portrait showing all the trajectories spiral away from* $(0,0)$, *an unstable spiral or an unstable focus.*

Nature of the eigenvalues λ, μ	Nature of the steady state (0,0)	Stability of the steady state (0,0)
Real, unequal, same sign:		
(i) $\lambda < 0,\ \mu < 0$	Node	Asymptotically stable
(ii) $\lambda > 0,\ \mu > 0$	Node	Unstable
Real, unequal, opposite sign:		
(iii) $\lambda < 0,\ \mu > 0$	Saddle point	Unstable
Real, equal, same sign:		
(iv) $\lambda = \mu = \lambda^* < 0$	Node	Asymptotically stable
(v) $\lambda = \mu = \lambda^* > 0$	Node	Unstable
Complex conjugates (a+ib, a-ib):		
(vi) $a < 0$	Spiral point	Asymptotically stable
(vii) $a > 0$	Spiral point	Unstable
(viii) a=0	Center	Stable but not asymptotically stable

Thus, a steady state or an equilibrium point can be stable, asymptotically stable, or unstable. A point is stable if the orbit of the system is inside a bounded neighborhood to the point for all time t after some t_0. A point is asymptotically stable if it is stable and the orbit approaches the critical point as $t \to \infty$. If the equilibrium point is not stable, then it is unstable.

Example 3.4.1 Determine the type and stability of steady state at $(0,0)$ of the following systems:

$$(i)\ \frac{dx}{dt} = 2x + 7y,\quad \frac{dy}{dt} = -5x - 10y. \qquad (ii)\ \frac{dx}{dt} = 3x - 4y,\quad \frac{dy}{dt} = 2x - y.$$

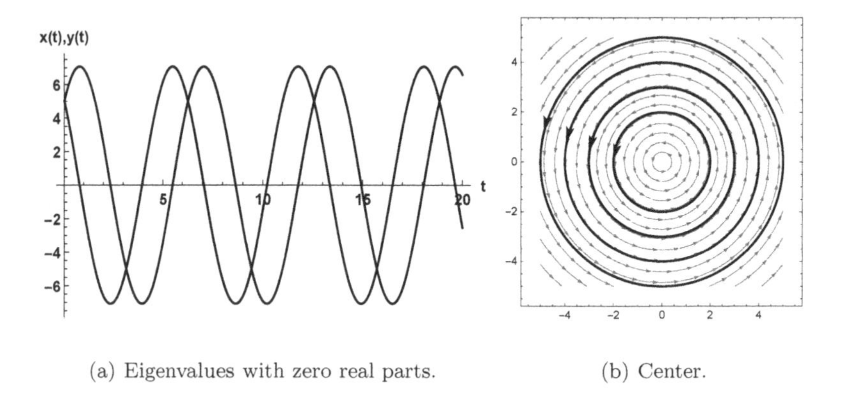

(a) Eigenvalues with zero real parts. (b) Center.

FIGURE 3.6: *Phase portrait showing all the trajectories are closed orbits about* $(0,0)$, *a center.*

Solution: (i) The characteristic equation is given by

$$\begin{vmatrix} 2-\lambda & 7 \\ -5 & -10-\lambda \end{vmatrix} = 0 \Rightarrow \lambda^2 + 8\lambda + 15 = 0 \Rightarrow \lambda = -3, -5.$$

The eigenvalues are real, unequal and negative, implying that the phase portrait is a stable node (asymptotically stable).

(ii) The characteristic equation is given by

$$\begin{vmatrix} 2-\lambda & 7 \\ -5 & -10-\lambda \end{vmatrix} = 0 \Rightarrow \lambda^2 - 2\lambda + 5 = 0 \Rightarrow \lambda = 1+2i, 1-2i.$$

The eigenvalues are imaginary with positive real part, implying that the phase portrait is an unstable focus.

3.5 Continuous Models

Different models are formulated and discussed in this section.

3.5.1 Carbon Dating

Carbon dating (Carbon 14 dating) is a method, developed by W.F. Libby at the University of Chicago in 1947 [97], that can be used to accurately date archaeological samples to determine the ages of the plant (wood fossil) or any material which got its carbon from the air. Carbon 14 (C^{14}), a radioactive isotope of carbon, is a result of constant bombardment by radiation from the sun in the atmosphere. During this bombardment, neutrons hit nitrogen 14 atoms and transmute them to carbon.

In a living organism, the absorption rate of C^{14} balances the disintegration rate of C^{14}. When the organism dies and the body is preserved, it does not absorb C^{14} but disintegration continues. As mentioned before, C^{14} is radioactive in nature and has a half-life (the time taken by a substance undergoing decay to decrease to half) of 5730 years. Scientists use this information. The method of carbon dating involves measuring the strength of C^{14} archaeological samples or fossils and then comparing it with the expected strength of C^{14} in the atmosphere, to calculate the accurate age.

Suppose an archaeological sample was found whose age needs to be determined. Let $A(t)$ be the amount of C^{14} present in the sample at any time t, then

$$\frac{dA}{dt} = -\lambda A \quad \text{(following radioactive decay law)}, \tag{3.13}$$

where λ is the decay constant of the sample. Integrating, we get,

$$A(t) = A_0 e^{-\lambda t},$$

where $A_0 = A(0)$ is the amount of C^{14} present in the sample when it was discovered. From equation (3.13), we can obtain the present ratio of disintegration of C^{14} in the archaeological sample, given by

$$M(t) = -\frac{dA}{dt} = \lambda A_0 e^{-\lambda t} \Rightarrow \frac{M(t)}{M(0)} = e^{-\lambda t} \Rightarrow t = \frac{1}{\lambda} \log \frac{M(0)}{M(t)}, \tag{3.14}$$

$M(0)$ $(= \lambda A_0)$, being the original rate of disintegration. From (3.14), we can determine the age, provided we can measure $M(t)$ and $M(0)$. The rate of disintegration of C^{14} in a comparable amount of archaeological sample or fossil (living wood) is equal to $M(0)$.

The half-life of the sample under investigation can be obtained from equation (3.14). Since the half-life is the amount of time required by the decreasing substance to reduce to half, we get

$$\frac{A_0}{2} = A_0 e^{-\lambda \tau} \Rightarrow \tau = \frac{1}{\lambda} \log_e 2 = \frac{0.6931}{\lambda}, \text{ where } \tau \text{ is the half-life.}$$

Example 3.5.1 A fossil is found that has 20% C^{14} compared to the living sample. How old is the fossil, knowing that the C^{14} half-life is 5730 years?

Solution: Let $A(t)$ be the amount of C^{14} present in the sample at any time t, then

$$A(t) = A_0 e^{-\lambda t} \quad \text{(following exponential decay)},$$

where A_0 is the initial amount of C^{14} present in the sample and λ is the decay constant of the sample. The fossil contains 20% of the original amount at time t, which implies

$$\frac{A(t)}{A_0} = \frac{20}{100} \Rightarrow e^{-\lambda t} = \frac{20}{100} \Rightarrow t = \frac{1}{\lambda}\ln(5). \tag{3.15}$$

Since the half-life is the amount of time required by the decaying substance to reduce to half, we get,

$$\frac{A_0}{2} = A_0 e^{-\lambda\tau}, \text{ where } \tau \text{ is the half-life.}$$

$$\Rightarrow \quad -\lambda\tau = \ln\left(\frac{1}{2}\right) \Rightarrow \frac{1}{\lambda} = \frac{\tau}{\ln(2)}. \tag{3.16}$$

From (3.15) and (3.16), we obtain,

$$t = \tau\frac{\ln(5)}{\ln(2)} = 5730 \times \frac{1.6094}{0.6931} \simeq 13305 \text{ years.}$$

3.5.2 Drug Distribution in the Body

The study of the movement of drugs in the body is called *pharmacokinetics*. The science of pharmacokinetics uses mathematical equations and utilizes them to describe the movement of the drug through the body [65].

We now study a simple problem in pharmacology, where we will be dealing with the dose-response relationship of a drug. In this problem, the drug present in the system follows certain laws. Let us assume that the rate of decrease of the concentration of the drug is directly proportional to the square of its amount present in the body and C_0 be the initial dose of the drug given to the patient at time $t = 0$. The mathematical model that captures this dynamics is given by

$$\frac{dC(t)}{dt} = -kC^2, \tag{3.17}$$

where k is a constant depending on the drug used and its value can be obtained from the experiment. Solving (3.17), we get

$$C(t) = \frac{C_0}{1 + C_0 kt}, \text{ where } C(0) = C_0.$$

Let an equal dose of drug C_0 be given to the body at equal time intervals, T. Then, immediately after the second dose, the concentration of the drug inside the body is

$$C_1 = C_0 + \frac{C_0}{1 + C_0 kT}.$$

Immediately after the third dose, the concentration of the drug inside the body is

$$C_2 = C_0 + \frac{C_1}{1 + C_1 kT}.$$

In a similar manner, we can conclude that

$$C_n = C_0 + \frac{C_{n-1}}{1 + C_{n-1} kT}, \tag{3.18}$$

which is a non-linear difference equation. Now,

$$C_{n+1} - C_n = \frac{C_n - C_{n-1}}{(1 + kTC_n)(1 + kTC_{n-1})}. \tag{3.19}$$

From (3.18), we conclude that $C_n > C_0$, which implies $C_{n+1} - C_n$ and $C_n - C_{n-1}$ have the same sign. Noting that C_n is an increasing function of n, we attempt to find the limiting value of the concentration by taking limits on both sides of (3.18), that is,

$$\lim_{t \to \infty} C_n = \lim_{t \to \infty} \left(C_0 + \frac{C_{n-1}}{1 + C_{n-1} kT} \right),$$

$$\Rightarrow \lim_{t \to \infty} C_n = C_0 + \frac{\lim_{t \to \infty} C_{n-1}}{1 + kT \lim_{t \to \infty} C_{n-1}},$$

$$\Rightarrow C_\infty = C_0 + \frac{C_\infty}{1 + C_\infty kT} \quad \text{where} \quad C_\infty = \lim_{t \to \infty} C_n = \lim_{t \to \infty} C_{n-1},$$

$$\Rightarrow kTC_\infty^2 - kTC_0 C_\infty - C_0 = 0,$$

$$\Rightarrow C_\infty = \frac{kTC_0 \pm \sqrt{k^2 T^2 C_0^2 + 4C_0 kT}}{2kT},$$

$$\Rightarrow C_\infty = \frac{C_0}{2} + \frac{C_0}{2} \sqrt{1 + \frac{4}{C_0 kT}} \quad \text{(taking positive sign only)}.$$

This implies $C_0 < C_n < C_\infty$ (since C_n is an increasing function), that is, the concentration is bounded.

3.5.3 Growth and Decay of Current in an L-R Circuit

We consider an L-R circuit where L is the inductance of the coil and R is the resistance. The coil is connected to a battery of voltage V through a key K (fig. 3.7).

In the **ON** position, the current flows through the coil. When the current $i(t)$ starts to flow, the negative lines of force move outward from the coil and an electromotive force (e.m.f.) will induce across L. According to the law of electromagnetic induction, this e.m.f. will oppose the voltage, as a result of

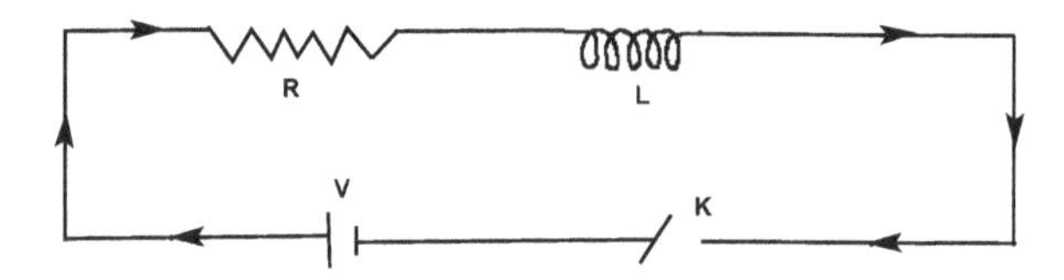

FIGURE 3.7: *The inductance-resistance (L-R) circuit, connected to a battery of voltage V through a key K.*

which there will be a voltage drop across R, which will also oppose the applied voltage. Let, at any time t, i be the current in the circuit increasing from 0 to a maximum value at a rate of increase $\frac{di}{dt}$. Now, the potential difference across the inductor is $V_1 = L\frac{di}{dt}$ and across the resistor is $V_2 = iR$. The differential equation modeling of this scenario is given by

$$V = L\frac{di}{dt} + iR \text{ (since } V = V_1 + V_2),$$

$$\Rightarrow \int_0^i \frac{di}{i - \frac{V}{R}} = -\frac{R}{L}\int_0^t dt \Rightarrow \log_e \left(i - \frac{i}{V/R}\right) = -\frac{R}{L}t,$$

$$\Rightarrow i(t) = \frac{V}{R}\left(1 - e^{-\frac{R}{L}t}\right),$$

which shows that the current grows exponentially. As $t \to \infty, i \to \frac{V}{R} = I$ (say), a steady value (fig. 3.8(a)).

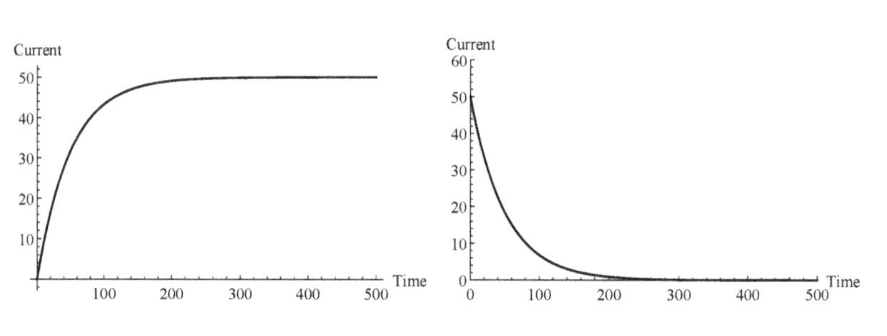

(a) Current grows and reaches a steady value.

(b) Current decays to zero.

FIGURE 3.8: *Graphs showing the (a) growth and (b) decay of current, with $L = 50$, $R = 1$ and $V = 50$.*

We now put the key in the **OFF** position. Initially, when the key was in the **ON** position, a steady current $I = \frac{V}{R}$ was flowing. With no current flowing in the circuit, the flux will reduce gradually, resulting in a voltage drop iR across the resistance R and the induced e.m.f. $L\frac{di}{dt}$ across the inductance L.

Now, since the key is **OFF**, the current becomes open, implying that the impressed voltage is zero.

The differential equation showing this scenario is given by

$$
\begin{aligned}
0 &= L\frac{di}{dt} + iR \text{ (since } V = V_1 + V_2 = 0), \\
\Rightarrow \int_I^i \frac{di}{i} &= -\int_0^t \frac{R}{L}dt \text{ (since at } t = 0, i = I), \\
\ln\left(\frac{i}{I}\right) &= -\frac{R}{L}t \Rightarrow i(t) = \frac{V}{R}\, e^{-\frac{R}{L}t}.
\end{aligned}
$$

Thus, the current decays exponentially as time increases and ultimately goes to zero (fig. 3.8(b)).

3.5.4 Rectilinear Motion under Variable Force

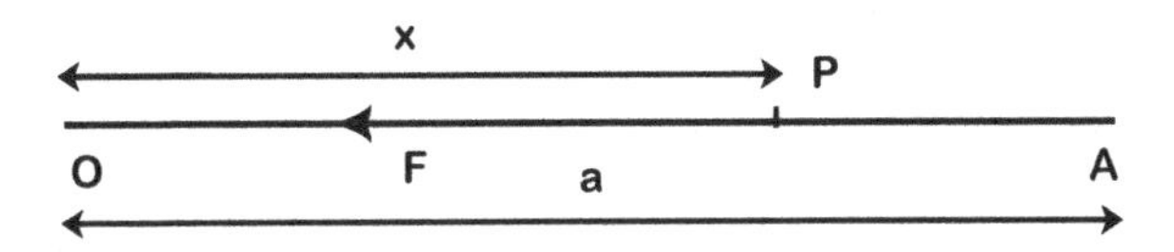

FIGURE 3.9: *A particle moving in a straight line towards the origin O (fixed) and acted upon by a force F directed towards O.*

Let a particle move in a straight line and be acted upon by a force $F = \frac{m\mu}{x}$, $\mu\ (> 0)$ being the constant of proportionality, which is always directed towards a fixed point O. Here, m is the mass of the particle and P is the position of the particle at any time t such that $OP = x$ (fig. 3.9). The equation of motion modeling the given scenario is given by

$$
mv\frac{dv}{dx} = -m\frac{\mu}{x},
$$

where $v\frac{dv}{dx}$ is the acceleration of the particle of mass m. Since the force is attractive, the sign of right-hand side is negative. Integrating, we get,

$$
\frac{v^2}{2} = -\mu \ln x + \text{constant}.
$$

Initially, let the particle start from rest (from A) at a distance a from the fixed point O (origin), then at time $t = 0, x = a, v = 0$,

$$
\Rightarrow \text{constant} = \mu \ln a \;\Rightarrow v^2 = 2\mu \ln\left(\frac{a}{x}\right).
$$

$$\Rightarrow v = \frac{dx}{dt} = -\sqrt{2\mu \ln\left(\frac{a}{x}\right)} \text{ (negative sign as distance decreases with time).}$$

$$\Rightarrow \sqrt{2\mu}\int_0^T dt = -\int_a^0 \frac{dx}{\sqrt{\ln\left(\frac{a}{x}\right)}},$$ where T is the time taken by the particle to reach the origin. Therefore,

$$\sqrt{2\mu}T = -\int_0^\infty \frac{ae^{-y^2}(-2y)dy}{y} \quad \left(\text{Substitute } \ln\left(\frac{a}{x}\right) = y^2\right)$$

$$= 2a\int_0^\infty e^{-y^2}dy = a\int_0^\infty \frac{e^{-z}}{\sqrt{z}}dz \text{ (Put } z = y^2)$$

$$= a\int_0^\infty e^{-z}z^{\frac{1}{2}-1}dz = a\,\Gamma\left(\frac{1}{2}\right) = a\sqrt{\pi} \Rightarrow T = a\sqrt{\frac{\pi}{2\mu}}.$$

3.5.5 Mechanical Oscillations

In this section, we discuss three different oscillations, namely, horizontal oscillations, vertical oscillations and damped forced oscillations.

3.5.5.1 Horizontal Oscillations

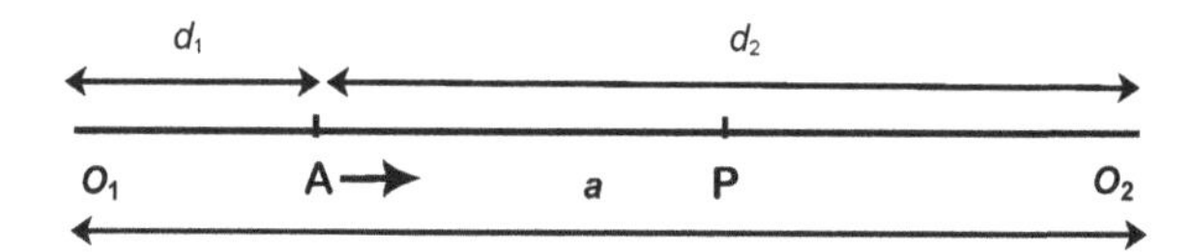

FIGURE 3.10: *A particle at rest is attracted by two forces towards two fixed centers O_1 and O_2, respectively.*

We consider a particle of mass m resting in equilibrium at a point A, being attracted by two forces equal to $m\mu_1^n \times (\text{distance})^n$ and $m\mu_2^n \times (\text{distance})^n$ respectively, towards two fixed centers O_1 and O_2 (fig. 3.10).

In this equilibrium position of the particle at A, the forces acting on it are equal and opposite, and hence O_1, A and O_2 are collinear.

Let $O_1O_2 = a, O_1A = d_1$ and $AO_2 = d_2$. Then $d_1 + d_2 = a$ and $\mu_1^n d_1^n = \mu_2^n d_2^n$, since the forces are equal and opposite at A.

$$\mu_1^n d_1^n = \mu_2^n d_2^n \Rightarrow \mu_1 d_1 = \mu_2 d_2,$$

$$\Rightarrow \frac{\mu_1}{d_2} = \frac{\mu_2}{d_1} = \frac{\mu_1 + \mu_2}{d_1 + d_2} = \frac{\mu_1 + \mu_2}{a},$$

$$\Rightarrow d_1 = \frac{a\mu_2}{\mu_1 + \mu_2} \text{ and } d_2 = \frac{a\mu_1}{\mu_1 + \mu_2}.$$

Now, let the particle be slightly displaced towards the fixed center O_2 and P be the new position of the particle at any time t, such that $AP = x$ (small).

The differential equation modeling of this physical scenario is given by

$$m\frac{d^2x}{dt^2} = -m\mu_1^n O_1P^n + m\mu_2^n O_2P^n. \tag{3.20}$$

When the particle is slightly displaced from the equilibrium position A to the position P, it tends to come back to its original position. Doing so, O_1 becomes the origin of attraction and O_2, the origin of repulsion. This explains the reason for the first term on the right-hand side of (3.20) to be negative and the second term to be positive.

$$\begin{aligned}
\text{Therefore, } \frac{d^2x}{dt^2} &= -\mu_1^n(d_1+x)^n + \mu_2^n(d_2-x)^n, \\
&= -\mu_1^n d_1^n\left(1+\frac{x}{d_1}\right)^n + \mu_2^n d_2^n\left(1-\frac{x}{d_2}\right)^n, \\
&= -\mu_1^n d_1^n\left\{1+\frac{nx}{d_1}+\frac{n(n-1)}{2!}\frac{x^2}{d_1^2}+\dots\right\} \\
&+ \mu_2^n d_2^n\left\{1-\frac{nx}{d_2}+\frac{n(n-1)}{2!}\frac{x^2}{d_2^2}+\dots\right\}, \\
\text{or, } \frac{d^2x}{dt^2} &= -\mu_1^n d_1^n - nx\mu_1^n d_1^{n-1} + \mu_2^n d_2^n - nx\mu_2^n d_2^{n-1}, \\
&\quad \text{(Since } x \text{ is small, neglecting higher powers of } x\text{)} \\
\Rightarrow \frac{d^2x}{dt^2} &= -n\left(\mu_1^n d_1^{n-1} + \mu_2^n d_2^{n-1}\right)x, \\
&= -n\left\{\mu_1^n\left(\frac{a\mu_2}{\mu_1+\mu_2}\right)^{n-1} + \mu_2^n\left(\frac{a\mu_1}{\mu_1+\mu_2}\right)^{n-1}\right\}x, \\
&= -\frac{n\mu_1^{n-1}\mu_2^{n-1}a^{n-1}}{(\mu_1+\mu_2)^{n-1}}(\mu_1+\mu_2)x, \\
\Rightarrow \frac{d^2x}{dt^2} &= -n\frac{(\mu_1\mu_2 a)^{n-1}}{(\mu_1+\mu_2)^{n-2}}\,x.
\end{aligned}$$

Therefore, we conclude that the motion of the particle is simple harmonic about A and the period of oscillation is

$$2\pi\sqrt{\frac{(\mu_1+\mu_2)^{n-2}}{(\mu_1\mu_2 a)^{n-1}}}.$$

3.5.5.2 Vertical Oscillations

Consider an elastic string of upstretched length AB $(= a)$, fixed at point A. A particle of mass m is attached to the end of the string so that the string is extended to the length AO $(= b)$ when the mass is at rest. In the equilibrium position, the tension in the string is balanced by the weight of the particle, that is,

$$mg = T_0 = \lambda\left(\frac{AO - AB}{AB}\right) \quad \text{(by Hooke's law)}$$

$$\Rightarrow mg = \lambda\left(\frac{b-a}{a}\right), \quad \lambda \text{ is the modulus of elasticity.} \tag{3.21}$$

Now, the particle is pulled to the point D and then released. Let P be the position of the particle at any time t such that $OP = x$ (fig. 3.11). The differential equation that models the scenario is given by

$$m\,\frac{d^2x}{dt^2} = mg - T, \text{ where } T \text{ is the tension of the string.}$$

$$\Rightarrow \; m\frac{d^2x}{dt^2} = mg - \lambda\frac{b+x-a}{a} = mg - \lambda\frac{b-a}{a} - \frac{\lambda x}{a},$$

$$\Rightarrow \; \frac{d^2x}{dt^2} = -\frac{\lambda x}{am} \quad \text{(using (3.21))}.$$

Hence, the motion is simple harmonic about the center O, the period of oscillation being $2\pi\sqrt{\frac{am}{\lambda}}$.

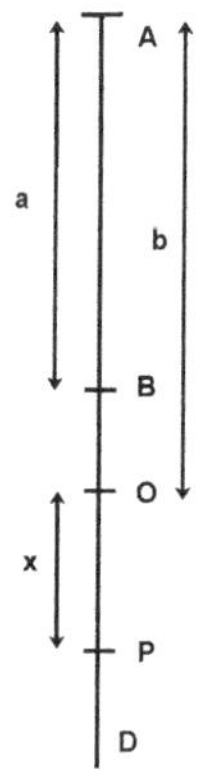

FIGURE 3.11: *Vertical oscillation.*

3.5.5.3 Damped and Forced Oscillations

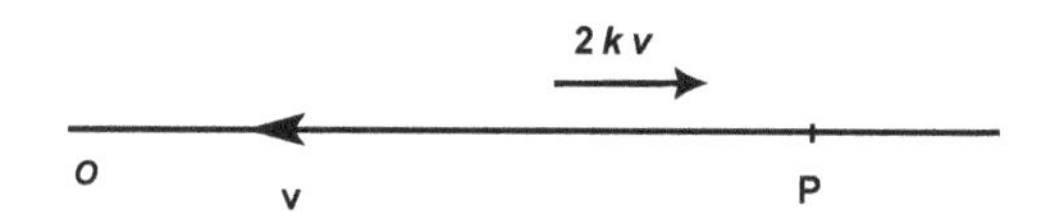

FIGURE 3.12: *A disturbing force 2kv acting on a particle moving in a straight line towards a fixed point O.*

We consider a particle moving in a straight line with an acceleration μ^2 (distance) towards a fixed point O in the line. A disturbing acceleration of $2k$ (velocity)($\mu > k$) also acts on the particle along with a periodic additional acceleration $F\cos(bt)$ [54] (fig. 3.12).

The equation of motion that models the scenario is given by

$$\frac{d^2x}{dt^2} = -\mu^2 x - 2k\frac{dx}{dt} + F\cos(bt).$$

Let $x = Ae^{mt}(A \neq 0)$ be a trial solution of

$$\frac{d^2x}{dt^2} + 2k\frac{dx}{dt} + \mu^2 x = 0,$$

then the required auxiliary equation is

$$m^2 + 2mk + \mu^2 = 0 \Rightarrow m = -k \pm \sqrt{k^2 - \mu^2} = -k \pm i\sqrt{\mu^2 - k^2} \ (\mu > k).$$

Complimentary function is $e^{-kt} A \cos(\sqrt{\mu^2 - k^2}t + \varepsilon_1)$ and the particular integral is given by

$$\frac{1}{D^2 + 2Dk + \mu^2} F \cos(bt) = F \frac{D^2 - 2Dk + \mu^2}{(D^2 + \mu^2)^2 - 4D^2k^2} \cos(bt)$$
$$= F \frac{(\mu^2 - b^2)\cos(bt) + 2kb\sin(bt)}{(\mu^2 - b^2)^2 + 4k^2b^2} = B\cos(bt - \varepsilon_2),$$
$$\text{where } B = \frac{F}{\sqrt{(\mu^2 - b^2)^2 + 4k^2b^2}} \text{ and } \tan\varepsilon_2 = \frac{2kb}{\mu^2 - b^2}.$$

Therefore, the general solution is

$$x = Ae^{-kt}\cos(\sqrt{\mu^2 - k^2}t + \varepsilon_1) + B\cos(bt - \varepsilon_2). \tag{3.22}$$

From (3.22), it is concluded that motion is the resultant of two oscillations,

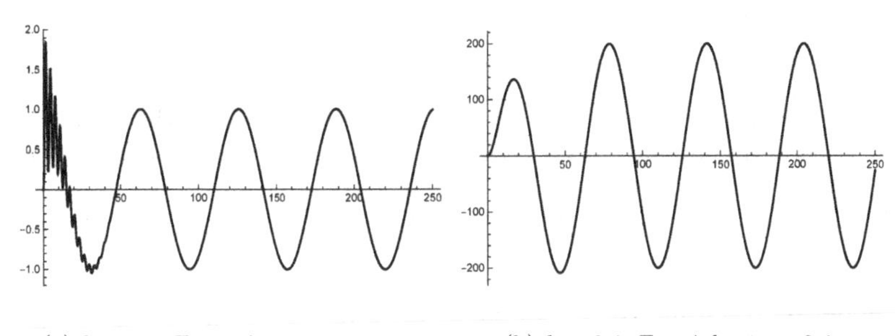

(a) $k = 0.1, F = 4, b = 0.1, \mu = 2.0$. (b) $k = 0.1, F = 4, b = \mu = 0.1$.

FIGURE 3.13: *A motion which is the resultant of two oscillations, namely, a free oscillation and a forced one.*

namely, the free oscillation (first part) and forced oscillation (second part). The arbitrary constants A and B can be obtained from the initial conditions. From the expression (3.22), it is clear that the amplitude of free oscillation decreases with time t because of the factor e^{-kt} and ultimately vanishes for large t. However, the amplitude of the forced oscillation persists as there is no diminishing factor, whose period of oscillation is $\frac{2\pi}{b}$. This is also evident from fig. 3.13(a).

Special Case: If the period of forced oscillation is equal to the period of free oscillation, that is, $\frac{2\pi}{b} = \frac{2\pi}{\mu} \Rightarrow b = \mu$, then the amplitude of the forced oscillation is $B = \dfrac{F}{2kb}$. If k is small, then the amplitude of the forced oscillation is very large (fig. 3.13(b)). This is the reason why a group of soldiers marching on a bridge are ordered to fall out. While marching in groups, the period of forced vibration may be equal to the natural period of the bridge structure. Then, a large amplitude of vibration may be generated, which may cause the bridge to crack and fall down.

3.5.6 Dynamics of Rowing

In rowing a boat, a person tries to push the boat forward against the water using the oar and thereby exerts a force, known as tractive force. We denote that force by T. Also, as the boat moves forward, the water adjacent to the sides of the boat exerts a force, resulting in losing its speed. We call this force a drag force and denote it by D. If $v(t)$ be the velocity of the boat at any time t, then the equation of motion is given by [62]

$$M\frac{dv}{dt} = T - D.$$

Let us now assume that the person has entered a race and let P be the effective power that the person can sustain for the entire length he has to row. Then, from physics, we get

$$P = T \times v \text{ (effective power = tractive force} \times \text{velocity).}$$

Also, from fluid dynamics, the drag force is proportional to the square of the velocity and to the surface area in contact with the water (wetted surface area). Thus,

$$D = kv^2S,$$

where S is wetted surface area and k is the constant of proportionality. Thus, the model representing the dynamics of rowing is given by [62]

$$M\frac{dv}{dt} = \frac{P}{v} - kv^2S = \frac{P - kv^3S}{v} = \frac{ks(\frac{P}{kS} - v^3)}{v},$$

$$\Rightarrow \int \frac{vdv}{a^3 - v^3} = \frac{kS}{M}\int dt, \text{ where } a^3 = \frac{P}{kS},$$

$$\Rightarrow \ln\frac{a^2 + av + v^2}{(a-v)^2} - 2\sqrt{3}\tan^{-1}\left(\frac{a+2v}{\sqrt{3}a}\right) = \frac{6akS}{M}t + \text{constant}.$$

Assuming at $t = 0, v = 0$, we get constant $= -\frac{\pi}{\sqrt{3}}$.

$$\text{Therefore, } \ln\left\{\frac{a^2 + av + v^2}{(a-v)^2}\right\} + \frac{\pi}{\sqrt{3}} - 2\sqrt{3}\tan^{-1}\left(\frac{a+2v}{\sqrt{3}a}\right) = \frac{6akS}{M}t,$$

where $a = \left(\frac{P}{kS}\right)^{1/3}$. This is more or less what we observe in the race, except the person rowing the boat may slow down at the end.

Alternative Solution:

$$Mv\frac{dv}{dx} = \frac{P}{v} - kSv^2 = \frac{P - kSv^3}{v}$$

$$\Rightarrow \int \frac{Mv^2}{P - kSv^3}dv = \int dx \Rightarrow -\frac{M}{3kS}\int \frac{dz}{z} = dx \ (z = P - kSv^3)$$

$$\Rightarrow -\frac{M}{3kS}\ln(z) = x + \text{Constant} \Rightarrow -\frac{M}{3PS}\ln(P - kSv^3) = x + \text{Constant}.$$

At time $t = 0, v = 0, x = 0 \Rightarrow$ Constant $= -\dfrac{MlnP}{3PS}$.

$$\Rightarrow \frac{M}{3PS}\left[\ln(P) - \ln(P - kSv^3)\right] = x \Rightarrow \ln\left(\frac{P - kSv^3}{P}\right) = -\frac{3PS}{M}x$$

$$\Rightarrow 1 - \frac{kSv^3}{P} = e^{-\frac{3PS}{M}x} \Rightarrow \frac{kS}{P}v^3 = 1 - e^{-\frac{3PS}{M}x}$$

$$\Rightarrow v^3 = \frac{P}{kS}\left[1 - e^{-\frac{3PS}{M}x}\right] \Rightarrow v = a\left[1 - e^{-\frac{3PS}{M}x}\right]^{\frac{1}{3}} \text{ where } a = \left(\frac{P}{kS}\right)^{\frac{1}{3}}.$$

3.5.7 Arms Race Models

We consider two neighboring countries A and B and let $x(t)$ and $y(t)$ be the expenditures on arms respectively by these two countries in some standardized monetary unit.

We construct a simple mathematical model by assuming the notion of mutual fear, that is, the more one country spends on arms, it encourages the other one to increase its expenditure on arms. Thus, we assume that each country spends on arms at a rate which is directly proportional to the existing expenditure of the other nation.

Mathematically, we can write [117]

$$\frac{dx}{dt} = \alpha y, \ \frac{dy}{dt} = \beta x \ (\alpha, \beta > 0), \tag{3.23}$$

$$\Rightarrow \frac{d^2x}{dt^2} = \alpha\frac{dy}{dt} = \alpha\beta x$$

$$\Rightarrow x = A_1e^{\sqrt{\alpha\beta}t} + A_2e^{-\sqrt{\alpha\beta}t}.$$

$$\text{Similarly, we get, } y = B_1e^{\sqrt{\alpha\beta}t} + B_2e^{-\sqrt{\alpha\beta}t}.$$

Thus, $x, y \to \infty$ as $t \to \infty$ and we conclude that both the countries A and B spend more and more money on arms with increasing time and no limits on the expenditure (fig. 3.14(a)). As the mathematical prediction of indefinitely large expenditure for both the countries is unrealistic, an improved model is desired.

In the modified model, other than the mutual fear, we also assume that the rate of change of one country's expenditure on arms will also be directly proportional to its own expenditure as the excessive expenditure on the arms puts the country's economy in the compromising position. Accordingly, the model (3.23) is modified as [117]

$$\begin{aligned} \frac{dx}{dt} &= \alpha y - \gamma x \ (\alpha, \beta, \gamma, \delta > 0), \\ \frac{dy}{dt} &= \beta x - \delta y. \end{aligned} \tag{3.24}$$

Clearly, $(0,0)$ is the only steady-state solution, provided $\gamma\delta - \alpha\beta \neq 0$. The characteristic equation is given by

$$\begin{vmatrix} -\gamma - \lambda & \gamma \\ \beta & -\delta - \lambda \end{vmatrix} = 0,$$
$$\Rightarrow \lambda^2 - (-\delta - \gamma)\lambda + \gamma\delta - \alpha\beta = 0.$$

Hence, the system is stable if $\gamma\delta - \alpha\beta > 0 \ \Rightarrow \ \gamma\delta > \alpha\beta$.

This implies if the product of the rates of depreciation $(\gamma\delta)$ on the expenditure of arms of both the countries A and B is greater than the product of rates of expenditure $(\alpha\beta)$ on arms of both the countries, the system will be stable and the countries will spend an allocated amount of money on arms, so that the economy of the country is not compromised.

A simple refinement of the model was made by Lewis F. Richardson (1881–1953), popularly known as the Richardson Arms Race model [130], where he assumed that the cause of the rate of increase of a country's armament, not only depend on mutual stimulation but also on the permanent underlying grievances of each country against the other. The modified model is [117, 127, 130]

$$\frac{dx}{dt} = \alpha y - \gamma x + r, \tag{3.25}$$
$$\frac{dy}{dt} = \beta x - \delta y + s, \tag{3.26}$$

where α, β, γ, δ are positive (as before) and r, s are constants which will be negative or positive, depending on the fact whether the country has overcome the grievance or not. The unique steady-state solution is obtained by solving

$$\alpha y^* - \gamma x^* + r = 0 \text{ and } \beta x^* - \delta y^* + s = 0,$$

provided $\gamma\delta - \alpha\beta \neq 0$. Solving, we obtain

$$x^* = \frac{r\delta + s\alpha}{\gamma\delta - \alpha\beta} \text{ and } y^* = \frac{r\beta + s\gamma}{\gamma\delta - \alpha\beta}.$$

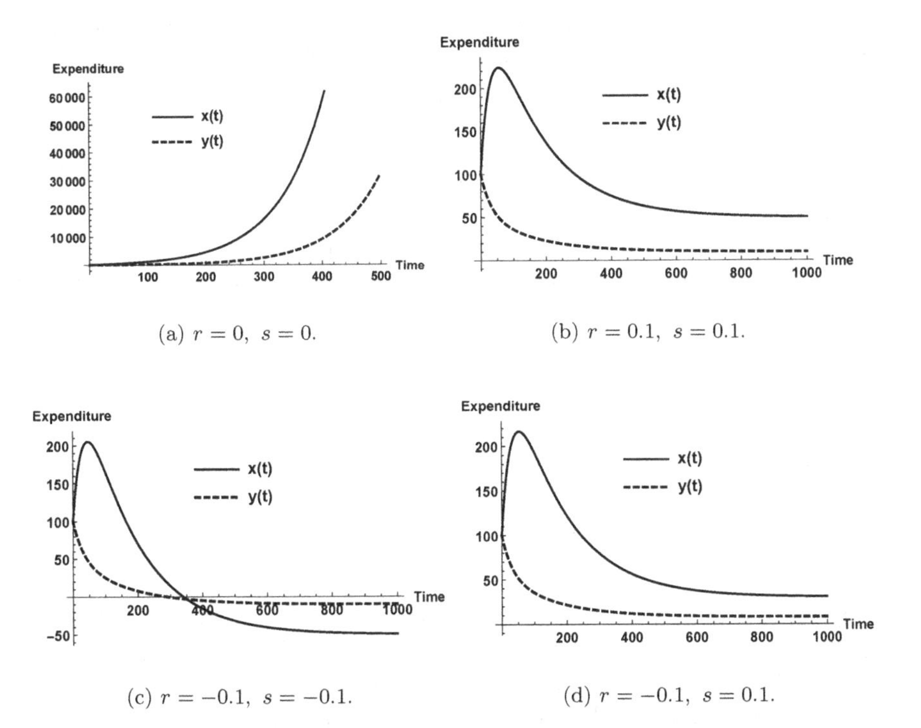

(a) $r = 0,\ s = 0.$

(b) $r = 0.1,\ s = 0.1.$

(c) $r = -0.1,\ s = -0.1.$

(d) $r = -0.1,\ s = 0.1.$

FIGURE 3.14: *The figures show the dynamics of the arms race model. (a) Unlimited expenditure on arms with increasing time, (b) money on arms are spent in a strategic manner, (c) complete disarmament, and (d) mutual goodwill.*

The characteristic equation is

$$\lambda^2 + (\gamma + \delta)\lambda + \gamma\delta - \alpha\beta = 0.$$

Case I: $\gamma\delta - \alpha\beta > 0, r > 0, s > 0$. In this case, the system is stable. This means both the countries spent on arms in a strategic manner so that the economy of the country is not compromised (fig. 3.14(b)).

Case II: $\gamma\delta - \alpha\beta > 0, r < 0, s < 0$. In this case, though the system is stable, the equilibrium solution becomes negative. However, expenditures cannot be negative in reality. Suppose $(x_0, y_0) > 0$ be the initial expenditures, then $x(t) \to x^*$ and $y(t) \to y^*$ at $t \to \infty$, and for that, it has to pass through zero values. Thus, as $x(t)$ becomes zero, (3.26) reduces to

$$\frac{dy}{dt} = -\delta y + s \Rightarrow y(t) = \frac{s}{\delta} + C_1 e^{-\delta t}.$$

Since $s < 0, y(t)$ decreases till it reaches the value zero. A similar argument is valid for $x(t)$. Thus, in this case, both the countries will stop spending on arms [73], and then this will result in a complete disarmament (fig. 3.14(c)).

Case III: $\gamma\delta - \alpha\beta > 0, r > 0, s < 0$. In this case, one of the countries has overcome the grievance ($s < 0$). The system is stable with positive equilibrium solution if $\delta r + \alpha s > 0$ and $\beta r + \gamma s > 0$ ($s < 0$). The system approaches an equilibrium value which is less than the previous cases when both $r, s > 0$ (fig. 3.14(d)). That means when one country has overcome the grievance and started spending less on armaments, this will have an effect on the other country, who will also start spending less in order to develop mutual goodwill [73].

Case IV: $\gamma\delta - \alpha\beta < 0, r > 0, s > 0$. The system becomes unstable, which will lead to a runaway arms race ($x \to \infty, y \to \infty$) as one of the eigenvalues is positive and the other is negative.

Case V: $\gamma\delta - \alpha\beta < 0, r < 0, s < 0$. The system is unstable, but the equilibrium solution is positive. Though one of the eigenvalues is positive and the other is negative, there is a possibility of disarmament as well as a runaway arms race, depending on the initial expenditure on arms by both the countries [73].

3.5.8 Epidemic Models

Mathematical epidemiology is the use of mathematical models to predict the course of infectious disease and to compare the effects of different control strategies. In epidemic models, the population is divided into three main classes, namely, a susceptible class, denoted by $S(t)$ (persons who are vulnerable to the disease or who can be easily infected by the disease), infected class denoted by $I(t)$ (persons who already have the disease), and recovered class, denoted by $R(t)$ (persons who have recovered from the disease). One can define more classes if the situation demands, for modifications in the models.

Susceptible-Infective (SI) Model: Let a population consist of (n+1) persons of which n persons are in susceptible group and only one is infected, so that $S(t) + I(t) = n + 1, S(0) = n, I(0) = 1$. A susceptible person gets infected when he comes in contact with an infected one. Mathematically, we can say that the rate of increase of the infected class is proportional to the product of the susceptible and infected persons. Hence, the susceptible class also decreases at the same rate. The system modeling the described situation is given by [73]

$$\frac{dS}{dt} = -\beta SI,$$
$$\frac{dI}{dt} = \beta SI \ (\beta > 0).$$

$$\text{Now, } \frac{dS}{dt} + \frac{dI}{dt} = 0 \ \Rightarrow \ S(t) + I(t) = \text{constant}.$$
$$\Rightarrow \ \text{constant} = S(0) + I(0) = n + 1 \ \Rightarrow \ S(t) + I(t) = n + 1.$$
$$\Rightarrow \ \frac{dS}{dt} = -\beta S(n + 1 - S) \text{ and } \frac{dI}{dt} = \beta I(n + 1 - I).$$

Integrating the first differential equation, we obtain

$$\int \frac{dS}{S(n+1-S)} = -\int \beta dt \ \Rightarrow \frac{1}{n+1} \int \left[\frac{1}{n+1-S} + \frac{1}{S} \right] dS = -\int \beta dt,$$
$$\Rightarrow -\ln(n + 1 - S) + \ln(S) = -(n + 1)\beta t + A \text{ (constant)}.$$
$$\text{At } t = 0, S(0) = n \ \Rightarrow \ \text{A} = \ln(n).$$
$$\Rightarrow \ \ln \left[\frac{S}{n(n+1-S)} \right] = -(n + 1)\beta t \ \Rightarrow \ \frac{S}{n(n+1-S)} = e^{-(n+1)\beta t},$$
$$\Rightarrow \ S(t) = \frac{n(n+1)}{n + e^{(n+1)\beta t}}. \text{ Therefore,}$$
$$I(t) = (n + 1) - S(t) = (n + 1) - \frac{n(n+1)}{n + e^{(n+1)\beta t}} = \frac{n+1}{1 + ne^{-(n+1)\beta t}}.$$

As $t \to \infty, S(t) \to 0$ and $I(t) \to (n + 1)$. Therefore, we conclude that as time increases, all the susceptible persons will become infected (fig. 3.15(a)).

Susceptible-Infective-Susceptible (SIS) Model: A simple refinement of the previous model has been made and named as the SIS model, where it is assumed that the infected person has the ability to recover and move to the susceptible class at a rate α (say). Initially $S(0) = n, I(0) = 1$, then, we get the required model as [73]

$$\frac{dS}{dt} = -\beta SI + \alpha I,$$
$$\frac{dI}{dt} = \beta SI - \alpha I, \ (\beta, \alpha > 0).$$

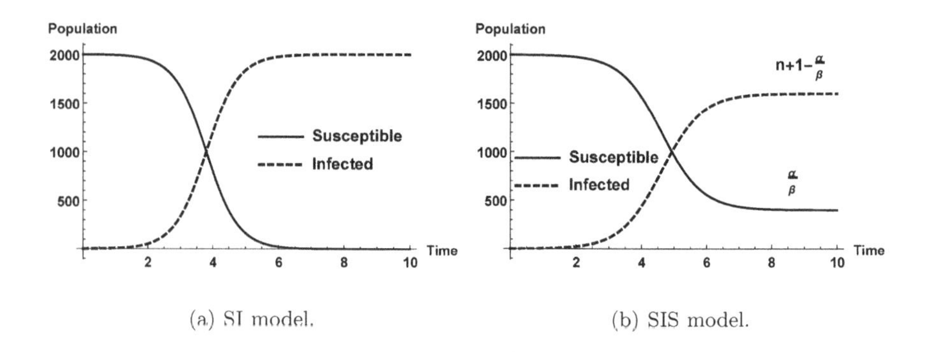

(a) SI model. (b) SIS model.

FIGURE 3.15: *The figures show the dynamics of epidemic models. (a) SI model with* $\beta = 0.001, S(0) = 2000, I(0) = 1$, *(b) SIS model with* $\beta = 0.001, \alpha = 0.4, S(0) = 2000, I(0) = 1$.

$$\text{Now, } \frac{dS}{dt} + \frac{dI}{dt} = 0 \Rightarrow S(t) + I(t) = K \text{ (constant)}.$$
$$\Rightarrow K = S(0) + I(0) = n+1 \Rightarrow S(t) + I(t) = n+1.$$
$$\frac{dS}{dt} = -[(n+1)\beta + \alpha]S + \beta S^2 + (n+1)\alpha,$$
$$\frac{dI}{dt} = [(n+1)\beta - \alpha]I - \beta I^2 = cI - \beta I^2, \text{ where } c = (n+1)\beta - \alpha.$$
$$\Rightarrow \frac{dI}{I(1-\frac{\beta}{c}I)} = c\,dt \Rightarrow \frac{\frac{c}{\beta}dI}{I(\frac{c}{\beta} - I)} = c\,dt \Rightarrow \left[\frac{1}{I} + \frac{1}{\frac{c}{\beta} - I}\right] = c\,dt.$$

Integrating we obtain,

$$\ln(I) - \ln\left(\frac{c}{\beta} - I\right) = ct + B \text{ (constant). Now, } I(0) = 1 \Rightarrow B = -\ln\left(\frac{c}{\beta} - 1\right).$$
$$\Rightarrow \ln(I) - \ln\left(\frac{c}{\beta} - I\right) + ln\left(\frac{c}{\beta} - 1\right) = c\,t \Rightarrow \frac{I(\frac{c}{\beta} - 1)}{\frac{c}{\beta} - I} = e^{ct},$$
$$\Rightarrow I(t) = \frac{\frac{c}{\beta}}{1 + (\frac{c}{\beta} - 1)e^{-ct}} = \frac{(n+1) - \frac{\alpha}{\beta}}{1 + (n+1-\frac{\alpha}{\beta} - 1)e^{-[(n+1)\beta - \alpha]t}}.$$

$$S(t) = n+1 - I(t) = n+1 - \frac{n+1-\frac{\alpha}{\beta}}{1 + (n+1-\frac{\alpha}{\beta} - 1)e^{-[(n+1)\beta-\alpha]t}},$$
$$\Rightarrow S(t) = \frac{(n+1)(n+1-\frac{\alpha}{\beta} - 1)e^{-[(n+1)\beta-\alpha]t} + \frac{\alpha}{\beta}}{1 + (n+1-\frac{\alpha}{\beta} - 1)e^{-[(n+1)\beta-\alpha]t}}.$$

AS $t \to \infty, S \to \frac{\alpha}{\beta}$ and $I \to n+1-\frac{\alpha}{\beta}$; provided $(n+1)\alpha - \beta > 0$. Hence, in this case, a fraction of susceptible persons will be there, which have not been infected or a fraction of infected persons have recovered and becomes susceptible again (fig. 3.15(b)).

Susceptible-Infective-Recovered (SIR) Model: This model was developed by Kermack and McKendrick [81] and is given by the set of differential equations as follows [42, 81]:

$$\begin{aligned}
\frac{dS}{dt} &= -\beta SI, \\
\frac{dI}{dt} &= \beta SI - \alpha I, \\
\frac{dR}{dt} &= \alpha I, \ (\beta, \alpha > 0).
\end{aligned}$$

It is assumed that the susceptibles become infected when they come in contact with one another (βSI) and a fraction of the infected class (αI) recovers from the disease and moves to the recovered class. Now,

$$\frac{dS}{dR} = \frac{dS}{dt}\frac{dt}{dR} = \frac{-\beta SI}{\alpha I} = -\frac{\beta}{\alpha}S,$$

$$\Rightarrow \frac{dS}{S} = -\frac{\beta}{\alpha}dR \ \Rightarrow \ln(S) = -\frac{\beta}{\alpha}R + \text{constant}.$$

Initially, $S(0) = n, R(0) = 0 \ \Rightarrow$ constant $= \ln(n)$,

$$\Rightarrow \ \ln(S) = -\frac{\beta}{\alpha}R + \ln(n) \ \Rightarrow \ S = ne^{-\frac{\beta}{\alpha}R}.$$

Again,

$$\frac{dI}{dS} = \frac{dI}{dt}\frac{dt}{dS} = \frac{\beta SI - \alpha I}{-\beta SI} = -1 + \frac{\alpha}{\beta}\frac{1}{S},$$

$$\Rightarrow \int dI = -\int dS + \frac{\alpha}{\beta}\int\frac{dS}{S} \ \Rightarrow \ I(t) = -S + \frac{\alpha}{\beta}\ln(S) + \text{ constant}.$$

Initially at $t = 0, S(0) = n, I(0) = 1 \ \Rightarrow$ constant $= 1 + n - \frac{\alpha}{\beta}\ln(n)$.

$$\Rightarrow I(t) = -S + \frac{\alpha}{\beta}\ln(S) + n + 1 - \frac{\alpha}{\beta}\ln(n) = n + 1 - S + \frac{\alpha}{\beta}\ln\left(\frac{S}{n}\right).$$

Since $S = ne^{-\frac{\beta}{\alpha}R}$, we have

$$\frac{dR}{dt} = \alpha I = \alpha(n + 1 - S - R) = \alpha\left(n + 1 - ne^{-\frac{\beta}{\alpha}R} - R\right),$$

$$\Rightarrow \frac{dR}{dt} = \alpha\left[n + 1 - n\left(1 - \frac{\beta}{\alpha}R + \frac{\beta^2}{2\alpha^2}R^2\right) - R\right], \text{(Assuming } \frac{R}{\frac{\alpha}{\beta}} \text{ is small)},$$

$$\Rightarrow \frac{dR}{dt} = \alpha\left[1 - \frac{n\beta^2}{2\alpha^2}\left\{R^2 - \frac{2\alpha^2}{n\beta^2}\left(\frac{n\beta}{\alpha} - 1\right)R\right\}\right],$$

$$= \alpha\left[1 - \frac{n\beta^2}{2\alpha^2}\left\{R - \frac{\alpha^2}{n\beta^2}\left(\frac{n\beta}{\alpha} - 1\right)\right\}^2 + \frac{\alpha^2}{2n\beta^2}\left(\frac{n\beta}{\alpha} - 1\right)^2\right],$$

$$= \frac{n\beta^2}{2\alpha}\left[\frac{2\alpha^2}{n\beta^2} + \frac{\alpha^4}{n^2\beta^4}\left(\frac{n\beta}{\alpha} - 1\right)^2 - \left\{R - \frac{\alpha^2}{n\beta^2}\left(\frac{n\beta}{\alpha} - 1\right)\right\}^2\right],$$

$$= \frac{n\beta^2}{2\alpha}\left[B^2 - (R - A)^2\right] \text{ where } A = \frac{\alpha^2}{n\beta^2}\left(\frac{n\beta}{\alpha} - 1\right), \text{ and}$$

$$B^2 = \frac{2\alpha^2}{n\beta^2} + \frac{\alpha^4}{n^2\beta^4}\left(\frac{n\beta}{\alpha} - 1\right)^2. \text{ Integrating we get,}$$

$$\int \frac{dR}{B^2 - (R - A)^2} = \int \frac{n\beta^2}{2\alpha}dt \Rightarrow \frac{1}{B}\tanh^{-1}\left(\frac{R - A}{B}\right) = \frac{n\beta^2}{2\alpha}t + \text{constant.}$$

Initially at $t = 0, R(0) = 0 \Rightarrow \text{constant} = \frac{1}{B}\tanh^{-1}(\frac{-A}{B}) = -\frac{1}{B}\tanh^{-1}(\frac{A}{B})$.

$$\Rightarrow \frac{1}{B}\tanh^{-1}\left(\frac{R - A}{B}\right) = \frac{n\beta^2}{2\alpha}t - \frac{1}{B}\tanh^{-1}\left(\frac{A}{B}\right),$$

$$\Rightarrow \frac{R - A}{B} = \tanh\left[\frac{n\beta^2}{2\alpha}t - \frac{1}{B}\tanh^{-1}\left(\frac{A}{B}\right)\right],$$

$$\Rightarrow R(t) = A + B\tanh\left[B\frac{n\beta^2}{2\alpha}t - \frac{1}{B}\tanh^{-1}\left(\frac{A}{B}\right)\right]. \text{ Therefore,}$$

$$S(t) = ne^{-\frac{\beta}{\alpha}\left[A + B\tanh\left\{B\frac{n\beta^2}{2\alpha}t - \frac{1}{B}\tanh^{-1}\left(\frac{A}{B}\right)\right\}\right]} \text{ and}$$

$$I(t) = n + 1 - S(t) - R(t).$$

The numerical solution of the model shows that both susceptible and infected goes to zero and there is a full recovery (fig. 3.16(a)).

Susceptible-Infective-Removed-Susceptible (SIRS) Model: A refinement of the SIR model can be made by assuming that the recovered person becomes susceptible again due to loss of immunity at a rate proportional to the population in recovery class R, with proportionality constant γ. The following differential equations describe the model [30]:

$$\frac{dS}{dt} = -\beta SI + \gamma R,$$

$$\frac{dI}{dt} = \beta SI - \alpha I,$$

$$\frac{dR}{dt} = \alpha I - \gamma R, \ (\beta, \alpha, \gamma > 0).$$

Fig. 3.16(b) shows the dynamics of SIRS model for $\beta = 0.001$, $\alpha = 0.4$ and $\gamma = 0.01$.

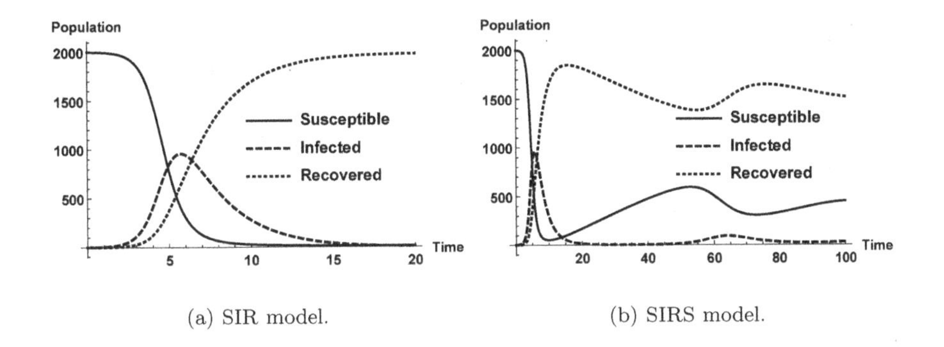

(a) SIR model. (b) SIRS model.

FIGURE 3.16: *The figures show the dynamics of epidemic models. (a) SIR model with* $\beta = 0.001, \alpha = 0.4, S(0) = 2000, I(0) = 1, R(0) = 0$, *(b) SIRS model with* $\beta = 0.001, \alpha = 0.4, \gamma = 0.01, S(0) = 2000, I(0) = 1, R(0) = 0$.

We can use the SIRS model to capture the dynamics of COVID-19. The susceptible population becomes infected by COVID-19 at a rate β (per-capita effective contact rate), which is the number of effective contacts made by a given individual per unit time. We are trying to minimize the value of β by practicing social distancing. Once infected, the susceptible population (βSI) moves to the infected class. The infected class recovers from the virus by hard immunity of individual (since no vaccine is available) but have the chance to reinfection. Hence, from the infected class, αI has moved to the recovered class, and γR has moved to the susceptible class. We want the see the dynamics of the spread of COVID-19 with $\beta = 0.00002856, \alpha = 0.19819303, \gamma = 0.001$ [146] and initial condition $S(0) = 15000, I(0) = 1, R(0) = 0$. The figure shows that the initial spread is high, which then decays over time (fig. 3.17(a)). The susceptible as well as recovered class also show "ups and down" behaviour before reaching a steady value (fig. 3.17(b)).

3.5.9 Combat Models

F.W. Lanchester was a British Engineer, who formulated a series of mathematical problems used to predict the outcome and number of soldiers surviving a given battle during the first world war. Lanchester developed a system of differential equations, which have been utilized throughout conventional military combat situations. The models helped in better planning, prediction of battles and their possible outcomes. Additionally, these equations have been manipulated to apply to multiple battle scenarios such as two conventional armies, two guerrilla armies, one guerrilla army, and one conventional army. With a proper amount of variables, these equations can be used to analyze extremely complex battles. Lanchester mathematical models

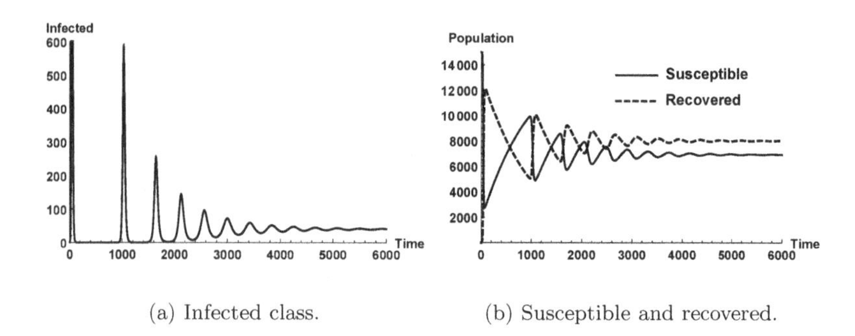

(a) Infected class. (b) Susceptible and recovered.

FIGURE 3.17: *The figures show the dynamics of the spread of COVID-19 along with the dynamics of susceptible and recovered classes.*

were built on equations, very similar to the law of mass action:

$$\textit{rate of change} = \textit{rate in} - \textit{rate out},$$

where *rate in* denotes the number of new troops supplied (reinforcement) and *rate out* denotes the number of troops not available to fight the battle, either due to attacks from the opposing army (combat loss) or other factors like desertion, sickness, etc. (operational loss).

3.5.9.1 Conventional Combat Model

Let $x(t)$ and $y(t)$ denote the number of troops or combatants for army A and army B, respectively. In conventional combat, the two opposite armies must directly interact with one another in the open, using conventional means (like knives, guns, etc.) with the exclusion of any chemical, biological and nuclear weapons. The following assumptions are made to formulate the model:

- The operational losses are neglected.
- There is no new supply of troops (no reinforcement).
- The combat loss rate of a conventional army A is proportional to the size of the opposing army B.

The system of differential equations describing the model is

$$\frac{dx}{dt} = -\alpha y, \quad \frac{dy}{dt} = -\beta x,$$

where α, β (> 0) are the fighting effectiveness coefficients of the armies B and A, respectively. Suppose, the initial number of troops for army A and army B are x_0 and y_0, respectively. Then, we have an initial value problem

$$\frac{dx}{dt} = -\alpha y, \qquad \frac{dy}{dt} = -\beta x, \qquad x(0) = x_0, \qquad y(0) = y_0,$$

which has a unique solution.

3.5.9.2 Guerrilla Combat Model

Let $x(t)$ and $y(t)$ denote the number of troops or combatants for the army A and army B, respectively. In Guerrilla combat, troops are deployed in small groups that are hidden (covert). If the troops are large, they are more likely to get caught. However, if the opposing army is large, they are more likely to find the guerrilla troops. The following assumptions are made to formulate the model:

- The operational losses are neglected.
- There is no new supply of troops (no reinforcement).
- The combat loss rate of a guerrilla troop army is proportional to the product of the sizes of both the armies A and B.

The system of differential equations describing the model is

$$\frac{dx}{dt} = -\alpha xy, \quad \frac{dy}{dt} = -\beta xy,$$

with initial conditions

$$x(0) = x_0, \qquad y(0) = y_0,$$

where α, β (> 0) are fighting effectiveness coefficients of the forces B and A, respectively.

3.5.9.3 Mixed Combat Model

Let $x(t)$ and $y(t)$ denote the number of troops or combatants for a conventional army A and a guerrilla force B. The following assumptions are made to formulate the model:

- The operational losses are neglected.
- There is no new supply of troops (no reinforcement).
- The combat loss rate for the conventional army A is proportional to the size of the opposing guerrilla force B and the combat loss rate for guerrilla force B is proportional to the product of the sizes of both the armies A and B.

The system of differential equations describing the model is

$$\frac{dx}{dt} = -\alpha y, \quad \frac{dy}{dt} = -\beta xy,$$

with initial conditions

$$x(0) = x_0, \qquad y(0) = y_0,$$

where α, β (> 0) are fighting effectiveness coefficient of the forces B and A, respectively.

3.5.9.4 Guerrilla Combat Model (Revisited)

The mathematical model for Guerrilla combat is described by the initial value problem

$$\frac{dx}{dt} = -\alpha xy, \quad \frac{dy}{dt} = -\beta xy \qquad (\alpha, \beta > 0),$$

with initial conditions

$$x(0) = x_0, \qquad y(0) = y_0,$$

which has a unique solution. Now,

$$\frac{dy}{dx} = \frac{dy}{dt}\frac{dt}{dx} = \frac{-\beta xy}{-\alpha xy} = \frac{\beta}{\alpha} \Rightarrow \beta x(t) - \alpha y(t) = \text{constant}$$

Using the initial conditions, we obtain constant $= \beta x_0 - \alpha y_0$. Therefore,

$$\beta\, x(t) - \alpha\, y(t) = \beta\, x_0 - \alpha\, y_0, \; \forall t \geq 0. \tag{3.27}$$

From (3.27) we get

$$y = \frac{\beta x - (\beta x_0 - \alpha y_0)}{\alpha} \tag{3.28}$$

Substituting (3.28) in $\frac{dx}{dt} = -\alpha xy$, we get the initial value problem as

$$\begin{aligned} \frac{dx}{dt} &= -x[\beta x - (\beta x_0 - \alpha y_0)], \; x(0) = 0 \\ \Rightarrow \frac{1}{x^2}\frac{dx}{dt} &= (\beta x_0 - \alpha y_0)\frac{1}{x} - \beta. \end{aligned} \tag{3.29}$$

We put $z = \frac{1}{x}$ $\left(\frac{dz}{dt} = -\frac{1}{x^2}\frac{dx}{dt}\right)$ in (3.29) and obtain

$$\frac{dz}{dt} + (\beta x_0 - \alpha y_0)z = \beta. \tag{3.30}$$

This is a first-order linear differential equation. Integrating factor is

$$e^{\int(\beta x_0 - \alpha y_0)dt} = e^{(\beta x_0 - \alpha y_0)t}.$$

Multiplying both sides of (3.30) by the integrating factor we obtain,

$$\frac{d}{dt}[ze^{(\beta x_0 - \alpha y_0)t}] = \beta e^{(\beta x_0 - \alpha y_0)t} \;\Rightarrow\; ze^{(\beta x_0 - \alpha y_0)t} = \frac{\beta e^{(\beta x_0 - \alpha y_0)t}}{\beta x_0 - \alpha y_0} + \text{constant},$$

$$\frac{1}{x}e^{(\beta x_0 - \alpha y_0)t} = \frac{\beta e^{(\beta x_0 - \alpha y_0)t}}{\beta x_0 - \alpha y_0} + \text{constant}.$$

Since $x(0) = x_0$, we get constant $= \dfrac{1}{x_0} - \dfrac{\beta}{\beta x_0 - \alpha y_0}$. Therefore,

$$\begin{aligned}
\frac{1}{x} e^{(\beta x_0 - \alpha y_0)t} &= \frac{\beta e^{(\beta x_0 - \alpha y_0)t}}{\beta x_0 - \alpha y_0} + \frac{1}{x_0} - \frac{\beta}{\beta x_0 - \alpha y_0} \\
&= \frac{x_0 \beta e^{(\beta x_0 - \alpha y_0)t} - \alpha y_0}{x_0(\beta x_0 - \alpha y_0)}, \\
\Rightarrow x(t) &= \frac{x_0(\beta x_0 - \alpha y_0)}{\beta x_0 - \alpha y_0 e^{-(\beta x_0 - \alpha y_0)t}}, \quad \forall t \geq 0. \qquad (3.31)
\end{aligned}$$

Substituting (4.21) in (3.28), we obtain

$$\begin{aligned}
\alpha y &= \frac{\beta x_0(\beta x_0 - \alpha y_0)}{\beta x_0 - \alpha y_0 e^{-(\beta x_0 - \alpha y_0)t}} - (\beta x_0 - \alpha y_0) \\
\Rightarrow y &= \frac{(\beta x_0 - \alpha y_0) y_0}{-\alpha y_0 + \beta x_0 e^{(\beta x_0 - \alpha y_0)t}}, \quad \forall t \geq 0.
\end{aligned}$$

***CaseI* : $\boldsymbol{\beta x_0 - \alpha y_0 > 0}$.** In this case, both $x(t)$ and $y(t)$, are strictly decreasing functions for $t \geq 0$ (as $\frac{dx}{dt}$ and $\frac{dy}{dt} < 0$) and

$$\lim_{t \to \infty} x(t) = \frac{\beta x_0 - \alpha y_0}{\beta} > 0, \quad \lim_{t \to \infty} y(t) = 0.$$

Therefore, army A wins and army B loses (fig. 3.18(a)).

***CaseII* : $\boldsymbol{\beta x_0 - \alpha y_0 < 0}$.** In this case also , both $x(t)$ and $y(t)$, are strictly decreasing functions for $t \geq 0$ and

$$\lim_{t \to \infty} x(t) = 0, \quad \lim_{t \to \infty} y(t) = \frac{\alpha y_0 - \beta x_0}{\alpha} > 0.$$

Hence, in this case, army B wins and army A loses (fig. 3.18(b)).

***CaseIII* : $\boldsymbol{\beta x_0 - \alpha y_0 = 0}$.** In this case, the initial condition problem becomes (using 3.29)

$$\frac{dx}{dt} = -\beta x^2, \quad x(0) = 0,$$

which has a unique solution

$$x(t) = \frac{x_0}{1 + \beta x_0 t}, \quad \forall t \geq 0.$$

Substituting this value of $x(t)$ in (3.28), we obtain

$$\begin{aligned}
y(t) &= \frac{\beta x_0}{\alpha(1 + \beta x_0 t)} = \frac{\alpha y_0}{\alpha(1 + \alpha y_0 t)} \text{ (since } \beta x_0 = \alpha y_0) \\
\Rightarrow y(t) &= \frac{y_0}{(1 + \alpha y_0 t)}, \quad \forall t \geq 0.
\end{aligned}$$

Therefore, when $\beta x_0 - \alpha y_0 = 0$, both $x(t)$ and $y(t)$, are strictly decreasing functions for $t \geq 0$ and we have

$$\lim_{t\to\infty} x(t) = \lim_{t\to\infty} y(t) = 0.$$

Also,

$$\frac{x(t)}{y(t)} = \frac{x_0}{y_0} = \frac{\alpha}{\beta},$$

at any time t, implying that the ratio between the number of troops of the two armies in battle remains constant in time. This is the case of a tie situation.

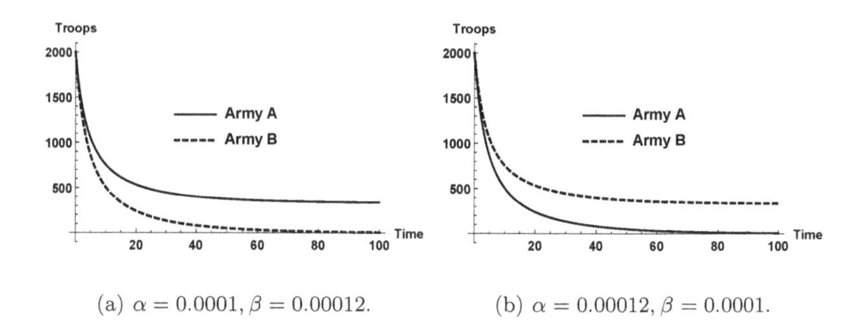

(a) $\alpha = 0.0001, \beta = 0.00012.$ (b) $\alpha = 0.00012, \beta = 0.0001.$

FIGURE 3.18: *The figures show the dynamics of a guerrilla combat model. (a) Army A wins. (b) Army B wins. Initial conditions: $x(0) = y(0) = 2000$, in both the cases.*

Note: Another effective tool to analyze the behavior of solutions in time for the given system of differential equations is given by the notion of conservative law, which states that a particular measurable property of an isolated physical system does not change as the system evolves over time. Basically, we will try to obtain such a conservation law for this Guerrilla combat model, which essentially is an expression depending on the unknowns and remaining constants along with any given solution of the system. Let us consider the function

$$L(x, y) = \beta x(t) - \alpha y(t).$$

Now,

$$\begin{aligned}
\frac{d}{dt}(\beta x - \alpha y) &= \beta\frac{dx}{dt} - \alpha\frac{dy}{dt}, \\
&= \beta(-\alpha xy) - \alpha(-\beta xy), \\
&= -\alpha\beta xy + \alpha\beta xy = 0 \\
\Rightarrow \beta x(t) - \alpha y(t) &= \text{constant} = \beta x_0 - \alpha y_0, \quad \forall t \geq 0,
\end{aligned}$$

which is the same as the solution (3.27). Therefore, the expression

$$L(x, y) = \beta x(t) - \alpha y(t),$$

is a conservation law of the system and is known as *Lanchester Linear Law* for the guerrilla combat model. Please note that there is no particular method to find the expression of $L(x, y)$. It is just a trial and error method. You need to choose $L(x, y)$ in such a manner that its derivative becomes zero.

3.5.10 Mathematical Model of Love Affair

The dynamics of love affair was introduced by Strogatz [154] with the help of a linear system, where he simply considered the simplest model of the ill-fated romance between Romeo and Juliet. Here, we consider a modified version of Strogatz's linear model given by Rinaldi [131]:

$$\frac{dM}{dt} = aM + bL + f,$$
$$\frac{dL}{dt} = cM + dL + g.$$

The variables M, L can be named after any famous lovers like Romeo and Juliet, Cleopatra and Mark Antony, Lancelot and Guinevere, Paris and Helena, and many more. We call this pair Majnun and Layla, taken from a romantic poem by Nizami of Ganje, inspired by an Arab legend.

Here, $M(t)$ is the love of Majnun for Layla and $L(t)$ is the love of Layla for Majnun. The constants a and b specify the romantic style of Manjun, implying that the parameter a signifies the extent to which Majnun is encouraged by his own feeling and b signifies the extent to which Majnun is encouraged by the feelings of Layla. In a similar manner, the parameter d gives the growth of love for Layla by her own feelings and c gives the extent to which Layla is encouraged by the feelings of Majnun. f expresses how Layla appeals to Majnun and g expresses how Majnun appeals to Layla.

The equilibrium point is $\left(\frac{bg - df}{ad - bc}, \frac{ag - cf}{ad - bc}\right)$ $(ad \neq bc)$. The characteristic equation is given by

$$\begin{vmatrix} a - \lambda & b \\ c & d - \lambda \end{vmatrix} = 0 \Rightarrow \lambda^2 - (a + d)\lambda + ad - bc = 0.$$

For stability, we must have $a + d < 0, ad - bc > 0$.

Case I: $a = 0, b = 3, f = 0, c = 2, d = 0, g = 0$. The dynamics of love is given by

$$\frac{dM}{dt} = 3L \text{ and } \frac{dL}{dt} = 2M,$$

which means that the love of Majnun is encouraged by the feelings of Layla and vice-versa. The characteristic equation is $\lambda^2 - 6 = 0 \Rightarrow \lambda = \pm\sqrt{6}$.

Therefore,

$$M(t) = c_1 e^{\sqrt{6}t} + c_2 e^{-\sqrt{6}t} \text{ and } L(t) = \frac{\sqrt{6}}{3}[c_1 e^{\sqrt{6}t} - c_2 e^{-\sqrt{6}t}].$$

As t becomes large, $e^{-\sqrt{6}t} \to 0$, implying

$$M(t) \sim c_1 e^{\sqrt{6}t} \text{ and } L(t) \sim \frac{\sqrt{6}}{3} c_1 e^{\sqrt{6}t}.$$

Thus, each of them can tell the other that *I love you exponentially* (fig. 3.19(a)).

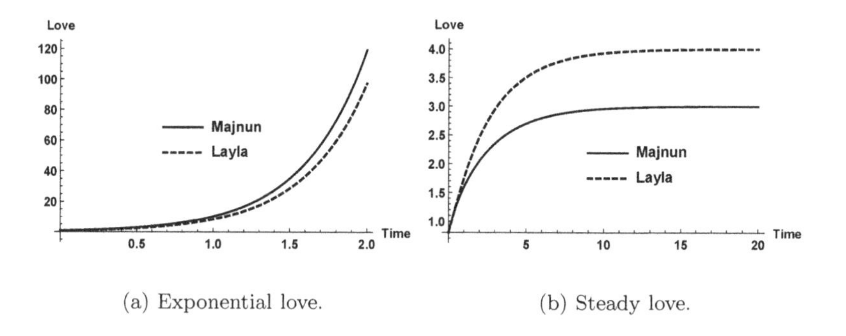

(a) Exponential love. (b) Steady love.

FIGURE 3.19: *The figures show the dynamics of love affair of a couple, namely, the exponential love and the steady love.*

Case II: $a = -2, b = 1, f = 2, c = 1, d = -1, g = 1$. The model equations are

$$\frac{dM}{dt} = -2M + L + 2 \text{ and } \frac{dL}{dt} = -L + M + 1.$$

The equilibrium solution of the model is $(3, 4)$. Also, $a + d = -3 < 0$ and $ad - bc = 2 - 1 = 1 > 0$, implying that the system is stable. The model shows a forgetting process, which gives rise to a loss of interest in the partners (because of the terms $-2M$, $-L$). This may happen due to long distance love affairs, where Majnun and Layla are staying in different cities. The return of love occurs with the love of the partner (+L and +M), while the instinct is sensitive only to the partner's appeal (physical, financial, intellectual, etc.). Thus, the model describes the romantic relationship of cautious lovers (a most common type of people) with a positive appeal ($f = 2, g = 1$), which are capable of establishing a steady love relationship [162]. In the beginning, the relationship starts with complete indifference, then love grows continuously until a plateau is reached (fig. 3.19(b)).

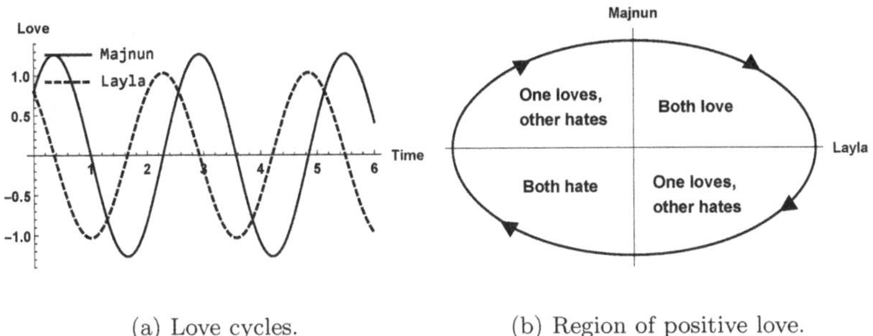

(a) Love cycles.

(b) Region of positive love.

FIGURE 3.20: *The figures show the love–hate relationship of a couple as the graphs take both the positive and negative values.*

Case III: $a = 0, b = 3, f = 0, c = -2, d = 0, g = 0$. The dynamics of love is given by

$$\frac{dM}{dt} = 3L \text{ and } \frac{dL}{dt} = -2M,$$

which means that the love of Majnun is encouraged by the feelings of Layla, he warms up when she loves him and grows cold when she does not. On the other hand, Layla turns out to be a fickle lover [154], the more Majnun loves her, the more she runs away and hides. But, Layla finds him strangely attractive, when Majnun gets discouraged and backs off. The characteristic equation is $\lambda^2 + 6 = 0 \Rightarrow \lambda = \pm i\sqrt{6}$. Therefore,

$$M(t) = c_1 \cos(\sqrt{6}t) + c_2 \sin(\sqrt{6}t),$$

$$L(t) = \frac{\sqrt{6}}{3}\left[-c_1 \sin(\sqrt{6}t) + c_2 \cos(\sqrt{6}t)\right].$$

Assuming $M(0) = 1$ and $L(0) = 0$, we get $c_1 = 1$ and $c_2 = 0$. Hence,

$$M(t) = \cos(\sqrt{6}t) \text{ and } L(t) = -\frac{\sqrt{6}}{3}\sin(\sqrt{6}t). \tag{3.32}$$

The love functions of Majnun and Layla can take both positive and negative values (which may be stated as hate). When Majnun's love for Layla is at a maximum, Layla's love for him vanishes, as a result of which her love rate becomes negative, resulting in converting her love to hate (fig. 3.20(a)). An interesting question to look into is when $M(t) > 0$ and $L(t) > 0$, that is, what is the time duration when both Majnun and Layla will love each other?
To answer this question, we obtain the geometrical representation of (3.32) as

$$M^2(t) + \frac{3}{2}L^2(t) = 1,$$

which represents an ellipse. The relationship starts at $t = 0$, when $M(0) = 1$

and $L(0) = 0$, that is, at the top point of the ellipse. It is observed that one-quarter of the time both of them are in love, one-quarter of the time they hate each other, and the rest of the time one loves and the other hates (fig. 3.20(b)). Thus, three-quarters of the time, the couple is frustrated because of Layla's fickleness [32].

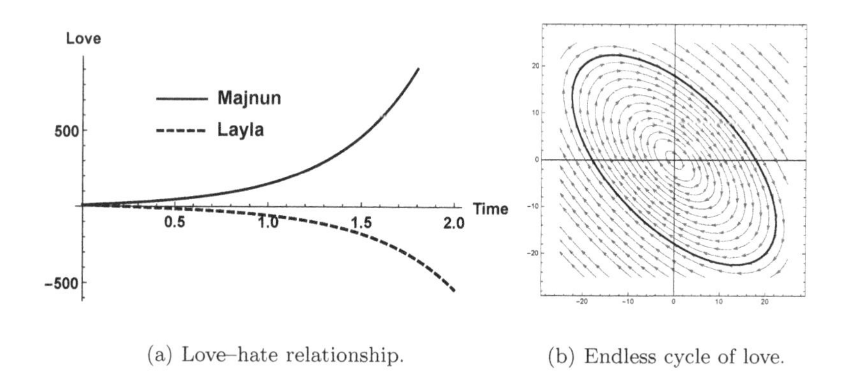

(a) Love–hate relationship. (b) Endless cycle of love.

FIGURE 3.21: *The figures show the dynamics of fire–ice relationship as well as endless cycle of love, of a couple.*

Case IV: $a = 3, b = 2, c = -2, d = -3, f = 0, g = 0$. In this case, the model equations are

$$\frac{dM}{dt} = 3M + 2L \text{ and } \frac{dL}{dt} = -2M - 3L.$$

Here, $a > 0, b > 0$, that is, Majnun is encouraged by his own feelings as well as Layla's (Eager beaver); however, Layla retreats from her own feelings as well as Majnun's (Hermit). The characteristic equation is

$$\begin{vmatrix} 3-\lambda & 2 \\ -2 & -3-\lambda \end{vmatrix} = 0 \Rightarrow \lambda = \pm\sqrt{5}.$$

Therefore, the system is a saddle (unstable). This means if two people are exactly opposite (fire and ice), they end up in odds, one loves but the other hates (fig. 3.21(a)). There is an endless cycle of love and hate with about a quarter of time of mutual love and a quarter of time of mutual hate [118].

Case V: $a = 3, b = 5, c = -5, d = -3, f = 0, g = 0$. In this case, the model equations are

$$\frac{dM}{dt} = 3M + 5L \text{ and } \frac{dL}{dt} = -5M - 3L.$$

In this case, $|b| > |a|$ and $\lambda = \pm\sqrt{4i}$. Therefore, the system is a centre (fig. 3.21(b)).

3.6 Bifurcations

Consider an autonomous system of the form

$$\frac{dx}{dt} = f(x, \mu), \text{ where } x \in \mathbb{R}^n \text{ and } \mu \in \mathbb{R}.$$

The word *bifurcation* was introduced by *Poincare*, a French mathematician, in the field of non-linear dynamics. He used this word to indicate qualitative changes in the behavior of systems, where one or more system parameters are varied. Mathematically, this means when the parameter μ crosses some point $\mu = \mu^*$, the phase portrait of the system for $\mu < \mu^*$ is topologically different from the phase portrait of the system for $\mu > \mu^*$. The point $\mu = \mu^*$ is called a bifurcation point at which the system undergoes a bifurcation. If the phase portrait changes its topological structure (loses topological equivalence) as a parameter is varied, we say that a bifurcation has occurred [76]. Examples include changes in the number or stability of fixed points, closed orbits, or saddle connections, as a parameter is varied.

Different types of local bifurcations will be discussed now. By local bifurcations, it is meant that the qualitative changes occur in the neighborhood of equilibria (fixed points) or periodic orbits as the system parameter passes through the critical threshold $\mu = \mu^*$. The local bifurcations of fixed points (equilibrium points) are classified into static bifurcations or dynamic bifurcations. Saddle-node bifurcation, pitchfork bifurcation, and transcritical bifurcation are examples of static bifurcations as only branches of fixed points or static solutions meet. Hopf bifurcation is an example of dynamic bifurcation.

3.6.1 Bifurcations in One-Dimension

The qualitative dynamical behavior of a one-dimensional continuous dynamical system is determined by its equilibria and their stability, so all bifurcations are associated with bifurcations of equilibria.

3.6.1.1 Saddle-Node Bifurcation

In saddle-node bifurcation, fixed points are created and destroyed. As the parameter varies, the two equilibria existing on one side of the bifurcation disappear on the other side of the bifurcation. This means that as the parameter varies, two equilibria move towards each other, coincide and are destroyed. A saddle-node bifurcation of a fixed point (equilibrium point or steady state or critical point) of the system

$$\frac{dx}{dt} = f(x, \mu), \text{ where } f : \mathbb{R} \times \mathbb{R} \to \mathbb{R},$$

is a smooth function and $\mu \in R$, occurs at (x^*, μ^*) if

(i) there is an equilibrium at $x = x^*$ for $\mu = \mu^*$, that is, $f(x^*, \mu^*) = 0$,

(ii) the Jacobian matrix $D_x f(x^*, \mu^*) = \dfrac{\partial f(x^*, \mu^*)}{\partial x} = 0$,

(iii) $\dfrac{\partial^2 f(x^*, \mu^*)}{\partial x^2} \neq 0$, (iv) $\dfrac{\partial f(x^*, \mu^*)}{\partial \mu} \neq 0$.

Example 3.6.1 Consider the system [154]

$$\frac{dx}{dt} = \mu - x^2, \text{ where } \mu \text{ is the parameter.}$$

Let $f(x, \mu) = \mu - x^2$. The system does not have any equilibrium point for $\mu < 0$ and for $\mu > 0$, it has two nontrivial equilibrium points, namely, $-\sqrt{\mu}$ and $+\sqrt{\mu}$. The Jacobian matrix (in this case, a single element $-2x$) has a single eigenvalue $\lambda = -2x$. Clearly, the equilibrium point $\sqrt{\mu}$ is a stable node $(f'(\sqrt{\mu}) = -2\sqrt{\mu} < 0)$ and $-\sqrt{\mu}$ is an unstable node $(f'(-\sqrt{\mu}) = 2\sqrt{\mu} > 0)$. Now, it is noted that at $(0, 0)$,

(i) $f(0,0) = (\mu - x^2)_{(0,0)} = 0$, (iii) $\left(\dfrac{\partial^2 f}{\partial x^2}\right)_{(0,0)} = -2 \neq 0$,

(ii) $\left(\dfrac{\partial f}{\partial x}\right)_{(0,0)} = (-2x)_{(0,0)} = 0$, (iv) $\dfrac{\partial f}{\partial \mu}_{(0,0)} = 1 \neq 0$.

As μ passes through $\mu = 0$ from positive to negative, the number of equilibrium points changes from two to zero. Hence, $\mu = 0$ is a saddle-node bifurcation point at the origin (fig. 3.22(a)).

3.6.1.2 Transcritical Bifurcation

In transcritical bifurcation, the fixed points change their stability as the bifurcation parameter is varied. The fixed points of the system exist for all parameter values and can never be destroyed. A transcritical bifurcation of a fixed point of the system

$$\frac{dx}{dt} = f(x, \mu), \text{ where } f : \mathbb{R} \times \mathbb{R} \to \mathbb{R}$$

is a smooth function and $\mu \in R$, occurs at (x^*, μ^*) if

(i) $f(x^*, \mu^*) = 0$, (ii) $\dfrac{\partial f(x^*, \mu^*)}{\partial x} = 0$,

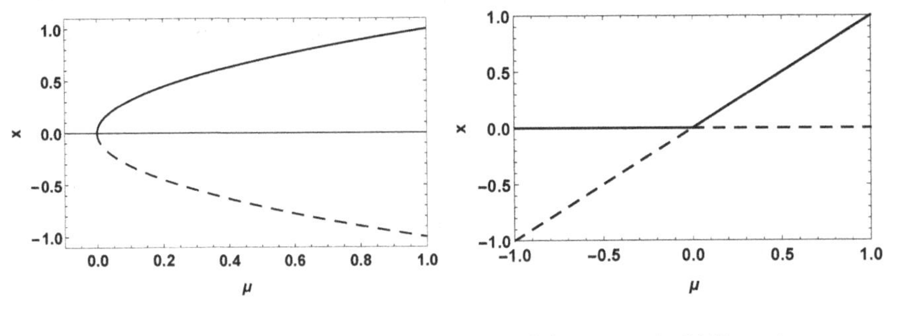

(a) Saddle-node bifurcation. (b) Transcritical bifurcation.

FIGURE 3.22: *The figures show (a) a saddle-node bifurcation as μ passes through $\mu = 0$ from positive to negative, (b) a transcritical bifurcation at the equilibrium point $(0,0)$ as the parameter μ passes through $\mu = 0$.*

$$\text{(iii)} \quad \frac{\partial f(x^*, \mu^*)}{\partial \mu} = 0, \qquad \text{(iv)} \quad \frac{\partial^2 f(x^*, \mu^*)}{\partial x^2} \neq 0, \qquad \text{(v)} \quad \frac{\partial^2 f(x^*, \mu^*)}{\partial x \partial \mu} \neq 0.$$

Example 3.6.2 Consider the system [154]

$$\frac{dx}{dt} = \mu x - x^2, \text{ where } \mu \text{ is the parameter.}$$

Let $f(x, \mu) = \mu x - x^2$. The system has two equilibrium points, namely, $x^* = 0,\ \mu$ and $f'(x) = \mu - 2x$. For $\mu < 0, x^* = 0$ is a stable equilibrium point and $x^* = \mu$ is an unstable one. As μ increases, the unstable equilibrium point approaches the origin and coincides with it when $\mu = 0$. For $\mu > 0, x^* = 0$ becomes unstable whereas $x^* = \mu$ is now stable. Thus, there is an exchange of stabilities between the equilibrium points as the parameter μ is varied. Now, it is noted that at $(x^*, \mu^*) = (0, 0)$,

$$f(0,0) = 0, \ \frac{\partial f(0,0)}{\partial x} = (\mu - 2x)|_{(0,0)} = 0, \ \frac{\partial f(0,0)}{\partial \mu} = x|_{(0,0)} = 0,$$
$$\frac{\partial^2 f(0,0)}{\partial x^2} = -2 \neq 0, \ \frac{\partial^2 f(0,0)}{\partial x \partial \mu} = 1 \neq 0.$$

Hence, a transcritical bifurcation occurs at $\mu = 0$ (fig. 3.22(b)).

3.6.1.3 Pitchfork Bifurcation

A pitchfork bifurcation is a particular type of local bifurcation (possible in dynamical systems) that has symmetry. In such cases, equilibrium points appear and disappear in symmetrical pairs. The bifurcation diagram looks like a pitchfork, hence the name pitchfork bifurcation. There are two types of pitchfork bifurcations, namely supercritical and subcritical. A pitchfork bifurcation is called supercritical if a stable solution branch bifurcates into two new stable branches as the parameter μ is increased. It is called subcritical if two unstable and one stable equilibria collapse to produce one stable one.

A pitchfork bifurcation of a fixed point of the system

$$\frac{dx}{dt} = f(x, \mu), \text{ where } f : \mathbb{R} \times \mathbb{R} \to \mathbb{R}$$

is a smooth function and $\mu \in R$, occurs at (x^*, μ^*) if

$$\text{(i) } f(x^*, \mu^*) = 0, \qquad \text{(ii) } \frac{\partial f(x^*, \mu^*)}{\partial x} = 0, \qquad \text{(iii) } \frac{\partial f(x^*, \mu^*)}{\partial \mu} = 0,$$

$$\text{(iv) } \frac{\partial^2 f(x^*, \mu^*)}{\partial x^2} = 0, \qquad \text{(v) } \frac{\partial^2 f(x^*, \mu^*)}{\partial x \partial \mu} \neq 0, \qquad \text{(vi) } \frac{\partial^3 f(x^*, \mu^*)}{\partial x^3} \neq 0.$$

The pitchfork bifurcation is supercritical if $\frac{\partial^3 f(x^*, \mu^*)}{\partial x^3} < 0$ and subcritical if $\frac{\partial^3 f(x^*, \mu^*)}{\partial x^3} > 0$.

Example 3.6.3 *Consider the dynamical system*

$$\frac{dx}{dt} = \mu x - x^3.$$

For $\mu < 0$, the system has one asymptotically stable fixed point x=0. For $\mu > 0$, the system has three fixed points, namely, $x = 0, \pm\sqrt{\mu}$, for which $\pm\sqrt{\mu}$ are stable and $x = 0$ is unstable. As μ passes from negative to positive values, the equilibrium point $x = 0$ loses its stability at $\mu = 0$ (bifurcation point), and two new stable equilibria appear. Also, at $(x^*, \mu^*) = (0, 0)$,

$$f(0,0) = 0, \quad \frac{\partial f(0,0)}{\partial x} = 0, \quad \frac{\partial f(0,0)}{\partial \mu} = 0, \quad \frac{\partial^2 f(0,0)}{\partial x^2} = 0, \quad \frac{\partial^2 f(0,0)}{\partial x \partial \mu} = 1 \neq 0.$$

Since, $\frac{\partial^3 f(0,0)}{\partial x^3} = -6 < 0$, the bifurcation is supercritical (fig. 3.23(a)). For the dynamical system $\frac{dx}{dt} = \mu x + x^3$, it can be easily shown that the pitchfork bifurcation is subcritical (fig. 3.23(b)).

3.6.2 Bifurcation in Two-Dimensions

Bifurcations of the equilibrium point in one-dimension have analogs in two-dimensions (and higher). As we switch from one-dimensional systems to two-dimensional systems, we still observe that equilibrium points can change stability or created/destroyed as parameters are varied (saddle-node, transcritical, pitchfork). In contrast to one-dimensional, however, now oscillations can be also switched on and off (Hopf).

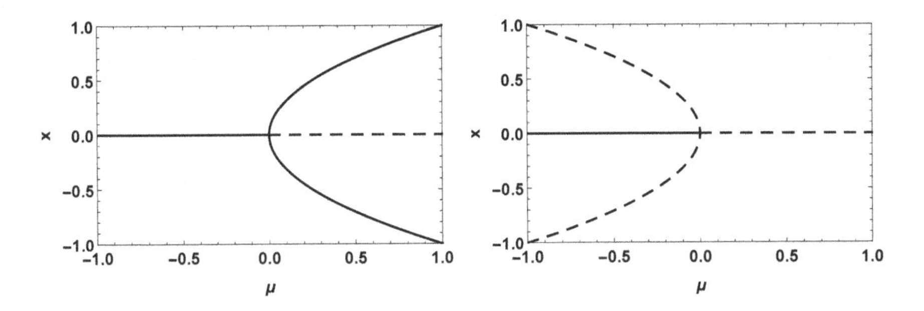

(a) Supercritical pitchfork bifurcation (b) Subcritical pitchfork bifurcation

FIGURE 3.23: *Pitchfork bifurcations at the equilibrium point* $(0, 0)$ *as the parameter* μ *passes through* $\mu = 0$.

3.6.2.1 Sotomayar's Theorem

Consider a system of the form

$$\frac{d\mathbf{x}}{dt} = \mathbf{f}(x, \mu), \text{ where } \mathbf{x} \in \mathbb{R}^n \text{ and } \mu \in \mathbb{R}.$$

Let $A_{n \times n}$ has a simple eigenvalue $\lambda = 0$ with eigenvector $\mathbf{v}$ and that A^T has an eigenvector $\mathbf{w}$ corresponding to the eigenvalue $\lambda = 0$. Furthermore, suppose that A has k eigenvalues with negative real part and $(n - k - 1)$ eigenvalues with positive real part. Then [123],

(a) the system experiences a saddle-node bifurcation at the equilibrium point $x = x^*$ as the parameter μ passes through the bifurcation values $\mu = \mu^*$ if
(i) $\mathbf{w}^T \mathbf{f}_\mu(x^*, \mu^*) \neq 0$, (ii) $\mathbf{w}^T [D^2\mathbf{f}(x^*, \mu^*)(\mathbf{v}, \mathbf{v})] \neq 0$.

(b) The system experiences a transcritical bifurcation at the equilibrium point $x = x^*$ as the parameter μ passes through the bifurcation values $\mu = \mu^*$if
(i) $\mathbf{w}^T \mathbf{f}_\mu(x^*, \mu^*) = 0$, (ii) $\mathbf{w}^T [D\mathbf{f}_\mu(x^*, \mu^*)\mathbf{v}] \neq 0$,
(iii) $\mathbf{w}^T [D^2\mathbf{f}(x^*, \mu^*)(\mathbf{v}, \mathbf{v})] \neq 0$.

(c) The system experiences a pitchfork bifurcation at the equilibrium point $x = x^*$ as the parameter μ passes through the bifurcation values $\mu = \mu^*$ if
(i) $\mathbf{w}^T \mathbf{f}_\mu(x^*, \mu^*) = 0$,
(ii) $\mathbf{w}^T [D\mathbf{f}_\mu(x^*, \mu^*)\mathbf{v}] \neq 0$,
(iii) $\mathbf{w}^T [D^2\mathbf{f}(x^*, \mu^*)(\mathbf{v}, \mathbf{v})] = 0$,
(iv) $\mathbf{w}^T [D^3\mathbf{f}(x^*, \mu^*)(\mathbf{v}, \mathbf{v}, \mathbf{v})] \neq 0$.

3.6.2.2 Saddle-Node Bifurcation

The saddle-node bifurcation is the basic mechanism for creation and destruction of fixed points, which shows that nothing really new happens when more dimensions are added. All the action is confined to a one-dimensional subspace along which the bifurcations occur, while in the extra dimensions, the flow is either simple attraction or repulsion from that subspace [76].

Example 3.6.4 Consider a two-dimensional dynamical system given by

$$\begin{aligned}\frac{dx}{dt} &= \mu - x^2,\\ \frac{dy}{dt} &= -y.\end{aligned}$$

Let $f_1(x,y,\mu) = \mu - x^2$ and $f_2(x,y,\mu) = -y$. For $\mu > 0$, the system has two fixed points (equilibrium points), namely, $(\pm\sqrt{\mu}, 0)$. The Jacobian matrix is given by

$$D\mathbf{f} = \begin{pmatrix} \frac{\partial f_1}{\partial x} & \frac{\partial f_1}{\partial y} \\ \frac{\partial f_2}{\partial x} & \frac{\partial f_2}{\partial y} \end{pmatrix} = \begin{pmatrix} -2x & 0 \\ 0 & -1 \end{pmatrix}.$$

It has a zero eigenvalue at $\mu = 0$, the other eigenvalue is -1. Clearly, the fixed point $(-\sqrt{\mu}, 0)$ is a stable node and $(\sqrt{\mu}, 0)$ is a saddle. As μ passes through $\mu = 0$, from positive to negative, the number of equilibrium points changes from two to zero. Now,

$$A = D\mathbf{f}(0,0) = \begin{pmatrix} 0 & 0 \\ 0 & -1 \end{pmatrix}, A^T = \begin{pmatrix} 0 & 0 \\ 0 & -1 \end{pmatrix}, \text{ and } \mathbf{f}_\mu = \begin{pmatrix} \frac{\partial f_1}{\partial \mu} \\ \frac{\partial f_2}{\partial \mu} \end{pmatrix} = \begin{pmatrix} 1 \\ 0 \end{pmatrix}.$$

Let $\mathbf{v} = \mathbf{w} = (1,0)^T$, be the respective eigen vectors of A and A^T then,

$$\mathbf{w}^T \mathbf{f}_\mu(0,0) = \begin{pmatrix} 1 & 0 \end{pmatrix} \begin{pmatrix} 1 \\ 0 \end{pmatrix} = 1 \neq 0 \text{ and}$$

$$\mathbf{w}^T [D^2\mathbf{f}(0,0)(\mathbf{v},\mathbf{v})] = \begin{pmatrix} 1 & 0 \end{pmatrix} \begin{pmatrix} -2 \\ 0 \end{pmatrix} = -2 \neq 0, \text{ where}$$

$$D^2\mathbf{f}(x_0)(\mathbf{v},\mathbf{v}) = \begin{pmatrix} \frac{\partial^2 f_1(x_0)}{\partial x^2}\mathbf{v}_1\mathbf{v}_1 + \frac{\partial^2 f_1(x_0)}{\partial x \partial y}\mathbf{v}_1\mathbf{v}_2 + \frac{\partial^2 f_1(x_0)}{\partial x \partial y}\mathbf{v}_2\mathbf{v}_1 + \frac{\partial^2 f_1(x_0)}{\partial y^2}\mathbf{v}_2\mathbf{v}_2 \\ \frac{\partial^2 f_2(x_0)}{\partial x^2}\mathbf{v}_1\mathbf{v}_1 + \frac{\partial^2 f_2(x_0)}{\partial x \partial y}\mathbf{v}_1\mathbf{v}_2 + \frac{\partial^2 f_2(x_0)}{\partial x \partial y}\mathbf{v}_2\mathbf{v}_1 + \frac{\partial^2 f_2(x_0)}{\partial y^2}\mathbf{v}_2\mathbf{v}_2 \end{pmatrix}.$$

Therefore, by Sotomayor's theorem [123], the system experiences a saddle-node bifurcation at the equilibrium point $(0,0)$ as the parameter μ passes through $\mu = 0$ (fig. 3.22(a)).

3.6.2.3 Transcritical Bifurcation

In transcritical bifurcation, the fixed points change their stability as the bifurcation parameter is varied. The fixed points of the system exist for all parameter values and can never be destroyed. It just interchanges its stability with another fixed point.

Example 3.6.5 Consider a system given by [154]

$$\begin{aligned}\frac{dx}{dt} &= \mu x - x^2,\\ \frac{dy}{dt} &= -y.\end{aligned}$$

Let $f_1(x, y, \mu) = \mu x - x^2$ and $f_2(x, y, \mu) = -y$. The system has two fixed points, namely, $(0, 0)$ and $(\mu, 0)$. For $\mu < 0$, the equilibrium point $(\mu, 0)$ is unstable and $(0, 0)$ is stable with increasing μ. The unstable fixed point $(\mu, 0)$ approaches the origin and coalesces with it when $\mu = 0$. And, when $\mu > 0$, the fixed point $(\mu, 0)$ becomes stable and the origin becomes unstable. The Jacobian matrix is given by

$$D\mathbf{f} = \begin{pmatrix} \frac{\partial f_1}{\partial x} & \frac{\partial f_1}{\partial y} \\ \frac{\partial f_2}{\partial x} & \frac{\partial f_2}{\partial y} \end{pmatrix} = \begin{pmatrix} \mu - 2x & 0 \\ 0 & -1 \end{pmatrix}. \text{ Now,}$$

$$A = D\mathbf{f}(0,0) = \begin{pmatrix} \mu & 0 \\ 0 & -1 \end{pmatrix}, A^T = \begin{pmatrix} \mu & 0 \\ 0 & -1 \end{pmatrix}, \ \mathbf{f}_\mu = \begin{pmatrix} \frac{\partial f_1}{\partial \mu} \\ \frac{\partial f_2}{\partial \mu} \end{pmatrix} = \begin{pmatrix} x \\ 0 \end{pmatrix}, \text{ and}$$

$$D\mathbf{f}_\mu(0,0)\mathbf{v} = \begin{pmatrix} \frac{\partial^2 f_1}{\partial x \partial \mu} & \frac{\partial^2 f_1}{\partial y \partial \mu} \\ \frac{\partial^2 f_2}{\partial x \partial \mu} & \frac{\partial^2 f_2}{\partial y \partial \mu} \end{pmatrix} \begin{pmatrix} v_1 \\ v_2 \end{pmatrix} = \begin{pmatrix} 1 & 0 \\ 0 & 0 \end{pmatrix} \begin{pmatrix} 1 \\ 0 \end{pmatrix} = \begin{pmatrix} 1 \\ 0 \end{pmatrix}.$$

Clearly, A has a simple eigenvalue at $\mu = 0$ and let $\mathbf{v}=(1, 0)^T$ and $\mathbf{w}=(1, 0)^T$ be the eigenvectors of A and A^T respectively, corresponding to the eigenvalue $\lambda = 0$. Then,

$$\mathbf{w}^T \mathbf{f}_\mu(0,0) = \begin{pmatrix} 1 & 0 \end{pmatrix} \begin{pmatrix} 0 \\ 0 \end{pmatrix} = 0, \ \mathbf{w}^T [D\mathbf{f}_\mu(0,0)\mathbf{v}] = \begin{pmatrix} 1 & 0 \end{pmatrix} \begin{pmatrix} 1 \\ 0 \end{pmatrix} = 1 \neq 0, \text{ and}$$

$$\mathbf{w}^T [D^2\mathbf{f}(0,0)(\mathbf{v},\mathbf{v})] = \begin{pmatrix} 1 & 0 \end{pmatrix} \begin{pmatrix} -2 \\ 0 \end{pmatrix} = -2 \neq 0.$$

Therefore, by Sotomayor's theorem [123], the system experiences a transcritical bifurcation at the equilibrium point $(0, 0)$ as the parameter μ passes through $\mu = 0$ (fig. 3.22(b)).

3.6.2.4 Pitchfork Bifurcation

Pitchfork bifurcations occur in systems with symmetry. There are two classes of pitchfork: supercritical (the bifurcating pitchfork branches are stable fixed points) and subcritical (the bifurcating pitchfork branches are unstable fixed points).

Example 3.6.6 Consider the dynamical system [154]

$$\begin{aligned}\frac{dx}{dt} &= \mu x - x^3,\\ \frac{dy}{dt} &= -y.\end{aligned}$$

Let $f_1(x, y, \mu) = \mu x - x^3$ and $f_2(x, y, \mu) = -y$. There are three fixed points, namely, $(0, 0)$, $(\pm\sqrt{\mu}, 0)$.

For $\mu < 0$, the origin is the only equilibrium point and is stable. For $\mu > 0$, the origin $(0, 0)$ becomes unstable. Therefore, there is a change of stability of the trivial fixed point as μ passes through $\mu = 0$. For $\mu > 0$, two new stable equilibrium points appear on either side of the origin, points symmetrically located at $(-\sqrt{\mu}, 0)$ and $(+\sqrt{\mu}, 0)$. The Jacobian matrix

$$D\mathbf{f} = \begin{pmatrix} \frac{\partial f_1}{\partial x} & \frac{\partial f_1}{\partial y} \\ \frac{\partial f_2}{\partial x} & \frac{\partial f_2}{\partial y} \end{pmatrix} = \begin{pmatrix} \mu - 3x^2 & 0 \\ 0 & -1 \end{pmatrix} \text{ has the eigenvalue } \lambda = \mu \text{ at } (0,0)$$

and $\lambda = -2\mu$ at $(\pm\sqrt{\mu}, 0)$. Now,

$$A = D\mathbf{f}(0,0) = \begin{pmatrix} \mu & 0 \\ 0 & -1 \end{pmatrix}, A^T = \begin{pmatrix} \mu & 0 \\ 0 & -1 \end{pmatrix}, \text{ and } \mathbf{f}_\mu = \begin{pmatrix} \frac{\partial f_1}{\partial \mu} \\ \frac{\partial f_2}{\partial \mu} \end{pmatrix} = \begin{pmatrix} x \\ 0 \end{pmatrix}.$$

Clearly, A has a simple eigenvalue at $\mu = 0$ and let $\mathbf{v}=(1, 0)^T$ and $\mathbf{w}=(1, 0)^T$ be the eigenvectors of A and A^T respectively, corresponding to the eigenvalue $\lambda = 0$ (since, $\mu = 0$). Then,

$$\mathbf{w}^T\mathbf{f}_\mu(0,0) = \begin{pmatrix} 1 & 0 \end{pmatrix}\begin{pmatrix} 0 \\ 0 \end{pmatrix} = 0, \ \mathbf{w}^T[D\mathbf{f}_\mu(0,0)\mathbf{v}] = \begin{pmatrix} 1 & 0 \end{pmatrix}\begin{pmatrix} 1 \\ 0 \end{pmatrix} = 1 \neq 0,$$

$$\mathbf{w}^T[D^2\mathbf{f}(0,0)(\mathbf{v},\mathbf{v})] = \begin{pmatrix} 1 & 0 \end{pmatrix}\begin{pmatrix} 0 \\ 0 \end{pmatrix} = 0, \text{ and}$$

$$\mathbf{w}^T[D^3\mathbf{f}(0,0)(\mathbf{v},\mathbf{v},\mathbf{v})] = \begin{pmatrix} 1 & 0 \end{pmatrix}\begin{pmatrix} -6 \\ 0 \end{pmatrix} = -6 \neq 0.$$

Therefore, by Sotomayor's theorem [123], the system experiences a pitchfork bifurcation at the equilibrium point $(0, 0)$ as the parameter μ passes through $\mu = 0$. This is an example of a supercritical pitchfork bifurcation (fig. 3.23(a)).

The term "pitchfork" is due to the fact that the bifurcating non-trivial branches have the geometry of a pitchfork at $(0, 0)$. The characteristic of a supercritical pitchfork bifurcation is that there is a branch of stable equilibrium points (locally) on one side of the bifurcation point ($\mu = 0$, in this case) and two branches of stable equilibrium points and a branch of unstable equilibrium points on the other side of the bifurcation point.

Note: Readers may look into the system $\dot{x} = \mu x + x^3, \dot{y} = -y$ for subcritical pitchfork bifurcation. The characteristic of subcritical pitchfork bifurcation is that there are two branches of unstable equilibrium points and a branch of stable equilibrium points on one side of the bifurcation point and a branch of unstable equilibrium points on the other side of the bifurcation point (fig. 3.23(b)).

3.6.2.5 Hopf Bifurcation

We consider a two-dimensional system of the form

$$\frac{dx}{dt} = f(x, y, \mu), \quad \frac{dy}{dt} = g(x, y, \mu), \tag{3.33}$$

where μ is a parameter. Before describing Hopf bifurcation, it is important to know about the limit cycle.

Limit Cycle: A limit cycle is an isolated closed trajectory. By "isolated", we mean that the neighboring trajectories are not closed. Mathematically, we define the limit cycle as a closed path C of the system (3.33), which is approached spirally from either inside or the outside by a non-closed path C_1 of the system (3.33), either as $t \to +\infty$ or as $t \to -\infty$ [136]. The limit cycle is called stable or unstable according as the nearby curve spirals towards C or away from C. Scientifically, stable limit cycles are very important as they model systems which oscillate even in the absence of an external driving force (self-sustained oscillations, for example, heart beat, daily rhythms in body temperature, hormone secretion, etc.). If the system is slightly perturbed, it always returns to the stable limit cycle. The limit cycle occurs only in non-linear systems.

The system (3.33) has a fixed point (x^*, y^*), which may depend on μ. Let the eigenvalues of the linearized system about this fixed point be given by

$$\lambda(\mu) = \alpha(\mu) + i\beta(\mu) \text{ and } \bar{\lambda}(\mu) = \alpha(\mu) - i\beta(\mu).$$

A Hopf bifurcation is a local bifurcation, which occurs when a complex conjugate pair of eigenvalues $\lambda_{\pm}$ of the linearized system about the equilibrium point crosses from $Re(\lambda_{\pm}) < 0$ (the left half-plane) to the $Re(\lambda_{\pm}) > 0$ (the right half-plane) and a limit cycle emerges from the fixed point. If the limit cycle that emerges is stable, we call it a supercritical Hopf bifurcation and the emergence of an unstable limit cycle is called a subcritical Hopf bifurcation. Thus, a Hopf bifurcation occurs when we have a change in stability of a fixed point from one type of focus to another. Along with the change in stability, a limit cycle emerges in the phase plane.

Hopf Bifurcation Theorem: A Hopf bifurcation of the fixed point of the two-dimensional system occurs at some critical value of the parameter, $\mu = \mu_c$ (say), if the following conditions are satisfied:

(i) $f(x^*, y^*, \mu_c) = 0, \; g(x^*, y^*, \mu_c) = 0,$

(ii) The Jacobian matrix $\begin{pmatrix} \frac{\partial f}{\partial x} & \frac{\partial f}{\partial y} \\ \frac{\partial g}{\partial x} & \frac{\partial g}{\partial y} \end{pmatrix}_{(x^*, y^*)}$ has a pair of purely imaginary eigenvalues $\pm i\beta$ at (x^*, y^*, μ_c), that is, $\alpha(\mu_c) = 0, \beta(\mu_c) = \beta \neq 0,$

(iii) $\dfrac{d\alpha(\mu)}{d\mu} \neq 0$ at $\mu = \mu_c$.

To find whether the Hopf bifurcation is supercritical or subcritical, we calculate [30]

$$s = \frac{3\pi}{4\beta}\left[f_{xxx} + f_{xyy} + g_{xxy} + g_{yyy}\right] +$$
$$\frac{3\pi}{4\beta^2}\left[f_{xy}(f_{xx} + f_{yy}) + g_{xy}(g_{xx} + g_{yy}) + f_{xx}g_{xx} - f_{yy}g_{yy}\right],$$

at the steady state (x^*, y^*) when $\mu = \mu_c$.

(i) If $s < 0$, the Hopf bifurcation is supercritical and a stable limit cycle emerges for $\mu > \mu_c$.

(ii) If $s > 0$, the Hopf bifurcation is subcritical and an unstable limit cycle emerges for $\mu > \mu_c$.

(iii) If $s = 0$, no definite conclusions can be obtained, further investigations are necessary.

Example 3.6.7 Consider the system

$$\frac{dx}{dt} = \mu x - y + \left(x + \frac{3}{2}y\right)(x^2 + y^2),$$
$$\frac{dy}{dt} = x + \mu y + \left(-\frac{3}{2}x + y\right)(x^2 + y^2).$$

Clearly, the origin $(0, 0)$ is the equilibrium point. The linearization about the origin gives

$$\frac{dx}{dt} = \mu x - y, \quad \frac{dy}{dt} = x + \mu y,$$

whose Jacobian matrix is $\begin{pmatrix} \mu & -1 \\ 1 & \mu \end{pmatrix}$. The eigenvalues are $\mu \pm i$. Thus, we get a stable focus if $\mu < 0$ and an unstable one if $\mu > 0$. Also, $\frac{d(\mu)}{d\mu} = 1\ (\neq 0)$ at $\mu = 0$. The dynamics of the system change from a stable focus to an unstable one, and the real part of the eigenvalues changes from negative to positive as μ passes through $\mu = 0$, which is the bifurcation point. It is to be noted that all the three conditions for Hopf bifurcation are satisfied, namely,
(i) $f(0, 0, 0) = 0, g(0, 0, 0) = 0$,
(ii) the Jacobian matrix has purely imaginary eigenvalues $\pm i$ at $\mu = 0$.
(iii) $\frac{d(\mu)}{d\mu} = 1 \neq 0$ at $\mu = 0$.
Hence, we conclude that the system undergoes Hopf bifurcation at the point $\mu = 0$.

To find if the Hopf bifurcation is supercritical or subcritical, we calculate

$$s = \frac{3\pi}{4\beta}\left[f_{xxx} + f_{xyy} + g_{xxy} + g_{yyy}\right] +$$
$$\frac{3\pi}{4\beta^2}\left[f_{xy}(f_{xx} + f_{yy}) + g_{xy}(g_{xx} + g_{yy}) + f_{xx}g_{xx} - f_{yy}g_{yy}\right].$$

at the steady state $(0,0)$ when $\mu = 0$. Here, $\beta = 1$, $f_{xxx}(0,0) = 6$, $f_{xyy}(0,0) = 2$, $g_{xxy}(0,0) = 2$, $g_{yyy}(0,0) = 6$, $f_{xx}(0,0) = f_{xy}(0,0) = f_{yy}(0,0) = 0$, $g_{xx}(0,0) = g_{xy}(0,0) = g_{yy}(0,0) = 0$. Hence,

$$\begin{aligned} s &= \frac{3\pi}{4\beta}\left[f_{xxx}(0,0) + f_{xyy}(0,0) + g_{xxy}(0,0) + g_{yyy}(0,0)\right] + 0 \\ &= \frac{3\pi}{4\beta}\left[6 + 2 + 2 + 6\right] = 12\pi > 0. \end{aligned}$$

Therefore, the Hopf bifurcation is subcritical (fig. 3.24(a)) and an unstable limit cycle emerges for $\mu > 0$ (fig. 3.24(b), fig. 3.24(c), fig. 3.24(d)).

3.7 Estimation of Model Parameters

When a model is formulated on the basis of various logical assumptions, it contains parameters which need to be estimated. In this section, we will discuss about a process to estimate model parameters. A proper model needs to be formulated to fit a set of given data and that requires the use of empirical evidence in the data, knowledge of the process and some trial-and-error experimentation. Wide knowledge of the behavior of different mathematical functions is quite helpful in a proper model selection. To have an initial guess of the model, it is better to plot the data. Even if you have a good idea of what the form of the regression function will be, plotting allows a preliminary check of the underlying assumptions required for the model fitting to succeed [112]. The key is to choose a simplest function that can model the data (there is no point in choosing a complex function unnecessarily). Once the basic form of the functional part of the model has been selected, we next estimate the unknown parameters in the function. In theory, there are as many different ways of estimating parameters, here we use the method of least squares.

3.7.1 Least Squares Method

The principle of Least Squares states that for the best-fitting curve, the sum of the squares of differences between the observed and the corresponding estimated values should be the minimum. The method consists of finding an analytical expression of a curve on the basis of the given set of data so that the

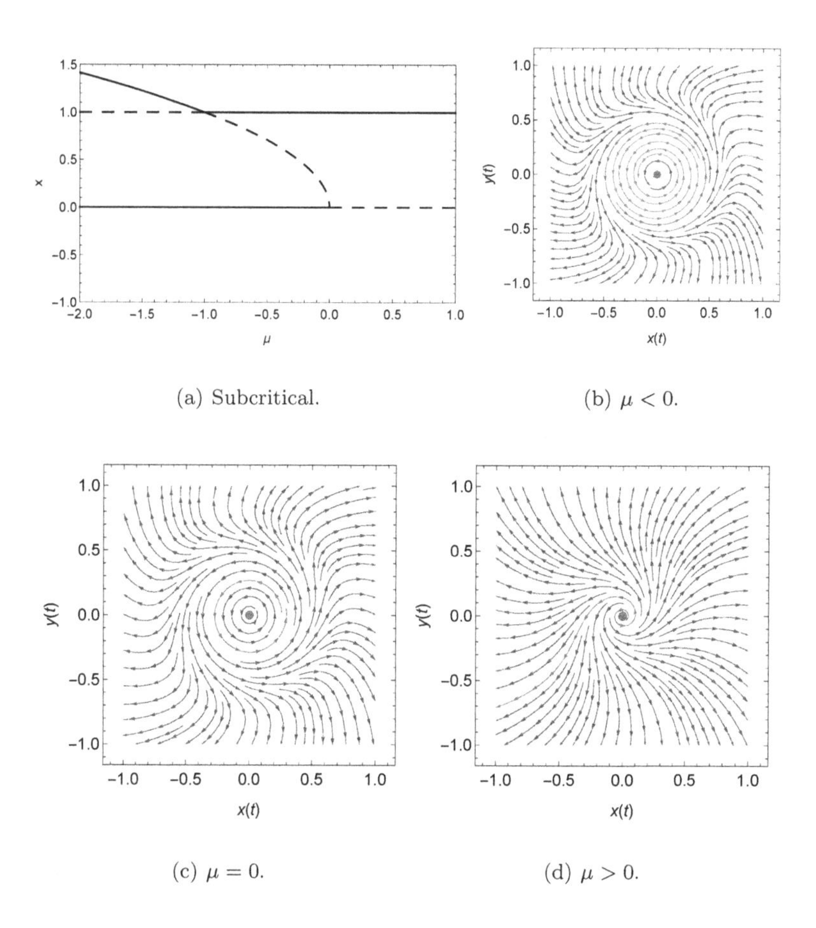

(a) Subcritical. (b) $\mu < 0$.

(c) $\mu = 0$. (d) $\mu > 0$.

FIGURE 3.24: *The system exhibits a subcritical Hopf bifurcation as the parameter μ passes through $\mu = 0$.*

curve is a **best-fit**. The term **best-fit** is interpreted in accordance with the principle of least squares, which means minimizing the sum of the squares of the deviations of the actual values from its estimated values. Mathematically, in the least square method, the parameters $\alpha_1, \alpha_2, \ldots$ in the function $\mathbf{f}(\mathbf{x}; \boldsymbol{\alpha})$ are estimated by minimizing the function (bold denotes vector quantity)

$$E = \sum_{i=1}^{n} [y_i - \mathbf{f}(\mathbf{x}; \boldsymbol{\alpha})]^2,$$

where there is a collection of n data points.

3.7.2 Fitting a Suitable Curve to the Given Data

It is desired to fit a straight line $y = A + Bx$ to the given set of n observations (data) (x_i, y_i); $(i = 1, 2, ..., n)$. The equation $y = A + Bx$ represents a family of straight lines for different values of arbitrary constants A and B. The problem is to determine A and B so that the straight line is a line of best-fit.

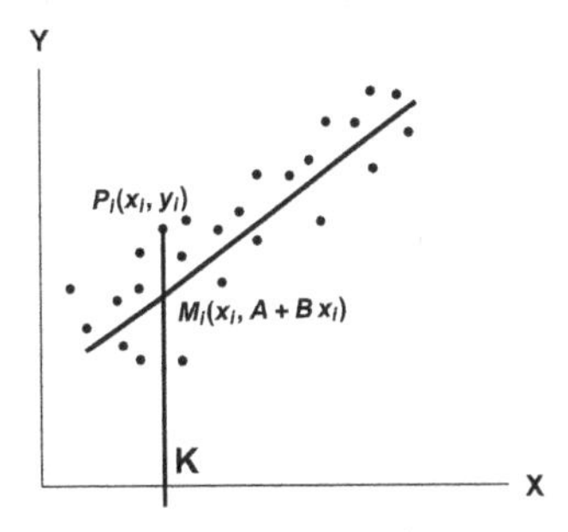

Let $P_i(x_i, y_i)$ be any arbitrary point in the scatter diagram. A perpendicular is dropped from P_i on the x-axis, which cuts the straight line $y = A + Bx$ at M_i. The x-coordinate (abscissa) of M_i is x_i and since M_i lies on the straight line $y = A + Bx$, its y-coordinate (ordinate) is $A + Bx_i$. Therefore coordinates of M_i are $(x_i, A + Bx_i)$.

The difference between the observed values and expected values, which is known as **residuals**, is given by

$$P_iM_i = P_iK - M_iK = y_i - (A + Bx_i).$$

Since these differences may be positive in some cases and negative in others, it is more convenient to make the sum of squares of these residuals, a minimum. Thus, by the principle of Least Square, A and B are to be determined so that

$$E = \sum_{i=1}^{n} (P_iM_i)^2 = \sum_{i=1}^{n} (y_i - A - Bx_i)^2 \text{ is minimum.}$$

By the principle of maxima and minima, the partial derivatives of E with respect to A and B should vanish separately, that is,

$$\frac{\partial E}{\partial A} = 0 \Rightarrow -2\sum_{i=1}^{n}(y_i - A - Bx_i) = 0 \Rightarrow \sum_{i=1}^{n} y_i = An + B\sum_{i=1}^{n} x_i, \qquad (3.34)$$

and

$$\frac{\partial E}{\partial B} = 0 \Rightarrow -2\sum_{i=1}^{n} x_i(y_i - A - Bx_i) = 0 \Rightarrow \sum_{i=1}^{n} x_iy_i = A\sum_{i=1}^{n} x_i + B\sum_{i=1}^{n} x_i^2. \qquad (3.35)$$

Equations (3.34) and (3.35) so obtained are called **normal equations.**

The values $\sum\limits_{n=1}^{n} x_i$, $\sum\limits_{n=1}^{n} y_i$, $\sum\limits_{n=1}^{n} x_i^2$, $\sum\limits_{n=1}^{n} x_iy_i$ can be easily calculated from the given set of observations. Substituting these values in the normal equations, the values of A and B are obtained. With these values of A and B, the straight line $y = A + Bx$ is the line of best-fit to the given set of observations.

In a similar manner, parameters of any suitable function can be estimated using the least square method to fit a given data. In practice, codes are written in any programming language (like Fortran, C, C++) or in any computing system (like MATHEMATICA, MATLAB) to estimate the parameters.

Example 3.7.1 Consider the data set, data1 $= \{(12, 5.3), (18, 5.7), (24, 6.3), (30, 7.2), (36, 8.0), (42, 8.7), (48, 8.9)\}$. We need to find the curve of best-fit for this data set. We first plot the points to get a rough idea of the function we need to choose. The plot shows a linear path (straight line), and hence we choose the function $f(x) = a + bx$, where the parameters a and b are to be estimated.

We use MATHEMATICA's LinearModelFit or FindFit functions to estimate the values of the parameter, such that the straight line is the line of best-fit. These functions have in-built programs which minimizes the sum of square of the residuals to estimate the parameters.

Mathematica Code (Using LinearModelFit):

```
data1 = {{12,5.3},{18,5.7},{24,6.3},{30,7.2},{36,8.0},{42,8.7},{48,8.9}};
l1=ListPlot[data1,PlotStyle→ {Black,PointSize[Large]}]
m1=LinearModelFit[data1,x,x]
Show[l1,Plot[m1[x],{x,0,50}]]
```

Mathematica Code (Using FindFit):

```
data1 = {{12,5.3},{18,5.7},{24,6.3},{30,7.2},{36,8.0},{42,8.7},{48,8.9}};
l1=ListPlot[data1,PlotStyle→ {Black,PointSize[Large]}]
fn[x_]:=a+b x
fit=FindFit[data1,fn[x],{a,b},x];
fn[x] /. fit
Show[l1,Plot[fn(x)/. fit,{x,0,50}],PlotRange→ All]
```

The first two lines of the code plot the give data to give a rough guess of the function to be chosen (fig. 3.25(a)). Rest of the lines execute the method of least squares through in-built functions LinearModelFit/FindFit to find the line of best-fit (fig. 3.25(b)).

Example 3.7.2 Consider the data set, data2 $= \{(-5, 48), (-4, 35), (-3, 24), (-2, 15), (-1, 8), (0, 3), (1, 0), (2, -1), (3, 0), (4, 3), (5, 8), (6, 15), (7, 24), (8, 35), (9, 48), (10, 63)\}$. We need to find the curve of best-fit for this data set. As stated before, we first plot the points to get a rough guess of the function, which shows a parabolic path. Hence, we choose the function $f(x) = a + bx + cx^2$, where the parameters a, b and, c are to be estimated.

We use MATHEMATICA's NonLinearModelFit or FindFit functions to estimate the values of the parameter, such that the parabola is the curve of best-fit.

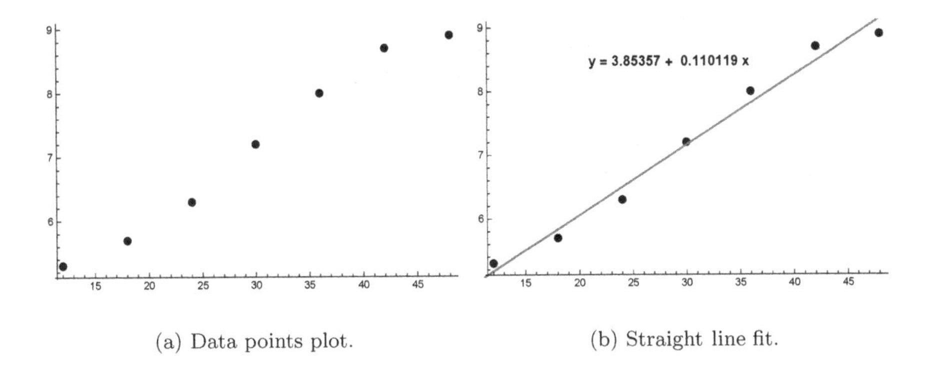

(a) Data points plot. (b) Straight line fit.

FIGURE 3.25: *The figures show (a) plot of data points for the initial guess of the function, (b) the line of best-fit,* $y = 3.85357 + 0.110119x$*, for the given set of data points.*

Mathematica Code (Using NonLinearModelFit):

```
data2 = {{-5,48},{-4,35},{-3,24},{-2,15},{-1,8},{0,3},{1,0},{2,-1},
{3,0},{4,3},{5,8},{6,15},{7,24},{8,35},{9,48},{10,63}};
l2=ListPlot[data1,PlotStyle→ {Black,PointSize[Large]}]
m2=NonLinearModelFit[data2,a + b x + c x^2,{a,b,c},x]
Show[l2,Plot[m2[x],{x,-5,10}]]
```

The first two lines of the code plot the given data to give an initial guess of the function to be chosen (fig. 3.26(a)). The rest of the lines execute the method of least squares through the in-built function NonLinearModelFit to find the line of best-fit (fig. 3.26(b)). In-built function Findfit can also be used.

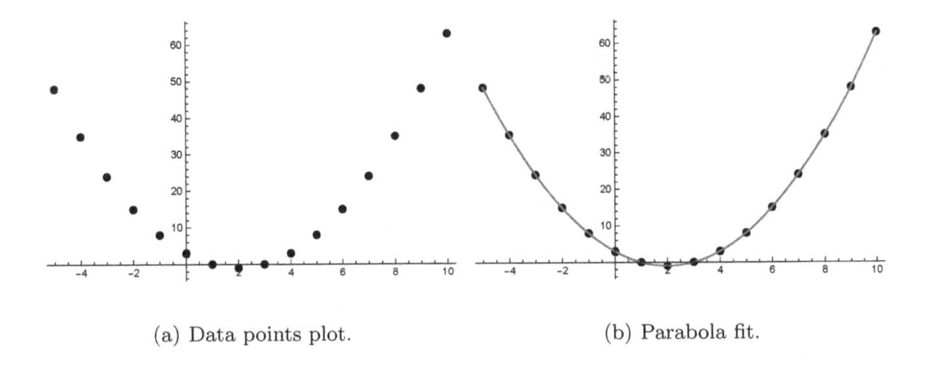

(a) Data points plot. (b) Parabola fit.

FIGURE 3.26: *The figures show (a) plot of data points for the initial guess of the function, (b) the curve of best-fit,* $y = 3 - 4.0x + 1.0x^2$*, for the given set of data points.*

Example 3.7.3 Consider the data set, data3 = {(35., 0.001), (35.5, 0.001), (36., 0.010), (36.5, 0.044), (37.,0.111), (37.5, 0.214), (38.,0.258), (38.5, 0.205), (39.,0.111),(39.5, 0.043), (40., 0.010)}. We first plot the points to get an initial guess of the function, which looks like a bell curve or normal curve (probability density function of Normal Distribution). Hence, we choose the function of the form $f(x) = ae^{-(x-b)^2}$, where the parameters a and, b are to be estimated.

Mathematica Code (Using NonLinearModelFit):

```
data3 = {{35.0,0.001},{35.5,0.001},{36,0.010},{36.5,0.044},{37.0,0.111},
{37.5,0.214},{38.0,0.258},{38.5,0.205},{39,0.111},{39.5,0.043},{40.,0.010}};
l3=ListPlot[data3,PlotStyle→ {Black,PointSize[Large]}]
m3=NonLinearModelFit[data3,a e^-(x-b)^2,{a,b},x]
Show[l3,Plot[m3[x],{x,5,10}]]
```

The first two lines of the code plot the give data to give an initial guess of the function to be chosen (fig. 3.27(a)). The rest of the lines execute the method of least squares through the in-built function NonLinearModelFit to find the line of best-fit (fig. 3.27(b)). However, many times, this may not work (fig. 3.27(a)). Then, we have to provide an initial guess of the parameters also. So, we modify the fourth line of the Mathematica code to get the best-fit curve (fig. 3.27(b)).

```
m3=NonLinearModelFit[data3,a e^-(x-b)^2,{{a,0.9},{b,45}},x].
```

Needless to say, this initial guess of the parameters is a trial and error method.

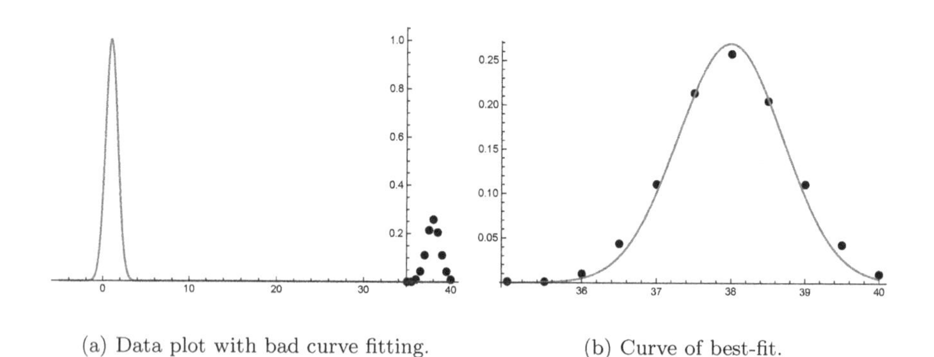

(a) Data plot with bad curve fitting.

(b) Curve of best-fit.

FIGURE 3.27: *The figures show (a) plot of data points for the initial guess of the function along with a bad-fit curve, (b) the curve of best-fit,* $y = 0.269513e^{-(x-37.988)^2}$, *for the given set of data points.*

Example 3.7.4 Consider the data set, Data = {(0, 44.5), (2, 45.3), (4, 52.6), (6, 60.4), (8, 70.2), (10,75.9), (12, 79.8), (14, 79.1), (16, 72.8), (18, 63.5), (20,52.5), (22, 44.6)}. Fit the function $A\ \sin(Bx + C) + D$ to the given data set by estimating the parameters A, B, C, D.

Solution: Mathematica code to find the estimation of the parameters A, B, C, and D are given as follows:

Mathematica Code (Using FindFit):

```
Data = {{0,44.5},{2,45.3},{4,52.6},{6,60.4},{8,70.2},{10,75.9},
{12,79.8},{14,79.1},{16,72.8},{18,63.5},{20,52.5},{22,44.6}}.
l4=ListPlot[data4,PlotStyle→ {Black,PointSize[Large]}]
fn(x_):=A Sin[B x+C]+D
fit=FindFit[data4,fn[x],{{A,2}, {B,0.5},{C,1},{D,-2}},x];
fn[x]/. fit
Show[l4,Plot[fn[x]/. fit,{x,0,25}],PlotRange→ All]
```

The estimated values are $A = 18.3506$, $B = -0.266222$, $C = 17.393$, and $D = 62.0586$. Fig3.28(a) shows the curve of best-fit for the function $A\ \sin(Bx + C) + D$.

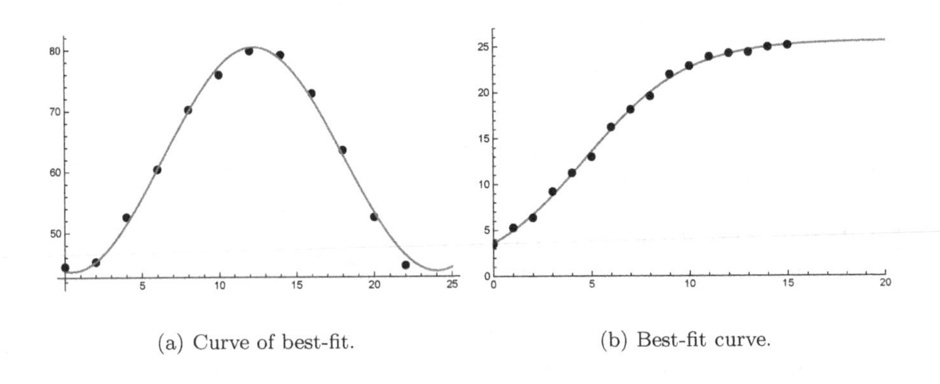

(a) Curve of best-fit. (b) Best-fit curve.

FIGURE 3.28: *The figures show (a) the curve of best-fit for the function* $18.3506 \sin(17.393 - 0.266222x) + 62.0586$, *(b) the curve of best-fit for the given set of data points, satisfying logistic growth.*

3.7.3 Parameter Estimation for ODE

In this section, we discuss the estimation of parameters in an ODE model, where we use the method of least squares directly with ODE. Consider a system of ODE of the form

$$\frac{d\mathbf{x}}{dt} = \mathbf{f}(t, \mathbf{x}; \boldsymbol{\alpha}), \quad \mathbf{x} \in \mathbb{R}^n, \ \mathbf{f} \in \mathbb{R}^n, \ \boldsymbol{\alpha} \in \mathbb{R}^m.$$

A collection of k data points is given by $(t_1, \mathbf{x_1}), (t_2, \mathbf{x_2}), ..., (t_k, \mathbf{x_k})$. We can now determine the optimal values for the parameter $\boldsymbol{\alpha}$ by minimizing the error

$$E(\boldsymbol{\alpha}) = \sum_{j=1}^{k} |\mathbf{x_j} - \mathbf{x}(t_j; \boldsymbol{\alpha})|^2,$$

where $|\cdot|$ denotes standard Euclidean norm $\left|\begin{pmatrix} x_1 \\ x_2 \\ x_3 \end{pmatrix}\right| = \sqrt{x_1^2 + x_2^2 + x_3^2}$.

Example 3.7.5 Consider the data set, data1 = {(0,3.493), (1,5.282), (2,6.357), (3,9.201), (4,11.224), (5,12.964), (6,16.226), (7, 18.137), (8, 19.590), (9, 21.955), (10,22.862), (11, 23.869), (12, 24.243), (13, 24.344), (14, 24.919), (15,25.108)}, which shows the population (in thousands), of harbor seals in the Wadden Sea over the years 1997 to 2012 [1]. We need to fit the given data to the logistic growth equation, given by

$$\frac{dx}{dt} = rx\left(1 - \frac{x}{k}\right), \quad x(0) = a,$$

by estimating the parameters r, k, and a.

We use MATHEMATICA's FindFit or NonlinearModelFit functions to estimate the values of the parameter, such that the solution of the given differential equation is the curve of best-fit.

Mathematica Code (Using FindFit):

```
data1 = {{0, 3.493}, {1, 5.282}, {2, 6.357}, {3, 9.201}, {4, 11.224},
{5, 12.964}, {6, 16.226}, {7, 18.137}, {8, 19.590}, {9, 21.955}, {10, 22.862},
{11, 23.869}, {12, 24.243}, {13, 24.344}, {14, 24.919}, {15, 25.108}};
model = ParametricNDSolveValue[{x'[t] == r x[t](1 - x[t]/k), x[0] == a},
x, {t, 0, 20}, {r, k, a}]
fit = FindFit[data1, model[r, k, a][t], {r, k, a}, t]
Plot[model[r, k, a][t]/. fit, {t, 0, 20}, Epilog -> {Black, PointSize[Large],
Point[data1]}, PlotRange -> {{0, 20}, {0, 27}}]
```

The estimated values of the parameters are $r = 0.385215$, $k = 25.6567$, and $a = 3.60666$. Fig. 3.28(b) shows the curve of best-fit for the given set of data.

Example 3.7.6 Consider the following data sets, which shows the interaction (predator-prey type) between lynx (a type of wildcat) and hare (mammals in the same biological family as rabbits), as measured by pelts (The pelt of an animal is its skin, which can be used to make clothing or rugs) collected by the Hudson Bay Company between 1900 and 1920 [60]:
dataset1 = {(0, 30), (1, 47.2), (2, 70.2), (3, 77.4), (4,36.30), (5,20.6), (6, 18.1), (7, 21.4), (8, 22), (9, 25.4), (10, 27.1), (11,40.3), (12, 57), (13, 76.6),

(14, 52.3),(15, 19.5), (16,11.2), (17, 7.6), (18, 14.6), (19, 16.2), (20, 24.7)};
dataset2 = {(0, 4), (1, 6.1), (2, 9.8), (3, 35.2), (4,59.4), (5,41.7), (6, 19), (7, 13), (8, 8.3), (9, 9.1), (10, 7.4), (11,8), (12, 12.3), (13, 19.5), (14, 45.70), (15, 51.1), (16,29.7), (17, 15.8), (18, 9.7), (19, 10.1), (20, 8.6)}; We try to fit the given data to the famous Lotka-Volterra predator-prey model, given by

$$\frac{dx}{dt} = a_1 x - b_1 xy, \ \frac{dy}{dt} = -a_2 y + b_2 xy, \ x(0) = 30, \ y(0) = 4,$$

by estimating the parameters a_1, b_1, a_2 and b_2.

We use MATHEMATICA's FindFit function to estimate the values of the parameter, such that the solution of the given differential equations is the curves of best-fit.

Mathematica Code (Using FindFit):

```
dataset1 = {{0, 3.493}, {1, 5.282}, {2, 6.357}, {3, 9.201}, {4, 11.224},
{5, 12.964}, {6, 16.226}, {7, 18.137}, {8, 19.590}, {9, 21.955}, {10, 22.862},
{11, 23.869}, {12, 24.243}, {13, 24.344}, {14, 24.919}, {15, 25.108}};
model1 = ParametricNDSolveValue[{x'(t) = a1x(t) - b1x(t)y(t),
y'(t) = b2x(t)y(t) - a2y(t), x(0) = 30, y(0) = 4}, x, {t, 0, 20}, {a1, b1, a2, b2}]
fit1 = FindFit[dataset1, model1[a1, b1, a2, b2][t], {{a1, 0.4}, {b1, 0.05}, {a2, 1.9},
{b2, 0.05}}, t]
Plot[model1[a1, b1, a2, b2][t]/. fit1, {t, 0, 20}, Epilog → {Black, PointSize[Medium],
Point[data1]}, PlotRange → {{0, 20}, {0, 80}}]
```

The estimated values of the parameters are $a_1 = 0.535891$, $b_1 = 0.0288135$, $a_2 = 0.861979$ and $b_2 = 0.0263876$. Fig. 3.29(a) shows the curve of best-fit for the given set of Hare-data and fig. 3.29(b) shows the curve of best-fit for Lynx-data.

Note: Sometimes, it is expected to provide an initial guess of the parameters to be estimated (see line 6 of the code). This guess is totally on a trial and error basis, and very often, the code will give numerical errors if the initial guess is not correct. Modify the code accordingly to obtain fig. 3.28(b).

3.8 Chaos in Continuous Models

Chaos, in an autonomous deterministic system, is defined as an aperiodic long-term behaviour (phase space trajectories do not converge to a point or a periodic orbit) that is sensitive to initial conditions (trajectories that start nearby initially, separate exponentially fast) [154]. Thus, chaos requires these

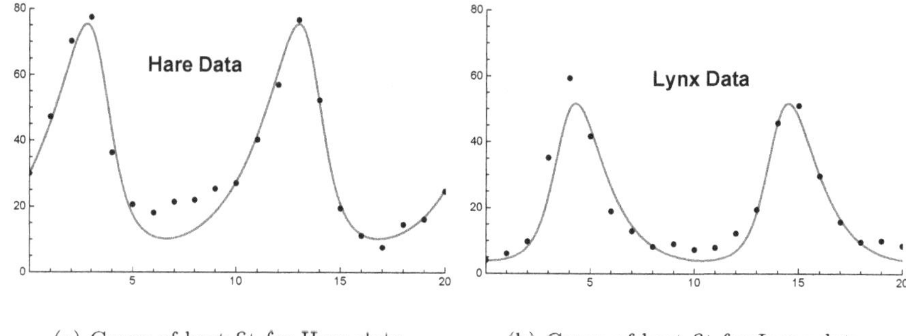

(a) Curve of best-fit for Hare data. (b) Curve of best-fit for Lynx data.

FIGURE 3.29: *The figures show the raw data match collected by the Hudson Bay Company, of the interaction of lynx and hare (as measured by pelts) with the Lotka-Volterra predator-prey model, by estimating the system parameters.*

three ingredients, namely, a periodic long-term behaviour, a deterministic system and sensitive dependence on initial conditions. The last condition implies that nearby trajectories diverge exponentially (positive Lyapunov exponent). Continuous systems in a two-dimensional phase space cannot experience such a divergence, hence chaotic behaviours can only be observed in a deterministic continuous systems with a phase space of dimension 3, at least. However, non-autonomous two-dimensional systems are equivalent to autonomous three-dimensional systems, which can exhibit chaotic behavior. Similarly, delay systems are infinite-dimensional, therefore, chaos can be observed in such systems. Discrete systems can exhibit chaotic behaviors even if they are one-dimensional.

3.8.1 Lyapunov Exponents

Lyapunov exponents measures the exponential attraction or separation in time of two adjacent orbits in the phase space with close initial conditions. The system indicates chaos if it has at least one positive Lyapunov exponent. If the largest Lyapunov exponent is negative, then the orbit converges in time and the system is insensitive to initial conditions. If it is positive, then the distance between adjacent orbits grows exponentially and the system exhibits sensitive dependence on initial conditions, and hence the system is chaotic. Mathematically, suppose x_0 and $x_0 + \delta$ be two neighbouring points at time t, where δ is a tiny separation vector of initial length. It can be shown that $\delta = \delta_0 e^{\lambda t}$. If $\lambda > 0$, neighboring trajectories separate exponentially rapid. So, positive λ indicates the sensitivity to initial conditions. Let x_0 be the initial condition chosen arbitrarily, then the Lyapunov exponents are given by [120]

$$\lambda_i(x_0) = \lim_{t \to 0} \frac{1}{t} ln|m_i(t)|, \ (i = 1, 2, .., n),$$

and m is the eigenvalues of the Jacobian matrix of the system.

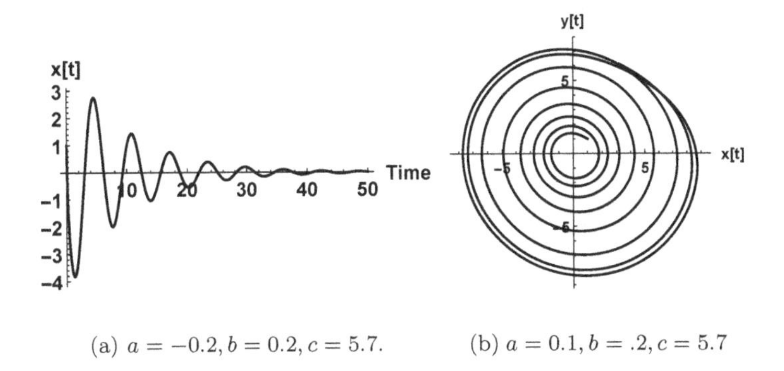

(a) $a = -0.2, b = 0.2, c = 5.7.$ (b) $a = 0.1, b = .2, c = 5.7$

FIGURE 3.30: *The figures show the dynamics of the Rössler system: (a) the system converges to the equilibrium point* $(-0.007, -0.035, 0.035)$*, (b) a unit cycle of period* 1 *is observed.*

3.8.2 Rössler Systems: Equations for Continuous Chaos

In the 1970s, Rössler systems were introduced as prototype equations with the minimum ingredients for continuous-time chaos. The three-dimensional Rössler system was originally conceived as a simple model for studying chaos. The Rössler system has only one non-linear term and can be thought as a minimal model for continuous-time chaos and as simplification of the well-known Lorenz system [99] (E. N. Lorenz was the first scientist to report continuous in a model of turbulence, under the name of deterministic non-periodic flow). In 1976, Rössler considered the system of ordinary differential equations [137]

$$\frac{dx}{dt} = -(y+z), \ \frac{dy}{dt} = x + ay, \ \frac{dz}{dt} = b + z(x-c),$$

where a, b, c are real parameters. The original values of the parameters studied by Rössler were respectively $a = 0.2$, $b = 0.2$, and $c = 5.7$, and x, y, z are three variables, which evolve with time. The linear terms in the first two equations create oscillations in the variables x and y and the expected chaotic behavior appears from the lone non-linear term (xz) of the system (third equation) [63]. Rössler system has two equilibrium points, namely,

$$P_1(\frac{c+\sqrt{c^2-4ab}}{2}, \frac{-c-\sqrt{c^2-4ab}}{2a}, \frac{c+\sqrt{c^2-4ab}}{2a}) \text{ and}$$
$$P_2(\frac{c-\sqrt{c^2-4ab}}{2}, \frac{-c+\sqrt{c^2-4ab}}{2a}, \frac{c-\sqrt{c^2-4ab}}{2a}).$$

These equilibria exist iff $c^2 - 4ab \geq 0$. The stability of these equilibrium points can be determined by obtaining eigenvalues of the Jacobian matrix

$$\begin{pmatrix} 0 & -1 & -1 \\ 1 & a & 0 \\ z & 0 & x-c \end{pmatrix},$$

which in turn can be obtained by solving the cubic equation

$$\lambda^3 - (a + x - c)\lambda^2 + (ax + 1 + z - ac)\lambda + c - x - az = 0.$$

For the parameter values, $a = 0.2, b = 0.2, c = 5.7$, the eigenvalues are $\lambda_{1,2} = 0.0971028 \pm i0.995786, \lambda_3 = -5.68718$.

The dynamics of the Rössler's model mainly depends on the parameters a, b, and c. By varying the parameters a, b, c, it is possible to make the system coverage towards a periodic orbit, equilibrium point, or escape towards infinity. We now display different dynamics of Rössler's model by varying its parameter.
(i) For $a = -0.2$, $b = 0.2$, $c = 5.7$, the system converges to the centrally located equilibrium point (fig. 3.30(a)).
(ii) For $a = 0.1$, $b = 0.2$, $c = 5.7$, we get a unit cycle of period 1 (fig. 3.30(b)).

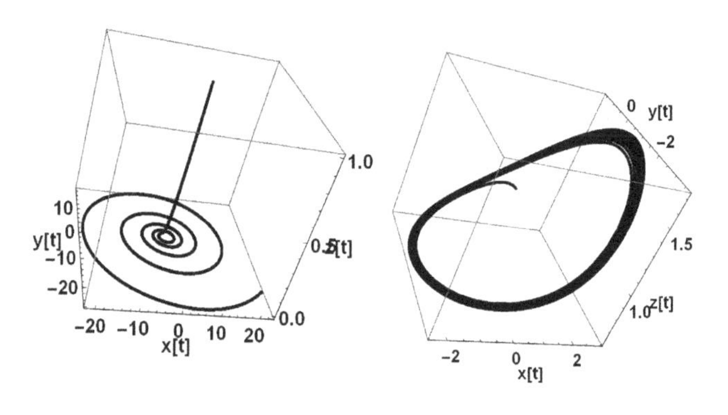

(a) $a = 0.2, b \approx 0, c = 5.7$. (b) $a = 0.2, b = 5, c = 5.7$.

FIGURE 3.31: *The figures show the dynamics of the Rössler system: (a) the attractor approaches infinity, (b) the system converges to period* 1.

(iii) For $a = 0.2$, $b = 0.00000000001$, $c = 5.7$, we observe that the attractor approaches infinity (fig. 3.31(a)) for small b (approaching zero).
(iv) For $a = 0.2$, $b = 5.0$, $c = 5.7$, the system converges to period 1 (not to a chaotic state) (fig. 3.31(b)).

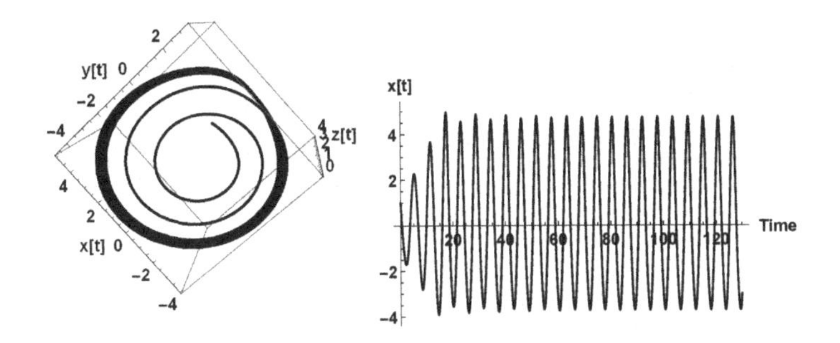

FIGURE 3.32: *The figures show limit cycle for the Rössler system for* $a = 0.2, b = 0.2, c = 2.5$.

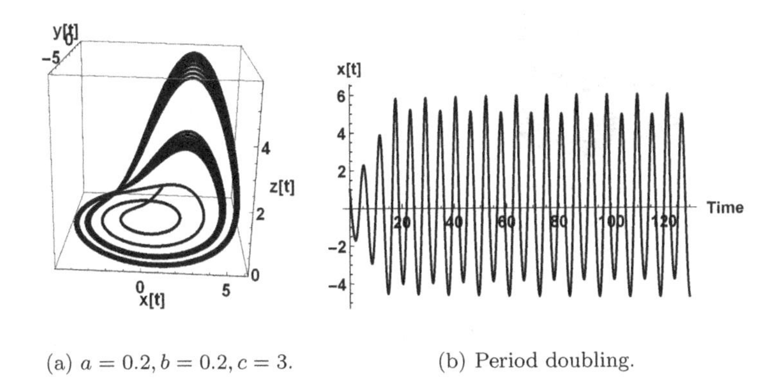

(a) $a = 0.2, b = 0.2, c = 3$. (b) Period doubling.

FIGURE 3.33: *The figures show period doubling for the Rössler system with two distinct sets of peaks and troughs.*

(v) For $a = 0.2$, $b = 0.2$, $c = 2.5$, we obtain a limit cycle (fig. 3.32).

(vi) For $a = 0.2, b = 0.2, c = 3.0$, the attractor undergoes a period doubling in which the attractor consists of two loops which are both traversed before the motion is repeated (fig. 3.33(a)). This behavior can be clearly seen from the time series plot. There are two distinct sets of peaks and troughs (fig. 3.33(b)) and the period of the motion is approximately twice that of the limit cycle.

(vii) For $a = 0.2$, $b = 0.2$, $c = 4.0$, we have a further period doubling with four distinct sets of peaks and troughs (fig. 3.34(a)).

(viii) For a=0.2, b=0.2, c=5.7, we get chaotic behavior (fig. 3.34(b), fig. 3.34(c)). The Lyapunov exponent is positive in this case (fig. 3.34(d)).

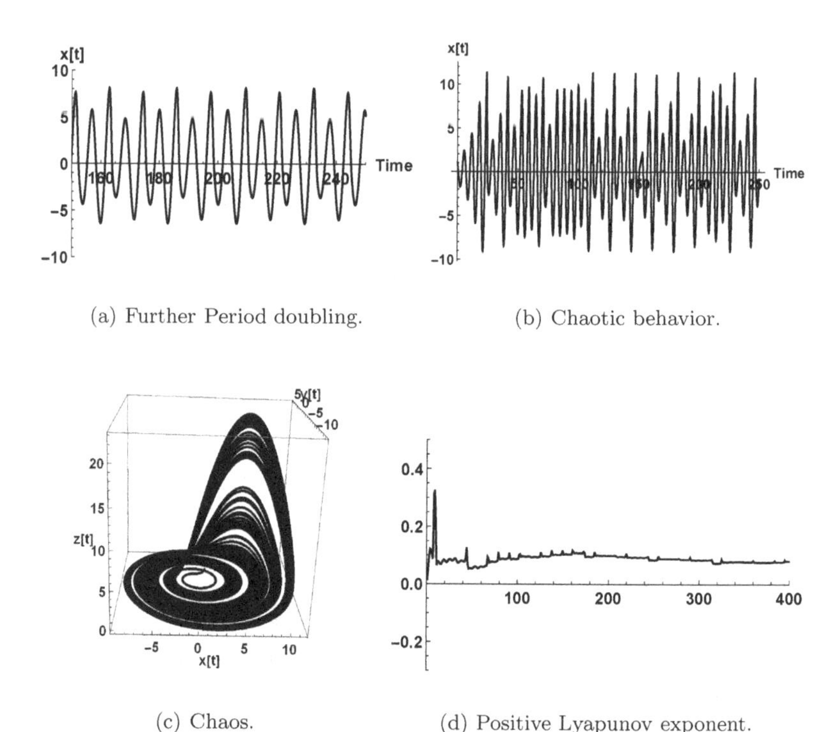

(a) Further Period doubling. (b) Chaotic behavior.

(c) Chaos. (d) Positive Lyapunov exponent.

FIGURE 3.34: *The figures show the dynamics of the Rössler system and its route to chaos for $a = 0.2$, $b = 0.2$, $c = 5.7$.*

3.9 Miscellaneous Examples

Problem 3.9.1 *The generalized Verhulst population model, with crowding effect of the form $(-\beta N^\alpha)$ is given by*

$$\frac{dN}{dt} = rN - \beta N^\alpha \qquad (r > 0, \beta > 0, \alpha > 1).$$

Obtain the equilibrium and check for stability. Also, solve the model for N and predict the behavior of the population for a long period of time.

Solution: The equilibrium solution is given by

$$\frac{dN}{dt} = 0 \Rightarrow N(r - \beta N^{\alpha-1}) = 0 \Rightarrow N^* = 0 \text{ and } N^* = \left(\frac{r}{\beta}\right)^{\frac{1}{\alpha-1}}.$$

Let $f(N) = rN - \beta N^\alpha$, then $f'(N) = r - \beta\alpha N^{\alpha-1}$.

Now, $f'(0) = r > 0 \Rightarrow$ the system is unstable about $N^* = 0$.
Also, $f'\left[\left(\frac{r}{\beta}\right)^{\frac{1}{\alpha-1}}\right] = r(1-\alpha) < 0$ (since $\alpha > 1$), implying that the system is stable about $N^* = \left(\frac{r}{\beta}\right)^{\frac{1}{\alpha-1}}$. We put $z = N^{\alpha-1}$ and we obtain

$$\frac{dz}{z(r-z\beta)} = (\alpha-1)dt \Rightarrow \left(\frac{1}{z} + \frac{\beta}{r-\beta z}\right)dz = r(\alpha-1)dt$$

Integrating, we get $\ln(k) - \ln(r-\beta z) = rt(\alpha-1) + \ln c$

$$\Rightarrow \frac{z}{r-\beta z} = ce^{rt(\alpha-1)} \Rightarrow N^{\alpha-1} = \frac{rc}{e^{-rt(\alpha-1)} + \beta c},$$

$$\Rightarrow N(t) = \left[\frac{rc}{e^{-rt(\alpha-1)} + \beta c}\right]^{\frac{1}{\alpha-1}}. \text{ As } t \to \infty, N \to \left(\frac{r}{\beta}\right)^{\frac{1}{\alpha-1}}.$$

Thus, the population attains its equilibrium value for large time (fig. 3.35(a)).

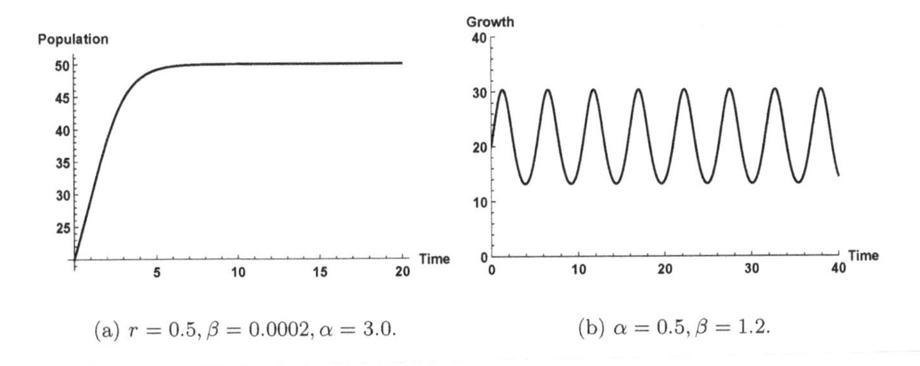

(a) $r = 0.5, \beta = 0.0002, \alpha = 3.0$.

(b) $\alpha = 0.5, \beta = 1.2$.

FIGURE 3.35: *The figures show (a) the population reaches its equilibrium value for large time, (b) sinusoidal behavior with increasing amplitude.*

Problem 3.9.2 *A seasonal growth model is given by*

$$\frac{dS}{dt} = \alpha S\ \cos(\beta t),\ \alpha, \beta \text{ are positive constants.}$$

Obtain the solution and comment on the behavior of the solution $S(t)$ of this model.

Solution: Given that

$$\frac{dS}{dt} = \alpha S \cos(\beta t) \Rightarrow \frac{dS}{S} = \alpha \cos(\beta t) dt,$$

$$\Rightarrow \ln S = \frac{\alpha}{\beta}\sin(\beta t) + \ln c \text{ (arbitrary constant)} \Rightarrow S(t) = ce^{\frac{\alpha}{\beta}\sin(\beta t)}.$$

Thus, $S(t)$ will be an exponential function, the power of which varies as a sinusoidal function with increasing amplitude (fig. 3.35(b)).

Problem 3.9.3 *Let $G(t)$ be the amount of glucose in the bloodstream at any time t, α (> 0) is the constant rate of infusion, and β (> 0) is the removal rate of glucose from the bloodstream. Using the differential equation, construct a mathematical model of the infusion of glucose into the bloodstream. By solving the differential equation, predict the glucose level in the bloodstream at any time t. Also, find the limiting value of $G(t)$ for large t.*

Solution: The equation governing the required model is

$$\frac{dG}{dt} = \alpha - \beta G \Rightarrow \frac{dG}{\alpha - \beta G} = dt \Rightarrow \frac{\ln(\alpha - \beta G)}{-\beta} = t + c,$$

$\Rightarrow G(t) = \dfrac{\alpha}{\beta} - \dfrac{c_1 e^{-\beta t}}{\beta}$, ($c_1 = e^{-\beta c}$, arbitrary constant), which gives the glucose level in the bloodstream at any time t.

At $t \to \infty, G \to \dfrac{\alpha}{\beta}$ (since $e^{-\beta t} \to 0$), which gives its limiting value for large t.

Problem 3.9.4 *Suppose in a chemical reaction two substances, M_1 and M_2, react in equal amounts to form a compound M_3. Let $C(t)$ be the concentration of the compound M_3 at time t, which satisfies the differential equation*

$$\frac{dC}{dt} = r(a - C)(b - C),$$

where r is a positive constant; a and b are initial concentrations of M_1 and M_2 at time $t = 0$. Obtain the concentration of M_3 as a function of time for $t > 0$, assuming $C(0) = 0$. Also, determine the limiting concentration of $C(t)$ for large t.

Solution: Given that,

$$\frac{dC}{dt} = r(a - C)(b - C) \Rightarrow \frac{1}{a - b}\left[\frac{dC}{C - a} - \frac{dC}{C - b}\right] = rdt.$$

Integrating, we get, $\dfrac{1}{a - b}\left[\ln\left(\dfrac{C - a}{C - b}\right) - \ln\left(\frac{a}{b}\right)\right] = rt$ (since, $C(0) = 0$),

$$\Rightarrow \frac{C - a}{C - b} = \frac{a}{b} e^{(a-b)rt} \Rightarrow C(t) = ab\left[\frac{1 - e^{(a-b)rt}}{b - ae^{(a-b)rt}}\right],$$

which gives the concentration of the compound M_3 at any time t. As t becomes large ($t \to \infty$), the limiting value of C(t) is given by

$$C_\infty = ab \lim_{t\to\infty}\left[\frac{1-e^{(a-b)rt}}{b-ae^{(a-b)rt}}\right] = \frac{ab}{b} = a, \text{ if } a < b,$$

$$= ab \lim_{t\to\infty}\left[\frac{1-e^{(a-b)rt}}{b-ae^{(a-b)rt}}\right] = ab \lim_{t\to\infty}\left[\frac{e^{-(a-b)rt}-1}{be^{-(a-b)rt}-a}\right] = \frac{ab}{a} = b, \text{ if } a > b.$$

Problem 3.9.5 *A spherical raindrop of radius a falls from a height h and accumulates moisture from the atmosphere as it descends, thereby increasing the radius of the spherical raindrop at a rate* λa. *Show that the radius of the raindrop is* $\lambda a\sqrt{\frac{2h}{g}}(1+\sqrt{1+\frac{g}{2h\lambda^2}})$, *when it hits the ground, g being the acceleration due to gravity.*

Solution: Let M be the mass of the raindrop, then $M = \frac{4}{3}\pi a^3\rho$, a is the radius and ρ is the density. Now, $\frac{dr}{dt} = \lambda a \Rightarrow \quad r = a(1+\lambda t),\ r(0) = a.$ Therefore, mass at time $t = \frac{4}{3}\pi r^3\rho = M(1+\lambda t)^3$. Equation of motion is

$\frac{d}{dt}\left(M(1+\lambda t)^3\dot{x}\right) = M(1+\lambda t)^3 g$. Integrating we get,

$$(1+\lambda t)^3\ \dot{x}(t) = \frac{g}{4\lambda}\left(1+\lambda t - \frac{1}{(1+\lambda t)^3}\right) \ \text{(since, } \dot{x}(0) = 0).$$

Integrating again, $x(t) = \frac{g}{8\lambda^2}\left[(1+\lambda t)^2 + \frac{1}{(1+\lambda t)^2} - 2\right)$ (since, $x(0) = 0$).

$$\Rightarrow \quad x(t) = \frac{g}{8\lambda^2}\left(1+\lambda t - \frac{1}{1+\lambda t}\right)^2 = \frac{g}{8\lambda^2}\left(\frac{r}{a}-\frac{a}{r}\right)^2.$$

When the raindrop reaches the ground, $x = h$. This implies

$$\frac{g}{8\lambda^2}\left(\frac{r}{a}-\frac{a}{r}\right)^2 = h \ \Rightarrow\ r^2 - 2\lambda a\sqrt{\frac{2h}{g}}r - a^2 = 0,$$

$$\Rightarrow\ r = \lambda a\sqrt{\frac{2h}{g}} + \sqrt{\frac{2h}{g}\lambda^2 a^2 + a^2} = \lambda a\sqrt{\frac{2h}{g}}\left(1+\sqrt{1+\frac{g}{2h\lambda^2}}\right).$$

(negative sign is inadmissible as the value of r becomes negative).

Problem 3.9.6 *An elastic string of unstretched length a and modulus of elasticity* λ *is fixed to a point on a smooth horizontal table and the other end is tied to a particle of mass m, which is lying on the table. The particle is pulled to a distance, where the extension of the string is b and then let go. Find the time of complete oscillation.*

Solution: We consider an elastic string, whose unstretched length is $OA_1 = a$ (fig. 3.36) and modulus of elasticity is λ. One end of the elastic

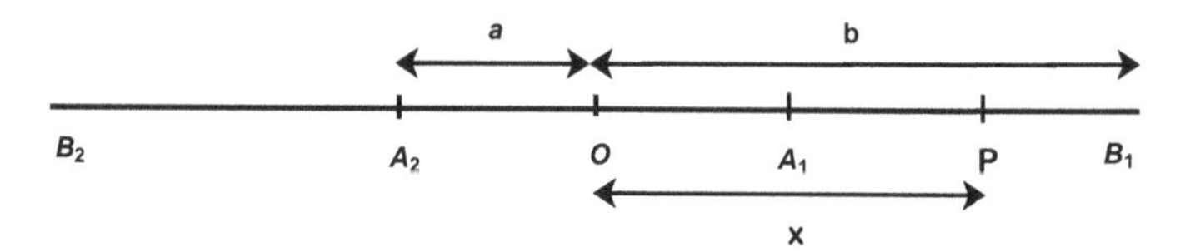

FIGURE 3.36: *The motion of an elastic string lying on a smooth horizontal table.*

string is fixed to a point on a smooth horizontal table and the other end is attached to a particle of mass m, which is lying on the table. The particle is now pulled to a distance where the extension of the string is b (A_1B_1) and then let go. Let P be the position of the particle at any time t, such that $OP = x$.

Then, the differential equation modeling the situation is given by

$$m\frac{d^2x}{dt^2} = -T,$$

$$\begin{aligned} \text{where } T &= \text{tension in the string } \ (a < x < a+b) \\ &= (\text{modulus of elasticity}) \times \frac{\text{increase of length}}{\text{original length}} \\ &= \lambda\left(\frac{x-a}{a}\right) \quad \text{(by Hooke's law).} \end{aligned}$$

$$\text{Therefore, } \frac{d^2(x-a)}{dt^2} = -\frac{\lambda}{am}(x-a), \tag{3.36}$$

which shows that the motion is simple harmonic about the center A_1, with amplitude b. The solution of (3.36) is

$$x - a = K_1\cos\left(\sqrt{\frac{\lambda}{am}}t\right) + K_2\sin\left(\sqrt{\frac{\lambda}{am}}t\right),$$

where K_1 and K_2 are arbitrary constants.

$$\frac{dx}{dt} = -K_1\sqrt{\frac{\lambda}{am}}\sin\left(\sqrt{\frac{\lambda}{am}}t\right) + K_2\sqrt{\frac{\lambda}{am}}\cos\left(\sqrt{\frac{\lambda}{am}}t\right).$$

When $t = 0$, the particle was at B_1, where $x = a + b$ and $\frac{dx}{dt} = 0$

$$\Rightarrow K_1 = b \quad \text{and} \quad K_2 = 0.$$

$$\text{Therefore, } x - a = b\cos\left(\sqrt{\frac{\lambda}{am}}t\right) \tag{3.37}$$

and

$$\frac{dx}{dt} = -b\sqrt{\frac{\lambda}{am}} \sin\left(\sqrt{\frac{\lambda}{am}} t\right). \tag{3.38}$$

Let T_1 be the time taken by the particle from the point B_1 to A_1. Then, from (3.37) we get (putting $x = a$),

$$0 = b \cos\left(\sqrt{\frac{\lambda}{am}} T_1\right) \Rightarrow \sqrt{\frac{\lambda}{am}} T_1 = \frac{\pi}{2} \Rightarrow T_1 = \sqrt{\frac{am}{\lambda}} \frac{\pi}{2}.$$

The velocity of the particle at the point A_1 is given by (3.38) as

$$\frac{dx}{dt} = -b\sqrt{\frac{\lambda}{am}} \sin\left(\sqrt{\frac{\lambda}{am}} T_1\right) = -b\sqrt{\frac{\lambda}{am}} \sin\left(\frac{\pi}{2}\right) = -b\sqrt{\frac{\lambda}{am}}.$$

(Negative sign because distance decreases with time.)

Therefore, the particle reaches its maximum velocity at point A_1. At this point, the elastic string becomes slack and the tension ceases. The equation of motion given by equation (3.36) does not hold any more, and the particle moves with uniform speed $b\sqrt{\frac{\lambda}{am}}$ from the point A_1, until it reaches the point A_2. Once it crosses the point A_2, the string again becomes taut and tension now acts in the direction A_2O. The velocity slowly decreases and goes to zero at the point B_2 such that $OB_2 = OB_1$. The particle then retraces its path and reaches B_1 and this cycle of motion goes on.

$$\text{Now, time taken from } A_1 \text{ to } O = \frac{\text{Distance}}{\text{Speed}} = \frac{a}{b\sqrt{\frac{\lambda}{am}}}.$$

$$\text{Therefore, time taken from } B_1 \text{ to } O = \sqrt{\frac{am}{\lambda}} \frac{\pi}{2} + \frac{a}{b\sqrt{\frac{\lambda}{am}}} = \left(\frac{\pi}{2} + \frac{a}{b}\right) \sqrt{\frac{am}{\lambda}}.$$

Hence, the time for complete oscillation is four times the time from B_1 to O

$$= 4\left(\frac{\pi}{2} + \frac{a}{b}\right) \sqrt{\frac{am}{\lambda}}.$$

Problem 3.9.7 *Consider a vertical air column of cross-section α. Let an element of this column be bounded by two planes at a height h and $h+\delta h$ above the sea level. Let ρ be the average density of the element, p and $p+\delta p$ be the pressures at height h and $h+\delta h$, respectively. With the help of a differential equation, model the situation connecting atmospheric pressure p kg/m^2 and height h meters above the sea level.*

Solution: Column δh of air is in equilibrium under the following forces:

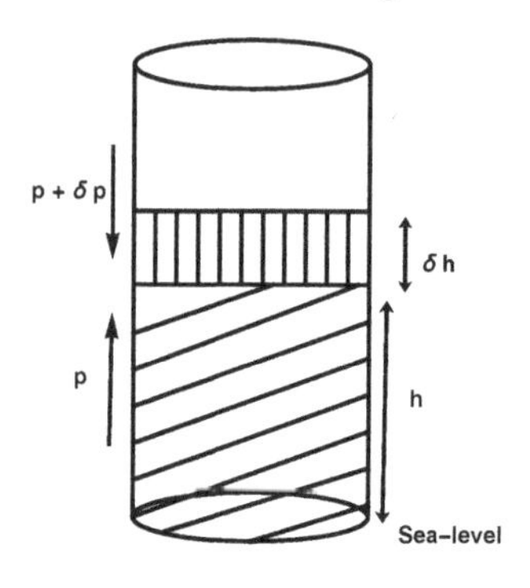

(i) pkg/m^2 in the upward direction,
(ii) $(p+\delta p)kg/m^2$ in the downward direction.

$$p = p + \delta p + \frac{\rho g \delta h \alpha}{\alpha} \Rightarrow \frac{\partial p}{\partial h} = -\rho g.$$

As $\delta h \to 0$, we obtain $\dfrac{dp}{dh} = -\rho g$,

which is the governing differential equation for the atmospheric pressure at any height h above the sea level.

Case I: Temperature is constant
From Boyle's law (For a gas of a given mass m, kept at a constant temperature t, volume V of the gas is inversely proportional to the pressure p), we obtain

$$V \propto \frac{1}{p} \Rightarrow pV = k \text{ (a constant)}.$$

If ρ be the density of the gas, we can write,

$$V = \frac{m}{\rho} \Rightarrow p\frac{m}{\rho} = k \Rightarrow p = \frac{k}{m}\rho \Rightarrow p = K\rho \Rightarrow \rho = \frac{p}{K}, \text{ where } K = \frac{k}{m}.$$

Therefore, $\dfrac{dp}{dh} = -\dfrac{pg}{K} \Rightarrow \ln p = -\dfrac{g}{K}h +$ constant.

At the sea level, $h = 0,\ p = p_0 \Rightarrow$ constant $= \ln p_0$.

Hence, $\ln p = -\dfrac{g}{K}h + \ln p_0 \Rightarrow p = p_0 e^{-\frac{g}{K}h}$.

Case II: Temperature is a variable.
Let $p = K\rho^n\ (n \neq 1) \Rightarrow \rho = \left(\dfrac{p}{K}\right)^{1/n}$, then

$$\text{then, } \frac{dp}{dh} = -\left(\frac{p}{K}\right)^{1/n} g \Rightarrow \frac{n}{n-1}\left(p_0^{\frac{n-1}{n}} - p^{\frac{n-1}{n}}\right) = ghk^{-1/n}.$$

Problem 3.9.8 *The price of sugar, initially* 50 *per kg, is* $p(t)$ *per kg. After* t *weeks, the demand is* $D = 120 - 2p + 5\frac{dp}{dt}$ *and the supply* $S = 3p - 30 + 50\frac{dp}{dt}$ *thousand kilograms per week. Show that, for* $D = S$*, the price of sugar must vary over time accordingly to the law* $p = 30 + 20e^{-\frac{t}{9}}$*. Predict the prices of the sugar after* 15 *weeks and* 60 *weeks. Draw a graph to show the approach of the price to the equilibrium value.*

Solution: For $D = S$, we get,

$$120 - 2p + 5\frac{dp}{dt} = 3p - 30 + 50\frac{dp}{dt} \Rightarrow 9\frac{dp}{dt} + p = 30,$$
$$\Rightarrow p(t) = Ae^{-\frac{t}{9}} + 30 \Rightarrow p(t) = 20e^{-\frac{t}{9}} + 30 \text{ (since } P(0) = 50).$$

Clearly, $p(15) = 33.78$ and $p(60) = 30.03$. Both analytically and graphically, we see that as $t \to \infty$, $p(t) \to 30$ (fig. 3.37(a)).

Problem 3.9.9 *We consider two trees that are growing independently, supported by independent fixed supplies of the substrate. Let $p_1(t)$ and $p_2(t)$ be the dry weights of the trees at time t, respectively. The system of differential equations governing the scenario is given by*

$$\frac{dp_1}{dt} = \alpha_1 p_1(k_1 - p_1), \quad \frac{dp_2}{dt} = \alpha_2 p_2(k_2 - p_2), \quad (\alpha_1, k_1, \alpha_2, k_2 > 0).$$

(i) Explain the parameters α_1, α_2, k_1 and k_2.
(ii) Find the equilibrium points and show that the system is stable. Solve the differential equations and obtain the solutions for $p_1(t)$ and $p_2(t)$.
(iii) Taking $\alpha_1 = \alpha_2 = 0.1, k_1 = 7, k_2 = 10$ and with different initial conditions $(p_1(0), p_2(0))$, plot the time series graphs and comment.

Solution (i) The given equations can be written as

$$\frac{dp_1}{dt} = \alpha_1 k_1 p_1 - \alpha_1 p_1^2, \quad \frac{dp_2}{dt} = \alpha_2 k_2 p_2 - \alpha_2 p_2^2.$$

Here, α_1 is the rate at which the dry weight p_1 of Tree I decreases on its own, k_1 being its carrying capacity, and $\alpha_1 k_1$ is the rate of increase of the weight of Tree I or the natural growth rate. Similarly, α_2 is the rate at which the dry weight p_2 of Tree II decreases on its own, that is, the growth is restricted and k_2 is the carrying capacity. $\alpha_2 k_2$ is the rate of increase of weight of Tree II or the natural growth rate.

(ii) The equilibrium solutions (p_1^*, p_2^*) are obtained by solving $\alpha_1 p_1(k_1 - p_1) = 0, \ \alpha_2 p_2(k_2 - p_2) = 0 \Rightarrow \ (p_1^*, p_2^*) = (0, 0); (k_1, 0); (0, k_2)$ and (k_1, k_2). The Jacobian matrix about (p_1^*, p_2^*) is

$$\begin{pmatrix} \alpha_1 k_1 - 2\alpha_1 p_1^* & 0 \\ 0 & \alpha_2 k_2 - 2\alpha_2 p_2^* \end{pmatrix},$$

whose eigenvalues are $\alpha_1 k_1 - 2\alpha_1 p_1^*$ and $\alpha_2 k_2 - 2\alpha_2 p_2^*$. For (k_1, k_2), the eigenvalues are $-\alpha_1 k_1 (< 0)$ and $-\alpha_2 k_2 (< 0)$. Therefore, the system is always asymptotically stable about (k_1, k_2). It can be easily shown that the system is unstable about the other three equilibrium points.

Given that $\dfrac{dp_1}{dt} = \alpha_1 p_1(k_1 - p_1) \ \Rightarrow \ \dfrac{dp_1}{p_1(k_1 - p_1)} = \alpha_1 dt,$

$$\left[\frac{1}{p_1} + \frac{1}{k_1 - p_1}\right] = k_1 \alpha_1 dt \ \Rightarrow \ \ln\left(\frac{p_1}{k_1 - p_1}\right) = k_1 \alpha_1 t + \ln c_1,$$

$\Rightarrow \ p_1 = \dfrac{k_1 c_1}{c_1 + e^{-k_1 \alpha_1 t}}$. In a similar manner, $p_2 = \dfrac{k_2 c_2}{c_2 + e^{-k_2 \alpha_2 t}}$.

(iii) Figs.(3.37(b), 3.37(c), 3.37(d)) show the growth of the dry weights of two independent trees. Fig.(3.37(b)) shows that even though Tree II starts with more dry weights, it soon becomes less Tree I. Both the trees reach their equilibrium dry weights even if they start from negligible values (fig. 3.37(c), fig. 3.37(d)).

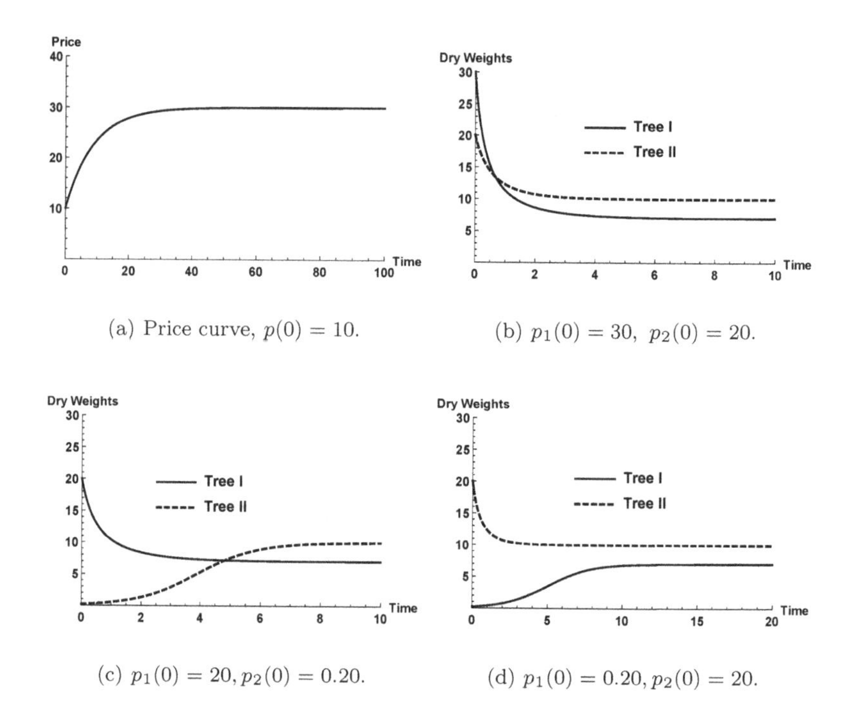

(a) Price curve, $p(0) = 10$.

(b) $p_1(0) = 30,\ p_2(0) = 20$.

(c) $p_1(0) = 20, p_2(0) = 0.20$.

(d) $p_1(0) = 0.20, p_2(0) = 20$.

FIGURE 3.37: *The figures show (a) the price of the sugar approaching a steady value of* 30 *as time increases, (b)(c)(d) the dynamics of the dry weights of two independent trees for different initial conditions.*

Problem 3.9.10 *Consider the pricing policy of edible oil, where the manufacturers stock the product to meet any sudden unexpected demand from customers. Let $S(t)$ and $Q(t)$ be the sales forecast and production forecast, respectively, and $p(t)$ be the price of edible oil (dollar per barrel) at any time t. Then, the general pricing policy is given by*

$$S(t) = \alpha_1 - \beta_1 p - \gamma_1 \frac{dp}{dt},\ Q(t) = \alpha_2 - \beta_2 p - \gamma_2 \frac{dp}{dt},\ \frac{dp}{dt} = -\delta(L(t) - L_0).$$

Here, $\alpha_1, \alpha_2, \beta_1, \beta_2, \gamma_1, \gamma_2, \delta$ are positive constants, L is the inventory level and L_0 the desired optimum inventory level. The changes in inventory follow the law

$$\frac{dL}{dt} = Q - S.$$

Show that the equation

$$\frac{d^2p}{dt^2} + \delta(\gamma_1 - \gamma_2)\frac{dp}{dt} + \delta(\beta_1 - \beta_2)p = \delta(\alpha_1 - \alpha_2)$$

gives the forecast price. Hence, deduce that if $\gamma_1 > \gamma_2,\ \beta_1 > \beta_2$, the price tends to a steady value as t increases.

Solution: Given that, $\dfrac{dL}{dt} = Q - S = (\alpha_2 - \alpha_1) - p(\beta_2 - \beta_1) - (\gamma_2 - \gamma_1)\dfrac{dp}{dt}$.

Also, $\dfrac{dp}{dt} = -\delta[L(t) - L_0]$, (differentiating we get), $\dfrac{d^2p}{dt^2} = -\delta\dfrac{dL(t)}{dt}$

$$\Rightarrow \quad \frac{d^2p}{dt^2} = -\delta\left[(\alpha_2 - \alpha_1) - p(\beta_2 - \beta_1) - (\gamma_2 - \gamma_1)\frac{dp}{dt}\right]$$

$$\Rightarrow \quad \frac{d^2p}{dt^2} + \delta(\gamma_1 - \gamma_2)\frac{dp}{dt} + \delta(\beta_1 - \beta_2)p = \delta(\alpha_1 - \alpha_2),$$

which is a second-order ordinary differential equation with constant coefficients, gives the forecast price. The complementary function of this second-order ordinary differential equation is

$$p_c = Ae^{m_1 t} + Be^{m_2 t}, \quad A, \ B \text{ are arbitrary constants, and}$$

$$m_1, m_2 = -\delta(\gamma_1 - \gamma_2) \pm \delta(\gamma_1 - \gamma_2)\sqrt{1 - 4\frac{(\beta_1 - \beta_2)}{\delta(\gamma_1 - \gamma_2)^2}},$$

and both are negative as $\gamma_1 > \gamma_2$, $\beta_1 > \beta_2$, implies

$$\delta(\gamma_1 - \gamma_2) > \delta(\gamma_1 - \gamma_2)\sqrt{1 - 4\frac{(\beta_1 - \beta_2)}{\delta(\gamma_1 - \gamma_2)^2}}.$$

The particular integral is

$$p_I = \frac{1}{D^2 + \delta(\gamma_1 - \gamma_2)D + \delta(\beta_1 - \beta_2)}\delta(\alpha_1 - \alpha_2) = \frac{\delta(\alpha_1 - \alpha_2)}{\delta(\beta_1 - \beta_2)} = \frac{\alpha_1 - \alpha_2}{\beta_1 - \beta_2}.$$

Therefore, the general solution is $(p_c + p_I)$

$$p(t) = Ae^{-m_1 t} + Be^{-m_2 t} + \frac{\alpha_1 - \alpha_2}{\beta_1 - \beta_2}.$$

As $t \to \infty$, p(t) tends to a steady value $\frac{(\alpha_1 - \alpha_2)}{(\beta_1 - \beta_2)}$, as both m_1 and m_2 are negative (fig. 3.38(a)).

Problem 3.9.11 *Consider a model of species competing for food. The governing equation is given by*

$$\frac{dx}{dt} = \alpha x - \beta y, \quad \frac{dy}{dt} = \gamma y - \delta x,$$

where x and y are two competing species, $\alpha, \beta, \gamma, \delta$ are positive constants.
(i) Show that

$$\frac{d^2x}{dt^2} - (\alpha + \gamma)\frac{dx}{dt} + (\alpha\gamma - \beta\delta)x = 0$$

and solve for x. Also, find the solution for y.
(ii) If, at $t = 0$, $x = 100$ and $y = 200$, obtain graphically the time when one species is eliminated (take $\alpha = 0.05, \beta = 0.1, \gamma = 0.2, \delta = 0.2$).

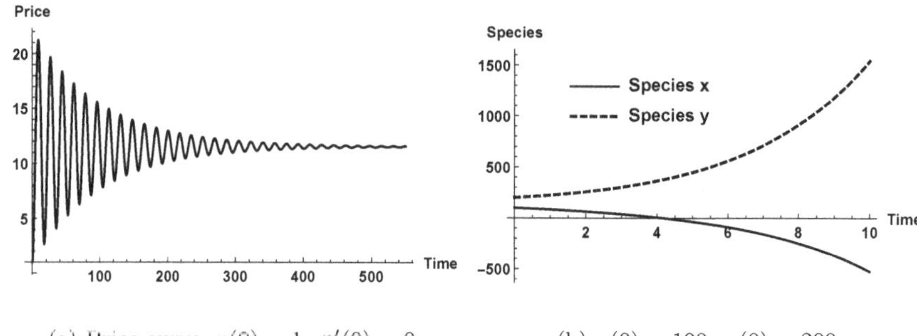

(a) Price curve, $p(0) = 1,\ p'(0) = 0$.

(b) $x(0) = 100,\ y(0) - 200$.

FIGURE 3.38: *The figures show (a) the price of the edible oil reaches a steady value of \$11.54 as time increases ($\alpha_1 = 40, \alpha_2 = 25, \beta_1 = 3.4, \beta_2 = 2.1, \gamma_1 = 0.5, \gamma_2 = 0.3$), (b)the behavior of two species competing for food, where the species x goes to extinction ($\alpha = 0.05, \beta = 0.1, \gamma = 0.2, \delta = 0.2$).*

Solution: (i) $\dfrac{d^2x}{dt^2} = \alpha\dfrac{dx}{dt} - \beta\dfrac{dy}{dt} = \alpha\dfrac{dx}{dt} - \beta(\gamma y - \delta x),$

$$\Rightarrow \frac{d^2x}{dt^2} = \alpha\frac{dx}{dt} - \gamma(-\frac{dx}{dt} - \alpha x) + \delta\beta x = \alpha\frac{dx}{dt} + \gamma\frac{dx}{dt} - \alpha\beta x + \delta\beta x,$$

$$\Rightarrow \frac{d^2x}{dt^2} - (\alpha + \gamma)\frac{dx}{dt} + (\alpha\gamma - \beta\delta)x = 0.$$

This is a second-order ordinary differential equation with constant coefficients, whose auxiliary equation is

$$m^2 - (\alpha + \gamma)m + (\alpha\gamma - \beta\delta) = 0.$$

Solving, we get, $m_1, m_2 = \dfrac{(\alpha+\gamma) \pm \sqrt{\gamma^2 + \alpha^2 + 4\beta\delta - 2\alpha\gamma}}{2}$.

Therefore, the required solution is

$x(t) = Ae^{m_1 t} + Be^{m_2 t}$, where A and B are arbitrary constants.

$$y(t) = \frac{1}{\beta}\left(\alpha x - \frac{dx}{dt}\right) = \frac{A(\alpha - m_1)}{\beta}e^{m_1 t} + \frac{B(\alpha - m_2)}{\beta}e^{m_2 t}.$$

(ii) The model is solved numerically with $\alpha = 0.05, \beta = 0.1, \gamma = 0.2, \delta = 0.2$ and fig. 3.38(b) shows that species x goes to extinction after 4.0 units of time.

Problem 3.9.12 *A mathematical model for epidemics consisting of susceptible (S), infected (I), and removals (R) is given by*

$$\frac{dS}{dt} = -\beta S^2 I,\ \frac{dI}{dt} = \beta S^2 I - \gamma I,\ \frac{dR}{dt} = \gamma I,$$

where β and γ are positive constants.
(i) Find the threshold density of susceptible.
(ii) Show that

$$\frac{dR}{dt} = \gamma \left\{ n - R - \frac{S_0}{1 + \frac{\beta S_0 R}{\gamma}} \right\},$$

where n is the total population size, $S(0) = S_0, I(0) = I_0, R(0) = 0$ and $I_0 < S_0$.

Solution: (i) The threshold density of the susceptible is by putting

$$\frac{dI}{dt} = 0 \Rightarrow \beta S^2 I - \gamma = 0 \Rightarrow S^{\dagger} = \sqrt{\frac{\gamma}{\beta}}.$$

(ii) Adding the given three equations we get,

$$\frac{dS}{dt} + \frac{dI}{dt} + \frac{dR}{dt} = 0 \Rightarrow S + I + R = n \text{ (constant)}.$$

Now, $\frac{dS}{dt} = -\beta S^2 I = -\beta S^2 \frac{1}{\gamma}\frac{dR}{dt} \Rightarrow \frac{dS}{S^2} = -\frac{\beta}{\gamma} dR.$

Integrating we get, $\frac{1}{S} = \frac{\beta}{\gamma} R + \frac{1}{S_0}$, since $S(0) = S_0$ and $R(0) = 0$.

$$\Rightarrow S(t) = \frac{S_0}{1 + \frac{\beta S_0 R}{\gamma}}.$$

Now, $\frac{dR}{dt} = \gamma I = \gamma(n - R - S) = \gamma \left\{ n - R - \frac{S_0}{1 + \frac{\beta S_0 R}{\gamma}} \right\}.$

Problem 3.9.13 *The British Museum was authorized in 1988 by the Vatican to date a cloth relic known as the* Shroud of Turin, *which contains the negative image of a human body, widely believed to be that of Jesus. The British Museum's report confirmed that the cloth fibers contained between 92% and 93% of their original C^{14}. Estimate the approximate age of the Shroud, using the method of carbon dating. Assume that the half-life of radioactive C^{14} is 5730 years.*

Solution: Let $A(t)$ be the amount of C^{14} present in the sample at any time t, then

$$\frac{dA}{dt} = -\lambda A \Rightarrow A(t) = A_0 e^{-\lambda t},$$

where λ is the decay constant of the sample and is given by

$$\lambda = \frac{1}{\tau} \log_e 2, \quad \tau \text{ being the half-life of } C^{14} = \frac{\log_e 2}{5730} \approx 0.000121.$$

Therefore, the fraction of the original C^{14} present after t-years is

$$\begin{aligned} A(t) &= A_0 e^{-0.000121t} \\ \Rightarrow t &= -\frac{1}{0.000121} \log_e \left(\frac{A(t)}{A_0} \right), \end{aligned}$$

where $A_0 = A(0)$, is the units of C^{14} present at time t $= 0$. Taking $\frac{A(t)}{A_0} =$ 0.92 and 0.93 respectively, we get,

$$\begin{aligned} t_1 &= -\frac{1}{0.000121} \log_e (0.92) \approx 689 \text{ and} \\ t_2 &= -\frac{1}{0.000121} \log_e (0.93) \approx 600. \end{aligned}$$

Therefore, from the test conducted by the British Museum in 1988, it was concluded that the Shroud was between 600 and 689 years old, thereby placing its origin between 1299 A.D. and 1388 A.D.

Problem 3.9.14 *We consider a competition model of the form*

$$\begin{aligned} \frac{dx}{dt} &= 0.05x \left(1 - \frac{x}{250000}\right) - axy, \\ \frac{dy}{dt} &= 0.08y \left(1 - \frac{y}{400000}\right) - axy, \end{aligned}$$

where $x(t)$ denotes the population of blue whales and $y(t)$ denotes the population of fin whales. Here, 0.05, 0.08 *are intrinsic growth rates;* 250000, 400000 *are carrying capacities of the blue whales and fin whales respectively, a is the rate of inter-specific competition (for food).*

(i) Find the condition for stability for the given system at the non-zero equilibrium point (x^*, y^*).

Solution: The equilibrium points are $(0,0)$, $(250000, 0)$, $(0, 400000)$ and (x^*, y^*), where $x^* = \dfrac{250000. - 2. \times 10^{12} a}{1. - 4. \times 10^{13} a^2}$, $y^* = \dfrac{400000. - 2. \times 10^{12} a}{1. - 4. \times 10^{13} a^2}$. The Jacobian matrix at the point (x^*, y^*) is given by

$$\begin{pmatrix} 0.05\left(1 - \frac{x^*}{250000}\right) - \frac{2}{10^7} x^* - ay^* & -ax^* \\ -ay^* & 0.08\left(1 - \frac{y^*}{400000}\right) - \frac{2}{10^7} y^* - ax^* \end{pmatrix}$$

$$= \begin{pmatrix} -\frac{2}{10^7} x^* & -ax^* \\ -ay^* & -\frac{2}{10^7} y^* \end{pmatrix} \text{ since, } 0.05\left(1 - \frac{x^*}{250000}\right) - ay^* = 0, \text{ and}$$

$$0.08\left(1 - \frac{y^*}{400000}\right) - ax^* = 0.$$

The characteristic equation of the Jacobian matrix is given by

$$\begin{vmatrix} -\frac{2}{10^7}x^* - \lambda & -ax^* \\ -ay^* & -\frac{2}{10^7}y^* - \lambda \end{vmatrix} = 0,$$
$$\Rightarrow \lambda^2 + \frac{2}{10^7}\lambda(x^* + y^*) + \left(\frac{4}{10^{14}} - a^2\right)x^*y^* = 0.$$

By Routh Hurwitz criteria, the system will be stable if

$$\frac{4}{10^{14}} - a^2 > 0 \Rightarrow \ -2 \times 10^{-7} < a < 2 \times 10^{-7}.$$

(ii) Solve the model numerically, assuming $x(0) = 6000$ and $y(0) = 60,000$ for $a = 10^{-8}$ and 10^{-6} and conclude on the dynamics of the system.

Solution: For $a = 10^{-8}$, the equilibrium points are $(0, 0)$, $(250000, 0)$, $(0, 400000)$ and $(230924, 381526)$. The Jacobian matrix at $(230576, 388471)$ is $\begin{pmatrix} -0.0461151 & -0.00230576 \\ -0.00388471 & -0.0776942 \end{pmatrix}$, whose eigenvalues are -0.078 and -0.046. Hence, the system is stable about $(230576, 381471)$. It can be easily shown that the system is unstable for the other three equilibrium points.

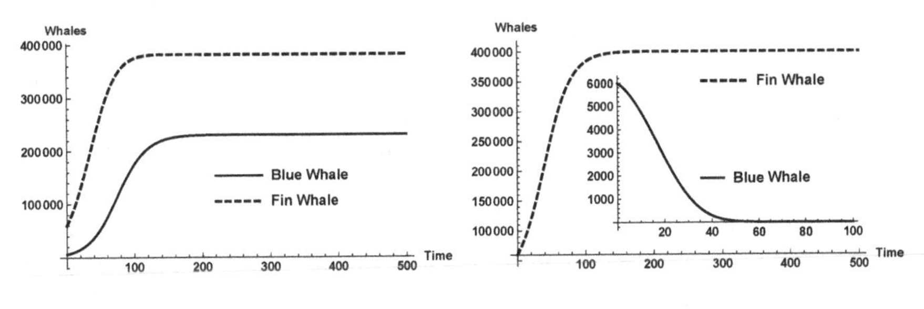

(a) Both whales coexist.

(b) Blue whale becomes extinct.

FIGURE 3.39: *The dynamics of blue whales and fin whales population competing for food.*

For $a = 10^{-6}$, the equilibrium points are $(0, 0)$, $(250000, 0)$, $(0, 400000)$ and $(72917, 35417)$. Since, $a = 10^{-6}$ does not lie in $-2 \times 10^{-7} < a < 2 \times 10^{-7}$, the system is unstable about $(72917, 35417)$. The system is also unstable for $(0, 0)$ but stable for both $(250000, 0)$ and $(0, 400000)$. The competition model of blue whales and fin whales is solved numerically. For $a = 10^{-8}$, both whales coexist (fig. 3.39(a)) and for $a = 10^{-6}$, the blue whale population dies out (fig. 3.39(b)).

(iii) Assuming that $x(0) = 6000, y(0) = 60000$ and $a = 10^{-7}$, numerically find the dynamics what happens to both the species of whales.

Solution: Proceed as in *(ii)*.

(iv) Consider the intrinsic growth rate of both the species of whales to be 5% per year. Assuming $a = 10^{-8}$ and intrinsic growth rates of blue whales to be 2%, 4%, 8% and 20% per year, solve the model numerically and comment on the dynamics of the both the species of whales.

Solution: Fig. 3.40(a) shows the dynamics of growth of blue whales with increasing growth rate $r = 0.02$, 0.04, 0.08 and 0.20, respectively. The blue whales start reaching the steady-state value at a faster rate, once it crossed the growth rate of 5%, and at one point, the number of blue whales exceeds the number of fin whales, for a short period of time, as observed from the graph.

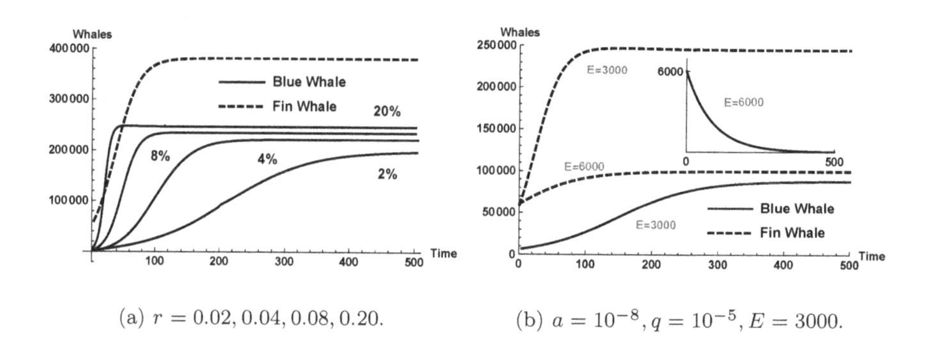

(a) $r = 0.02, 0.04, 0.08, 0.20.$ (b) $a = 10^{-8}, q = 10^{-5}, E = 3000.$

FIGURE 3.40: *The figures show (a) the behavior of blue whales with increasing intrinsic growth rates, (b) the behavior of both the species of whales with increasing harvesting.*

(v) The effects of harvesting are now considered on the two whale populations. Assuming $a = 10^{-8}$ and $E = 3000$ boat days per year, which results in the annual harvest of qEx blue whales and qEy fin whales (q = catch ability coefficient $= 10^{-5}$), rewrite the model and find (through numerical simulation), what happens to both the whale species. Repeat the same process for $E = 6000$ boat days per year. Find the range of E for which the number of whales of both species approaches a non-zero equilibrium.

Solution: With harvesting taken into account, the modified model is given by

$$\frac{dx}{dt} = 0.05x\left(1 - \frac{x}{250000}\right) - axy - qEx,$$
$$\frac{dy}{dt} = 0.08y\left(1 - \frac{y}{400000}\right) - axy - qEy.$$

It can be easily checked that the stability condition does not depend on q or E. Fig. 3.40(b) shows the effect of harvesting on both the whale species. When the number of boat days per year is 3000, both the whale species coexist (survive), but their numbers are reduced, compared to the non-harvesting case. When the number of boat days per year is increased to 6000, the fin whales survive but all the blue whales die out.

The non-zero equilibrium point (with E) is given by $(230576 - 47.619E, 388471 - 47.619E)$ and both need to be positive, which implies $E < min(4842.1, 8157.9) \approx min(4842, 8158)$. Therefore, the range of E for which the number of whales of both species approaches a non-zero equilibrium is $0 < E < 4842$.

(vi) Taking E = 500, 1000, 1500, 2000, 2500, 3000, 4000, 4500, 5500, 6000, 7000 boat days, plot time series graph to find which case results in the highest sustainable yield.

Solution: Proceed as in *(ii)*.

(vii) Suppose the model is modified to

$$\begin{aligned}\frac{dx}{dt} &= 0.05x\left(\frac{x - C_x}{x + C_x}\right)\left(1 - \frac{x}{250000}\right) - axy,\\ \frac{dy}{dt} &= 0.08y\left(\frac{y - C_y}{y + C_y}\right)\left(1 - \frac{y}{400000}\right) - axy,\end{aligned}$$

where C_x and C_y are the minimum viable population levels below which the growth rates of the blue whale and fin whale are negative. Use numerical simulation to comment on the coexistence of the two species of whales by assuming $a = 10^{-8}$, $C_x = 35000$, and $C_y = 16000$. Find the equilibrium points and classify each of them as stable or unstable. What does the model predict about the future of the two whale populations, assuming $x(0) = 6000$ and $y(0) = 60000$? Suppose $C_x = 1500$ instead of 35000, what does the model predicts now?

Solution: With initial populations $x(0) = 6000, y(0) = 60000$ and $a = 10^{-8}$, $C_x = 35000, C_y = 16000$, the model predicts that the fin whales survive and reach a steady-state value while the blue whales goes on extinction. (fig. 3.41(a)). But, with $C_x = 1500$ and $C_y = 16000$, both whales survive and reach non-zero equilibria (fig. 3.41(b)). Linear stability analysis is similar to *(i)*.

Problem 3.9.15 *We consider a battle between two forces A and B. The military strategy of wearing down the enemy by continued losses in personnel (attrition) is due to infantry (direct fire) and artillery (area fire). We assume that the loss in personnel due to direct fire is proportional to the number of enemy infantry. We also assume that the attrition rate due to artillery is*

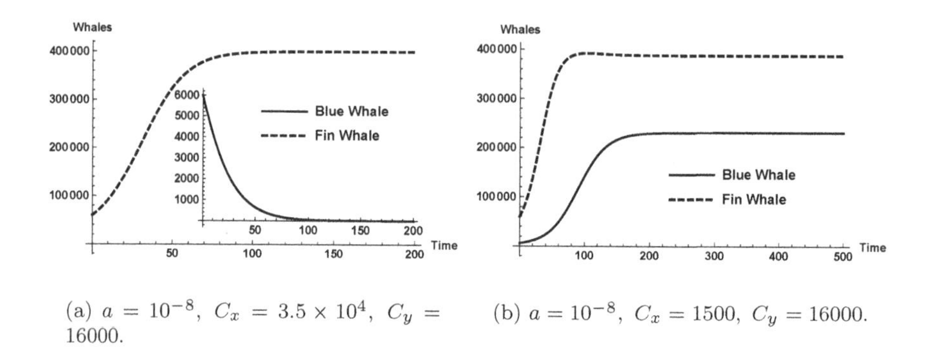

(a) $a = 10^{-8}$, $C_x = 3.5 \times 10^4$, $C_y = 16000$.

(b) $a = 10^{-8}$, $C_x = 1500$, $C_y = 16000$.

FIGURE 3.41: *The dynamics of the whales with minimum viable population levels C_x and C_y, below which the growth rates of the blue whale and fin whale are negative.*

proportional to the product of the two forces A and B. Also, force B has greater weapon effectiveness than force A. The mathematical model of the above scenario is given by

$$\frac{dx}{dt} = -\lambda ay - \lambda bxy,$$
$$\frac{dy}{dt} = -ax - bxy,$$

where x and y are the number of personnel in force A and force B, respectively; $a, b, \lambda\ (> 1)$ are positive constants. Since the weapon effectiveness of force B is more than force A, they cause more harm. Therefore, the rates in the first equation are multiplied by some constant $\lambda\ (> 1)$. Solve the model numerically by taking $a = 0.05$ and $b = 0.005$ for $\lambda = 1.5, 2, 3$ and 5, starting with $x(0) = 50$ and $y(0) = 30$ and comment on the result.

Solution: The model is solved numerically for different λ. For $\lambda = 1.5$, the number of personnel $y(t)$ of force B die within 10 units of time despite having the greater weapon effectiveness than force A (fig. 3.42(a)), whereas, for $\lambda = 2.0$, the number of personnel $x(t)$ of force B die in 30 units of time (fig. 3.42(b)). For $\lambda = 3.0$ and 5.0, we observe similar dynamics, only, personnel $x(t)$ of force B die at a much faster rate with respect to units of time (fig. 3.42(c) and fig. 3.42(d)).

Problem 3.9.16 *Analyze the stability of the origin for the system*

$$\dot{x} = \frac{dx}{dt} = -x + y^2, \quad \dot{y} = \frac{dy}{dt} = -y.$$

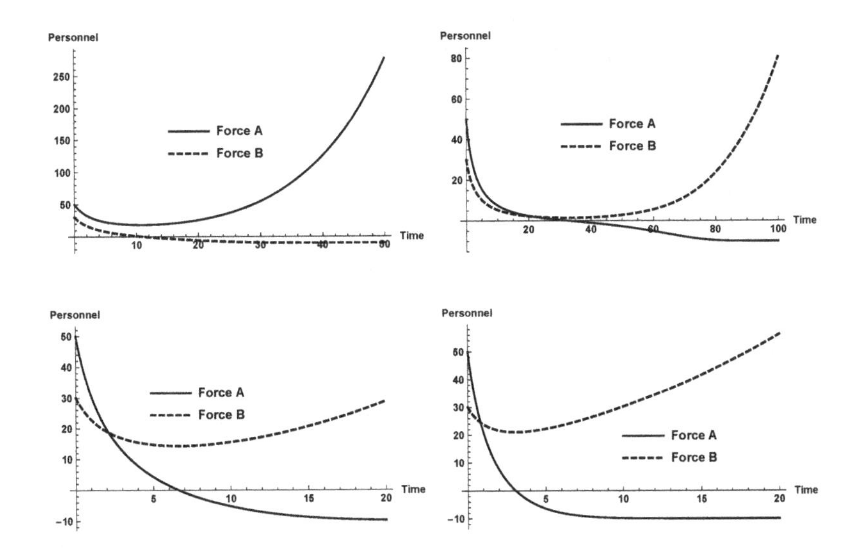

FIGURE 3.42: *The dynamics of the personnel of force A and force B during the battle.*

Solution: Clearly, $(0,0)$ is the only equilibrium point (steady state or critical point) of the given system. Linearizing about the steady state $(0,0)$, we obtain the system as

$$\begin{pmatrix} \dot{x} \\ \dot{y} \end{pmatrix} = \begin{pmatrix} \frac{dx}{dt} \\ \frac{dy}{dt} \end{pmatrix} = \begin{pmatrix} -1 & 0 \\ 0 & -1 \end{pmatrix} \begin{pmatrix} x \\ y \end{pmatrix}.$$

The variational matrix $\begin{pmatrix} -1 & 0 \\ 0 & -1 \end{pmatrix}$ has eigen-values $-1, -1$ (negative). Hence, by Lyapunov's indirect method, the system is locally asymptotically stable (LAS) about $(0,0)$.

To make further conclusions about the stability of the origin $(0,0)$, (namely, its region of attraction), we use a function $V(x,y) = ax^2 + by^2$, where a and b are some positive constants to be determined. Clearly, the function $V(x,y)$ is positive definite as $V(x,y) > 0 \ \forall \ (x,y) \neq (0,0)$ and $V(0,0) = 0$. Now,

$$\dot{V}(x,y) = \frac{dV}{dt} = \frac{\partial V}{\partial x}\dot{x} + \frac{\partial V}{\partial y}\dot{y} = -2ax^2 + 2axy^2 - 2by^2,$$

which is not negative definite for any non-zero values of a and b.

One may try to find another form of Lyapunov function for the global stability of the given system, but here we give an alternative method of proving global asymptotic stability of the given system about the origin $(0, 0)$. Since, y is independent of x, it is possible to find a closed-form solution. Thus,

$$\frac{dy}{dt} = -y \Rightarrow y(t) = y_0\, e^{-t}.$$

$$\text{Therefore, } \frac{dx}{dt} = -x + y^2 = -x + y_0^2\, e^{-2t},$$

$$\Rightarrow \frac{dx}{dt} + x = y_0^2\, e^{-2t} \Rightarrow x(t) = x_0\, e^{-t} + e^{-t}\left(1 - e^{-t}\right) y_0^2.$$

So, as $t \to \infty$, $x(t) \to 0$ and $y(t) \to 0$ for any initial condition x_0 and y_0. Hence, the system is globally asymptotically stable about the origin $(0, 0)$.

Problem 3.9.17 *Consider the autonomous non-linear system*

$$\frac{dx}{dt} = x - xy, \quad \frac{dy}{dt} = xy - y.$$

Compute the non-zero equilibrium point of the system and study its stability.

Solution: The equilibrium points of the model is given by

$$x - xy = x(1 - y) = 0, \quad xy - y = y(x - 1) = 0.$$

Solving, we get the non-zero equilibrium points as (1,1).

For stability analysis, we define the function

$$V(x, y) = x + y - \ln(x) - \ln(y) - 2.$$

Clearly, the function $V(x, y)$ is positive definite as $V(x, y) > 0\ \forall\ (x, y) \neq (1, 1)$ and $V(1, 1) = 0$. Now,

$$\dot{V}(x, y) = \frac{dV}{dt} = \frac{\partial V}{\partial x}\dot{x} + \frac{\partial V}{\partial y}\dot{y} = \left(1 - \frac{1}{x}\right)(x - xy) + \left(1 - \frac{1}{y}\right)(xy - y) = 0.$$

Thus, $V(x, y)$ is negative semi-definite, implying that $V(x, y)$ is a Lyapunov function and hence the system is stable (not asymptotically) about the equilibrium point (1,1).

Problem 3.9.18 *Find the critical point of the non-homogeneous linear system and determine the type and stability of the critical point.*

$$\frac{dx}{dt} = x + 2y - 6, \quad \frac{dy}{dt} = 6x - 3y + 24.$$

Solution: The critical point is given by

$$x + 2y - 6 = 0, \quad 6x - 3y + 24 = 0.$$

Solving we get $x = -2$ and $y = 4$. Putting $x = X - 2$ and $y = Y + 4$ in the given differential equation, we obtain,

$$\frac{dx}{dt} = X + 2Y, \quad \frac{dy}{dt} = 6X - 3Y.$$

The characteristic equation is given by

$$\begin{vmatrix} 1-\lambda & 2 \\ 6 & -3-\lambda \end{vmatrix} = 0 \Rightarrow \lambda^2 + 2\lambda - 15 = 0 \Rightarrow \lambda = -5, 3.$$

The eigenvalues are real, unequal and of opposite signs, implying that the phase portrait is a saddle, which is unstable.

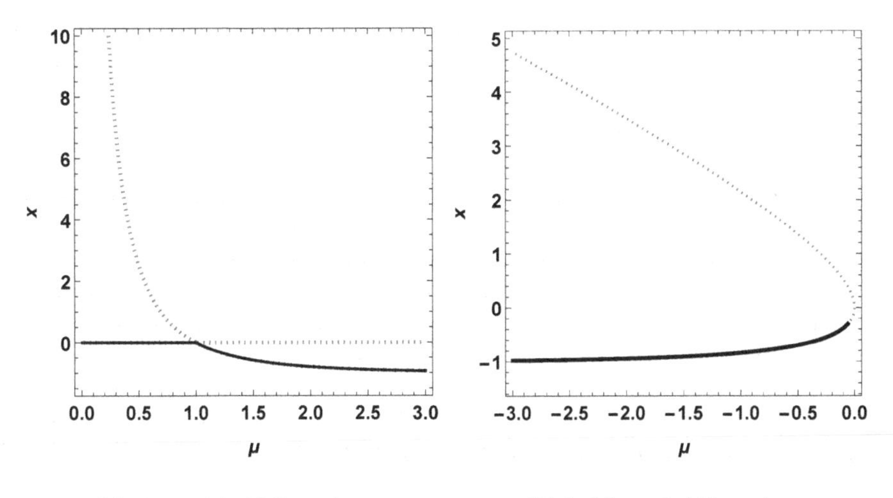

(a) Transcritical bifurcation. (b) Saddle-node bifurcation.

FIGURE 3.43: *The figures show (a) transcritical bifurcation as μ passes through $\mu = 1$; (b) saddle-node bifurcation as μ passes through $\mu = 0$.*

Problem 3.9.19 *Show that the system $\frac{dx}{dt} = \mu x - ln(1+x)$ undergoes a transcritical bifurcation as μ is varied, and find the value of μ at the bifurcation point.*

Solution: Clearly, $x = 0$ is a fixed point for all μ (by inspection). Transcritical bifurcation occurs when $f(x^*) = f'(x^*) = 0$, where $f(x, \mu) = \mu x - ln(1+x)$ and $f'(x) = \mu - \frac{1}{1+x}$. Here, $f(0) = 0$ and $f'(0) = \mu - 1$, which vanishes at $x = 0$ for $\mu = 1$. Now,

$$f(0,0)=0,\ \frac{\partial f(0,1)}{\partial x}=\left(\mu-\frac{1}{1+x}\right)|_{(0,1)}=0,$$
$$\frac{\partial f(0,1)}{\partial \mu}=x|_{(0,1)}=0,$$
$$\frac{\partial^2 f(0,1)}{\partial x^2}=\frac{1}{(1+x)^2}|_{(0,1)}=1\neq 0,\ \frac{\partial^2 f(0,1)}{\partial x\partial \mu}=1\neq 0.$$

Therefore, the system exhibits transcritical bifurcation at $\mu=1$ and the bifurcation occurs at $x=0$ (fig. 3.43(a)). For small x,

$$\frac{dx}{dt}=\mu x-[(x-\frac{x^2}{2})+O(x^3)]=(\mu-1)x+\frac{x^2}{2}+O(x^3).$$

Hence, a transcritical bifurcation occurs at $\mu=1$, which is also the equation for the bifurcation curve. The non-zero fixed point is given by the solution of

$$\mu-1+\frac{x}{2}=0\ \Rightarrow\ x=2(1-\mu).$$

Note: It is to be noted that the transcritical bifurcation arises in systems where there is some basic trivial solution branch ($x=0$, in this case), there exists for all values of the parameter μ. At the bifurcation point $(x,\mu)=(0,1)$, a second solution branch $x=\mu$ crosses the first one. As the branches cross, one solution becomes stable from unstable while the other goes from unstable to stable, which is termed as the exchange of stability.

Problem 3.9.20 *Show that the system* $\frac{dx}{dt}=\mu+x-ln(1+x)$ *undergoes a saddle-node bifurcation as* μ *is varied and find the value of* μ *at the bifurcation point.*

Solution: The bifurcation point x^* is obtained by imposing the condition that the graphs of $\mu+x$ and $\ln(1+x)$ intersect tangentially, which implies that the functions and their derivatives and equal, that is, $\mu+x=\ln(1+x)$ and $\frac{d}{dx}(\mu+x)=\frac{d}{dx}(\ln(1+x))$. From the second equation, $1=\frac{1}{1+x}\ \Rightarrow x=0$. Hence, from the first equation, $\mu=0$. Therefore, the bifurcation occurs at the point $x=0$ for $\mu=0$. Now,

$$f(0,0)=0,\ \left(\frac{\partial f}{\partial x}\right)_{(0,0)}=(1-\frac{1}{1+x})_{(0,0)}=0,$$
$$\left(\frac{\partial^2 f}{\partial x^2}\right)_{(0,0)}=(\frac{1}{(1+x)^2})_{(0,0)}=1\neq 0,\ \left(\frac{\partial f}{\partial \mu}\right)_{(0,0)}=1\neq 0.$$

Therefore, the system exhibits saddle-node bifurcation at $\mu=0$ and the bifurcation occurs at $x=0$ (fig. 3.43(b)).

3.10 Mathematica Codes

Mathematica codes of selected figures.

3.10.1 Stable Node (Figure 3.1(a))

```
s1 = NDSolve[{x'[t] = −3x[t], y'[t] = −2y[t], x[0] = 5, y[0] = 5}, {x, y}, {t, 20}];
Plot[Evaluate[{x[t], y[t]}/. s1, {t, 0, 5}], PlotRange → All, PlotStyle → {Black,}
AxesLabel → {"t", "x(t),y(t)"}, BaseStyle → {FontWeight → "Bold",
FontSize → 12}]
```

Note: All time-series graphs in this chapter can be generated by modifying this code.

3.10.2 Stable Node (Figure 3.1(b))

```
data1 = Table[{{x, y}, {−3x, −2y}}, {x, −5, 5, 0.1}, {y, −5, 5, 0.1}];
plot1 = ListStreamPlot[data1]
eq[x0_, y0_]:={x'[t] = −3x[t], y'[t] = −2y[t], x[0] = x0, y[0] = y0};
graph[x0_, y0_]:=Module[{xans, yans},
{xans, yans} = {x[t], y[t]}/. Flatten[NDSolve[eq(x0, y0), {x[t], y[t]}, {t, 0, 5}]];
ParametricPlot[{xans, yans}, {t, 0, 5}, PlotRange → {{−5, 5}, {−5, 5}},
AxesLabel → {"x", "y"}, DisplayFunction → Identity, PlotStyle → Black]];
points = {{5, 5}, {−5, 5}, {−5, −5}, {5, −5}, {5, 2.5}, {2.5, 5}, {−2.5, 5}, {−5, 2.5},
{−5, −2.5}, {−2.5, −5}, {2.5, −5}, {5, −2.5}, {5, 0}, {0, 5}, {−5, 0}, {0, −5}};

    Module[{i, x0temp, y0temp, pair, newgraph}, graphs = {};
    Do[pair = points[[i]];
    x0temp = First[pair];
    y0temp = Last[pair];
    newgraph = graphz1[x0temp, y0temp];
    graphs = Append[graphs, newgraph], {i, 1, 16}]]
    graph1 = Show[graphs, DisplayFunction → $DisplayFunction]
    Show[plot1, graph1, ImageSize → 250]
```

Note: All phase-portraits in this chapter can be generated by modifying this code.

3.10.3 One-Dimensional Bifurcation (Figure 3.43(a))

```
f[y_, a_]:=ay − ln(y + 1)
amin = 0;
amax = 3;
ymin = −1.5;
ymax = 10;
sols = NSolve[f(y, a) = 0, y];
p1 = Plot[Evaluate[Table[y/. sols[[i]], {i, 1, Length[sols]}]], {a, amin, amax},
PlotStyle → {{AbsoluteThickness[2], Black}}, DisplayFunction → Identity];
p2 = ContourPlot[f[y, a], {a, amin, amax}, {y, ymin, ymax}, Contours → {0},
ContourShading → False, FrameLabel → {μ, x}, PlotPoints → 50,
ContourStyle → {{Dotted, AbsoluteThickness[3]}},
BaseStyle → {FontWeight → Bold, FontSize → 14];
Show[p2, p1, PlotRange → {ymin, ymax}]
```

Note: All one-dimensional bifurcations (saddle-node, transcritical, pitchfork) can be generated by modifying this code.

3.10.4 Chaotic Behavior (Figure 3.34(c))

```
s3 = NDSolve[{X'(t) = −(Y(t) + Z(t)), Y'(t) = aY(t) + X(t),
Z'(t) = b − cZ(t) + X(t)Z(t), X(0) = 1, Y(0) = 1, Z(0) = 1}/.
{a → 0.2, b → 0.2, c → 5.7}, {X, Y, Z}, {t, 0, 500}];
```

```
ParametricPlot3D[Evaluate[{X(t), Y(t), Z(t)}/. s3], {t, 0, 500},
BoxRatios → {1, 1, 1}, PlotRange → All, PlotPoints → 1500,
PlotStyle → {Black}, AxesLabel → {x[t], y[t], z[t]}, BaseStyle → {FontWeight
→ Bold, FontSize → 12}, AxesLabel → TraditionalForm/@l]
```

Note: All parametric plots in 3D can be generated by modifying this code.

3.11 Matlab Codes

Matlab codes of selected figures.

3.11.1 Stable Node (Figure 3.1(a))

```
function stable_node
t = 0 : 0.001 : 5; initial_x = [5 5];
[t, x] = ode45(@rhs, t, initial_x);
plot(t, x(:, 1), t, x(:, 2));
xlabel('t'); ylabel('x(t), y(t)');
function dxdt = rhs(t, x)
dxdt_1 = -3 * x(1); dxdt_2 = -2 * x(2);
dxdt = [dxdt_1; dxdt_2];
end
end
```

3.11.2 Stable Node (Figure 3.1(b))

```
xp = @(t, x)[-3 0; 0 -2] * x;
[t x] = ode45(xp, [0, 5], [5; 5]); plot (x(:, 1), x(:, 2),'black'), hold on
[t x] = ode45(xp, [0, 5], [-5; 5]); plot (x(:, 1), x(:, 2),'black'), hold on
[t x] = ode45(xp, [0, 5], [-5; -5]); plot (x(:, 1), x(:, 2),'black'), hold on
[t x] = ode45(xp, [0, 5], [5; 2.5]); plot (x(:, 1), x(:, 2),'black'), hold on

[t x] = ode45(xp, [0, 5], [2.5; 5]); plot (x(:, 1), x(:, 2),'black'), hold on
[t x] = ode45(xp, [0, 5], [-2.5; 5]); plot (x(:, 1), x(:, 2),'black'), hold on
[t x] = ode45(xp, [0, 5], [-5; -2.5]); plot (x(:, 1), x(:, 2),'black'), hold on
[t x] = ode45(xp, [0, 5], [2.5; -5]); plot (x(:, 1), x(:, 2),'black'), hold on
[t x] = ode45(xp, [0, 5], [5; -2.5]); plot (x(:, 1), x(:, 2),'black'), hold on
[t x] = ode45(xp, [0, 5], [-5; 2.5]); plot (x(:, 1), x(:, 2),'black'), hold on
[t x] = ode45(xp, [0, 5], [5; 0]); plot (x(:, 1), x(:, 2),'black'), hold on
[t x] = ode45(xp, [0, 5], [0; 5]); plot (x(:, 1), x(:, 2),'black'), hold on
[t x] = ode45(xp, [0, 5], [-5; 0]); plot (x(:, 1), x(:, 2),'black'), hold on
[t x] = ode45(xp, [0, 5], [0; -5]); plot (x(:, 1), x(:, 2),'black'), hold on
[x1, x2] = meshgrid(-5 : 0.5 : 5, -5 : 0.5 : 5);
x1dot = -3 * x1; x2dot = -2 * x2;
quiver(x1, x2, x1dot, x2dot)
xlabel('x1') ylabel('x2')
```

3.11.3 Saddle Node Bifurcation (Figure 3.22(a))

```
syms mu x
eqn = mu − x^2 = 0;  S = solve(eqn, x);
r = −0.1 : 0.01 : 1; f = subs(S, r); fprimo = inline(' − 2 ∗ x');
for i = 1 : numel(S)
segnodif = fprimo(f(i, :));
for k = 1 : numel(segnodif);
if isreal(segnodif)(k) == 0;
segnodif)(k) = NaN;
end
end
I1 = find(realsegnodif > 0); I2 = find(realsegnodif < 0);
I3 = find(realsegnodif == 0);   holdon
plot(r(I1), real(f(i, I1)),' −')
plot(r(I2), real(f(i, I2)),' b')
plot(r(I3), real(f(i, I3)),' *')
axis([r(1) r(end)  − 1.0 1.0])
xlabel('mu') ylabel('x') title(Saddle-node bifurcation')
end
```

3.11.4 Chaotic Behavior (Figure 3.34(c))

```
function Rossler
a = 0.2; b = 0.2; c = 5.7;
t = 0 : 0.01 : 500;  initial_x = [1 1 1];
[t, x] = ode45(@rhs, t, initial_x);
plot3(x(:, 1), x(:, 2), x(:, 3));
xlabel('x(t)'); ylabel('y(t)'); zlabel('z(t)');
function  dxdt = rhs(t, x)
dxdt_1 = −(x(1) + x(3)); dxdt_2 = x(1) + a ∗ x(2);
dxdt_3 = b − c ∗ x(3) + x(1) ∗ x(3);
dxdt = [dxdt_1; dxdt_2; dxdt_3];
end
end
```

3.12 Exercises

1. A system with a non-linear force is given by $\frac{da}{dt} = -kt + \alpha t^3$, $(\alpha, k > 0)$. Find the steady-state solutions and check for stability about each of them.

2. A system satisfies the equation $\frac{dy}{dx} = \alpha(e^{\beta x} - 1)$ $(\alpha, \beta > 0)$. Determine all the equilibrium points and check for stability about each of the equilibrium points.

3. Smith's model of population growth is given by $\frac{dN}{dt} = \frac{\alpha N(f-\beta N)}{1+\alpha r N}$. Find the equilibrium solutions and check for stability.

4. Show that the explicit solution of the logistic equation $\frac{dN}{dt} = aN - bN^2$ is given by

   $N(t) = \frac{\frac{a}{b}}{1 + (\frac{a-bN_0}{bN_0})e^{-at}}$, where $N(0) = N_0$. Hence, deduce

   $N(t) = \alpha + \beta \tanh\left(\frac{at}{2}\right)$, where α and β are constants to be determined.

5. A growth model of a population is given by

   $$\frac{dN}{dt} = aN^2 - bN.$$

   (i) Obtain the exact solution with $N(0) = N_0$.
   (ii) Find the equilibrium solutions of the population and check their stability.
   (iii) What will happen to the population for large t, $(t \to \infty)$, when $N_0 > \frac{b}{a}$, $N_0 < \frac{b}{a}$, and $N_0 = \frac{b}{a}$?

6. Analyze the stability (both local and global) of the following systems about the origin $(0, 0)$:

   (i) $\frac{dx}{dt} = y - x - xy^2$, $\frac{dy}{dt} = -2x - y - x^2y$.

   (ii) $\frac{dx}{dt} = -y - x^3$, $\frac{dy}{dt} = x - y^3$.

   (iii) $\frac{dx}{dt} = -x + 4y$, $\frac{dy}{dt} = -x - y^3$.

   (iv) $\frac{dx}{dt} = (x - y)\,(x^2 + y^2 - 1)$, $\frac{dy}{dt} = (x + y)\,(x^2 + y^2 - 1)$.

7. Discuss the type and stability of steady state at $(0,0)$ of the system

$$\frac{dx}{dt} = -5x + ky, \ \frac{dy}{dt} = -x + y \ \ (k > 0),$$

for different values of k.

8. (i) Suppose an archaeologist excavates a bone and measures its content for radiocarbon C^{14}. If the result is 25% of the carbon present in the bones of a living organism, what can be said about the age of the bone? The half-life of C^{14} is 5730 years.
(ii) There are 8 grams of C^{14} remaining in a fossil after 17830 years. Calculate the initial amount of C^{14} (in grams) in the fossil. Half-life of C^{14} is 5730 years.
(iii) A fossil is found after 33450 years. What percentage of the initial sample of C^{14} remains, after the given time has passed? Half-life of C^{14} is 5730 years.

9. The rate of decrease of concentration of medicine in the bloodstream is directly proportional to the amount present in the bloodstream. Equal doses D of the medicine are given to a patient at times $t = 0, T, 2T, ..., nT, ...$ If $x(t)$ be the concentration of the medicine at time t, use a differential equation to formulate a continuous model. Find an expression for $x(nT)$ and hence evaluate $\lim_{x \to \infty} x(nT)$.

10. A drug of dose y_0 is given to a patient at regular intervals of time T. The concentration of the drug present in the system follows the law

$$\frac{dc}{dt} = -ke^c \ (k > 0), \ c(0) = y_0.$$

Show that the concentration of the drug $c(2T)$, just before the third dosage, is given by

$$c(2T) = -\log\left\{kT\left(1 + e^{-y_0}\right) + e^{-2y_0}\right\}. \text{ Hence, prove that}$$
$$c(nT) = -\log\left\{kT\left(1 + e^{-y_0} + - - - + e^{-(n-1)y_0}\right) + e^{-ny_0}\right\}.$$

Also, find the required time interval T_c, if the concentration of the drug in the system tends to the value c_T as the number of doses increases.

11. In a series RLC circuit, the differential equation governing the current i and the circuit parameters is given by

$$V - L\frac{di}{dt} - \frac{Q}{C} = iR,$$

where V is the impressed constant voltage from the battery, $\frac{Q}{C}$ is the

magnitude of the voltage developed across the capacitor, when a voltage is consumed in establishing the charge Q across the capacitor and $L\frac{di}{dt}$ is the induced emf which opposes the impressed voltage.
(i) Show that the current satisfies the differential equation

$$L\frac{d^2i}{dt^2} + R\frac{di}{dt} + \frac{i}{c} = 0.$$

(ii) Find the solution of the differential equation stated in (i), considering the cases (a) $R^2 > \frac{4L}{C}$, (b) $R^2 = \frac{4L}{C}$ and (c) $R^2 < \frac{4L}{C}$. What can you say about the behavior of the current in all these cases?

12. Consider an RLC circuit with impressed AC voltage $V = V_0 \sin wt$, where V_0 is the peak value of the voltage and $w = 2\pi f$, f being the frequency of supplied voltage. Show that the current $i(t)$ satisfies the differential equation

$$L\frac{d^2i}{dt^2} + R\frac{di}{dt} + \frac{i}{e} = wV_0\cos wt,$$

whose solution is

$$\begin{aligned} i &= e^{-\frac{R}{2L}t}\left[A_1\cos\sqrt{\left(\frac{1}{LC} - \frac{R^2}{4L^2}\right)}t + A_2\sin\sqrt{\left(\frac{1}{LC} - \frac{R^2}{4L^2}\right)}t\right] \\ &+ I_0\sin(wt - \theta), \end{aligned}$$

where

$$I_0 = \frac{V_0}{\sqrt{R^2 + \left(wL - \frac{1}{wC}\right)^2}} \text{ and } \theta = \tan^{-1}\frac{\left(wL - \frac{1}{wC}\right)}{R}.$$

13. If the charges on a plate are $+Q$, $-Q$, and V give the voltage between the plates, then the capacitance is given by $C = \frac{Q}{V}$. The ratio of flow of electric charge is the current flowing in a circuit at any time t, that is, $i(t) = \dfrac{dQ}{dt}$. We consider a circuit where a resistance R is connected with a capacitor of capacitance C and a battery (Volt V) is connected in series through a key K. The differential equation governing the charging of the capacitor is given by

$$V - \frac{Q}{C} = iR, \text{ where } i = \frac{dQ}{dt}.$$

(i) Show that the charge stored in the capacitor at any time t is given by $Q = Q_0\left(1 - e^{-\frac{t}{CR}}\right)$ and the current $i(t) = \frac{V}{R}e^{-\frac{t}{CR}}$.
(ii) If a charged condenser of Capacitance C is allowed to discharge through a resistance R, then $Q = Q_0e^{-\frac{t}{CR}}$.

14. A linear spring-mass system satisfies the differential equation

$$m\frac{d^2x}{dt^2} = -kx.$$

Show that

$$m\left(\frac{dx}{dt}\right)^2 + kx^2 = A \text{ (constant)}.$$

If $x(0) = x_0$ and $\frac{dx}{dt}\mid_{t=0} = v_0$, evaluate A. Also, find the velocity of the mass when it passes through its equilibrium solution.

15. A particle of mass m moving with initial velocity u is retarded by air resistance, which is proportional to the square of the velocity at that instant. Show that $v = \frac{u}{1+kut}$ and $v = ue^{-kx}$, where v is the velocity at any time t, at a distance x from the starting point and k is the constant of the proportionality.

16. A particle of mass m is projected vertically upwards under gravity with a velocity u, the air resistance being Kv per unit mass, where v is the velocity of the particle at any time t and K is a constant.
(i) Write down the equation of motion.
(ii) Show that the particle comes to rest at a height $\frac{u}{K} - \frac{g}{K^2}\log(1 + \frac{Ku}{g})$ above the point of projection, g being the acceleration due to the gravity and u is the initial velocity of projection.
(iii) Suppose the particle falls downwards from rest, instead of being projected upward, then show that the distance covered by the particle in time t is $\frac{gt}{K} + \frac{g}{K^2}(e^{-Kt} - 1)$.

17. A truck of mass M, whose engine works at a constant rate R, runs on a level road. If the maximum attainable velocity be W, show that the distance at which the truck (starting from rest) acquires a velocity V is $\frac{MW^3}{R}(\log(\frac{W}{W-V}) - \frac{V}{W} - \frac{1}{2}\frac{V^2}{W^2})$. Assume that there is a frictional resistance F, and P is the pull of the truck.

18. A steamer of mass M requires a horsepower H at its maximum speed V. The engine of the steamer exerts a constant propeller thrust at all speeds. As the steamer moves, there is a resistance that is proportional to the square of the speed. If the steamer acquires a velocity of v in time t from rest, show that $t = \frac{MV^2}{H+g}\log(\frac{V+v}{V-v})$, g being the acceleration due to gravity.

19. (i) A particle of mass m moves in a straight line with an acceleration $\mu(x + \frac{c^4}{x^3})$, which is directed towards the origin. Show that the particle will arrive at the origin in time $\frac{\pi}{4\sqrt{\mu}}$, if it starts from rest at a distance c.
(ii) A particle of mass m moves in a straight line with an acceleration $\mu x^{-5/3}$, which is directed towards the origin. Show that the particle will

arrive at the origin in time $\frac{2c^{4/3}}{\sqrt{3\mu}}$, if it starts from rest at a distance c. What will be its velocity there?

(iii) A particle of mass m moves in a straight line with an acceleration $\left(\frac{c^5}{x^2}\right)^{1/3}$, which is directed towards the origin. Show that the particle will arrive at the origin with a velocity $c\sqrt{6\mu}$ in time $\frac{8}{15}\sqrt{\frac{6}{\mu}}$, if it starts from rest at a distance c.

20. A particle is projected vertically upwards with a speed u from a point on the earth's surface. Let v be the speed of the particle in any position x, R be the radius of the earth, and g the acceleration due to gravity.
(i) Assuming that the acceleration due to gravity varies inversely as the square of the distance x from the center of the earth and neglecting air resistance, show that

$$v^2 = u^2 - 2gR^2\left(\frac{1}{R} - \frac{1}{x}\right).$$

(ii) If H be the greatest height reached by the particle, then show that

$$H = \frac{2gR^2}{2gR - u^2} - R.$$

21. The acceleration due to gravity varies inversely as the square of the distance from the center, when the attracted particle is outside the surface $\left(\frac{\mu_1}{x^2}\right)$ and, inside the earth, the acceleration at any point varies as its distance from the center of the earth $(\mu_2 x)$. Let a be the radius of the earth and g be the acceleration on the surface of the earth.
(i) If b be the distance from the center of the earth to the point from which the particle falls, and v_1 be the velocity on reaching the surface, then show that $v_1^2 = 2ag\left(1 - \frac{a}{b}\right)$.
(ii) Also, show that $v_2^2 = ag\left(3 - \frac{2a}{b}\right)$, where v_2 is the velocity on reaching the center.

22. A particle moves in a straight line under an acceleration μ^2 (distance) towards a fixed point in the line. A periodic disturbing force $F\cos(bt)$ is also acting on the particle.
(i) Write the differential equation that models this situation and find its general solution.
(ii) If the particle starts from rest at a distance a from the center, then show that

$$x = \left(a - \frac{F}{\mu^2 - b^2}\right)\cos(\mu t) + \frac{F}{\mu^2 - b^2}\cos(bt).$$

(iii) What will be the solution when $\mu = b$, assuming the same initial condition stated in (ii)?

23. Consider a light elastic string of natural length a and modulus of elasticity 2 mg. One end of the string is attached to a fixed point O and the other end to a particle of mass m. Initially, the particle is held at rest at O and released. (i) Write the differential equation which models this scenario.
(ii) Find the greatest extension of the string during the motion.
(iii) Show that the particle will reach back to O again after a time

$$\sqrt{\frac{2a}{g}}\left[\frac{\pi}{2}+2+\sin^{-1}\frac{1}{\sqrt{5}}\right].$$

24. (a) A particle of mass m falling from rest, is subjected to the gravitational force and air-resistance kv^2 $(k > 0)$, prove that $v^2 = a^2(e^{\frac{-2kx}{m}} - 1)$, where $mg = ka^2$.

(b) A raindrop of mass m falls from a cloud with an initial velocity, the air resistance being $\frac{gv^2}{k^2}$ per unit mass. Assuming that the raindrop does not gather further moisture as it proceeds, show that the velocity of the raindrop when it has traveled a distance S is $v^2 = k^2 + (u^2 - k^2)e^{-\frac{2gS}{k^2}}$, g being the acceleration due to gravity.

25. (a) Newton's law of cooling states that the rate of change of the temperature of an object is proportional to the difference between it's own temperature and the temperature of it's surroundings. Suppose a dead-body with a temperature 85 °F was discovered at mid-night, the surrounding temperature was 70 °F (constant). Almost instantly, the body was removed to the morgue, where the surrounding temperature was maintained at 40 °F. After one hour, the body temperature was found to be 60 °F. Using Newton's law of cooling, find the time of death.

(b) A cheesecake baking in the oven is removed with an internal temperature of 165 °F and is placed inside a refrigerator maintaining 35 °F. The cheesecake has cooled to 150 °F after 10 minutes. How long must one wait so that the cheesecake cools down to 70 °F, before it can be eaten?

(c) A cup of coffee with initial temperature T_0 is put in a room at the temperature of T_{S0}. The coffee cools according to Newton's law with the constant k. However, the temperature of the room slowly increases by the linear law $T_S = T_{S0} + \beta t$, where β is the known parameter. Find an expression for the time τ, when the temperature of the coffee and the surrounding environment temperature become equal.

26. (a) Two chemicals A and B react to produce another chemical C. The rate at which C is produced varies as the product of the amounts of A and B simultaneously. The formation requires 2 kgs of A for every 3 kgs of B. If 10 kgs of A and 30 kgs of B are present initially, then 5 kgs

of C are produced in 1 hour. Using the differential equation, obtain an expression for the amount of C at any time t.

(b) Two chemical substances x and y combine in the ratio 3:4 to form a third substance z. In 5 minutes, 50 gms of z are formed, when 40 gms of x and 70 gms of y are mixed together. Calculate how many gms of z will be formed in 210 minutes.

27. (a) A tank contains 1000 liters of fresh water. Salt water which contains 150 gms of salt/liter runs into the tank at the rate of 5 liters/min and the well-stirred mixture runs out of it at the same rate. When will the tank contain 5000 gms of salt?

    (b) A tank contains 60 liters of a solution of which 90% water and 10% alcohol. At time $t = 0$, a second solution consist of 50% water and 50% alcohol is poured into the tank at a rate of 2 liters per minute. The solution is drained out of the tank at the rate of 3 liters per minute. Assuming that the tank is continuously stirred, formulate a mathematical model to express the rate of change of $S(t)$, the amount of salt in the tank at time t. Hence, find the amount of salt in the tank between $t = 0$ and $t = 50$.

28. Mr. Gomes borrowed A amount of money from a bank at a daily interest r. He also makes a micro-payment of k amount to the bank each day.
    (i) If x be the amount of money owed by the bank at the end of each day, then model the situation with the help of a differential equation, and solve it.
    (ii) If the amount k is very small, Mr. Gomes will never pay off the loan. Obtain the critical value of k such that the amount of loan remains the same.
    (iii) If Mr. Gomes wishes to return the loan amount in $\frac{n}{2}$ days, how much amount he has to pay each day?

29. A tumor ceases to grow as its interior does not have the oxygen supply. This situation is modeled by the Gompertz growth law $\frac{dM}{dt} = r(t)M(t)$, $M(0) = M_0$, where $\frac{dr}{dt} = -\alpha r(t)$, $r(0) = r_0$, r is the effective growth rate of tumor decrease exponentially. Find an expression for $M(t)$ and show that $\ln(\frac{M}{\theta}) = \frac{-r}{\alpha}e^{-\alpha t}$, where θ is a constant whose value needs to be determined.

30. A population of fish follows logistic growth given by $\frac{dF}{dt} = rF(1 - \frac{F}{K})$, where r is the intrinsic growth rate and K is the carrying capacity. Also, given $F(t_1) = K_1$, $F(t_1 + T) = K_2$ and $F(t_2 + T) = K_3$, show that the carrying capacity is given by

$$K = \frac{\frac{1}{K_1} + \frac{1}{K_2} + \frac{1}{K_3}}{\frac{1}{K_1 K_3} - \frac{1}{K_2}^2}.$$

31. Levins [93, 94] proposed the basic metapopulation model as

$$\frac{dp}{dt} = cp(1-p) - mp,$$

where p $(0 < p < 1)$ is the fraction of patches occupied by a species. (i) Explain the model. Find the non-trivial steady-state solution and interpret the result biologically. Also, find the stability condition.
(ii) Hanksi [56] estimated colonization and extinction probabilities for a large population of butterflies in Finland. Put $c = c_0 e^{-\alpha D}$, where D is the average distance between the habitats and $m = m_0 e^{-\beta A}$, where A is the average area of the preferred habitats. Explain the model. Find the non-trivial steady-state solution and interpret the result biologically. Also, find the stability condition.

32. A prey-predator model with constant cover k (> 0) is given by

$$\frac{dx}{dt} = \alpha x - \beta(x-k)y, \; \frac{dy}{dt} = -\gamma y + \delta(x-k)y, \; (\alpha, \beta, \gamma, \delta > 0).$$

The predators (y) do not have access to k of the prey (x). Find the equilibrium points and check for their stability. For $\alpha = 0.1, \beta = 0.002, \gamma = 0.05, \delta = 0.001, k = 50, x(0) = y(0) = 40$, solve the model numerically and plot the graph. What conclusion do you obtain from the graph?

33. Let $N(t)$ be the number of tiger population at any time t. The quotient of birth rate and death rate by the population size N are respectively by, $\frac{Birthrate}{N} = \frac{3}{2} + \frac{1}{1000}N$ and $\frac{Deathrate}{N} = \frac{1}{2} + \frac{1}{3000}N$. Formulate a model (using differential equation) that describes the growth and regulation of this tiger population. Solve for $N(t)$, assuming $N(0) = 100$ and describe the long-term behavior of this tiger population $t \to \infty$.

34. A population follows a generalized logistic growth given by

$$\frac{dx}{dt} = \frac{rx}{\alpha}\left[1 - \left(\frac{x}{K}\right)^{\alpha}\right], \; x(0) = x_0, \; \alpha > 0.$$

(i) Find the exact solution of this differential equation and show that the limiting population is K.
(ii) What happens to the model if $\alpha \to 0$ and $\alpha \to -1$?

35. The favorite food of the tiger shark is the sea turtle. A two-species prey-predator model is given by

$$\frac{dP}{dt} = P(a - bP - cS), \; \frac{dS}{dt} = S(-k + \lambda P), \; P(0) = P_0, \; S(0) = S_0,$$

where P is the sea turtle (prey), S is the tiger shark (predator), and a, b, c, k, λ are positive constants.

(i) Let $b = 0$ and the value of k is increased. Ecologically, what is the interpretation of increasing k and what is its effect on the non-zero equilibrium populations of sea turtles and sharks?
(ii) Obtain all the equilibrium solutions for $b = 0$ and $b \neq 0$.
(iii) Obtain the linearized system about the equilibrium point (P^*, S^*) and find the condition of its stability. Interpret the result in the context of the model ecologically.
(iv) Obtain time-series solution for $a = 0.1, b = 0.0005, c = 0.002, k = 0.05, \lambda = 0.001, P(0) = 20, S(0) = 15$. What do you expect to happen to the dynamics of the model if c $= 0$? Repeat the same with $c = 0$ and $\lambda = 0$.
(v) The two species prey-predator model is now modified as

$$\frac{dP}{dt} = P(a - bP - cS), \ \frac{dS}{dt} = S(-k + \lambda P - \sigma S), P(0) = P_0, S(0) = S_0.$$

Describe the model by pointing out the difference between this and the previous one. Obtain the equilibrium points and check for their stability.
(vi) For $a = 0.1, b = 0.0005, c = 0.002, k = 0.05, \lambda = 0.001, \sigma = 0.01$, check numerically the stability condition. Hence, obtain the time-series solution as well as phase portrait for $P(0) = 20, S(0) = 15$.

36. May's prey-predator model is given by

$$\frac{dx}{dt} = rx\left(1 - \frac{x}{k}\right) - \frac{\beta xy}{x + \alpha}, \frac{dy}{dt} = sy\left(1 - \frac{y}{\gamma x}\right), \ (r, k, \alpha, \beta, s, \gamma > 0).$$

(i) Explain the model. Find the equilibrium points and obtain the condition(s) for stability.
(ii) For $r = \gamma = \beta = \alpha = 1.0, s = 1.2, k = 10$, obtain numerically the equilibrium points, check the stability condition(s), find time-series solution and phase portrait for $x(0) = y(0) = 10$, and comment on the dynamics of the two species. How the dynamics of the model changes if $s = 0.2$?

37. Malaria is a disease spread by carriers. Let $S(t)$ and $C(t)$ be the number of susceptibles and carriers in a population. Let the carriers, after identification, be removed from the population at a rate β. It is also assumed that malaria spreads at a rate proportional to the product of $S(t)$ and $C(t)$. Formulate a mathematical model using differential equations for susceptibles and carriers, and explicitly solve for $C(t), S(t)$ assuming $C(0) = C_0, S(0) = S_0$. Also, Find the number of susceptibles that escape the epidemic.

38. Consider a model that tries to capture the buying behavior of the consumer towards a branded mouthwash (or any product). Let $L(t)$ be the level of buying of the consumer and $A(t)$ be the attitude of the

consumer towards the product. Then, the differential equation governing the model is

$$\frac{dL}{dt} = \alpha A - \beta L, \ \frac{dA}{dt} = \gamma L - \delta A + cV,$$

where $V = V(t)$ is the advertising policy, $\alpha, \beta, \gamma, \delta$ are positive parameters.
(i) Show that $L(t)$, the level of buying of the consumer, satisfies the equation

$$\frac{d^2L}{dt^2} + (\beta + \delta)\frac{dL}{dt} + (\beta\delta - \alpha\gamma)L = \alpha cV.$$

Also, show that for constant advertising V_0, the buying level tends to a limiting value.
(ii) Predict the buying behavior when $\alpha = \gamma = c = 1, \beta = \delta = 2, L(0) = 0, A(0) = 0$ and

$$V(t) = \begin{cases} 100 \text{ units for } 0 < t < 10, \\ 0, \text{ for } t > 10. \end{cases}$$

39. Space docking is a technique by which two spacecraft are prevented from colliding by some mechanism. The differential equation modeling the space docking mechanism is given by

$$\frac{dx}{dt} = -kwx - kcy, \ \frac{dy}{dt} = x - y, \ x(0) = x_0, y(0) = y_0, \ (k, w, c > 0).$$

Find the condition of stability of the model about the equilibrium point $(0, 0)$. Taking $w = 10, c = 5$ and $k = 0.02$, verify numerically, the stability of the system about the equilibrium solution $(0, 0)$. Hence, draw the time-series solution and phase portrait for this model for $x(0) = y(0) = 10$.

40. F.W. Lanchester developed a conventional combat model between two armies during World War I. Let $x(t)$ and $y(t)$ be the number of troops for army A and army B, respectively. Lanchester assumed that the combat loss rate of a conventional army is proportional to the size of the opposing army. He also assumed that the operational losses are neglected and there is no reinforcement.
(i) Obtain the differential equations that incorporate Lanchester's assumptions.
(ii) Solve the equation explicitly and comment on the long term behavior of the model.

41. F.W. Lanchester developed a mixed combat model between two armies (conventional army and guerrilla army) during World War I. Let $x(t)$ and $y(t)$ be the number of troops for conventional army A and a guerrilla army B, respectively. Lanchester assumed that the combat loss rate for

the conventional army A is proportional to the size of the opposing guerrilla force B and the combat loss rate for the guerrilla force B is proportional to the product of the sizes of both the armies A and B. Also, he assumed that the operational losses are neglected and there is no reinforcement.
(i) Obtain the differential equations that incorporate Lanchester's assumptions.
(ii) Solve the equation explicitly and comment on the long-term behavior of the model.

42. (a) Let $D(t)$ and $N(t)$ denote the national debt and total national income, respectively. Domer's first debt model assumes that the rate of the national debt is proportional to the national income and the rate of increase of national income is constant.
(i) Formulate the mathematical model using differential equations from the given assumptions.
(ii) Assuming $D(0) = D_0$ and $N(0) = N_0$, solve the differential equation and comment on the dynamics of the model.
(iii) If the rate of increase of national income is proportional to the income (instead of being constant), obtain the modified Domar debt model (Domar's second debt model).
(iv) Solve the differential equations obtained in (iii) and find the ratio $\frac{D(t)}{N(t)}$ as $t \to \infty$. Interpret the result in the context of the model.

(b) A modified Domar Debt model is given by $D^{'}(t) = aN(t)$; $N^{'}(t) = bN^n(t)$, where $D(t)$ denotes the national debt and $N(t)$ denotes the national income.
(i) Solve the differential equation and deduce Domar's first and second debt models by letting $n \to 0$ and $n \to 1$.
(ii) Discuss the behavior of $\frac{D(t)}{N(t)}$ as $t \to \infty$, for a general value of n.

(c) Let $S(t)$, $I(t)$, and $N(t)$ be the Savings, Investment, and National Income at time t. Domar Macro model assumes that Savings are proportional to the National Income, and that all Savings are invested, and Investment is proportional to the rate of increase of national income.
(i) Formulate a mathematical model using differential equations with the given assumptions.
(ii) Solve the differential equations and comment on the dynamics of the model, assuming $N(0) = N_0$.

43. (a) The three-species ecological system is given by

$$\frac{dx}{dt} = x(a - cy), \quad \frac{dy}{dt} = y(-k + fx - mz), \quad \frac{dz}{dt} = z(-e + sy),$$

where $a, c, k, f, m, e, s > 0$.

(i) Explain the model by describing the role of each species in the ecological system.
(ii) Obtain the equilibrium points and the condition(s) under which the linearized system is stable.
(iii) For $a = 0.5, c = 0.5, k = 1.5, f = 0.5, m = 0.6, e = 0.5, s = 0.75$, find the non-zero equilibrium point and check its stability. Draw the time-series graph and phase-portrait for this set of parameters and comment on the dynamics of the model.
(iv) Repeat the same for $a = 0.5, c = 0.5, k = 1.5, f = 1.5, m = 0.6, e = 0.5, s = 0.75$, and $a = 0.5, c = 1.5, k = 1.5, f = 1.5, m = 0.6, e = 0.5, s = 0.75$. What differences do you observe? Compare the change in dynamics of the model.

(b) The three species ecosystem is given by [157]

$$\frac{dx}{dt} = a_1 x - b_1 xy, \quad \frac{dy}{dt} = -a_2 y + b_2 xy - \frac{c_1 z y^2}{D^2 + y^2}, \quad \frac{dz}{dt} - a_3 z + \frac{c_2 z y^2}{D^2 + y^2}.$$

where all the parameters $a_1, b_1, a_2, b_2, c_1, D, c_2, a_3$ are positive.
(i) Explain the model by taking into account the interactions between the species x, y and z.
(ii) Obtain the equilibrium points and the condition(s) under which the linearized system is stable.
(iii) For $a_1 = 0.5, b_1 = 0.5, a_2 = 0.5, b_2 = 0.5, c_1 = 0.6, D = 10, c_2 = 0.75, a_3 = 0.0065$, find the non-zero equilibrium point and check its stability. Draw the time-series graph and phase-portrait for this set of parameters and comment on the dynamics of the model. Obtain the time-series graphs for $a_3 = 0.007426, 0.01, 0.05$ and compare the results.

44. An arms race model between two countries A and B has the form

$$\frac{dx}{dt} = ay^2 - mx + r, \quad \frac{dy}{dt} = bx^2 - ny + s,$$

where a, b, m, n are positive constants, and r and s can take any sign.
(i) Explain the model in detail.
(ii) Find the equilibrium solution(s) of the model for $r = 0, s = 0$ and obtain the condition for their stability. Now, numerically check the result for $a = 0.1, b = 0.15, m = 2.0, n = 1.5, r = 0, s = 0$. Draw the time-series graph and the phase-portrait of the model. What conclusion can be obtained from the figure in the context of the model?
(iii) Repeat the same for (a) $r = 2, s = 2$; (b) $r = 2, s = -2$; (c) $r = -2, s = 2$; (d) $r = -2, s = -2$. What conclusion can you draw about the outcomes of such an arms race for the following combinations of r and s. Which one gives the favorable result?

45. A two-mode laser model is given by

$$\frac{dn_1}{dt} = G_1(N_0 - \alpha_1 n_1 - \alpha_2 n_2)n_1 - k_1 n_1 \ (G_1, N_0, \alpha_1, \alpha_2, k_1 > 0),$$
$$\frac{dn_2}{dt} = G_2(N_0 - \alpha_1 n_1 - \alpha_2 n_2)n_2 - k_2 n_2 \ (G_2, k_2 > 0),$$

where n_1 and n_2 are two different kinds of photons that are produced.
(i) Find all the equilibrium points of the model and discuss the stability of the linearized system about the equilibrium points.
(ii) Draw different phase portraits for the following set of parameters:
(a) $N_0 = 5, G_1 = G_2 = \alpha_1 = \alpha_2 = 1, k_1 = 7.5, k_2 = 7.5$;
(b) $N_0 = 5, G_1 = G_2 = \alpha_1 = \alpha_2 = 1, k_1 = 7.5, k_2 = 2.5$;
(c) $N_0 = 5, G_1 = G_2 = \alpha_1 = \alpha_2 = 1, k_1 = 2.5, k_2 = 7.5$;
(d) $N_0 = 5, G_1 = G_2 = \alpha_1 = \alpha_2 = 1, k_1 = 7.5, k_2 = 7.5$.
What does the model predicts about the long-term behavior of the laser?

46. In a forest, the fox population grows at the rate of 20% per year and the wolf population at the rate of 50% per year. The species compete for the same resources and the forest can support 100 foxes and 100 wolves.
(i) By taking $F(t)$ and $W(t)$ to be the fox and the wolf populations, respectively, at any time t, formulate a mathematical model. Find the solution for $W(t)$ assuming $W(0) = W_0$.
(ii) Assuming that the competition among the foxes and the wolves decreases the growth rates by an amount proportional to the product of the two populations, modify the model to show the interaction of the foxes and the wolves, 0.0002 and 0.0005 being the rates of decrease for the foxes and the wolves respectively. Find the equilibrium point(s) of the modified model.
(iii) Perform linear stability analysis about all the equilibrium points of the modified model and comment on the stability of the system. Draw time-series graphs and phase portrait for $F(0) = 40, W(0) = 20$.
(iv) Modify the model further by considering an additional term to model the hunting of both foxes and wolves, E being the measure of amount of hunting. At time $t = 0$ (when hunting of both the species started), $F(0) = 40$ and $W(0) = 20$ and at time $t = 10$, $W(5) = 50$. Find the value of E for this to happen. Obtain the graph for long-term behavior of the two populations if the same level of hunting continues.

47. A model for interaction of messenger RNA-M and protein E is given by

$$\frac{dM}{dt} = \frac{E^k}{1+E^k} - \alpha M, \quad \frac{dE}{dt} = M - \beta E.$$

(i) For $k = 1$, interpret the model. Find the steady-state(s) of the model for $k = 1$. Check the stability of the model about the steady state(s). For $\alpha = 0.1, \beta = 0.2, x(0) = 10, y(0) = 10$, draw the time-series graph

and phase-portrait of the model.
(ii) Find the steady state solution(s) for $k = 2$. Check the stability of the model about the steady state(s). What happens where (a) $\alpha\beta < \frac{1}{2}$ (b) $\alpha\beta = \frac{1}{2}$ and (c) $\alpha\beta > \frac{1}{2}$? Draw the time-series graph and phase-portrait of the model for $\alpha = 0.1, \beta = 0.2, x(0) = 10, y(0) = 10$.

48. In biological pattern formation (zebra stripes and butterfly wing patterns), Lewis [95] proposed a simple model involving a biochemical switch, where a gene G is activated by a biochemical signal subsystem S. The model is given by

$$\frac{dG}{dt} = k_1 s_0 - k_2 G + \frac{k_3 G^2}{k_4^2 + G^2} \quad (k_1, k_2, k_3, k_4 > 0),$$

where $G(t)$ is the gene product concentration, the concentration S_0 of S is fixed.
(i) Describe the model and put it in the dimensionless form

$$\frac{dx}{d\tau} = s - rx + \frac{x^2}{1 + x^2}$$

by suitably determining dimensionless quantities.
(ii) Find the fixed points for s=0. Is there any condition which needs to be fulfilled? If yes, what is the condition?
(iii) If $x(0) = 0$, what happens to $x(\tau)$ if s is slowly increased from zero? What happens if s then goes back to zero?

49. (a) A particle moves in a plane with an acceleration which is always directed towards and perpendicular to the axis of x. If initially the particle is projected with an initial velocity $\sqrt{\frac{\mu}{a}}$ parallel to the x-axis from the point (0,2a), show that the path of the particle is a cycloid.

(b) A particle moves in a plane with an acceleration
(i) which is always directed towards a fixed point and varies directly as the distance from the fixed point.
(ii) which is always directed away from a fixed point and varies directly as the distance from the fixed point.
If initially the particle is projected from the point (a,0) with a velocity V along the y-axis (increasing direction), obtain the paths of the particle for both cases.

50. (a) A smooth tube of length L is capable of rotating in a vertical plane about one of its ends, which is fixed. A particle is placed at the other end of the tube when it is in a horizontal position and is made to rotate counterclockwise with a constant angular velocity Ω. If Ω is small, show that the particle will reach the fixed end in time $\left(\frac{6a}{gw}\right)^{\frac{1}{3}}$ approximately.

(b) A smooth tube of length L rotates in a horizontal plane with a constant angular velocity Ω about one of its ends, which is fixed. A particle, that is placed at the other end of the tube is projected with a velocity $L\Omega$ towards the fixed end of the tube. Using differential equations, write the equation of motion. Show that the time taken by the particle to reach half the length of the tube is $\frac{1}{\Omega}\log 2$.

51. Consider the Lorenz system defined by the following set of differential equations:

$$\frac{dx}{dt} = \sigma(y - x), \quad \frac{dy}{dt} = rx - y - xz, \quad \frac{dz}{dt} = xy - bz,$$

where σ, r, and b are positive constants. Analyze the stability of the origin under the condition $0 < r < 1$ (which is known not to lead to any complex behavior).

52. The breadth of a river is L. O and A are two points on opposite sides of the river bank such that OA is perpendicular to the direction of flow of the river. A boat starts from point A to reach the other side and rows with constant velocity v in such a manner that it is always directed towards O. Show that the path of the boat is a parabola, assuming that the river flows with the same velocity v.

53. Consider a pendulum of mass M at the end of a rigid but massless rod of length L. Let θ be the angle, which the pendulum makes in the downward direction and g be the acceleration due to gravity. Formulate a model (using differential equation) that captures the dynamics of the pendulum. Also, discuss the stability of the system about the equilibrium point $(0, 0)$.

54. Consider the Richardson's arms race model

$$\frac{dx}{dt} = \alpha y - \gamma x + r, \quad \frac{dy}{dt} = \beta x - \delta y + s,$$

where x and y are the expenditures on arms by nations A and B respectively.
(i) Taking $\alpha = 1.0, \beta = 1.2, \gamma = 0.9, \delta = 0.8, r = 1.0, s = -2.0$, find the equilibrium solution of the model and check its stability. What can you say about the values of r and s? Draw the time-series graph and the phase portrait by choosing suitable initial conditions. What conclusion do you obtain from them?
(ii) The model is now modified by placing constraints upon the nations A and B by introducing carrying capacity terms [91] and the modified model is

$$\frac{dx}{dt} = (\alpha y - \gamma x + r)\left(1 - \frac{x}{k_1}\right), \quad \frac{dy}{dt} = (\beta x - \delta y + s)\left(1 - \frac{y}{k_2}\right),$$

where k_1 and k_2 are the maximum possible arms expenditures for nations A and B, respectively. Taking $\alpha = 1.0, \beta = 1.2, \gamma = 0.9, \delta = 0.8, r = 1.0, s = -2.0, k_1 = 7$ and $k_2 = 9$, find the equilibrium solutions of the model and check their stabilities. Compare the models by drawing time series graphs and the phase portrait by choosing the same initial conditions.

55. (a) Show that the non-linear autonomous system

$$\frac{dx}{dt} = \mu - x - e^{-x}$$

undergoes a saddle-node bifurcation as μ is varied and find the value of μ at the bifurcation point.

(b) Consider the non-linear autonomous system

$$\frac{dx}{dt} = 1 + x\mu + x^2 \ (\mu > 0).$$

Show that the system undergoes a saddle-node bifurcation as μ passes through $\mu = 2$.

(c) A hypothetical reaction in the study of isothermal autocatalytic reactions was considered by Gray and Scott (1985), whose kinetics in dimensionless form is given as follows:

$$\frac{dx}{dt} = a(1-x) - xy^2, \ \frac{dy}{dt} = xy^2 - (a+k)y,$$

where a and k are positive parameters. Show that the saddle-node bifurcation occurs at $k = -a \pm \frac{\sqrt{a}}{2}$.

56. Consider a model for a vegetation V in a desert, where the growth of the vegetation is limited in the amount of water W in the soil:

$$\frac{dW}{dt} = a_1 - b_1 WV - c_1 W,$$
$$\frac{dV}{dt} = b_2 WV - c_2 W,$$

where a is the rainfall dependent water uptake in the soil, b is the extra water uptake and evaporation by the vegetation, c is the normal evaporation, d is the water dependent growth of the vegetation, and e is the death rate of the vegetation.
(i) How much water does the soil contain if there is no vegetation? Suppose the rainfall increases two-fold because of a change in climate. How much water would the soil contain if there is still no vegetation?

How much water would the soil contain if there is a vegetation? How much water would the soil contain if the rainfall increases two-fold in the presence of vegetation?
(ii) Obtain the equilibrium points of the model and find the condition for its stability.
(iii) For $a_1 = 4.0, b_1 = 0.1, c_1 = 0.02, b_2 = 0.02, c_2 = 0.1$, check numerically the analytical results obtained in (ii). Hence, obtain the time-series solution as well as phase portrait for $W(0) = 10, V(0) = 5$.

57. (a) Show that the non-linear autonomous system

$$\frac{dx}{dt} = x - \mu x(1 - x) \ (\mu > 0),$$

undergoes a transcritical bifurcation as μ is varied, and find the value of μ at the bifurcation point.

(b) The dynamics of the number of photons $n(t)$ in a laser field is given by

$$\frac{dn}{dt} = (GN_0 - k)n - \alpha Gn^2,$$

where G is the gain coefficient for stimulated emission, k is the decay rate due to photon loss by scattering, α is the rate at which atoms drop back to their ground states and, in the absence of a laser field, the number of exited atoms is kept fixed at N_0. Find the equilibrium points of the system and check their stability. Show that the system undergoes a transcritical bifurcation at $N_0 = k/G$.

58. The yearly yield of fish in a lake with constant harvesting is represented by

$$\frac{dx}{dt} = rx(1 - x) - H,$$

where $r(> 0)$ is the rate of the logistic growth and H is the constant harvesting. Show that the system exhibits saddle-node bifurcation.

59. (a) Consider the two-dimensional non-linear autonomous system,

$$\frac{dx}{dt} = y + \frac{x}{\sqrt{x^2 + y^2}} \left[1 - (x^2 + y^2)\right],$$

$$\frac{dy}{dt} = -x + \frac{y}{\sqrt{x^2 + y^2}} \left[1 - (x^2 + y^2)\right].$$

Show that the system exhibits a limit cycle.

(b) Consider the non-linear autonomous system,

$$\frac{dx}{dt} = y + x\left(1 - x^2 - y^2\right), \ \frac{dy}{dt} = -x + y\left(1 - x^2 - y^2\right).$$

Show that there exists a limit cycle for the given system.

60. Consider the two-dimensional dynamical system,

$$\frac{dx}{dt} = y, \ \frac{dy}{dt} = x^2 - y - \mu.$$

Show that the system experiences a saddle-node bifurcation at $(0,0)$ as μ passes through $\mu = 0$.

61. (a) Consider the dynamical system,

$$\frac{dx}{dt} = \mu x + y - x(x^2 + y^2), \ \frac{dy}{dt} = -x + \mu y - y(x^2 + y^2).$$

Show that the system undergoes Hopf bifurcation at $\mu = 0$.

(b) Brusselator model of biochemical reaction for a certain chemical reaction is given by

$$\frac{dx}{dt} = a - (b+1)x + x^2 y, \ \frac{dy}{dt} = bx - x^2 y, \ a, b > 0,$$

where x, y are concentrations of the chemicals. Discuss the bifurcation of the model and find the condition for which the Brusselator model admits Hopf bifurcation [90].

62. A prey-predator model is given by

$$\frac{dx}{dt} = x(b - x - \frac{y}{1+x}), \ \frac{dy}{dt} = y(\frac{x}{1+x} - ay),$$

where $x, y > 0$ are populations and $a, b > 0$ are parameters. Show that a Hopf bifurcation occurs at the positive equilibrium point if $a = a_c = \frac{4(b-2)}{b^2(b+2)}$ and $b > 2$.

63. (a) Consider the dynamical system,

$$\frac{dx}{dt} = x(\mu - x^2 - y^2) - y, \ \frac{dy}{dt} = y(\mu - x^2 - y^2) + x.$$

Show that the system exhibits supercritical Hopf bifurcation as μ passes through $\mu = 0$.

(b) Show that the system,

$$\frac{dx}{dt} = -y + x(\mu + x^2 + y^2), \ \frac{dy}{dt} = x + y(\mu + x^2 + y^2),$$

undergoes subcritical Hopf bifurcation at the origin $(0,0)$ as μ passes through $\mu = 0$.

64. Formulate a competition model between two species x and y, where the nutrients for both species are limited, both the species compete with each other for the same nutrient and the species x migrates from somewhere else into the region of interest at a rate α per unit time. Find the equilibrium position of the system and check for stability of the linearized model.

65. (a) Consider the data set, dataset $=$ {(0, 3.93), (10, 5.31), (20, 7.24), (30, 9.64), (40, 12.87), (50,17.07), (60, 23.19), (70, 31.44), (80, 39.82), (90, 50.16), (100,62.95), (110, 75.99), (120, 91.97), (130, 105.71), (140,122.78), (150, 131.67), (160, 151.33), (170, 179.32), (180,203.21), (190, 226.5), (200, 249.63), (210, 281.42)}. Fit the function

$$\frac{mk}{m + (k - m)e^{-rx}}$$

to the given data by estimating the parameters m, k, r.

(b) Consider the data set, dataset $=$ {(0, 4.77), (0.05, 4.14), (0.1, 3.67), (0.15, 2.88), (0.2, 2.5), (0.25,2.15), (0.3, 0.89), (0.35, 0.41), (0.4, −0.38), (0.45,−0.74), (0.5,−1.71), (0.55,−1.95), (0.6,−2.69), (0.65,−2.94), (0.7,−2.81), (0.75,−3.53), (0.8,−3.09), (0.85,−3.64), (0.9, −3.71), (0.95,−3.69), (1,−3.41)}. Fit the function

$$A\ e^{-Bx}\ \cos(\omega x + \phi)$$

to the given data by estimating the parameters A, B, ω, ϕ.

66. Consider the model of interacting population

$$\frac{dx}{dt} = x\left(1 - \frac{x}{2} - y\right), \ \frac{dy}{dt} = y\left(x - 1 - \frac{y}{x}\right).$$

(i) Explain the model.
(ii) Obtain the equilibrium points and check their stability.
(iii) Numerically solve the model with initial condition $x(1) = 2$, $y(1) = 3$ and comment on its dynamics. Also, obtain the phase portrait.

67. The Holling-Tanner model for predator-prey interaction is

$$\frac{dx}{dt} = x\left(1 - \frac{x}{7}\right) - \frac{6xy}{7 + 7x}, \ \frac{dy}{dt} = 0.2y\left(1 - \frac{Ny}{x}\right) \ (N > 0).$$

(i) Explain the model.
(ii) Obtain the equilibrium points and check their stability.
(iii) Numerically solve the system for $N = 2.5, 0.5$ and comment on its dynamics. Also, obtain the phase portrait for $N = 2.5, 0.5$.

68. In a population of birds, the birth and death rates are respectively 0.45 and 0.65 per year. The constant rates of migration and emigrations respectively are 2000 and 1000 birds per year. Formulate a model for the bird population (using differential equation) and solve it. What will be the long term behavior of the model when the initial populations are (a) 4000 and (b) 7000?

69. Consider the data set, data1 = {(0,3.493), (1,5.282), (2,6.357), (3,9.201), (4,11.224), (5,12.964), (6,16.226), (7,18.137), (8,19.590), (9,21.955), (10,22.862), (11,23.869), (12,24.243), (13,24.344), (14,24.919), (15,25.108)}, which shows the population, in thousands, of harbor seals in the Wadden Sea over the years 1997 to 2012 [1]. Fit the given data, by estimating the parameters of the following growth equations, given by
   (i) $\dfrac{dx}{dt} = rx\left[\ln\left(\dfrac{k}{x}\right)\right]^{\gamma}$, $x(0) = 0.5$ (r, k, γ are parameters).
   (ii) $\dfrac{dx}{dt} = rx^{\frac{2}{3}}\left[1 - \left(\dfrac{x}{k}\right)^{\frac{1}{3}}\right]$, $x(0) = a$ (r, k, a are parameters).
   (iii) $\dfrac{dx}{dt} = rx\left[1 - \left(\dfrac{x}{k}\right)^{\beta}\right]$, $x(0) = 3.0$ (r, k, β are parameters).
   (iv) $\dfrac{dx}{dt} = rx^{1+\beta(1-\gamma)}\left[1 - \left(\dfrac{x}{k}\right)^{\beta}\right]^{\gamma}$, $x(0) = 3.0$ (r, β, γ, k are parameters).

70. A model between two interacting species x and y is given by

$$\frac{dx}{dt} = x(2 - 2x + y), \quad \frac{dy}{dt} = y(1 - y + x).$$

   (i) Interpret the model in ecological terms by taking into account of the interactions between the species.
   (ii) Find all the equilibrium points of the system and perform a linear stability analysis.
   (iii) Numerically solve the model for $x(0) = 20, y(0) = 10$ and comment on its dynamics. Also, obtain the phase portrait.

71. Consider the following data sets, which shows the competition of *S. cerevisiae* and the slower growing *S. kephir* that Gause performed in two experiments [41]:
   dataset1 = {(0, 0.375), (1.5, 0.92), (9, 3.08), (10, 3.99), (18, 4.69), (18,5.78), (23, 6.15), (25.5, 9.910), (27, 9.47), (38, 10.57), (42,7.27), (45.5, 9.88), (47, 8.3)};
   dataset2 = {(0, 0.29), (1.5, 0.37), (9, 0.63), (10, 0.98), (18, 1.47), (18,1.22), (23, 1.46), (25.5, 1.11), (27, 1.225), (38, 1.1), (42,1.71), (45.5, 0.96), (47, 1.84)};

Fit the given data to the competition model, given by

$$\frac{dx}{dt} = a_1x - b_1x^2 - c_1xy,$$
$$\frac{dy}{dt} = a_2y - b_2y^2 - c_2xy, \; x(0) = 0.42, \; y(0) = 0.63,$$

by estimating the parameters a_1, b_1, c_1, a_2, b_2 and c_2.

72. Discuss the bifurcation of the system

$$\frac{dx}{dt} = x^3 - 5x^2 - (\mu - 8)x + \mu - 4,$$

with μ as bifurcation parameter.

73. Show that the system

$$\frac{dx}{dt} = -y + x(\mu - x^2 - y^2), \; \frac{dy}{dt} = x + y(\mu - x^2 - y^2)$$

undergoes supercritical Hopf bifurcation at the origin $(0, 0)$ as μ passes through $\mu = 0$.

74. Show that the system

$$\begin{aligned} \frac{dx}{dt} &= x[\mu + (x^2 + y^2) - (x^2 + y^2)^2] - y, \\ \frac{dy}{dt} &= y[\mu + (x^2 + y^2) - (x^2 + y^2)^2] + x \end{aligned}$$

undergoes subcritical Hopf bifurcation at $(0, 0)$ as μ passes through $\mu = 0$.

75. Consider the model

$$\frac{dx}{dt} = xy - x, \; \frac{dy}{dt} = a + by - xy.$$

Determine the non-negative equilibrium point and check its stability. Taking $a = 5, b = 2$, draw the time-series graph of the model with initial condition $x(0) = 10, y(0) = 20$ and comment on its dynamics. Also, obtain the phase portrait of the model about its equilibrium points. Draw the bifurcation diagrams when the parameters a and b vary.

76. The following system of differential equations describes a prey–predator model that takes into account of the hunting of the two species:

$$\frac{dx}{dt} = 0.05x - 0.0004xy - C_1x,$$
$$\frac{dy}{dt} = -0.04y + 0.0001xy - C_2y, \; x(0) = 100, y(0) = 50.$$

(i) For $C_2 = 0.1$, find the equilibrium points in terms of C_1 and obtain the condition for stability of the model.
(ii) Obtain the values of C_1 such that the prey species are killed off before the predator (solve numerically).
(iii) For what values of C_1, the predators are killed off first?
(iv) For what values of C_1, neither of them are killed off?

77. A rumor propagation model (ISR Model), where I are the ignorant, S are the spreaders of rumor, R are the repressor is given by [125]

$$\frac{dI}{dt} = -\beta kIS,$$
$$\frac{dS}{dt} = \beta kIS - \alpha kS(S+R),$$
$$\frac{dR}{dt} = \alpha kS(S+R).$$

(i) Describe the model. Find the equilibrium point and perform linear stability analysis about it.
(ii) Solve the model numerically for $k = 0.8, \alpha = 0.01, \beta = 0.01$ with initial condition $I(0) = 1, S(0) = 0, R(0) = 0$. Repeat the process when (a) $\beta >> \alpha$ (b) $\beta << \alpha$.

78. When a patient takes a drug orally, it dissolves and releases the medications, which diffuses into the blood, and the bloodstream takes the medications to the site where it has a therapeutic effect. Let $C(t)$ be the concentration of drug in the compartment at time t, then the rate of change of $C(t)$ is $\frac{dC}{dt}$ = input rate of drug − output range of drug.
(i) A two compartmental model describing the rate of change in oral drug administration is given as follows [82]:

$$\frac{dC_1}{dt} = -k_1C_1,\ C_1(0) = c_0, \qquad \frac{dC_2}{dt} = -k_1C_1 - k_2C_2,\ C_2(0) = 0.$$

(i) Describe the model. Solve the model analytically and obtain the expressions for $C_1(t)$ and $C_2(t)$, respectively.
(ii) Numerically solve the model for the concentration of drug in the stomach $C_1(t)$ with an initial concentration of 500 units at different rate constants $k_1 = 0.9776/h, 0.7448/h, 0.3293/h, 0.2213/h$. Also, obtain the graph for variation of drug concentration within the body at different initial concentrations of $600, 250, 200$ and 100 units with $k_1 = 0.9776/h$.

79. The dynamics of the number of photons $p(t)$ in a laser field and the number of exited atoms $n(t)$ is given by a system of equations:

$$\frac{dp}{dt} = Gpn - kp, \qquad \frac{dn}{dt} = -Gpn - fn + \alpha \quad (G, k, f > 0),$$

where G is the given coefficient for stimulated emission, k is the decay

rate due to photon loss by scattering, f is the decay rate for spontaneous emission, and α is the pump strength and can take any sign.
(i) Non-dimensionalize the system and obtain all the equilibrium points. Perform linear stability analysis about all the equilibrium points and obtain the conditions for stability of the system.
(ii) Draw a time-series graph of the model by taking suitable values of the dimensionless parameters.
(iii) Draw different phase portraits depending on the values of various dimensionless parameters.

80. We consider a model of two species with mutual cooperation:

$$\frac{dS_1}{dt} = -\alpha S_1 + \beta S_2 \, , \frac{dS_2}{dt} = -\gamma S_2 + \delta S_1 .$$

(i) Show that $\frac{d^2 S_1}{dt^2} + (\alpha + \gamma)\frac{dS_1}{dt} + (\alpha\gamma - \beta\delta)S_1 = 0$.
(ii) Both the species goes to extinction unless $\beta\delta \geq \alpha\gamma$.
(iii) If $S_1(0) = 100, S_2(0) = 200$ initially, find the limiting population for both the species when $\alpha = 2, \beta = 4, \gamma = 2, \delta = 4$.

81. Odell [75] proposed a prey-predator system whose non-dimensionlize system is given as follows:

$$\frac{dx}{dt} = x[x(1-x) - y], \quad \frac{dy}{dt} = y(x - a),$$

where $x(t)$ and $y(t)$ are dimensionless population of the prey and predator, respectively and $a > 0$ is a control parameter.
(i) Find the equilibrium points and classify them. Perform linear stability analysis about all the equilibrium points and obtain the conditions for stability of the system.
(ii) Draw the time-series graph and phase portrait for $a > 1$ and comment on the long-term behavior of the model.
(iii) Show that a Hopf bifurcation occurs at $a_c = \frac{1}{2}$. Is it supercritical or subcritical?

82. Consider an SIS model with carriers (Carriers are those individuals who, although apparently healthy themselves, harbor infection which can be transmitted to others).
(i) Model the given scenario using differential equations by taking infection rate as β and recovery rate as α. Obtain the analytical solution of the model.
(ii) Suppose the infection is spread only by a constant number of carriers and the number of carriers decreases exponentially with time. Modify the model by taking decay rate of carrier as γ. Obtain the analytical solution of the model ($S(0) = S_0, I(0) = I_0$) and comment on its long-term behavior.

(iii) Taking infection rate $\beta = 0.00005$, recovery rate $\alpha = 0.0005$, and decay rate of carrier $\gamma = 0.02$, obtain the numerical solution of both the models and comment on the dynamics. Also, compare them with the analytical solutions.

83. The model for microbial growth in a chemostat is given by

$$\frac{dm}{dt} = k(c) - \rho m, \qquad \frac{dc}{dt} = \rho(c_0 - c) - \beta k(c)m,$$

where $m(t)$ is the population of micro-organisms per unit volume and $c(t)$ is the concentration of rate limiting nutrient in the medium at time t, and $k(c) = \frac{kc}{a+c}$. Find the steady-state solutions of the model and perform linear stability analysis. Obtain condition for stability for each steady-state solution. Numerically solve the model for $a = 10, \beta = 0.8, c_0 = 3.0, k = 0.2, \rho = 0.1$, draw time-series graph and comment on the dynamics of the model. Repeat the same analysis with $k(c) = k(1 - e^{\frac{c \ln 2}{a}})$.

84. A crab population, following a logistic growth, harvested at a constant rate E, is given by

$$\frac{dC}{dt} = rC\left(1 - \frac{C}{K}\right) - E.$$

Show that
(i) the ultimate population size is $\frac{r}{2}(1 + \sqrt{1 - \frac{4E}{rK}})$.
(ii) The limiting crab population is $\frac{K}{2}$. What is the condition that $C(t)$ approaches a constant limit as $t \to \frac{\pi}{2}\frac{k}{r^2}$.

85. A generalized measure of the forest's health is given by energy reserve E(t) and the average number of trees S(t). The model involving Spruce-Budworm and its effect on the balsam fir forest was proposed by Ludwig et al. [100]. In this problem, the dynamics of the forest was considered, given by the system of differential equations as follows:

$$\frac{dS}{dt} = r_s S\left(1 - \frac{S}{K_S}\frac{K_E}{E}\right),$$

$$\frac{dE}{dt} = r_E E\left(1 - \frac{K_E}{E}\right) - P\frac{B}{S},$$

where B is the constant Budworm population present and r_s, K_S,K_E, r_E, and P are all positive constants.
(i) Explain the model biologically. By defining suitable dimensionless quantities, obtain the dimensionless form of the system.
(ii) Find the steady-state solution and perform a linear stability analysis of the model. Obtain the condition for stability of the system.
(iii) Classify the bifurcation that occurs at the critical value of B.

(iv) For the large and small values of B, draw the phase portrait of the system.

86. A fishery model is given by,

$$\frac{dF}{dt} = rF(1 - \frac{F}{K}) - H\frac{F}{A+F},$$

where the fish population grows logistically in the absence of fishing with intrinsic growth rate r and carrying capacity K, the rate at which fishes are harvested decreases with F (when there is less fish, it is different to locate them and the daily harvest drops), $H(>0)$ being the coefficient of harvesting.
(i) Describe the model.
(ii) By defining suitable dimensionless quantities, obtain the dimensionless form of the model as $\frac{dx}{d\tau} = x(1-x) - h\frac{x}{a+x}$.
(iii) Find all the possible equilibrium points of the dimensionless model and perform linear stability analysis in each case.
(iv) Show that a bifurcation occurs when $h = a$ and classify this bifurcation.
(v) When $h = \frac{1}{2}(a+1)^2$, another bifurcation occurs for $a < a_c$. Obtain the value of a_c and classify the bifurcation.

87. The chlorine dioxide-iodine-malonic acid (CIO_2-I_2-$C_3H_4O_4$) reaction proposed by Lengyel et. al. [92] is given as follows:

$$\frac{dx}{dt} = a - x - \frac{4xy}{1+x^2}, \quad \frac{dy}{dt} = bx\left(1 - \frac{y}{1+x^2}\right),$$

where x and y are dimensionless concentrations of I^- and CIO_2^-.
(i) Find the steady-state solution and perform a linear stability analysis of the model.
(ii) Obtain the condition for stability of the system.
(iii) Show that a Hopf bifurcation occurs at $b = b_c$, which needs to be determined and check whether it is subcritical or supercritical.
(iv) Verify the results numerically for $a = 5, b = 1$ and $a = 10, b = 5$.

88. A sophisticated model for a laser is given by a system of equations:

$$\frac{dE}{dt} = K(P - E), \quad \frac{dP}{dt} = r_1(ED - P), \quad \frac{dD}{dt} = r_2(1 + \lambda - D - \lambda EP).$$

This is the Maxwell–Bloch equation where $E(t)$ is the electric field, $P(t)$ is the mean atomic-polarization density, $D(t)$ is the population inversion density, K is the rate of decay in the laser cavity due to beam transmission, r_1 and r_2 are the respective rates of decay due to atomic polarization and population invasion and λ is a pumping energy parameter.

(i) Find the equilibrium points of the system and perform linear stability analysis. Numerically check the analytical findings for $K = -3.4, r_1 = 3.54, r_2 = 2.92$, and $\lambda = 0.34$.
(ii) Draw time-series graph and phase portrait of the model for $K = -3.4, r_1 = 3.54, r_2 = 2.92, \lambda = 0.34$, with initial condition $E(0) = 0.62, P(0) = 0.9, D(0) = 1.45$. What can you say about the dynamics of the system from the graphs and phase portrait?
(iii) Repeat the process with the following sets of parameters:
(a) $K = -3.4, r_1 = 3.54, r_2 = 1.18, \lambda = 0.34$ with initial condition $E(0) = 0.62, P(0) = 0.90, D(0) = 1.25$.
(b) $K = -3.4, r_1 = 3.50, r_2 = 0.12, \lambda = 5.00$ with initial condition $E(0) = 0.40, P(0) = 0.50, D(0) = 1.45$.
(c) $K = -2.75, r_1 = 2.7, r_2 = 5.0, \lambda = 1.14$ with initial condition $E(0) = 0.27, P(0) = 0.455, D(0) = 2.11$.
(d) $K = -2.95, r_1 = 2.7, r_2 = 5.0, \lambda = 1.14$ with initial condition $E(0) = 0.27, P(0) = 0.455, D(0) = 2.11$.
(iv) Are period-doubling, chaotic behavior, strange attractors observed with the parameter sets mentioned above?

89. **The Zombie Model [98]:** In 1932, the first zombie film (*white zombie*) was released, which told the story of a woman being transformed into a zombie by a voodoo practitioner when she was visiting Haiti (North America). Later, in 1968, in the movie *Night of the living dead*, a different kinds of zombies were shown, feeding on human flesh. However, in the 2002 release of *28 days later*, a new generation of zombies were created, who were not truly dead but infected with a virus and it turns the victim into a violent creature very fast. The common mode of this virus transmission is via bite as shown in zombie fiction.

    Let normal live humans are considered as susceptibles (S), Z are zombies and R are dead (not permanently). The encounters between zombies and humans can result in humans becoming zombies or the zombie becoming dead. Also, sometimes the dead becomes zombies spontaneously. The differential equations modelling this scenario is given by

$$\frac{dS}{dt} = -\beta SZ, \quad \frac{dZ}{dt} = \beta SZ - \alpha SZ + \gamma R, \quad \frac{dR}{dt} = \alpha SZ - \gamma R.$$

    (i) Find the approximate analytical solution of the model, taking S(0)=n, Z(0)=1, R(0)=0.
    (ii) Taking $\beta = 1.0, \alpha = 4.0, \gamma = 0.2, S(0) = 0.99, Z(0) = 0.01, R(0) = 0$, obtain the numerical solution of the model for t=(0,150) and comment on the dynamics.

90. Consider the love affair model (Majnun-Layla model) described by

$$\frac{dM}{dt} = aM + bL + f, \quad \frac{dL}{dt} = cM + dL + g.$$

For the given values of a, b, c, d, g, f, characterize the romantic styles of Majnun and Layla, classify the fixed point and its implication for the affair, sketch $M(t)$ and $L(t)$ as functions of t (time-series graph), assuming $M(0) = m_0$, $L(0) = l_0$ (m_0, l_0 to be chosen by the readers, for example, $m_0 = 1, l_0 = 0$). Also, draw the phase portraits.
(i) $a = 3, b = 2, c = -1, d = 2, f = 0, g = 0$.
(ii) $a = -3, b = 2, c = -1, d = 2, f = 0, g = 0$.
(iii) $a = -3, b = 2, c = -2, d = 1, f = 0, g = 0$.
(iv) $a = -3, b = -2, c = -2, d = -1, f = 0, g = 0$.
(v) $a = -3, b = -2, c = -2, d = -1, f = 10, g = 10$. Change the values of f to $15, 20, 30$ and comment on the dynamics.
(vi) $a = 3, b = 2, c = 1, d = 2, f = 0, g = 0$. Put $f = -10, g = -10$ and comment on the change in dynamics.

3.13 Projects

1. Consider a predator–prey model where two prey species x, y are eaten by a single predator z:

$$\frac{dx}{dt} = x(1 - x) - \alpha_1 xy - \beta_1 xz,$$
$$\frac{dy}{dt} = y(1 - y) - \alpha_2 xy - \beta_2 xz,$$
$$\frac{dz}{dt} = c\beta_1 xz + c\beta_2 xz - z.$$

(i) Explain the model along with the parameters and its significance.
(ii) Find the equilibrium points. Use Routh Hurwitz's criteria to find the condition of stability about all the equilibrium points.
(iii) For $\alpha_1 = 1, \alpha_2 = 1.5, \beta_1 = 4.1, \beta_2 = 1, c = 0.5$, numerically verify the analytical results obtained in (ii).
(iv) Explain the dynamics of the model for $3.4 \leq \alpha_1 \leq 5.5$.
(iv) Show that (a) for $\alpha_1 = 3.4$, a transcritical bifurcation occurs that allows the second prey to invade, (b) for $\alpha_1 = 2.0$, a second transcritical bifurcation occurs when the predator can invade.
(v) Show that for $\alpha_1 = 5.5$, the system undergoes a hopf bifurcation. Is it supercritical or subcritical?
(vi) Obtain the phase-portrait for $\alpha_1 = 6.0, 8.0, 10.0$. Are limit cycles, period-doubling, chaotic behavior, strange attractors observed with the parameter sets mentioned above?

2. Consider the Lorenz system defined by the following set of differential equations:

$$\frac{dx}{dt} = \sigma(y - x), \ \frac{dy}{dt} = rx - y - xz, \ \frac{dz}{dt} = xy - bz,$$

where σ, r, and b are positive constants.
(i) For $\sigma = 10$, $r = 30$, $b = 3$, obtain the time series graph and 3D phase portrait with initial condition $x(0) = y(0) = z(0) = 0.01$.
(ii) Draw trajectories of the states of the Lorenz equations in a three-dimensional phase space for several different values of r while other parameters are kept constant. See how the dynamics of the Lorenz equations change as you vary r.

3. A simple resource consumer model based on a saturated functional response is given by

$$\frac{dx}{dt} = a_1 x(1 - \frac{x}{K}) - b_1 \frac{xy}{x + c_1}, \ \frac{dy}{dt} = -a_2 y + b_2 \frac{xy}{x + c_2},$$

(i) Give a biological interpretation and the dimension of all parameters. Is it biologically reasonable to choose $b_1 = b_2$? Give an interpretation for the following parameter choices $c_1 = c_2$, $c_1 > c_2$ and $c_1 < c_2$.
(ii) Find the equilibrium points. Use Routh Hurwitz's criteria to find the condition of stability about all the equilibrium points.
(iii) For $a_1 = 8$, $K = 1.6$, $b_1 = 1$, $c_1 = 1$, $a_2 = 0.2$, $b_2 = 0.95$, $c_2 = 1$, numerically verify the analytical results obtained in (ii).
(iv) Draw the time series graph and the phase portrait with the same set of parameters and initial condition $x(0) = y(0) = 1.5$.
(v) Suppose the consumer does not have access to m of its resources. How will you modify the model?
(vi) Find the equilibrium points of this new model and find the condition of stability about all the equilibrium points.
(vii) For $a_1 = 0.6$, $K = 0.89$, $b_1 = 0.57$, $c_1 = 0.56$, $a_2 = 0.056$, $b_2 = 0.006$, $c_2 = 0.56$, $m = 0.01$, numerically verify the analytical results obtained in (vi). Draw the time series graph and the phase portrait with the same set of parameters and initial condition $x(0) = 0.1$, $y(0) = 0.95$.
(viii) Discuss the bifurcation of the system with m as bifurcation parameter.
(ix) Consider sigmoid functional response $\frac{x^2}{x^2+c_1^2}$, and repeat the process from (i) to (viii).

Chapter 4

Spatial Models Using Partial Differential Equations

4.1 Introduction

Real-world modeling depends on many variables simultaneously. So, when we try to model some phenomena from the real world with the help of ordinary differential equations (ODE), we restrict our analysis to one independent variable (namely, time) only; hence, we only succeed in describing the dynamical behavior of the problem of interest with respect to that independent variable. Thus, using ODE models means that we are considering that independent variable, which is the most important factor affecting the problem of interest, and other factors are ignored. Because of this restriction, ODE models often fail to reflect the dynamics. Thus, a disparity between an ODE model and data may signify that its state variables depend on more than one independent variable (say, time and space). Hence, instead of using an ODE model in such cases, it may be appropriate to use a partial differential equation (PDE) model.

The advantage of PDE models is that they include derivatives of at least two independent variables, and hence, they can describe the dynamical behavior of the problem of interest in terms of two or more variables at the same time. For example, if we consider the flow of heat in a metal bar, it would be inappropriate NOT to model it with partial differential equations, to compute the temperature distribution with respect to time as well as space. Let us consider a simple predator-prey model (Lotka-Volterra)

$$\frac{dP_1(t)}{dt} = \alpha\ P_1(t) - \beta\ P_1(t)P_2(t),$$
$$\frac{dP_2(t)}{dt} = -\gamma\ P_2(t) + \delta\ P_1(t)P_2(t),$$

where we have used one independent variable (namely, time) to study the dynamics of the system. But, one can consider the effect of movement of the prey and the predator by adding a diffusion term to the equations, thereby making it a PDE model as

DOI: 10.1201/9781351022941-4

$$\frac{\partial P_1(t,x,y)}{\partial t} = \alpha\ P_1(t,x,y) - \beta P_1(t,x,y)P_2(t,x,y) + D_1\nabla^2 P_1(t,x,y),$$
$$\frac{\partial P_2(t,x,y)}{\partial t} = -\gamma\ P_2(t,x,y) + \delta P_1(t,x,y)P_2(t,x,y) + D_2\nabla^2 P_2(t,x,y),$$

where $\nabla^2 \cong \frac{\partial^2}{\partial x^2} + \frac{\partial^2}{\partial y^2}$.

This model is able to capture the spatial aspect of the model and can give a complete picture of the dynamics of the predator-prey system with respect to both time and space.

The general solution of the partial differential equation involves as many arbitrary functions as the order of the equation (order of the highest partial differential coefficient in the equation). Certain conditions are required in order to find these arbitrary functions. The standard notation for the space variables in applications are x, y, z, etc., and a solution may be required in some region Ω of space. In such a case, there will be some conditions to be satisfied on the boundary $\partial\Omega$, which are called boundary conditions (BCs). Similarly, in the case of the independent variables, one of them is generally taken as time (say, t), then there will be some initial conditions (ICs) to be satisfied. The conditions of partial differential equations are classified into two categories:

(i) Initial value problem (IVP): A partial differential equation with initial conditions, that is, dependent variable and an appropriate number of its derivatives are prescribed at the initial point of domain is called an initial value problem.

(ii) Boundary value problem (BVP): A partial differential with boundary conditions, that is, dependent variable and an appropriate number of its derivatives are prescribed at the boundary of domain is called a boundary value problem.

There are four broad categories of boundary conditions:

(a) Dirichlet boundary condition: On the boundary, the values of the dependent variable are specified.

(b) Newmann boundary condition: On the boundary, the normal derivative of the dependent variable is specified.

(c) Cauchy boundary condition: On the boundary, both the values of the dependent variable and its normal derivative are specified.

(d) Robin boundary condition: On the boundary, a linear combination of the dependent variable and its normal derivatives are specified.

A second-order partial differential equations of the form

$$Au_{xx} + Bu_{xy} + Cu_{yy} + Du_x + Eu_y + Fu = 0,$$

is elliptic, parabolic or hyperbolic when $B^2 - 4AC$ is respectively less than, equal to or greater than zero. This classification gives a better understanding of the choice of initial or boundary conditions. To have a unique, stable solution, each class of PDEs requires a different class of boundary conditions.

(i) Dirichlet or Neumann boundary conditions on a closed boundary surrounding the region of interest are the requirements to elliptic equations. Other boundary conditions are either insufficient to determine a unique solution, overly restrictive or lead to instabilities.

(ii) Cauchy boundary conditions on an open surface are the requirements to elliptic equations. Other boundary conditions are either too restrictive for a solution to exist, or insufficient to determine a unique solution.

(iii) Parabolic equations require Dirichlet or Neumann boundary conditions on an open surface. Other boundary conditions are too restrictive.

Example 4.1.1 *Initial value problem (IVP):*

$$\frac{\partial^2 u}{\partial t^2} - \frac{\partial^2 u}{\partial x^2} = 0, \ 0 \leq x \leq l, \ t > 0.$$

$$ICs: \ u(x,0) = 0, \ \frac{\partial u(x,0)}{\partial t} = x.$$

Example 4.1.2 *Boundary value problem (BVP):*

$$\frac{\partial u}{\partial t} = \frac{\partial^2 u}{\partial x^2} = 0, \ 0 \leq x \leq L, \ t > 0.$$

(i) BCs: $u(0,t) = 0, \ u(L,t) = 0$ *(Dirichlet boundary condition).*

(ii) BCs: $\frac{\partial u(0,t)}{\partial x} = 0, \ \frac{\partial u(L,t)}{\partial x} = 0$ *(Neumann boundary condition).*

(iii) BCs: $u(0,t) = 0, \ \frac{\partial u(L,t)}{\partial x} = 0$ *(Cauchy boundary condition).*

(iv) BCs: $u(0,t) = 1, \ \frac{\partial u(L,t)}{\partial x} + u(L,t) = 0$ *(Robin boundary condition).*

Example 4.1.3 *Initial boundary value problem (IBVP):*

$$\frac{\partial u}{\partial t} = \frac{\partial^2 u}{\partial x^2} = 0, \ 0 \leq x \leq l, \ t > 0.$$

$$BCs: \ u(0,t) = 0, \ u(l,t) = 0 \ (\textit{Dirichlet boundary condition}),$$
$$ICs: \ u(x,0) = x.$$

4.2 Heat Flow through a Small Thin Rod (One Dimensional)

We consider a thin rod of length L, made of homogenous material (material properties are translational invariant). We assume that the rod is perfectly insulated along its length so that heat can flow only through its ends (fig. 4.1).

FIGURE 4.1: *A thin homogenous rod of length L, perfectly insulated along its length.*

Let $u(x,t)$ be the temperature of this homogenous thin rod at a distance x at time t. We consider an infinitesimal piece from the rod with length $[x, x+\triangle x]$. Let A be the cross-section of the rod, ρ be the density of the material of the rod, then the infinitesimal volume is given by $\Delta V = A\Delta x$ and the corresponding infinitesimal mass is $\Delta m = \rho A \Delta x$. The amount of heat for the volume element is $Q = \sigma \Delta m\ u(x,t)$, where σ is the specific heat of the material of the rod.

At time $t + \Delta t$, the amount of heat is

$$\begin{aligned} Q_1 &= \sigma \Delta m u(x, t+\Delta t). \\ \text{Change in heat} &= Q_1 - Q = \sigma\Delta m\ u(x,t+\Delta t) - \sigma \Delta m\ u(x,t) \\ &= \sigma\rho A[u(x,t+\Delta t) - u(x,t)]\Delta x. \end{aligned}$$

Now, by the Fourier law of heat conduction, the heat flow is proportional to the temperature gradient, that is, $Q = -k\frac{\partial u}{\partial x} = -ku_x(x,t)$ (in one dimension), where k is the thermal conductivity of the solid and the negative sign denotes that the heat flux vector is in the direction of decreasing temperature. Therefore, the change in heat must be equal to the heat flowing in at x, minus the heat flowing out at $x + \Delta x$, during the time interval Δt, that is,

$$\sigma\rho A[u(x,t+\Delta t) - u(x,t)]\Delta x = [-ku_x(x,t) - (-ku_x(x+\Delta x, t))]\Delta t,$$

$$\frac{u(x,t+\Delta t) - u(x,t)}{\Delta t} = \left(\frac{k}{\sigma\rho A}\right)\frac{u_x(x+\Delta x,t) - u_x(x,t)}{\Delta x}.$$

$$\text{Taking } \Delta x, \Delta t \to 0, \text{ we obtain } \frac{\partial u}{\partial t} = c^2\frac{\partial^2 u}{\partial x^2},$$

which gives the required heat equation determining the heat flow through a small thin rod. $c^2 = \frac{k}{\sigma\rho A}$ is called the diffusivity of the material of the rod.

We use the separation of variables to solve the heat equation, which can also be termed as a one-dimensional diffusion equation.

Let $u(x,t) = X(x)\,T(t)$ be a solution of $\frac{\partial u}{\partial t} = c^2 \frac{\partial^2 u}{\partial x^2}$. Substituting, we get,

$$X(x)T^{'}(t) = c^2 X^{''}(x)T(t) \;\Rightarrow\; \frac{X^{''}}{X} = \frac{1}{c^2}\frac{T^{'}}{T} = \mu \;\text{ (a constant)}.$$

Both functions must be equal to some constant as one of them is a function of x only, and the other is a function of t.

$$\Rightarrow X^{''} = \mu X \quad \text{and} \quad T^{'} = \mu c^2 T. \tag{4.1}$$

Case I: μ is positive ($= \lambda^2$, say). From (4.1), we get $X(x) = A_1 \cosh(\lambda x) + A_2 \sinh(\lambda x)$ and $T(t) = A_3 e^{\lambda^2 c^2 t}$. Then, the solution of the heat equation is

$$u(x,t) = [C_1 \cosh(\lambda x) + C_2 \sinh(\lambda x)]\, e^{\lambda^2 c^2 t},\; C_1 = A_1 A_2,\; C_2 = A_2 A_3.$$

Case II: $\mu = 0$. From (4.1), we get, $X(x) = A_4 x + A_5$ and $T(t) = A_6$. Then, the solution of the heat equation is

$$u(x,t) = C_3 x + C_4.$$

Case III: μ is negative ($= -\lambda^2$, say). From (4.1), we get $X(x) = A_7 \cos(\lambda x) + A_8 \sin(\lambda x)$ and $T(t) = A_9 e^{-\lambda^2 c^2 t}$. Then, the solution of the heat equation is

$$u(x,t) = [C_5 \cos(\lambda x) + C_6 \sin(\lambda x)]\, e^{-\lambda^2 c^2 t}.$$

Combining, we can write the general solution of the heat equation as

$$u(x,t) = \begin{cases} [C_1 \cosh(\lambda x) + C_2 \sinh(\lambda x)]\, e^{\lambda^2 c^2 t}, & \mu = \lambda^2 > 0. \\ C_3 x + C_4, & \mu = 0. \\ [C_5 \cos(\lambda x) + C_6 \sin(\lambda x)]\, e^{-\lambda^2 c^2 t}, & \mu = -\lambda^2 < 0. \end{cases}$$

Notes: (i) All three solutions are not consistent.
(ii) The first solution indicates $t \to \infty, u \to \infty$. So, it is reasonable to assume that $u(x,t)$ is bounded as $t \to \infty$, from a realistic physical point of view.
(iii) The consistency of the third solution is always there; however, the second solution is consistent in some cases along with the first.
(iv) If the boundary conditions are of Dirichlet's type, homogeneous as well as periodic, that is, $u(0,t) = 0, u(L,t) = 0$, third solution is the only solution.
(v) If the boundary conditions are of Dirichlet's type, but non-homogeneous or non-periodic, that is, $u(0,t) = \alpha, u(L,t) = \beta$, second and third solution are the general solution of the given problem.
(vi) Second and third solutions constitute the general solution of the given problem, in the case of Neumann and Robin boundary conditions.

Example 4.2.1 *A uniform rod of length 20 cm with diffusivity D of the material of the rod, whose sides are insulated, is kept at initial temperature x, when $0 \leq x \leq 10$ and $20 - x$, when $10 \leq x \leq 20$. Both ends of the rod are suddenly cooled at $0\,^{\circ}$C and are kept at that temperature.*
(i) If $u(x,t)$ represents the temperature function at any point x at time t, formulate a mathematical model of the given situation, stating clearly the boundary and initial conditions.
(ii) Using the method of separation of variables, find the temperature function $u(x,t)$. Obtain the numerical solution of the problem for $D = 0.475$ and plot the graph.

Solution: *(i) The mathematical model of the given situation represents an initial boundary value problem of heat conduction and is given by*

$$\frac{\partial u}{\partial t} = D\frac{\partial^2 u}{\partial x^2}; \quad 0 \leq x \leq 20, t > 0 \tag{4.2}$$

Boundary Condition (BCs): $u(0,t) = 0 = u(20,t)$; $t > 0$ (since both ends of the rod are cooled suddenly at $0\,^{\circ}$C).
Initial Condition (ICs): $u(x,0) = \begin{cases} x, & 0 \leq x \leq 10, \\ 20 - x, & 10 \leq x \leq 20. \end{cases}$

(ii) Let $u(x,t) = X(x)T(t)$ be a solution of $\frac{\partial u}{\partial t} = D\frac{\partial^2 u}{\partial x^2}$. Then,

$$\frac{X''}{X} = \frac{1}{D}\frac{T'}{T} = -\lambda^2 \text{ (separation constant)}. \tag{4.3}$$

Since the boundary conditions are periodic and homogenous in x, the periodic solution of (4.2) exists if the separation constant is negative. One can also consider the other two cases, that is, the separation constant to be positive and zero but will arrive at the same conclusion. Basically, a negative separation constant gives a physically acceptable general solution. Solving (4.3) we get,

$$X(x) = A_1\cos(\lambda x) + A_2\sin(\lambda x) \quad \text{and} \quad T(t) = A_3 e^{-\lambda^2 Dt}.$$

Therefore, the complete solution of (4.2) is given by

$$u(x,t) = [C_1\cos(\lambda x) + C_2\sin(\lambda x)]\, e^{-\lambda^2 Dt}, \text{ where } C_1 = A_1A_2, C_2 = A_2A_3.$$

Applying the boundary conditions $u(0,t) = 0 = u(20,t)$, we obtain

$$(C_1\cos 0 + C_2\sin 0)e^{-\lambda^2 Dt} = 0 \text{ and } [C_1\cos(20\lambda) + C_2\sin(20\lambda)]e^{-\lambda^2 Dt} = 0,$$

$$\Rightarrow C_1 = 0 \text{ and } C_2\sin(20\lambda) = 0, \;\Rightarrow\; \sin(20\lambda) = 0 \;\Rightarrow\; \lambda = \frac{n\pi}{20}, n \text{ being}$$

an integer (for non-trivial solution $C_2 \neq 0$).

Therefore, the required solution is of the form

$$u(x,t) = C_2\sin\left(\frac{n\pi}{20}x\right) e^{\frac{-n^2\pi^2 D}{400}t}.$$

Noting that the heat conduction equation is linear, we use the principle of superposition to obtain its most general solution as

$$u(x,t) = \sum_{n=1}^{\infty} B_n \sin\left(\frac{n\pi}{20}x\right) e^{\frac{-n^2\pi^2 D}{400}t}.$$

Using the initial condition, we get

$$u(x,0) = \sum_{n=1}^{\infty} B_n \sin\left(\frac{n\pi}{20}x\right),$$

which is a half-range Fourier series, where

$$\begin{aligned} B_n &= \frac{2}{20}\int_0^{20} u(x,0)\ \sin\left(\frac{n\pi}{20}x\right) dx \\ &= \frac{2}{20}\int_0^{10} x\ sin\left(\frac{n\pi}{20}x\right) dx + \frac{2}{20}\int_{10}^{20} (20-x)\ \sin\left(\frac{n\pi}{20}x\right) dx \\ &= \frac{80}{n^2\pi^2}\sin\left(\frac{n\pi}{2}\right) = \begin{cases} 0, \ \mathit{if}\ n = 2m \ \mathit{and}\ m = 1,2,3,... \\ \dfrac{80}{(2m-1)^2\pi^2}, \ \mathit{if}\ n = 2m-1 \ \mathit{and}\ m = 1,2,3... \end{cases} \end{aligned}$$

Hence, the temperature function is given by (fig. 4.2)

$$u(x,t) = \sum_{n=1}^{\infty} \frac{80}{(2m-1)^2\pi^2} \sin\left(\frac{(2m-1)\pi x}{20}\right) e^{\frac{-(2m-1)^2\pi^2 D}{400}t}.$$

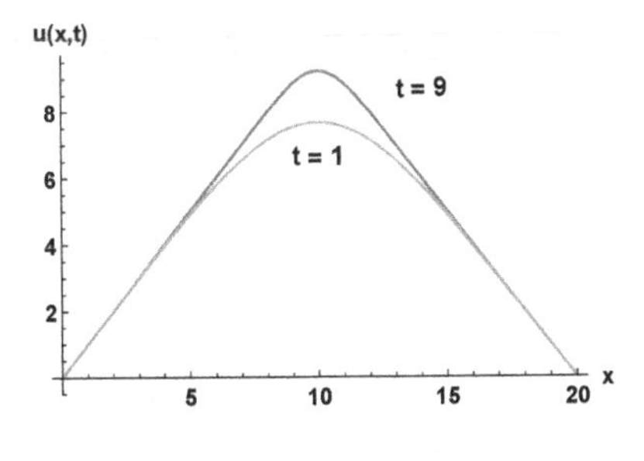

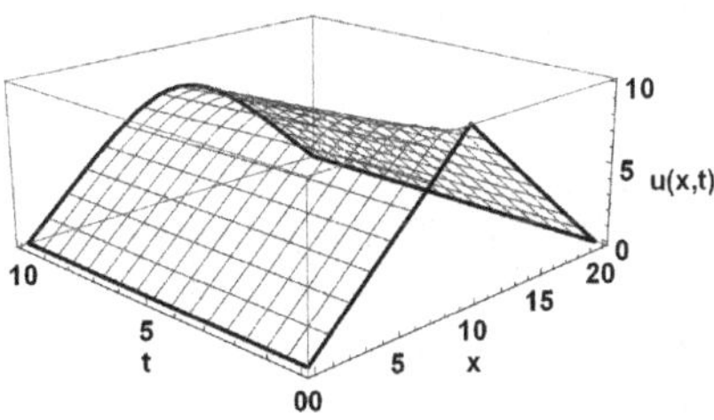

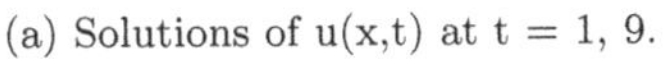
(a) Solutions of u(x,t) at t = 1, 9.

(b) Three-dimensional graph of u(x,t).

FIGURE 4.2: *The figures show the dynamics of heat flows with Dirichlet boundary conditions for $D = 0.475$. (a) Heat flow between $x = 0$ and $x = 20$, the value of $u(x,t)$ is zero at the boundaries. (b) Three-dimensional visualization of one-dimensional heat flow, which is smooth over time, even though the initial condition is a piecewise function.*

Example 4.2.2 *Consider a laterally insulated rod of length* 100 *cm with diffusivity* D *of the material of the rod, whose ends are also insulated. The initial temperature is* x, *when* $0 \le x \le 40$ *and* $100 - x$, *when* $40 \le x \le 100$.
(i) If $u(x,t)$ *represents the temperature function at any point* x *at time* t, *formulate a mathematical model of the given situation, stating clearly the boundary and initial conditions.*
(ii) Using the method of separation of variables, find the temperature function $u(x,t)$. *Obtain the numerical solution of the problem for* $D = 0.475$ *and plot the graph.*

Solution: *(i) The mathematical model of the given situation represents an initial boundary value problem of heat conduction and is given by*

$$\frac{\partial u}{\partial t} = D\frac{\partial^2 u}{\partial x^2};\ 0 \le x \le 100, t > 0. \tag{4.4}$$

Boundary Condition (BCs): $\dfrac{\partial u(0,t)}{\partial x} = 0,\ \dfrac{\partial u(100,t)}{\partial x} = 0;\ t > 0$ *(both ends of the rod are insulated).*

Initial Condition (ICs): $u(x,0) = \begin{cases} x, & 0 \le x \le 40, \\ 100 - x, & 40 \le x \le 100. \end{cases}$

(ii) Let $u(x,t) = X(x)T(t)$ *be a solution of* $\dfrac{\partial u}{\partial t} = D\dfrac{\partial^2 u}{\partial x^2}$. *Then,*

$$\frac{X^{''}}{X} = \frac{1}{D}\frac{T^{'}}{T} = \mu \Rightarrow X^{''} - \mu X = 0 \text{ and } T^{'} = \mu DT. \tag{4.5}$$

Case I: *Let* $\mu > 0\ (= \lambda^2\ )$. *From (4.5), we get* $X(x) = A_1 e^{\lambda x} + A_2 e^{-\lambda x} \Rightarrow X^{'}(x) = A_1\lambda e^{\lambda x} - A_2\lambda e^{-\lambda x}$ *and* $T(t) = A_3 e^{\lambda^2 Dt}$. *Using the given boundary conditions we get* $A_1 = 0, A_2 = 0$, *which gives trivial solution* $u(x,t) = 0$. *Hence, we reject* $\mu = \lambda^2$.

Case II: *Let* $\mu = 0$. *Then, the solution of (4.5) is* $X(x) = A_1 x + A_2 \Rightarrow X^{'}(x) = A_1$ *and* $T(t) = constant = A_3$ *(say). Using the boundary conditions, we get* $A_1 = 0$, *then* $X(x) = A_2$. *Therefore, corresponding to* $\mu = 0$, *a solution to the boundary value problem is given by*

$$u(x,t) = A_2 \times A_3 = B_0\ (say). \tag{4.6}$$

Case III: $\mu < 0\ (= -\lambda^2)$. *From (4.5), we obtain*

$$\begin{aligned} u(x,t) &= \left[C_1 \cos(\lambda x) + C_2 \sin(\lambda x)\right] e^{-\lambda^2 Dt}, \\ \Rightarrow \quad \frac{\partial u}{\partial x} &= \left[-C_1\lambda \sin(\lambda x) + C_2\lambda \cos(\lambda x)\right] e^{-\lambda^2 Dt}. \end{aligned}$$

Applying the boundary conditions $u_x(0,t) = 0 = u_x(100,t)$, *we get* $C_2 = 0$ *and* $-C_1\lambda \sin(100\lambda) = 0 \Rightarrow \sin(100\lambda) = 0 \Rightarrow \lambda = \frac{n\pi}{100}$, *n being an integer (for* *non-trivial solution $C_1 \neq 0$). Therefore, the required solution is of the form $u(x,t) = C_1 \cos\left(\frac{n\pi}{100}x\right) e^{\frac{-n^2\pi^2 D}{10000}t}$. Noting that the heat conduction equation is linear, we use the principle of superposition to obtain*

$$u(x,t) = \sum_{n=1}^{\infty} B_n \cos\left(\frac{n\pi}{100}x\right) e^{\frac{-n^2\pi^2 D}{10000}t}. \tag{4.7}$$

Equations (4.6) and (4.7) constitute a set of infinite solutions of (4.4). To obtain a solution, which will satisfy the initial condition, we consider a linear combination of these two solutions. Hence, the complete solution of (4.4) is of the form

$$u(x,t) = B_0 + \sum_{n=1}^{\infty} B_n \cos\left(\frac{n\pi}{100}x\right) e^{\frac{-n^2\pi^2 D}{10000}t}.$$

Using the initial condition, we get $u(x,0) = \frac{B_0}{2} + \sum_{n=1}^{\infty} B_n \cos\left(\frac{n\pi}{100}x\right)$, which is a half-range Fourier series, where

$$B_0 = \frac{1}{100}\left[\int_0^{40} x dx + \int_{40}^{100} (100-x)dx\right] = 25, \text{ and}$$

$$B_n = \frac{2}{100}\int_0^{100} u(x,0)\ \cos\left(\frac{n\pi}{100}x\right) dx = \frac{200}{n^2\pi^2}\left[2\cos\left(\frac{n\pi}{2}\right) - \cos(n\pi) - 1\right].$$

Hence, the temperature function is given by (fig. 4.3)

$$u(x,t) = 25 + \sum_{n=1}^{\infty} \frac{200}{n^2\pi^2}\left[2\cos\left(\frac{n\pi}{2}\right) - \cos(n\pi) - 1\right] \cos\left(\frac{n\pi x}{100}\right) e^{\frac{-n^2\pi^2 D}{10000}t}.$$

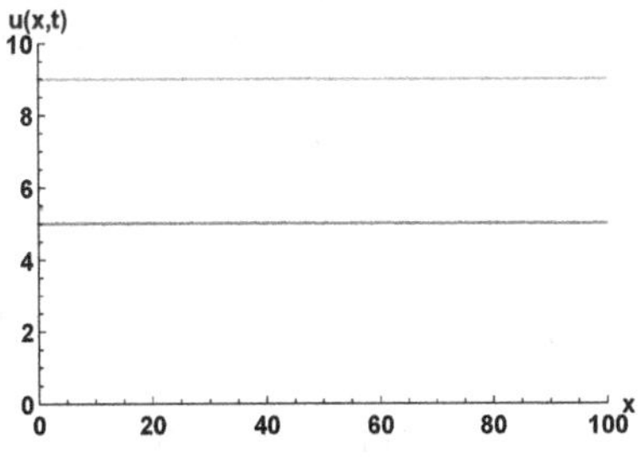

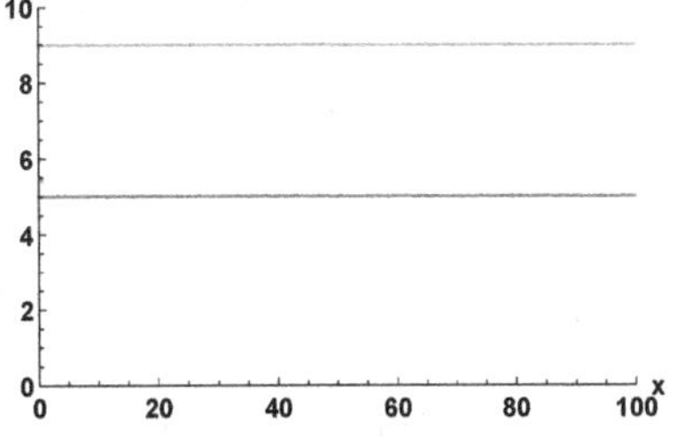

(a) Solutions of u(x,t) at t = 5, 9.

(b) Three-dimensional graph of $u(x,t)$.

FIGURE 4.3: *The figures show the dynamics of heat flows with Neumann boundary conditions ($D = 0.475$). (a) Constant heat flow for $t = 5, 9$ in $0 \leq x \leq 100$. (b) Three-dimensional visualization of one-dimensional heat flow, shows constant increase.*

4.3 Two-Dimensional Heat-Equation (Diffusion Equation)

Consider a thin rectangular plate made of some thermally conductive material, whose dimensions are $a \times b$. The plate is heated in some way and then insulated along its top and bottom. Our aim is to mathematically model the movement of thermal energy through the plate. Let $u(x, y, t)$ be the temperature of the plate at position (x,y) at time t. It can be shown that under ideal assumptions (say, uniform density, uniform specific heat, perfect insulation, no internal heat sources etc.), $u(x, y, t)$ satisfies the two-dimensional heat equation

$$\frac{\partial u}{\partial t} = k\left(\frac{\partial^2 u}{\partial x^2} + \frac{\partial^2 u}{\partial y^2}\right), \quad \text{for } 0 \leq x \leq a, \ 0 \leq y \leq b. \tag{4.8}$$

Suppose, the four edges of the plate $x = 0$, $x = a$, $y = 0$, $y = b$ are kept at zero temperature, which imposes some sort of boundary conditions, namely,

$$u(0, y, t) = u(a, y, t) = 0, \tag{4.9}$$
$$u(x, 0, t) = u(x, b, t) = 0. \tag{4.10}$$

The way the plate is heated initially is given by the initial condition

$$u(x, y, 0) = f(x, y), \quad (x, y) \in \Re, \quad \text{where } \Re = [0, a] \times [0, b]. \tag{4.11}$$

For a fixed t, the height of the surface $z = u(x, y, t)$ gives the temperature of the plate at time t and position (x,y). Our aim is to obtain a solution to the heat equation (4.8) subject to the boundary conditions (4.9), (4.10) and, initial condition (4.11). As before, we separate variables to produce simple solutions to (4.8), (4.9) and (4.10) and then use the principle of superposition to build up a solution that satisfies (4.11) as well.
Let $u(x, y, t) = X(x)Y(y)T(t)$ be a solution of (4.8). Substituting in (4.8), we obtain,

$$\frac{X''}{X} + \frac{Y''}{Y} = \frac{T''}{kT}. \tag{4.12}$$

Since x, y, t are independent variables, (4.12) will hold if each term on each side is equal to same separation constant, that is,

$$\frac{X'}{X} = \mu_1, \ \frac{Y'}{Y} = \mu_2, \ \frac{T'}{kT} = \mu_1 + \mu_2.$$

Let $\dfrac{X'}{X} = \mu_1 \Rightarrow X'' - \mu_1 X = 0$. Using boundary conditions (4.9) and (4.10) we get,

$$X(0)Y(y)T(t) = 0 \ \text{ and } \ X(a)Y(y)T(t) = 0.$$

$Y(y) = 0$ or $T(t) = 0$ will lead to the trivial solution $u = 0$, hence $Y(y) \neq 0$, $T(t) \neq 0$ and this implies $X(0) = 0$ and $X(a) = 0$.

Case I: If $\mu_1 = 0$, $X(x) = A_1 X + A_2$. Now, $X(0) = 0 \Rightarrow A_2 = 0$ and $X(a) = 0 \Rightarrow A_1 = 0 \Rightarrow X(x) = 0$. This leads to $u = 0$, which does not satisfy (4.11). So, we reject $\mu_1 = 0$.

Case II: If $\mu_1 > 0$ (say, ${\lambda_1}^2$), then $X(x) = A_1 e^{\lambda_1 x} + A_2 e^{-\lambda_1 x}$. Using $X(0) = 0, X(a) = 0$, we get $A_1 + A_2 = 0$, $A_1 e^{\lambda_1 a} + A_2 e^{-\lambda_1 a} = 0 \Rightarrow A_1 = 0$, $A_2 = 0$, which again leads to $u = 0$, and does not satisfy (4.11). So, we reject, $\mu_1 > 0$.

Case III: If $\mu_1 < 0$ (say, $-{\lambda_1}^2$), then $X(x) = A_1 \cos \lambda_1 x + A_2 \sin \lambda_1 x$. $X(0) = 0 \Rightarrow A_1 = 0$ and $X(a) = 0 \Rightarrow A_2 \sin(\lambda_1 a) = 0 \Rightarrow \sin(\lambda_1 a) = 0$ ($A_2 \neq 0$, otherwise $A_2 = 0$ gives trivial solution, not satisfying (4.11)) $\Rightarrow$ $\lambda_1 = \dfrac{n\pi}{a}$, $n = 1, 2, 3, \ldots$. Hence, the non-zero solution is given by

$$X(x) = A_2 \sin\left(\frac{m\pi x}{a}\right), \quad m = 1, 2, 3, \ldots \; .$$

In a similar manner,

$$Y(y) = B_2 \sin\left(\frac{n\pi y}{b}\right), \quad n = 1, 2, 3, \ldots \; .$$

Now,

$$\frac{T'}{kT} = \mu_1 + \mu_2 = -{\lambda_1}^2 - {\lambda_2}^2 = -\pi^2 \left(\frac{m^2}{a^2} + \frac{n^2}{b^2}\right),$$

$$\Rightarrow T' = -\lambda_{mn}^2 T, \text{ where } \lambda_{mn}^2 = k\pi^2 \left(\frac{m^2}{a^2} + \frac{n^2}{b^2}\right).$$

Solving we obtain,

$$T(t) = C\, e^{-\lambda_{mn}^2 t}, \quad m = 1, 2, \ldots \;, \quad n = 1, 2, \ldots \; .$$

We use the principle of superposition to obtain its most general solution as

$$u(x, y, t) = \sum_{n=1}^{\infty} \sum_{m=1}^{\infty} A_{mn} \sin\left(\frac{m\pi x}{a}\right) \sin\left(\frac{n\pi y}{b}\right) e^{-\lambda_{mn}^2 t}.$$

Putting $t = 0$ and using the initial condition (4.11), we get,

$$f(x, y) = \sum_{n=1}^{\infty} \sum_{m=1}^{\infty} A_{mn} \sin\left(\frac{m\pi x}{a}\right) \sin\left(\frac{n\pi y}{b}\right),$$

which is a double Fourier sine series, where

$$A_{mn} = \frac{4}{ab} \int_{x=0}^{a} \int_{y=0}^{b} f(x, y) \sin\left(\frac{m\pi x}{a}\right) \sin\left(\frac{n\pi y}{b}\right) dy\, dx.$$

Example 4.3.1 *The four edges of a thin rectangular plate of length a and breadth b kept at zero temperature and the faces are perfectly insulated. If the initial temperature of the plate is $x\ y\ (\pi - x)\ (\pi - y)$, find the temperature at any point in the plate.*

Solution: *Let $u(x, y, t)$ be the temperature of the plate at position (x, y) at time t. The model in the form of the initial boundary value problem is given by*

$$\frac{\partial u}{\partial t} = k\left(\frac{\partial^2 u}{\partial x^2} + \frac{\partial^2 u}{\partial y^2}\right), \quad \text{for } 0 \le x \le a,\ 0 \le y \le b.$$

Boundary Conditions : $u(0, y, t) = u(a, y, t) = 0, \quad u(x, 0, t) = u(x, b, t) = 0.$
Initial Condition : $u(x, y, 0) = x\ y\ (\pi - x)\ (\pi - y).$

Proceeding as in section(4.3), we obtain the solution as

$$u(x, y, t) = \sum_{n=1}^{\infty}\sum_{m=1}^{\infty} A_{mn}\ \sin\left(\frac{m\pi x}{a}\right)\ \sin\left(\frac{n\pi y}{b}\right)\ e^{-\lambda_{mn}^2 t}, \text{ where,}$$

$$A_{mn} = \frac{4}{ab}\int_{x=0}^{a}\int_{y=0}^{b} x\ y\ (\pi - x)\ (\pi - y)\ \sin\left(\frac{m\pi x}{a}\right)\ \sin\left(\frac{n\pi y}{b}\right) dy\ dx$$

$$\text{and } \lambda_{mn}^2 = k\ \pi^2\left(\frac{m^2}{a^2} + \frac{n^2}{b^2}\right), \quad m = 1, 2, 3, \ldots\ , \ n = 1, 2, 3, \ldots\ .$$

Therefore,

$$A_{mn} = \frac{4}{ab}\int_{x=0}^{a}(\pi x - x^2)\ \sin\left(\frac{m\pi x}{a}\right)\ dx\ \times \int_{y=0}^{b}(\pi y - y^2)\ \sin\left(\frac{n\pi y}{b}\right) dy$$
$$= \frac{16a^2b^2}{m^3n^3\pi^6}\ [1 - (-1)^m]\ [1 - (-1)^n],$$
$$= \begin{cases} 0, & \text{when } m = 2p \text{ or } n = 2q \text{ (even)}. \\ \dfrac{64a^3b^3}{m^3n^3\pi^6}, & \text{when } m = 2p - 1,\ n = 2q - 1 \text{ (odd)}. \end{cases}$$

Hence, the required solution is (fig. 4.4)

$$u(x, y, t) = \sum_{p=1}^{\infty}\sum_{q=1}^{\infty}\left[\frac{64a^3b^3}{\pi^3(2p-1)^3\ (2q-1)^3} \times \right.$$
$$\left. \sin\frac{(2p-1)\pi x}{a}\ \sin\frac{(2q-1)\pi y}{b}\ e^{-k\pi^2\left(\frac{(2p-1)^2}{a^2} + \frac{(2q-1)^2}{b^2}\right)t}\right].$$

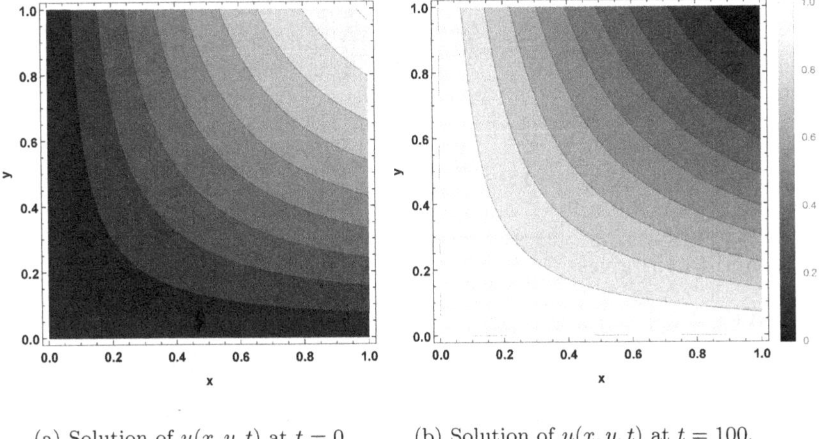

(a) Solution of $u(x, y, t)$ at $t = 0$. (b) Solution of $u(x, y, t)$ at $t = 100$.

FIGURE 4.4: *The figures show the dynamics of heat flows with Dirichlet boundary conditions and initial condition* $u(x, y, 0) = x\ y\ (\pi - x)\ (\pi - y)$.

4.4 Steady Heat Flow: Laplace Equation

In heat flow problems, sometimes, one has to deal with inhomogeneous boundary conditions, which require the study of steady heat flow (that is, time-independent solutions of the heat equation). These are called steady-state solutions, and they satisfy $\frac{\partial u}{\partial t} = 0$. In one-dimensional case, the heat equation for steady state becomes $\frac{\partial^2 u(x)}{\partial x^2} = 0$, whose solutions are straight lines. In a two-dimensional case, the heat equation for steady state becomes $\frac{\partial^2 u(x,y)}{\partial x^2} + \frac{\partial^2 u(x,y)}{\partial y^2} = 0$, which is the well-known Laplace equation. Solutions to the Laplace equation are called harmonic functions.

4.4.1 Laplace Equation with Dirichlet's Boundary Condition

To obtain the steady-state solutions to the Laplace equation

$$\nabla^2 u \equiv \frac{\partial^2 u(x, y)}{\partial x^2} + \frac{\partial^2 u(x, y)}{\partial y^2} = 0, \tag{4.13}$$

with Dirichlet's boundary condition:

$$u(0, y) = u(a, y) = 0,\ 0 \le y \le b, \tag{4.14a}$$

$$u(x, 0) = 0,\ 0 \le x \le a, \tag{4.14b}$$

$$u(x, b) = f(x),\ 0 \le x \le a, \tag{4.14c}$$

we assume the solution of the form

$$u(x, y) = X(x)\, Y(y). \tag{4.15}$$

Substituting we get,

$$\frac{X''}{X} = -\frac{Y''}{Y} = \mu \text{ (say) } \Rightarrow X'' - \mu X = 0, \tag{4.16a}$$

$$Y'' + \mu Y = 0. \tag{4.16b}$$

(4.14a) and (4.15) gives

$$X(0)\, Y(0) = 0 \text{ and } X(a)\, Y(y) = 0 \Rightarrow X(0) = 0 \text{ and } X(a) = 0$$
$$[Y(y) \neq 0 \text{ or else } u = 0, \text{ which does not satisfy (4.14c)}].$$

As before, $\mu = 0$ and $\mu > 0$ give trivial solution, which does not satisfy (4.14c), hence we reject both of them. For $\mu < 0$ $(= -\lambda^2)$, the solution of (4.16a) is

$$X(x) = A_1 \cos(\lambda x) + A_2 \sin(\lambda x).$$

$$\text{Now, } X(0) = 0 \Rightarrow A_1 = 0 \text{ and } X(a) = 0 \Rightarrow \lambda = \frac{n\pi}{a}, \quad n = 1, 2, 3, \ldots\ .$$

Hence, the non-zero solution of (4.16a) is given by

$$X(x) = A_2 \sin\left(\frac{n\pi x}{a}\right).$$

From (4.16b), we get

$$Y''(y) = \lambda^2\, Y = \frac{n^2\pi^2}{a^2}\, Y \Rightarrow Y(y) = B_1\, e^{\frac{n\pi}{a} y} + B_2\, e^{-\frac{n\pi}{a} y}.$$

(4.14b) and (4.15) give $u(x, 0) = X(x)\, Y(0) = 0 \Rightarrow Y(0) = 0$ $(X(x) \neq 0$ or else $u(0) = 0$, which does not satisfy (4.14c).

$$Y(0) = 0 \Rightarrow B_1 = -B_2 \Rightarrow Y(y) = 2B_1 \frac{e^{\frac{n\pi}{a} y} - e^{-\frac{n\pi}{a} y}}{2} = 2B_1 \sinh\left(\frac{n\pi y}{a}\right).$$

We use the principle of superposition to obtain the most general solution as

$$u(x, y) = \sum_{n=1}^{\infty} A_n \sin\left(\frac{n\pi x}{a}\right) \sinh\left(\frac{n\pi y}{a}\right). \tag{4.17}$$

The general solution (4.17) satisfies the Laplace equation (4.13) inside the rectangle, as well as the three homogeneous boundary conditions on three of its sides (left, right and bottom). We now use the boundary condition on the top of the rectangle to determine the values of A_n, which requires

$$u(x, b) = f(x) = \sum_{n=1}^{\infty} A_n \sin\left(\frac{n\pi x}{a}\right) \sinh\left(\frac{n\pi b}{a}\right).$$

This is a half range Fourier sine series of $f(x)$ in $0 \leq x \leq a$, where

$$\Rightarrow A_n = \frac{2}{a \sinh\left(\frac{n\pi b}{a}\right)} \int_0^a f(x) \sin\left(\frac{n\pi x}{a}\right) dx. \tag{4.18}$$

Therefore, the solution to the Laplace equation with Dirichlet's boundary conditions is (4.17), where A_n is given by (4.18).

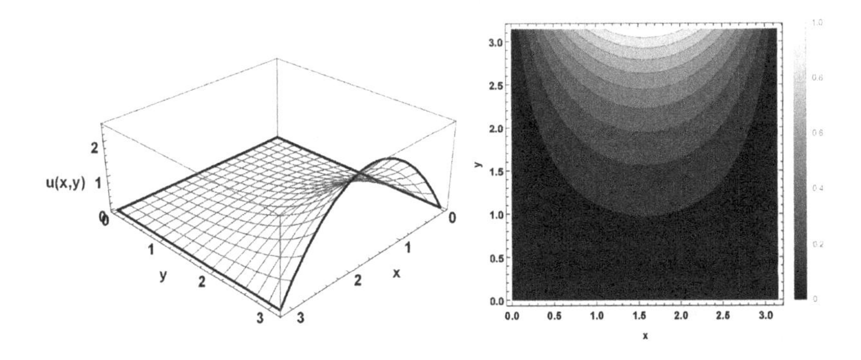

(a) Three-dimensional plot of steady heat flow.

(b) Contour plot of the flow.

FIGURE 4.5: *The figures show a steady heat flow in a two-dimensional rectangular plate with Dirichlet boundary condition.*

Example 4.4.1 *Find the steady-state temperature distribution in a rectangular plate of sides a and b, insulated at the lateral surface and satisfying the boundary conditions*

$$u(0, y) = 0, u(\pi, y) = 0 \text{ for } 0 \leq y \leq \pi,$$
$$u(x, 0) = 0, \text{ and } u(x, \pi) = x\ (\pi - x) \text{ for } 0 \leq x \leq \pi.$$

Solution: *We proceed as in section(4.4.1), and calculate*

$$A_n = \frac{2}{\pi \sinh(n\pi)} \int_0^\pi x(\pi - x)\sin(\pi x)dx = \frac{2(2 - 2\cos(n\pi) - n\pi\sin(n\pi))}{n^3\pi\sinh(n\pi)}$$

$$= \frac{4}{\pi n^3}\ [1 - (-1)^n]\ \cosh(n\pi) = \begin{cases} 0, & \text{when } n = 2m\ . \\ \dfrac{8\cosh((2m-1)\pi)}{(2m-1)^3\pi}, & \text{when } n = 2m - 1. \end{cases}$$

Hence, the required steady temperature u(x, y) is given by (fig. 4.5)

$$u(x, y) = \sum_{m=1}^{\infty} \frac{8\cosh(2m - 1)\pi}{\pi(2m - 1)^3}\ \sin(2m - 1)x\ \sin(2m - 1)y.$$

4.4.2 Laplace Equation with Neumann's Boundary Condition

To obtain the steady-state solution to the Laplace equation

$$\nabla^2 u \equiv \frac{\partial^2 u}{\partial x^2} + \frac{\partial^2 u}{\partial y^2} = 0,$$

with Neumann's boundary conditions:

$$u_x(0, y) = 0, \; u_x(a, y) = 0, \; 0 \le y \le b,$$
$$u_y(x, 0) = 0, \; u_y(x, b) = f(x) \; 0 \le x \le a.$$

We assume the solution the solution of the form $u(x, y) - X(x)\ Y(y)$. Proceeding as in (4.4.1), we get $X(x) = A_1 \cos\left(\frac{n\pi x}{a}\right)$ and $Y(y) = 2B_1 \cosh\left(\frac{n\pi y}{a}\right)$. Using the principle of superposition, the general solution can be written as

$$u(x, y) = \sum_{i=1}^{n} \frac{n\pi}{a} A_n \cos\left(\frac{n\pi x}{a}\right) \cosh\left(\frac{n\pi y}{a}\right).$$

$$u_y(x, b) = f(x) \Rightarrow f(x) = \sum_{i=1}^{n} \frac{n\pi}{a} \sinh\left(\frac{n\pi b}{a}\right) A_n \cos\left(\frac{n\pi x}{a}\right), \; 0 \le x \le a,$$

which is a Fourier Cosine series, where

$$A_n = \frac{2}{n\pi \sinh\left(\frac{n\pi b}{a}\right)} \int_0^a f(x) \cos\left(\frac{n\pi x}{a}\right) dx.$$

4.5 Wave Equation

The wave equation describes the propagation of oscillations and is represented by a linear second-order partial differential equation. Consider a homogeneous string of length L is tied at both ends. We assume that the string offers no resistance due to bending; that is, it is thin and flexible; the tension in the string is much greater than the gravitational force, and hence, it can be neglected; the motion of the string takes place in the vertical plane only (fig. 4.6).

Let ρ be the linear density of the string, and P and Q are two neighboring points on the string such that arc $PQ = \Delta s$. Let T_1 and T_2 be the tensions at points P and Q, which make angles α and β, respectively with the x-axis and let $u(x, t)$ be the displacement of the string at time t from its equilibrium state. Then, the equations of motion are

$$T_2 \cos\beta - T_1 \cos\alpha = 0 \quad \text{(along x-axis)}, \tag{4.19}$$
$$(\rho \Delta s) \frac{\partial^2 u}{\partial t^2} = T_2 \sin\beta - T_1 \sin\alpha \quad \text{(along y-axis)}.$$

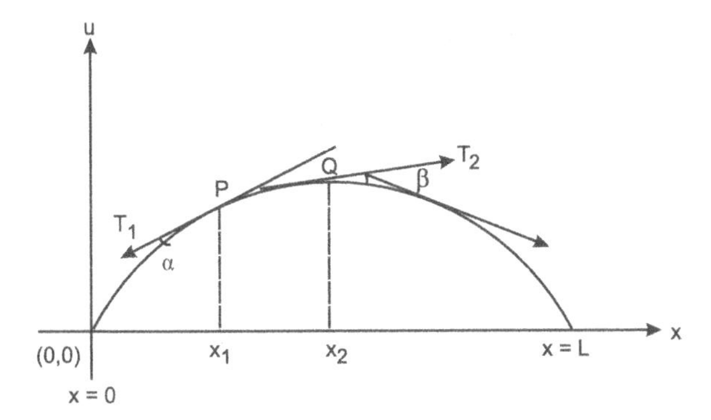

FIGURE 4.6: *A homogeneous string of length L, tied at both ends such that the string offers no resistance due to bending.*

From (4.19), we obtain, $T_1 \cos\alpha = T_2 \cos\beta = T$ (say), which implies

$$\frac{(\rho\Delta S)}{T}\frac{\partial^2 u}{\partial t^2} = \frac{T_2 \sin\beta}{T} - \frac{T_1 \sin\alpha}{T} = \frac{T_2 \sin\beta}{T_2 \cos\beta} - \frac{T_1 \sin\alpha}{T_1 \cos\alpha} = \tan\beta - \tan\alpha.$$

At the points P and Q, the slopes of the string are given by

$\tan\alpha = \frac{\partial u}{\partial x}|_{x_1}$ and $\tan\beta = \frac{\partial u}{\partial x}|_{x_2}$. Therefore,

$$\frac{(\rho\Delta S)}{T}\frac{\partial^2 u}{\partial t^2} = u_x(x_2, t) - u_x(x_1, t) \Rightarrow \frac{\rho}{T}\frac{\Delta S}{\Delta x}\frac{\partial^2 u}{\partial t^2} = \frac{u_x(x_1 + \Delta x, t) - u_x(x_1, t)}{\Delta x}.$$

As $\Delta x \to 0, \Delta s \to \Delta x$ and we get $\frac{\partial^2 u}{\partial t^2} = c^2 \frac{\partial^2 u}{\partial u^2}$, where $c^2 = \frac{T}{\rho}$,

which is the one-dimensional wave equation, c being the speed of the propagation of the wave.

4.5.1 Vibrating String

We consider a homogenous flexible string of length L, which is stretched between two fixed points $(0, 0)$ and $(L, 0)$. Initially, the string is released from a position $u = f_1(x)$ with a velocity $u_t = f_2(x)$ parallel to the y-axis.

Mathematically, we can formulate the model as follows:

$$\frac{\partial^2 u}{\partial t^2} = c^2 \frac{\partial^2 u}{\partial x^2}, \quad 0 \le x \le L, \ t > 0.$$

Boundary Conditions (BCs): $u(0, t) = 0, \ u(L, t) = 0$.
Initial Conditions (ICs): $u(x, 0) = f_1(x), \ \frac{\partial u(x,0)}{\partial t} = f_2(x)$.

We use separation of variables to solve the given wave equation. Let $u(x, t) = X(x)T(t)$ be a solution of $\frac{\partial^2 u}{\partial t^2} = c^2 \frac{\partial^2 u}{\partial x^2}$. Substituting, we obtain,

$$\frac{X''}{X} = \frac{T''}{c^2 T} = \text{constant}.$$

Since the boundary conditions are periodic and homogenous in x, $X(x)$ must be periodic, which is only possible if the constant is negative $(-\lambda^2)$.

$$\text{Solving } \frac{X''}{X} = -\lambda^2 \quad \text{and} \quad \frac{T''}{c^2T} = -\lambda^2, \text{ we obtain,}$$

$$X(x) = A_1\cos(\lambda x) + A_2\sin(\lambda x) \text{ and } T(t) = A_3\cos(c\lambda t) + A_4\sin(c\lambda t).$$

Therefore, the general solution is given by

$$u(x,t) = [A_1\cos\lambda x + A_2\sin\lambda x]\left[(A_3\cos(c\lambda t) + A_4\sin(c\lambda t)\right]. \tag{4.20}$$

Using the boundary conditions $u(0,t) = 0 = u(L,t)$ we obtain $A_1 = 0$ and $\sin(\lambda L) = 0$ $(A_2 \neq 0) \Rightarrow \lambda = \frac{n\pi}{L}$, n being an integer. Therefore equation (4.20) becomes

$$u(x,t) = A_2\sin\frac{n\pi x}{L}\left[A_3\cos\left(\frac{cn\pi t}{L}\right) + A_4\sin\left(\frac{cn\pi t}{L}\right)\right].$$

Noting that the wave equation is linear, we use the principle of superposition to obtain its most general solution as

$$u(x,t) = \sum_{n=1}^{\infty}\left[A_n\cos\left(\frac{cn\pi t}{L}\right) + B_n\sin\left(\frac{cn\pi t}{L}\right)\right]\sin\left(\frac{n\pi x}{L}\right).$$

Using the initial condition we get,

$$u(x,0) = f_1(x) = \sum_{n=1}^{\infty} A_n\sin\left(\frac{n\pi x}{L}\right), \text{ and}$$

$$\frac{\partial u(x,0)}{\partial t} = f_2(x) = \sum_{n=1}^{\infty}\frac{cn\pi}{L}B_n\sin\left(\frac{n\pi x}{L}\right).$$

Both are half-range Fourier sine series; therefore, we get,

$$A_n = \frac{2}{L}\int_0^L f_1(x)\sin\left(\frac{n\pi x}{L}\right)dx, \tag{4.21}$$

$$\frac{n\pi c}{L}B_n = \frac{2}{L}\int_0^L f_2(x)\sin\left(\frac{n\pi x}{L}\right)dx,$$

$$\Rightarrow B_n = \frac{2}{n\pi c}\int_0^L f_2(x)\sin\left(\frac{n\pi x}{L}\right)dx. \tag{4.22}$$

Hence, the displacement of the vibrating string is given by

$$u(x,t) = \sum_{n=1}^{\infty}\left[A_n\cos\left(\frac{cn\pi t}{L}\right) + B_n\sin\left(\frac{cn\pi t}{L}\right)\right]\sin\left(\frac{n\pi x}{L}\right),$$

where A_n and B_n are given by (4.21) and (4.22).

Corollary 1: If the homogenous flexible string of length L, stretched between two fixed points $(0,0)$ and $(L,0)$, is initially released from rest from a position $u = f_1(x)$, then its initial velocity is zero, that is, $\frac{\partial u(x,0)}{\partial t} = 0$. The solution in that case will be of the form

$$u(x,t) = \sum_{n=1}^{\infty} A_n \cos\left(\frac{cn\pi t}{L}\right) \sin\left(\frac{n\pi x}{L}\right), \quad \text{where}$$

$$A_n = \frac{2}{L}\int_0^L f_1(x) \sin\left(\frac{n\pi x}{L}\right) dx.$$

Corollary 2: If the homogenous flexible string of length L, stretched between two fixed points $(0,0)$ and $(L,0)$, is initially released with velocity $f_2(x)$, and the initial deflection of the string is zero, then $u(x,0) = 0$. The solution in that case will be of the form

$$u(x,t) = \sum_{n=1}^{\infty} B_n \sin\left(\frac{cn\pi t}{L}\right) \sin\left(\frac{n\pi x}{L}\right), \quad \text{where}$$

$$B_n = \frac{2}{cn\pi}\int_0^L f_2(x) \sin\left(\frac{n\pi x}{L}\right) dx.$$

Example 4.5.1 *A homogenous flexible string in a guitar is stretched between two fixed points $(0,0)$ and $(L,0)$, the length of the string being L units. The string of the guitar is initially plucked from rest from a position $\mu x(L-x)$. Find the displacement $u(x,t)$ of the string of the guitar at time t.*

Solution: *Mathematically, we can formulate the model as follows:*

$$\frac{\partial^2 u}{\partial t^2} = c^2 \frac{\partial^2 u}{\partial x^2}, \quad 0 \le x \le L, \ t > 0.$$

Boundary conditions (BCs): $u(0,t) = 0 = u(L,t)$.
Initial conditions (ICs): $u(x,0) = \mu x(L-x)$, $\frac{\partial u(x,0)}{\partial t} = 0$.
The solution is of the form

$$u(x,t) = (A_1 \cos\lambda x + A_2 \sin\lambda x)[A_3 \cos(c\lambda t) + A_4 \sin(c\lambda t)].$$

Applying boundary conditions, we get $A_1 = 0$, $\lambda = \frac{n\pi}{2}$ $(A_2 \neq 0,)$, *n being an integer. Using the principle of superposition, the possible solution is*

$$u(x,t) = \sum_{n=1}^{\infty} \left[A_n \cos\left(\frac{cn\pi t}{L}\right) + B_n \sin\left(\frac{cn\pi t}{L}\right)\right] \sin\left(\frac{n\pi x}{L}\right).$$

Now, $\dfrac{\partial u(x,0)}{\partial t} = 0$ *gives* $B_n = 0 \Rightarrow u(x,t) = \sum_{n=1}^{\infty} A_n \cos\left(\frac{cn\pi t}{L}\right) \sin\left(\frac{n\pi x}{L}\right)$.

The initial condition gives $u(x,0) = \mu x(L-x) = \sum_{n=1}^{\infty} A_n \sin\left(\frac{n\pi x}{L}\right)$, *which is a half-range Fourier sine series, where*

$$A_n = \frac{2}{L}\int_0^L \mu x(L-x)sin\left(\frac{n\pi x}{L}\right)dx = \frac{2\mu}{L}\left[2\left(\frac{L}{n\pi}\right)^3\{1-(-1)^n\}\right]$$

$$= \begin{cases} \frac{8\mu L^2}{n^3\pi^3}; & n = odd \\ 0, & n = even \end{cases}$$

Therefore, the required solution is (fig. 4.7)

$$u(x,t) = \sum_{n=1}^{\infty} \frac{8\mu L^2}{(2n-1)^3\pi^3} cos\left\{\frac{(2n-1)c\pi t}{L}\right\} sin\left\{\frac{(2n-1)\pi x}{L}\right\}.$$

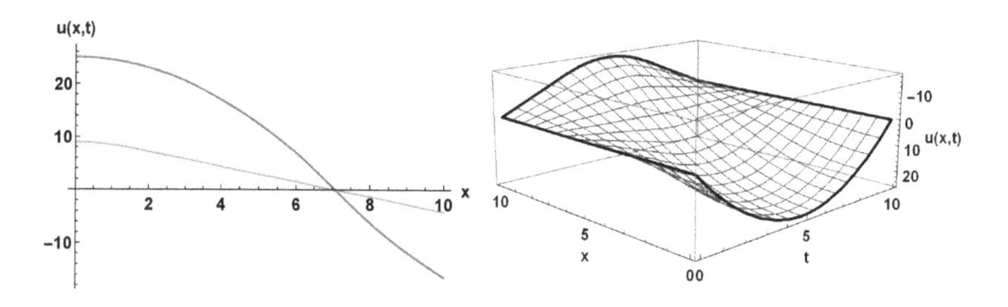

(a) Displacement $u(x,t)$ for $t = 5, 9$. (b) Three-dimensional view of $u(x,t)$.

FIGURE 4.7: *The figures show the displacement of the homogeneous string in a guitar dynamics between* $(0,0)$ *and* $(10,0)$.

4.6 Two-Dimensional Wave Equation

Consider a thin elastic membrane stretched tightly over a rectangular frame, of dimension $a \times b$, where the edges of the membrane are kept fixed to the frame. Some sort of vibration is observed if the membrane is perturbed from equilibrium. Our aim is to mathematically model the vibration of the membrane surface. Let $u(x,y,t)$ be the deflection of the membrane from equilibrium at position $(x,\ y)$ at time t. The surface $z = u(x,y,t)$ gives the shape of the membrane at time t (for a fixed t). It can be shown that under ideal assumptions (say, uniform membrane density, uniform tension, no resistance to motion, small deflection etc.), $u(x,y,t)$ satisfies the two dimensional wave equation

$$\frac{\partial^2 u}{\partial t^2} = c^2\left(\frac{\partial^2 u}{\partial x^2} + \frac{\partial^2 u}{\partial y^2}\right),\ 0 \le x \le a,\ 0 \le y \le b. \tag{4.23}$$

The edges of the membrane are kept fixed, which is expressed by the boundary conditions:

$$u(0,y,t)=0,\ u(a,y,t)=0,\ 0\le y\le b, t\ge 0, \tag{4.24}$$
$$u(x,0,t)=0,\ u(x,b,t)=0,\ 0\le x\le a, t\ge 0. \tag{4.25}$$

It is also specified how the membrane is initially deformed and set into motion, which is done through the initial conditions:

$$\text{Initial displacement: } u(x,y,0)=f(x,y),\ (x,y)\in\Re, \tag{4.26}$$

$$\text{Initial velocity: } \frac{\partial u(x,y,0)}{\partial t}=g(x,y),\ (x,y)\in\Re=[0,a]\times[0,b]. \tag{4.27}$$

Let $u(x,y,t)=X(x)\,Y(y)\,T(t)$ be a solution of (4.23). Substituting, we get

$$\frac{X''}{X}+\frac{Y''}{Y}=\frac{T''}{c^2T}. \tag{4.28}$$

Since x, y, t are independent variables, (4.28) will hold if each term on each side is equal to the same separation constant, that is,

$$\frac{X''}{X}=\mu_1,\ \frac{Y''}{Y}=\mu_2 \text{ and } \frac{T''}{c^2T}=\mu_1+\mu_2.$$

Now, $\frac{X''}{X}=\mu_1\Rightarrow X''-\mu_1X=0$. (4.24) gives

$$X(0)\,Y(y)\,T(t)=0 \text{ and } X(a)\,Y(y)\,T(t)=0,$$
$$\Rightarrow X(0)=0 \text{ and } X(a)=0$$
$$[Y(y)\neq 0 \text{ or else } u=0, \text{which does not satisfy (4.26)}].$$

As discussed before $\mu_1=0$ and $\mu_1>0$ give trivial solution, which does not satisfy (4.26), hence we reject both of them. For $\mu_1<0$ ($=-\lambda_1^2$, say), the solution of (4.28) is

$$X(x)=A_1\cos(\lambda_1x)+A_2\sin(\lambda_1x).$$

$X(0)=0\Rightarrow A_1=0$ and $X(a)=0\Rightarrow\lambda_1=\frac{m\pi}{a}, m=1,2,3,...$. Hence, non-zero solution of (4.28) is given by

$$X(x)=A_2\ \sin\left(\frac{m\pi x}{a}\right), m=1,2,3,... \ .$$

In a similar manner,

$$Y''-\mu_2Y=0\Rightarrow Y(y)=B_2\sin\left(\frac{n\pi y}{b}\right), n=1,2,3,...$$
$$\text{and } T''=c^2\ (\mu_1+\mu_2)\ T=c^2\ (-\lambda_1{}^2-\lambda_2{}^2)\ T.$$
$$\Rightarrow T''+\lambda_{mn}^2\ T=0, \text{where } \lambda_{mn}^2=c^2\ \pi^2\ (\frac{m^2}{a^2}+\frac{n^2}{b^2}).$$
$$\Rightarrow T(t)=A\ \cos(\lambda_{mn}t)+B\ \sin(\lambda_{mn}t).$$

We use the principle of superposition to obtain the most general solution as

$$u(x,y,t) = \sum_{m=1}^{\infty}\sum_{n=1}^{\infty} \left[A_{mn}\cos(\lambda_{mn}t) + B_{mn}\sin(\lambda_{mn}t)\right] \times \sin\left(\frac{m\pi x}{a}\right)\sin\left(\frac{n\pi y}{b}\right). \tag{4.29}$$

Differentiating (4.29) partially with respect to t, we obtain,

$$\frac{\partial u}{\partial t} = \sum_{m=1}^{\infty}\sum_{n=1}^{\infty} \left[-A_{mn}\,\lambda_{mn}\,\sin(\lambda_{mn}t) + B_{mn}\,\lambda_{mn}\,\cos(\lambda_{mn}t)\right] \times \sin\left(\frac{m\pi x}{a}\right)\sin\left(\frac{n\pi y}{b}\right) \tag{4.30}$$

Putting $t = 0$ in (4.29) and (4.30) and using (4.26) and (4.27), we obtain

$$f(x,y) = \sum_{m=1}^{\infty}\sum_{n=1}^{\infty} A_{mn}\,\sin\left(\frac{m\pi x}{a}\right)\sin\left(\frac{n\pi y}{b}\right),$$

$$\text{and } g(x,y) = \sum_{m=1}^{\infty}\sum_{n=1}^{\infty} B_{mn}\,\lambda_{mn}\,\sin\left(\frac{m\pi x}{a}\right)\sin\left(\frac{n\pi y}{b}\right),$$

which is a double Fourier sine series. Accordingly we get (n=1,2,..., m=1,2,...),

$$A_{mn} = \frac{4}{ab}\int_{x=0}^{a}\int_{y=0}^{b} f(x,y)\,\sin\left(\frac{m\pi x}{a}\right)\sin\left(\frac{n\pi y}{b}\right)dy\,dx$$

$$\text{and } B_{mn} = \frac{4}{ab}\int_{x=0}^{a}\int_{y=0}^{b} g(x,y)\,\sin\left(\frac{m\pi x}{a}\right)\sin\left(\frac{n\pi y}{b}\right)dy\,dx.$$

Corollary I: If the initial velocity $\frac{\partial u}{\partial t}(x,y,0) = g(x,y) = 0$, then $B_{mn} = 0$ and the solution reduces to

$$u(x,y,t) = \sum_{m=1}^{\infty}\sum_{n=1}^{\infty} A_{mn}\,\cos(\lambda_{mn}t)\,\sin\left(\frac{m\pi x}{a}\right)\sin\left(\frac{n\pi y}{b}\right).$$

Corollary II: If the initial displacement $u(x,y,0) = f(x,y) = 0$, then $A_{mn} = 0$ and the solution reduces to

$$u(x,y,t) = \sum_{m=1}^{\infty}\sum_{n=1}^{\infty} B_{mn}\,\sin(\lambda_{mn}t)\,\sin\left(\frac{m\pi x}{a}\right)\sin\left(\frac{n\pi y}{b}\right).$$

Example 4.6.1 *Model the deflection $u(x,y,t)$ of the square membrane of each side unity and $c = 1$, if the initial velocity is zero and the initial deflection is $10\sin(\pi x)\sin(2\pi y)$.*

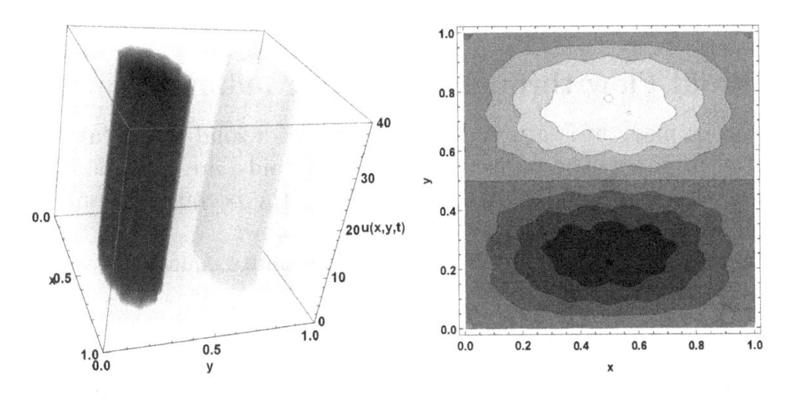

(a) Three-dimensional view of deflection $u(x, y, t)$.

(b) Density plot of deflection $u(x, y, t)$.

FIGURE 4.8: *The figures show the deflection $u(x, y, t)$ of the unit square membrane with Dirichlet boundary condition.*

Solution: *The deflection $u(x, y, t)$ of the given membrane is governed by the 2D wave equation*

$$\frac{\partial^2 u}{\partial t^2} = \frac{\partial^2 u}{\partial x^2} + \frac{\partial^2 u}{\partial y^2} \quad (c = 1),$$

with B.C.'s $u(0, y, t) = 0, u(1, y, t) = 0, u(x, 0, t) = 0, u(x, 1, t) = 0$, Initial deflection $u(x, y, 0) = 10\sin(\pi x)\sin(2\pi y)$ and, initial velocity $u_t(x, y, 0) = 0$. Following section(4.6), we get the general solution of the deflection as

$$u(x, y, t) = \sum_{m=1}^{\infty}\sum_{n=1}^{\infty} A_{mn}\cos(\lambda_{mn}t)\sin(m\pi x)\sin(n\pi y), \ \lambda_{mn}^2 = \pi^2(m^2 + n^2),$$

$$A_{mn} = 4\int_{x=0}^{1}\int_{y=0}^{1} 10\ \sin(\pi x)\ \sin(2\pi y)\ \sin(m\pi x)\ \sin(n\pi y)\ dy\ dx$$

$$= 10\int_{x=0}^{1} 2\sin(\pi x)\ \sin(m\pi x)\ dx \int_{y=0}^{1} 2\sin(2\pi y)\ \sin(n\pi y)\ dy.$$

Now, $$\int_{x=0}^{1} 2\sin(\pi x)\ \sin(m\pi x)\ dx == \begin{cases} 1, & m = 1, \\ 0, & m \neq 1, \end{cases}$$

and $$\int_{y=0}^{1} 2\sin(2\pi y)\ \sin(n\pi y)\ dy == \begin{cases} 1, & n = 2, \\ 0, & n \neq 2. \end{cases}$$

Hence, $A_{12} = 10 \times 1 \times 1 = 10$ and ${\lambda_{12}}^2 = \pi^2\ (1^2 + 2^2) = 5\ \pi^2$.

Required solution is $u(x, y, t) = 10\ \cos(\sqrt{5}\pi t)\ \sin(\pi x)\ \sin(2\pi y)$ *(fig. 4.8).*

4.7 Fluid Flow through a Porous Medium

A porous medium is a material consisting of a solid frame (also called a matrix) with pores. The pores, also known as voids, are interconnected and are filled with liquid or gas or both. To model the porous medium, we have to consider the three equations that govern the flow:

(i) Equation of conservation of mass or continuity equation:

$$\frac{\partial \rho}{\partial t} = -\vec{\nabla}.(\rho \vec{v}),$$

which states that the rate of mass entering a system is equal to the rate of mass leaving the system.

(ii) Darcy's law: $v = -a\vec{\nabla}p$, which states that velocity is proportional to the pressure gradient.

(iii) Equation of state: $\rho = p^{\gamma}$, where ρ is the fluid density, p is the pressure, $\vec{v}$ is the flow velocity vector field and γ is a constant, which is the ratio of specific heats.

The continuity equation gives

$$\begin{aligned}\frac{\partial \rho}{\partial t} &= -\vec{\nabla}.(\rho \vec{v}) = \vec{\nabla}.(\rho a \vec{\nabla} p) \ \ (\text{using } (ii)) = \vec{\nabla}.(\rho a \vec{\nabla} \rho^{1/\gamma}) \quad (\text{using } (iii)) \\ &= \vec{\nabla}.\left[\rho a \left(\frac{\partial}{\partial x}\hat{i} + \frac{\partial}{\partial y}\hat{j} + \frac{\partial}{\partial z}\hat{k}\right)\rho^{1/\gamma}\right] = \vec{\nabla}.(a\frac{\rho^{1/\gamma}}{\gamma}\vec{\nabla}\rho).\end{aligned}$$

$$\text{Now,} \ \vec{\nabla}\left(\rho^{1+\frac{1}{\gamma}}\right) = \left(\hat{i}\frac{\partial}{\partial x} + \hat{j}\frac{\partial}{\partial y} + \hat{k}\frac{\partial}{\partial z}\right)\rho^{1+\frac{1}{\gamma}} = \left(1 + \frac{1}{\gamma}\right)\rho^{\frac{1}{\gamma}}\vec{\nabla}\rho.$$

$$\text{Therefore,} \ \frac{\partial \rho}{\partial t} = \vec{\nabla}.\left(\frac{a\vec{\nabla}\rho^{1+\frac{1}{\gamma}}}{\gamma\left(1+\frac{1}{\gamma}\right)}\right) = \frac{a}{1+\gamma}\nabla^2\rho^{1+\frac{1}{\gamma}}.$$

Replacing ρ by u and rescaling t by $\frac{a}{1+\gamma}$, we get,

$$\frac{\partial u}{\partial t} = \Delta u^m,$$

where $m = 1 + \frac{1}{\gamma}$ and Δu^m is the linear diffusion term with $m > 1$. This equation governs the flow of fluids through a porous medium.

4.8 Traffic Flow

Traffic flow is defined as the total number of vehicles passing a given point in a given time and is expressed as vehicle per hour. It is the investigation of the interaction between vehicles, drivers, and infrastructure, which aims to understand and build an optimal road network with efficient traffic movement and minimal traffic congestion. The efficiency of the traffic flows depends on three basic characteristics, (i) flow (the number of vehicles passing a point in a given time period), (ii) velocity (vehicle's rate of motion in a particular direction) and, (iii) density (the number of vehicles occupying a unit length of the roadway at an instant in time).

There are two classes of models in the traffic flow problem: (a) Macroscopic, which deals with average behavior, such as traffic density, average speed and modular area, and (b) Microscopic, which is concerned with individual behavior, say, car-following models [10]. But, the computational cost for microscopic models is high as an ODE is linked to each car, which needs to be solved at each time step, and the size of the system increases with the increase of the number of cars. Therefore, a good macroscopic model is desirable as they are computationally less expensive due to fewer design details in terms of interaction among vehicles and between vehicles and their environment.

In this section, we will concentrate on the macroscopic modeling of the traffic flow, where the focus will be on modeling the density of cars and their flow, rather than modeling individual cars and their velocity. Let the number of cars per unit length at position x and in time t be $\rho(x,t)$ and the number of cars passing the position x per unit time at time t be $f(x,t)$, that is, $f(x,t)$ gives the rate of flow of the cars (traffic flux) and $\rho(x,t)$ denotes the density of cars (the state variable).

We now use the global conservation law on the statement that "cars are conserved" [5]. By the statement "cars are conserved", we mean that the number of cars in an arbitrary region of space $I = [x_1, x_2]$, called the cell, at the end of an arbitrary interval of time $T = [t_1, t_2]$, called the time-step, is equal to the number of cars in the cell at the beginning of the time-step (that is, t), plus the number of cars entering the cell, minus the number of cars going out of the cell, that is, the change of flux in the cell $[x_1, x_2]$.

Now, the number of cars entering the cell $[x_1, x_2]$ through the point x_1 during the time-step $[t_1, t_2]$ is $\displaystyle\int_{t_1}^{t_2} f(x_1, t)dt$.

The number of cars leaving the cell $[x_1, x_2]$ through the point x_2 during the time-step $[t_1, t_2]$ is $\displaystyle\int_{t_1}^{t_2} f(x_2, t)dt$.

The number of cars in the cell $[x_1, x_2]$ at time t_1 is $\displaystyle\int_{x_1}^{x_2} \rho(x, t_1)dx$

and the number of cars in the cell $[x_1, x_2]$ at time t_2 is $\int_{x_1}^{x_2} \rho(x, t_2)dx$.

Since "cars are conserved", we have

$$\int_{x_1}^{x_2} \rho(x, t_2)dx = \int_{x_1}^{x_2} \rho(x, t_1)dx + \int_{t_1}^{t_2} f(x_1, t)dt - \int_{t_1}^{t_2} f(x_2, t)dt,$$

$$\Rightarrow \int_{x_1}^{x_2} [\rho(x, t_2) - \rho(x, t_1)]dx + \int_{t_1}^{t_2} [f(x_2, t) - f(x_1, t)]dt = 0.$$

Using the fundamental theorem of Calculus (by assuming ρ and f to be a smooth function of x and t respectively), we get,

$$\int_{x_1}^{x_2} \int_{t_1}^{t_2} \frac{\partial \rho(x, t)}{\partial t} dtdx + \int_{t_1}^{t_2} \int_{x_1}^{x_2} \frac{\partial f(x, t)}{\partial x} dxdt = 0,$$

$$\Rightarrow \int_{x_1}^{x_2} \int_{t_1}^{t_2} \left[\frac{\partial \rho(x, t)}{\partial t} + \frac{\partial f(x, t)}{\partial x} \right] dtdx = 0.$$

(changing the order of integration for the second integral)

$$\Rightarrow \frac{\partial \rho(x, t)}{\partial t} + \frac{\partial f(x, t)}{\partial x} = 0. \tag{4.31}$$

(Assuming that the interval is piecewise continuous.)

Now, suppose there are 50 cars per kilometer on a road, and each car is traveling at 80 km/h, then to a person standing at one side of the road, 80 km worth of cars will pass in one hour, that is, $50 \times 80 = 4000$ cars per hour. In other words, the flux in this case is $\rho u = 50$ cars/km$\times 80$ km/h $= 4000$ cars/h. Therefore, we can express traffic flux as a product of traffic density and velocity, that is,

$$f(x, t) = \rho(x, t)u(x, t), \tag{4.32}$$

Using (4.32), (4.31) becomes

$$\frac{\partial \rho}{\partial t} + \frac{\partial (\rho u)}{\partial x} = 0. \tag{4.33}$$

Please note that the above continuity equation in one dimension has two unknowns, namely, ρ and u, which are again functions of x and t. Since we have only one partial differential equation, further information is necessary. A reasonable assumption by a traffic modeler may be $u = u(\rho)$, that is, velocity of the car is a function of traffic density alone (such functions are called equations of state or constitutive relations). Then (4.33) becomes

$$\frac{\partial \rho}{\partial t} + \frac{\partial (\rho u(\rho))}{\partial x} = 0,$$

which is a partial differential equation of the first order.

Corollary 1: For a single-lane open road, it is reasonable to take velocity as a function of traffic density only. The simplest relation between car velocity and traffic density is the linear relation

$$u(\rho) = u_{max}\left(1 - \frac{\rho}{\rho_{max}}\right), \qquad 0 \le \rho \le \rho_{max}, \tag{4.34}$$

where u_{max} is the maximum velocity with which an isolated car will travel, either due to speed limits or condition of the road or driver caution, such that $u(0) = u_{max}$ and ρ_{max} is the maximum traffic density (bumper-to-bumper traffic) where the velocity u is zero. Also, the velocity of the car diminishes with traffic density and hence $\frac{du}{d\rho} < 0, \quad \rho > 0$, which is true for the function (4.34).

Let $g(\rho) = \rho u(\rho) = u_{max}\left(\rho - \frac{\rho^2}{\rho_{max}}\right)$, then (4.33) becomes

$$\frac{\partial \rho}{\partial t} + \frac{\partial g(\rho)}{\partial x} = 0 \Rightarrow \frac{\partial \rho}{\partial t} + \frac{dg}{d\rho}\frac{\partial \rho}{\partial x} = 0 \Rightarrow \frac{\partial \rho}{\partial t} + g'(\rho)\frac{\partial \rho}{\partial x} = 0. \tag{4.35}$$

We next introduce a small perturbation in the traffic density

$$\rho = \rho_0 + \triangle\rho \ (\triangle\rho \ll \rho_0),$$

where ρ_0 is a constant traffic density in (4.35). Then

$$g'(\rho) = g'(\rho_0 + \triangle\rho) = g'(\rho_0) + \triangle\rho\, g''(\rho_0) + \ldots\ldots\ldots$$

and the linearized form of (4.35) is

$$\frac{\partial \rho}{\partial t} + g'(\rho_0)\frac{\partial \rho}{\partial x} = 0.$$

The partial derivatives can be written as $\frac{\partial}{\partial t}(\triangle\rho)$ and $\frac{\partial}{\partial x}(\triangle\rho)$, but since both the partial derivatives are of order $\triangle$, we have dropped $\triangle$ from both terms. Also, note that $g'(\rho) = \dfrac{dg}{d\rho}$ has the dimension of velocity, therefore $g'(\rho_0)$, also having the dimension of velocity, is a constant, say, v_0. Then the equation becomes

$$\frac{\partial \rho}{\partial t} + v_0\frac{\partial \rho}{\partial x} = 0. \tag{4.36}$$

The general solution of (4.36) is given by

$$\rho = h(x - v_0 t),$$

which represents linear traffic waves and the velocity v_0 is given by

$$v_0 = g'(\rho_0) = u_{max}\left(1 - \frac{2\rho_0}{\rho_{max}}\right).$$

We use the method of characteristics to solve (4.36). The characteristic base curves for this problem are solutions of

$$\frac{dx}{dt} = v_0 \Rightarrow x = v_0 t + x_0 \quad (\text{assuming } x(0) = x_0).$$

Clearly, this curve represents a straight line as both v_0 and x_0 are constants.

Let the initial traffic density be

$$\rho = \begin{cases} 150, & \text{for} \quad x < 0, \\ 150\left(1 - \frac{x_0}{2}\right), & \text{for} \quad 0 < x < 1, \\ 90, & \text{for} \quad x > 1, \end{cases}$$

and, let $\rho_{max} = \frac{1}{4}$ cars per meter $= \frac{1000}{4} = 250$ cars per km, $u_{max} = 80$ km/h. Then, for characteristics coming out of the negative x-axis

$$v_0 = u_{max}\left(1 - \frac{2\rho_0}{\rho_{max}}\right) = 80\left(1 - \frac{2 \times 150}{250}\right) = -16 \text{ km/h}.$$

For those emerging from $x > 1$, $v_0 = 80\left(1 - \frac{2 \times 90}{250}\right) = 22.4$ km.

For $0 < x < 1$, we have $v_0 = 80\left[1 - \frac{2 \times 150\left(1 - \frac{x_0}{2}\right)}{250}\right] \Rightarrow v_0 = -16 + 48x_0$,

which shows a general transition.

Summarizing, we can say that since the traffic density ρ is constant for each characteristic, we can calculate the traffic density in terms of x, for any time. Thus, for $x < -16t, \rho = 150$ and for $x > 1 + 22.4t, \rho = 90$. In between we can solve for x_0 as

$$x = (-16 + 48x_0)t + x_0 \Rightarrow x_0 = \frac{x + 16t}{1 + 48t}.$$

Therefore, the traffic density for the in-between region is given by

$$\rho(x,t) = 150\left(1 - \frac{x_0}{2}\right) = 150\left[1 - \frac{x + 16t}{2(1 + 48t)}\right] = 75\left(\frac{2 - x + 80t}{1 + 48t}\right).$$

4.9 Crime Model

Crimes occur in both urban and rural environments. Some areas are reasonably safe while others are dangerous, demonstrating that crime is not uniformly distributed. Criminals have their favorite zones and victim types, who are repeatedly targeted in a short time period [70, 71, 72, 77]. Some zones are commonly known as *crime hotspots* [151, 161]. Crime patterns depend on many factors. For example, the preference of a burglar to visit a previously burglarized house or an adjacent one will depend on the information about the schedules of the occupants or the types of valuables that may be stolen. The preference of the burglar may also depend on the choice of some favorable neighborhood where past successful burglaries have created an impression that the occupants are crime tolerant, resulting in the growth of more illegal activities. This is known as the *broken window effect* [151, 165].

Here we discuss a simple crime model, namely residential burglary, which is common in both urban and rural areas. The proposed model is a modification of that given in [151]. Let $A(x, y, t)$ denote the attractiveness to burglars and $C(x, y, t)$ be the criminal density for a location. The equations that model a simple residential burglary are

$$\begin{aligned} \frac{\partial A}{\partial t} &= \alpha\, CA - \beta\, A + D_1 \nabla^2 A, \\ \frac{\partial C}{\partial t} &= \gamma - \delta\, CA + D_2 \nabla^2 C, \quad \text{where, } \nabla^2 \equiv \frac{\partial^2}{\partial x^2} + \frac{\partial^2}{\partial y^2}. \end{aligned}$$

In the first equation, the term αCA shows the positive effect of successful burglaries on the attractiveness of location. A successful burglary at one location encourages the burglars to repeat the crime at the same location or adjacent to it. Thus, the density of burglars is increased due to the attractiveness of a location. The decay term $(-\beta A)$ implies that with time, the attractiveness of a particular location diminishes. Successful burglaries in the past may not encourage the burglars to commit the crime more recently, resulting in the decline of the attractiveness of a location. The diffusion term $D_1 \nabla^2 A$ measures the spread of attractiveness to the neighboring areas due to successful burglaries in that location, $D_1 (> 0)$ being the diffusion coefficient.

In the second equation, γ is the constant source of burglars per area at a given location. While some burglars leave the location after a successful burglary, new thieves may enter the location due to its attractiveness, and hence, a constant input of criminals is assumed. The term $(-\delta CA)$ implicates that due to the attractiveness of a location, a burglar will commit the burglary in that location rather than moving to another one. Hence, there will be a reduction in the criminal density. The term $D_2 \nabla^2 C$ gives the movement of the criminals. Due to some unavoidable circumstances, a burglar may also decide to move to a neighboring location, without committing a burglary in the present location.

The model has a unique equilibrium point, namely, $A^* = \frac{\alpha\gamma}{\beta\delta}, C^* = \frac{\beta}{\alpha}$ and it can be easily shown that the system is always stable without or with diffusion terms. The model is solved numerically for $\alpha = 1.2, \beta = 0.033, \gamma = 0.02, \delta = 1.0, D_1 = 0.25, D_2 = 25$ with boundary conditions

$$\frac{\partial A(0,y,t)}{\partial x} = 0, \quad \frac{\partial A(L,y,t)}{\partial x} = 0, \quad \frac{\partial A(x,0,t)}{\partial y} = 0, \quad \frac{\partial A(x,L,t)}{\partial y} = 0,$$
$$\frac{\partial C(0,y,t)}{\partial x} = 0, \quad \frac{\partial C(L,y,t)}{\partial x} = 0, \quad \frac{\partial C(x,0,t)}{\partial y} = 0, \quad \frac{\partial C(x,L,t)}{\partial y} = 0.$$

Fig. 4.9(a) shows no hotspots, fig. 4.9(b) shows transitory hot spots, where hot spots do not typically last forever, and will move about, deform, or disappear over time and fig. 4.9(c) is an example of stationary hot spots.

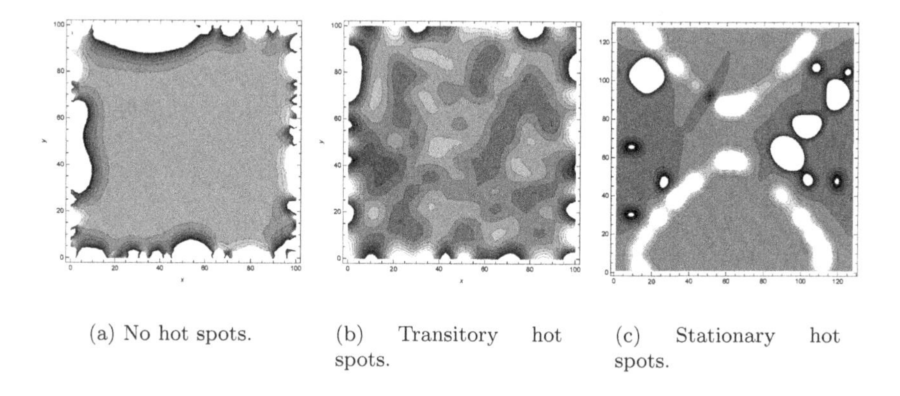

(a) No hot spots. (b) Transitory hot spots. (c) Stationary hot spots.

FIGURE 4.9: *The figures show criminal hotspots using parameters described in the text (L=100).*

4.10 Reaction-Diffusion Systems

Reaction-diffusion systems have established themselves as a powerful tool for describing the spatial distribution process. The typical form of reaction-diffusion equation is

$$\frac{\partial u(x,t)}{\partial t} = f(u) + \nabla^2 u,$$

which comprises of a reaction term and a diffusion term. Here, $u(x,t)$ is a state variable representing density/concentration of a substance, a population, etc. at position $x \in \Lambda \subset \mathbb{R}^n$ (Λ is an open set), $f(u) : \mathbb{R} \to \mathbb{R}$ is a smooth function that describes processes like birth, death, chemical reaction etc., that affects the present $u(x,t)$ (and not just diffuse in space), and $\nabla^2 u$ is the diffusion term with diffusion coefficient D, ∇^2 is the Laplacian.

4.10.1 Population Dynamics with Diffusion (Single Species)

We consider a reaction-diffusion equation of the form

$$\frac{\partial u(x,t)}{\partial t} = f(u(x,t)) + D\frac{\partial^2 u(x,t)}{\partial x^2}, \tag{4.37}$$

where $u(x,t)$ represents a population density, $f(u)$ shows growth (or decay) function of the population, and $D\frac{\partial^2 u(x,t)}{\partial x^2}$ denotes the spread of population in one-dimensional space x. Suppose, the population is confined to a region with length dimension L, that is, $0 \le x \le L$. This will imply the following boundary conditions (Neumann):

$$\frac{\partial u(0,t)}{\partial x} = 0, \quad \frac{\partial u(L,t)}{\partial x} = 0.$$

A homogenous steady-state solution of this model is given by $f(u_0) = 0$, as the solution is constant in time and space, implying

$$\frac{\partial u_0}{\partial t} = 0 \quad \text{and} \quad \frac{\partial^2 u_0}{\partial x^2} = 0. \tag{4.38}$$

Let $v(x,t) = u(x,t) - u_0$ be a small non-homogenous perturbation of the homogenous steady state, then

$$\frac{\partial v}{\partial t} = D\frac{\partial^2 v}{\partial x^2} + f(u_0 + v) = D\frac{\partial^2 v}{\partial x^2} + f(u_0) + vf'(u_0) + \frac{v^2}{2!}f''(u_0) + ... \quad .$$

Using $f(u_0) = 0$ and retaining linear terms, we get the linearized version of (4.37) as

$$\frac{\partial v(x,t)}{\partial t} = D\frac{\partial^2 v(x,t)}{\partial x^2} + f'(u_0)v(x,t). \tag{4.39}$$

Let $v(x,t) = Ae^{\lambda t}\cos(qx)$ be a solution of (4.39), then

$$A\left[\lambda - f'(u_0) + Dq^2\right] = 0 \text{ (since, } e^{\lambda t} \neq 0, \ \cos(qx) \neq 0).$$

Using the boundary condition $\frac{\partial u(L,t)}{\partial x} = \frac{\partial v(L,t)}{\partial x} = 0$, we get

$$\sin(qL) = 0 \Rightarrow qL = n\pi \Rightarrow q = \frac{n\pi}{L} \ (n = 1, 2, 3, ...).$$

If $A = 0$, then the solution is trivial. For a non-trivial solution, we must have

$$\lambda = f'(u_0) - Dq^2,$$

which gives the eigenvalue of the system. Hence, the given system is stable if

$$f'(u_0) - Dq^2 = f'(u_0) - D\frac{n^2\pi^2}{L^2} < 0.$$

Note: For an ODE, $\dot{u} = f(u)$, the system is stable if $f'(u_0) < 0$ and unstable if $f'(u_0) > 0$. Thus, the diffusion will maintain stability in a stable one-species system but it can stabilize an unstable one.

Corollary: For a reaction-diffusion equation of the form

$$\frac{\partial u(x,t)}{\partial t} = f(u(x,t)) + D\left[\frac{\partial^2 u(x,t)}{\partial x^2} + \frac{\partial^2 u(x,t)}{\partial x^2}\right],$$

with boundary conditions:

$$\frac{\partial u(0,y,t)}{\partial x} = 0, \quad \frac{\partial u(L,y,t)}{\partial x} = 0, \quad \frac{\partial u(x,0,t)}{\partial y} = 0, \quad \frac{\partial u(x,L,t)}{\partial y} = 0,$$

a solution of the linearized equation would be of the form

$$v(x,y,t) = Ae^{\lambda t}\cos(q_1 x)\cos(q_2 y).$$

Proceeding in a similar manner, we will obtain the stability condition as

$$f'(u_0) - D\frac{\pi^2}{L^2}\left[n_1^2 + n_2^2\right] < 0.$$

Example 4.10.1 *The spread of cancer cells is given by*

$$\frac{\partial C(x,t)}{\partial t} = rC(x,t)\left[1 - \frac{C(x,t)}{K}\right] + D\frac{\partial^2 C(x,t)}{\partial x^2}.$$

Suppose, the cancer cells are confined to a specific region with length dimension L, that is, $0 \leq x \leq L$. This will imply the following boundary conditions:

$$\frac{\partial C(0,t)}{\partial x} = 0, \quad \frac{\partial C(L,t)}{\partial x} = 0.$$

Let the initial density of cancer cells be 0.4 units, that is, $u(x,0) = 0.4$.
(a) For $r = 0.01$, $K = 1.0$ find the non-zero steady-state solutions. Check their stabilities for (i) $D = 0$ and (ii) $D = 10^{-5}, L = 1$.
(b) Plot figures for one-dimensional spread as well as two-dimensional spread.

Solution:*(a) The non-zero steady-state solution (C^*) is obtained by putting $rC\left[1 - \frac{C}{K}\right] = 0 \Rightarrow C^* = K$. For the given set of parameter values, the equilibrium solution is 1.0. Let $f(C) = rC\left[1 - \frac{C}{K}\right]$, then $f'(C^*) = r - \frac{2rC^*}{K}$.*
(i) When $D = 0$, $f'(1) = -0.01 < 0 \Rightarrow$ the system is stable about $C^* = 1$.
(ii) When $D = 10^{-5}(\neq 0)$, the condition for stability is $f'(C^*) - Dq^2 = f'(C^*) - D\frac{n^2\pi^2}{L^2} == -0.01 - n^2\pi^2 \times 10^{-5} < 0$. This shows that diffusion maintains stability in a stable cancer-growth system (fig. 4.10).

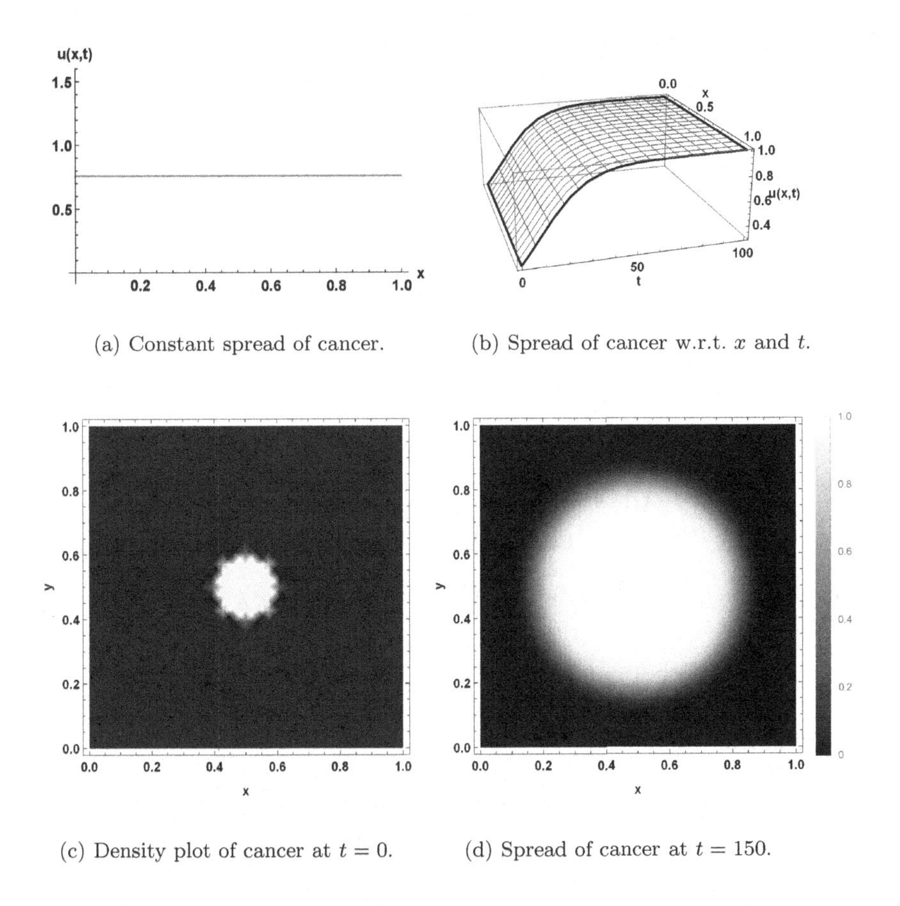

(a) Constant spread of cancer.

(b) Spread of cancer w.r.t. x and t.

(c) Density plot of cancer at $t = 0$.

(d) Spread of cancer at $t = 150$.

FIGURE 4.10: *The figures show the dynamics of the spread of cancer with Neumann boundary conditions.*

4.10.2 Population Dynamics with Diffusion (Two Species)

We consider a two species population with diffusion, given by

$$\begin{aligned}\frac{\partial u_1}{\partial t} &= f_1(u_1, u_2) + D_1 \frac{\partial^2 u_1}{\partial x^2},\\ \frac{\partial u_2}{\partial t} &= f_2(u_1, u_2) + D_2 \frac{\partial^2 u_2}{\partial x^2},\end{aligned}$$

where $f_1(u_1, u_2)$ and $f_2(u_1, u_2)$ gives the interaction terms between two species, and D_1 and D_2 are diffusion coefficients of first and second species respectively.

A spatially homogenous (uniform) steady state (u_1^*, u_2^*) of this model is given by

$$f_1(u_1^*, u_2^*) = 0, \quad f_2(u_1^*, u_2^*) = 0,$$

as the solution is constant in space and time implying

$$\frac{\partial u_i}{\partial t} = 0 \text{ and } \frac{\partial^2 u_i}{\partial x^2} = 0 \quad (i = 1, 2).$$

Let $v_1(x,t) = u_1(x,t) - u_1^*$, $v_2(x,t) = u_2(x,t) - u_2^*$ be small non-homogenous perturbations of the uniform steady state, then

$$\frac{\partial v_1}{\partial t} = f_1(u_1^* + v_1, u_2^* + v_2) + D_1\frac{\partial^2 v_1}{\partial x^2}, \quad \frac{\partial v_2}{\partial t} = f_2(u_1^* + v_1, u_2^* + v_2) + D_2\frac{\partial^2 v_2}{\partial x^2}.$$

$$\begin{aligned}
\text{Now, } \frac{\partial v_1}{\partial t} &= f_1(u_1^* + v_1, u_2^* + v_2) + D_1\frac{\partial^2 v_1}{\partial x^2} \\
&= D_1\frac{\partial^2 v_1}{\partial x^2} + f_1(u_1^*, u_2^*) + v_1\frac{\partial f_1}{\partial u_1}|_{(u_1^*,u_2^*)} + v_2\frac{\partial f_1}{\partial u_2}|_{(u_1^*,u_2^*)} \\
&+ \frac{v_1^2}{2!}\left(\frac{\partial^2 f_1}{\partial u_1^2}\right)_{(u_1^*,u_2^*)} + v_1 v_2\left(\frac{\partial^2 f_1}{\partial u_1 u_2}\right)_{(u_1^*,u_2^*)} + \frac{v_2^2}{2!}\left(\frac{\partial^2 f_1}{\partial u_2^2}\right)_{(u_1^*,u_2^*)} + \ldots
\end{aligned}$$

Using $f_1(u_1^*, u_2^*) = 0$ and retaining the linear terms, we obtain,

$$\frac{\partial v_1}{\partial t} = v_1\left(\frac{\partial f_1}{\partial u_1}\right)_{(u_1^*,u_2^*)} + v_2\left(\frac{\partial f_1}{\partial u_2}\right)_{(u_1^*,u_2^*)} + D_1\frac{\partial^2 v_1}{\partial x^2}. \tag{4.40}$$

In a similar manner,

$$\frac{\partial v_2}{\partial t} = v_1\left(\frac{\partial f_2}{\partial u_1}\right)_{(u_1^*,u_2^*)} + v_2\left(\frac{\partial f_2}{\partial u_2}\right)_{(u_1^*,u_2^*)} + D_2\frac{\partial^2 v_2}{\partial x^2}. \tag{4.41}$$

Let $v_1(x,t) = A_1 e^{\lambda t}\cos(qx)$ and $v_2(x,t) = A_2 e^{\lambda t}\cos(qx)$ be the solutions of (4.40) and (4.41). Substituting v_1 and v_2 in (4.40) and (4.41) we get,

$$\begin{aligned}
A_1(\lambda - a_{11} + D_1 q^2) - A_2 a_{12} &= 0, \\
-a_{21}A_1 + (\lambda - a_{22} + D_2 q^2)A_2 &= 0,
\end{aligned}$$

where

$$a_{11} = \left(\frac{\partial f_1}{\partial u_1}\right)_{(u_1^*,u_2^*)}, \quad a_{12} = \left(\frac{\partial f_1}{\partial u_2}\right)_{(u_1^*,u_2^*)}, \quad a_{21} = \left(\frac{\partial f_2}{\partial u_1}\right)_{(u_1^*,u_2^*)},$$
$$a_{22} = \left(\frac{\partial f_2}{\partial u_2}\right)_{(u_1^*,u_2^*)}.$$

Clearly, $A_1 = 0, A_2 = 0$ is a solution, which is trivial and is dismissed as we are interested in non-trivial solutions. The existence of a non-trivial solution implies

$$\begin{vmatrix} \lambda - a_{11} + D_1 q^2 & -a_{12} \\ -a_{21} & \lambda - a_{22} + D_2 q^2 \end{vmatrix} = 0, \text{ which implies}$$

$$\lambda^2 - \left(a_{11} + a_{22} - D_1 q^2 - D_2 q^2\right)\lambda + (a_{11} - D_1 q^2)(a_{22} - D_2 q^2) - a_{12}a_{21} = 0.$$

In the absence of diffusion, the characteristic equation is

$$\Rightarrow \lambda^2 - (a_{11} + a_{22})\lambda + (a_{11}a_{22} - a_{12}a_{21}) = 0.$$

The condition for stability is

$$a_{11} + a_{22} < 0, \qquad (4.42)$$

$$\text{and } a_{11}a_{22} - a_{12}a_{21} > 0. \qquad (4.43)$$

The conditions that the given system with diffusion have eigenvalues with $\mathrm{Re}\lambda < 0$ are

$$a_{11} + a_{22} - (D_1 + D_2)q^2 < 0, \qquad (4.44)$$

$$\text{and } (a_{11} - D_1 q^2)(a_{22} - D_2 q^2) - a_{12}a_{21} > 0. \qquad (4.45)$$

However, diffusion may have a destabilizing effect on the system, and violation of any one of the conditions given by (4.44) or (4.45) will lead to diffusive instability. Since D_1, D_2 and q^2 are all positive quantities, (4.44) is always true, provided (4.42) holds. Let

$$\begin{aligned} H(q^2) &= (a_{11} - D_1 q^2)(a_{22} - D_2 q^2) - a_{12}a_{21}, \\ &= D_1 D_2 (q^2)^2 - (D_1 a_{22} + D_2 a_{11})q^2 + (a_{11}a_{22} - a_{12}a_{21}). \end{aligned}$$

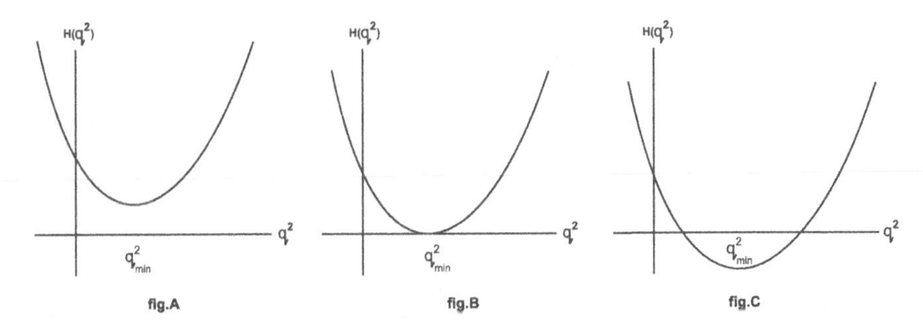

FIGURE 4.11: *The graph of $H(q^2)$, which represents a parabola and geometrically shows the region of diffusive instability.*

Clearly, the condition for diffusive instability is $H(q^2) < 0$. The graph of $H(q^2)$ is a parabola (as $D_1 > 0$, $D_2 > 0$) (fig. 4.11) and the function $H(q^2)$ has a minimum. Putting $\frac{dH(q^2)}{dq^2} = 0$ and solving, we obtain

$$q^2_{min} = \frac{1}{2}\left(\frac{a_{11}}{D_1} + \frac{a_{22}}{D_2}\right).$$

Hence, the minimal condition for diffusive instability is

$$H(q^2_{min}) < 0.$$

This implies

$$\begin{aligned}
\frac{D_1 D_2}{4}\left[\frac{a_{11}}{D_1}+\frac{a_{22}}{D_2}\right]^2 &- (D_1 a_{22}+D_2 a_{11})\frac{1}{2}\left[\frac{a_{11}}{D_1}+\frac{a_{22}}{D_2}\right] \\
&+ (a_{11}a_{22}-a_{12}a_{21})<0, \\
\Rightarrow\ D_1 D_2\left[\frac{(a_{11}D_2+a_{22}D_1)^2}{D_1^2 D_2^2}\right] &- 2\left(a_{11}a_{22}+\frac{a_{11}^2 D_2}{D_1}+\frac{a_{22}^2 D_1}{D_2}+a_{11}a_{22}\right) \\
&+ (a_{11}a_{22}-a_{12}a_{21})<0, \\
\Rightarrow\ D_1 D_2\left[\frac{(a_{11}D_2+a_{22}D_1)^2}{D_1^2 D_2^2}\right] &- 2\frac{(D_1 a_{22}+D_2 a_{11})^2}{D_1 D_2} \\
&+ 4(a_{11}a_{22}-a_{12}a_{21})<0, \\
\Rightarrow\ (D_1 a_{22}+D_2 a_{11})^2 &> 4D_1 D_2(a_{11}a_{22}-a_{12}a_{21}), \\
\Rightarrow\ (D_1 a_{22}+D_2 a_{11}) &> 2\sqrt{D_1 D_2}\sqrt{a_{11}a_{22}-a_{12}a_{21}}.
\end{aligned}$$

Upon summarizing the results, it can be concluded that the conditions for diffusive instability are
(i) $a_{11}+a_{22}<0$,
(ii) $a_{11}a_{22}-a_{12}a_{21}>0$,
(iii) $(a_{11}D_2+a_{22}D_1)>2\sqrt{D_1 D_2}\sqrt{a_{11}a_{22}-a_{12}a_{21}}$.

It is a well-known fact that an equilibrium point that is asymptotically stable in a non-spatial system may become unstable due to diffusion. Mathematical analysis has confirmed that the diffusive system first attains instability with respect to a spatially heterogeneous perturbation with a certain wave number, which results in the formation of the so-called dissipative patterns or regular spatial structure, known as Turing patterns.

Corollary: For a reaction-diffusion equation of the form

$$\begin{aligned}
\frac{\partial u_1}{\partial t} &= f_1(u_1,u_2)+D_1\left(\frac{\partial^2 u_1}{\partial x^2}+\frac{\partial^2 u_1}{\partial y^2}\right), \\
\frac{\partial u_2}{\partial t} &= f_2(u_1,u_2)+D_2\left(\frac{\partial^2 u_2}{\partial x^2}+\frac{\partial^2 u_2}{\partial y^2}\right),
\end{aligned}$$

with boundary conditions:

$$\begin{aligned}
&\frac{\partial u_1(0,y,t)}{\partial x}=0,\quad \frac{\partial u_1(L,y,t)}{\partial x}=0,\quad \frac{\partial u_1(x,0,t)}{\partial y}=0,\quad \frac{\partial u_1(x,L,t)}{\partial y}=0, \\
&\frac{\partial u_2(0,y,t)}{\partial x}=0,\quad \frac{\partial u_2(L,y,t)}{\partial x}=0,\quad \frac{\partial u_2(x,0,t)}{\partial y}=0,\quad \frac{\partial u_2(x,L,t)}{\partial y}=0,
\end{aligned}$$

a solution of the linearized equation would be of the form

$$v_1(x,y,t)=A_1 e^{\lambda t}\cos(q_1 x)\cos(q_2 y),\quad v_2(x,y,t)=A_2 e^{\lambda t}\cos(q_1 x)\cos(q_2 y).$$

Proceeding in a similar manner, we obtain

$$\lambda^2 - \left(a_{11} + a_{22} - D_1(q_1^2 + q_2^2) - D_2(q_1^2 + q_2^2)\right)\lambda$$
$$+(a_{11} - D_1(q_1^2 + q_2^2))(a_{22} - D_2(q_1^2 + q_2^2)) - a_{12}a_{21} = 0.$$

In the absence of diffusion, the characteristic equation is

$$\lambda^2 - (a_{11} + a_{22})\,\lambda + (a_{11}a_{22} - a_{12}a_{21}) = 0.$$

The condition for stability is

$$a_{11} + a_{22} < 0, \text{ and } a_{11}a_{22} - a_{12}a_{21} > 0. \tag{4.46}$$

The conditions that the given system with diffusion have eigenvalues with $\text{Re}\lambda < 0$ are

$$a_{11} + a_{22} - (D_1 + D_2)(q_1^2 + q_2^2) < 0, \text{ and} \tag{4.47}$$
$$(a_{11} - D_1(q_1^2 + q_2^2))(a_{22} - D_2(q_1^2 + q_2^2)) - a_{12}a_{21} > 0. \tag{4.48}$$

Violation of any one of the conditions given by (4.47) or (4.48) will lead to diffusive instability. Since D_1, D_2, q_1^2 and q_2^2 are all positive quantities, (4.47) is always true, provided (4.46) holds. Let

$$H(q_1^2, q_2^2) = (a_{11} - D_1(q_1^2 + q_2^2))(a_{22} - D_2(q_1^2 + q_2^2)) - a_{12}a_{21}.$$

Clearly, the condition for diffusive instability is $H(q_1^2, q_2^2) < 0$. (4.49)

We now find the minimum values of q_1^2 and q_2^2, say, (q_1^{*2}, q_2^{*2}) for which (4.49) holds.

$$\frac{\partial H}{\partial q_1^2} = 0 \text{ and } \frac{\partial H}{\partial q_2^2} = 0 \Rightarrow q_1^{*2} + q_2^{*2} = \frac{1}{2}\left(\frac{a_{11}}{D_1} + \frac{a_{22}}{D_2}\right).$$

And, $H_{q_1^2 q_1^2} H_{q_2^2 q_2^2} - (H_{q_1^2 q_2^2})^2 = (2D_1D_2)(2D_1D_2) - (2D_1D_2)^2 = 0.$

Hence, no conclusion about the maximum or minimum values of $H(q_1^2, q_2^2)$.

Now, it can be easily shown that

$$H(q_1^2, q_2^2) - H(q_1^{*2}, q_2^{*2}) = \left[q_1^2 + q_2^2 - \frac{1}{2}\left(\frac{a_{11}}{D_1} + \frac{a_{22}}{D_2}\right)\right]^2 > 0,$$

for all values of (q_1^2, q_2^2). Hence, $H(q_1^2, q_2^2)$ has a minimum value at (q_1^{*2}, q_2^{*2}).

Hence, the minimal condition for diffusive instability is $H(q_1^{*2}, q_2^{*2}) < 0$,

$$\Rightarrow (2a_{11}a_{22} - 4a_{12}a_{21})D_1D_2 - (a_{11}D_2)^2 - (a_{22}D_1)^2 < 0.$$

Example 4.10.2 *Consider the Brusselator equation with diffusion [122]:*

$$\frac{\partial u}{\partial t} = A - (B+1)u + u^2 v + D_1 \frac{\partial^2 u}{\partial x^2},$$
$$\frac{\partial v}{\partial t} = Bu - u^2 v + D_2 \frac{\partial^2 v}{\partial x^2},$$

The spatially homogeneous steady state (u^*, v^*) *is given by*

$$A - (B+1)u + u^2v = 0, Bu - u^2v = 0 \Rightarrow u^* = A \text{ and } v^* = \frac{B}{A}.$$

The characteristic equation is given by

$$\begin{vmatrix} \lambda - a_{11} + D_1q^2 & -a_{12} \\ -a_{21} & \lambda - a_{22} + D_2q^2 \end{vmatrix} = 0,$$

where $a_{11} = B - 1, a_{12} = A^2, a_{21} = -B, a_{22} = -A^2$. *Hence,*

$$\lambda^2 - \left(B - 1 - A^2 - D_1q^2 - D_2q^2\right)\lambda + (D_1q^2 + 1 - B)(D_2q^2 + A^2) + A^2B = 0.$$

The characteristic equation without diffusion is given by

$$\lambda^2 - \left(B - 1 - A^2\right)\lambda + A^2 = 0.$$

Without diffusion, the system is stable iff $B < 1 + A^2$.

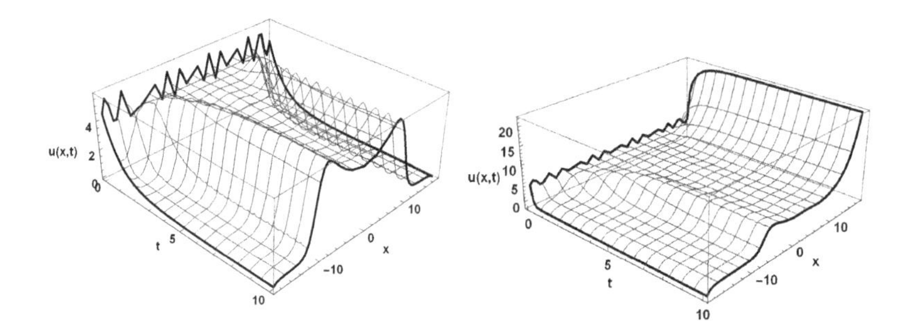

(a) Dirichlet boundary condition. (b) Neumann boundary condition.

FIGURE 4.12: *The figures show the behavior of the Brusselator equation with diffusion (one-dimensional).*

With diffusion, the system is stable iff $(D_1q^2+1-B)(D_2q^2+A^2)+A^2B > 0$. *Hence, the condition for diffusive instability is given by*

$$(D_1q^2 + 1 - B)(D_2q^2 + A^2) + A^2B < 0.$$

The minimum value of q^2 *for which the above inequality holds is*

$$q^2_{min} = \frac{1}{2}\left(\frac{a_{11}}{D_1} + \frac{a_{22}}{D_2}\right) = \frac{1}{2}\left(\frac{B-1}{D_1} - \frac{A^2}{D_2}\right).$$

Substituting this value and simplifying, we obtain the condition for diffusive instability as

$$D_2(B-1) - A^2D_1 > 2\sqrt{D_1D_2}A.$$

The model is solved numerically with $A = 4.5$, $B = 16.0$, $D_1 = 8.0$, $D_2 = 16.0$ *and* $L = 6.0\pi$, *with boundary conditions*
(i) $u(t,-L) = 0, u(t,L) = 0, v(t,-L) = 0, v(t,L) = 0$ *(Dirichlet),*
(ii) $\frac{\partial u(t,-L)}{\partial x} = 0, \frac{\partial u(t,L)}{\partial x} = 0, \frac{\partial v(t,-L)}{\partial x} = 0, \frac{\partial v(t,L)}{\partial x} = 0$ *(Neumann), and*
(iii) initial condition: $u(0,x) = A + \sin(2x), v(0,x) = \frac{B}{A} + \sin(2x)$.
Figs. 4.12(a), 4.12(b)) show the dynamics of the Brusselator model for the Dirichlet and Neumann boundary conditions respectively.

The model is now converted to a two-dimensional spatial model

$$\frac{\partial u}{\partial t} = A - (B+1)u + u^2 v + D_1 \left(\frac{\partial^2 u}{\partial x^2} + \frac{\partial^2 u}{\partial y^2} \right),$$

$$\frac{\partial v}{\partial t} = Bu - u^2 v + D_2 \left(\frac{\partial^2 v}{\partial x^2} + \frac{\partial^2 u}{\partial y^2} \right), \text{ with boundary conditions,}$$

$$\frac{\partial u(t,-L,y)}{\partial x} = 0, \frac{\partial u(t,L,y)}{\partial x} = 0, \frac{\partial u(t,x,-L)}{\partial y} = 0, \frac{\partial u(t,x,L)}{\partial y} = 0,$$

$$\frac{\partial v(t,-L,y)}{\partial x} = 0, \frac{\partial v(t,L,y)}{\partial x} = 0, \frac{\partial v(t,x,-L)}{\partial y} = 0, \frac{\partial v(t,x,L)}{\partial y} = 0.$$

Proceeding as before, the condition for diffusive instability is

$$2(B+1)A^2 D_1 D_2 - (B-1)^2 D_2^2 - A^4 D_1^2 < 0.$$

This diffusion-driven instability (also known as Turing instability) leads to Turing patterns (figs.4.13(a), 4.13(b)).

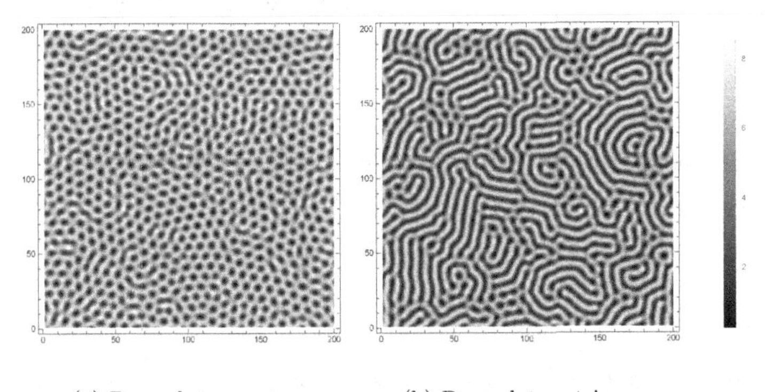

(a) Brusselator spots. (b) Brusselator stripes.

FIGURE 4.13: *The figures show numerical computation of Turing patterns. Parameter values* $A = 4.5$, $D_1 = 2.0$, $D_2 = 16.0$ *and* $L = 6.0\pi$. *(a) Spots (B=7.0), (b) Stripes (B=7.5).*

4.11 Mathematica Codes

Mathematica codes of selected figures.

4.11.1 Heat Equation with Dirichlet's Condition (Figure 4.2)

```
k = 0.475;
L = 20;
T = 10;
sol1 = NDSolve[{D[u[x, t], t] == kD[u[x, t], x, x], u[0, t] == 0, u[L, t] == 0,
u[x, 0] == Piecewise[{{x, 0 ≤ x ≤ 10}, {L − x, 10 ≤ x ≤ L}}, 0]}, u,
{x, 0, L}, {t, 0, T}]
t = {1, 9}
Plot[Evaluate[u[x, t]/. sol1], {x, 0, L}, PlotRange → All,
AxesLabel → {"x", "u(x,t)"}, BaseStyle → {FontWeight → "Bold",
FontSize → 14}]
Plot3D[u[x, t]/. sol1, {x, 0, L}, {t, 0, T}, AxesLabel → {x, t, u(x,t)},
BaseStyle → {FontWeight → "Bold", FontSize → 14},
BoundaryStyle → Directive[Black, Thick], PlotRange → All, PlotStyle → None]
```

4.11.2 Heat Equation with Neumann's condition (Figure 4.3)

```
k = 0.475; L = 100; T = 10;
sol2 = NDSolve[{D[u[x, t], t] == kD[u[x, t], x, x],
Derivative[0, 1][u][t, 0] == 0, Derivative[0, 1][u][t, 100] == 0,
u[x, 0] == Piecewise[{{x, 0 ≤ x ≤ 40}, {L − x, 40 ≤ x ≤ L}}, 0]}, u,
{x, 0, L}, {t, 0, T}]
t = {5, 9}
Plot[Evaluate[u[x, t]/. sol2], {x, 0, L}, PlotRange → All,
AxesLabel → {"x", "u(x,t)"}, BaseStyle → {FontWeight → "Bold",
FontSize → 14}]
Plot3D[u[x, t]/. sol2, {x, 0, L}, {t, 0, T}, AxesLabel → {x, t, u(x,t)},
BaseStyle → {FontWeight → "Bold", FontSize → 14},
BoundaryStyle → Directive[Black, Thick], PlotRange → All, PlotStyle → None]
```

4.11.3 One-Dimensional Wave Equation (Figure 4.7)

$c = \sqrt{0.5};$
$L = 10;$
$T = 10;$
$\text{sol3} = \text{NDSolve}[\{D[u[x,t],t,t] == c^2 D[u[x,t],x,x], u[t,0] == 0, u[t,L] == 0,$
$\text{Derivative}[1,0][u][0,x] == 0, u[0,x] == x(L-x)\}, u, \{x,0,L\}, \{t,0,T\}]$
$t = \{5,9\}$
$\text{Plot}[\text{Evaluate}[u[x,t]/.\,\text{sol3}], \{x,0,L\}, \text{AxesLabel} \to \{\text{“x”}, \text{“u(x,t)”}\},$
$\text{PlotRange} \to \text{All}, \text{BaseStyle} \to \{\text{FontWeight} \to \text{“Bold”}, \text{FontSize} \to 14\}]$
$\text{Plot3D}[u[x,t]/.\,\text{sol3}, \{x,0,L\}, \{t,0,T\}, \text{AxesLabel} \to \{\text{t}, \text{x}, \text{u(x,t)}\},$
$\text{BaseStyle} \to \{\text{FontWeight} \to \text{“Bold”}, \text{FontSize} \to 14\},$
$\text{BoundaryStyle} \to \text{Directive}[\text{Black}, \text{Thick}], \text{PlotRange} \to \text{All}, \text{PlotStyle} \to \text{None}]$

4.11.4 Two-Dimensional Heat Equation (Figure 4.4)

$a = 1; b = 1;$
$k = 0.475; T = 100;$
$\text{sol4} = \text{NDSolve}[\{D[u[x,y,t],t] == k(D[u[x,y,t],x,x] + D[u[x,y,t],y,y]),$
$u[0,y,t] == 0, u[a,y,t] == 0, u[x,0,t] == 0, u[x,b,t] == 0,$
$u[x,y,0] == xy(\pi - x)(\pi - y), u[x,y,t], \{x,0,a\}, \{y,0,b\}\{t,0,T\}]$
$\text{distribution}(\text{x\_}, \text{y\_}, \text{t\_}) = \text{First}[u(x,y,t)/.\,\text{sol4}]$
$t = 0;$
$\text{ContourPlot}[\text{distribution}(x,y,t), \{x,0,a\}, \{y,0,b\}, \text{Mesh} \to \text{False},$
$\text{FrameLabel} \to \{\text{x}, \text{y}\}, \text{BaseStyle} \to \{\text{FontWeight} \to \text{Bold}, \text{FontSize} \to 12\},$
$\text{ColorFunction} \to \text{GrayLevel}, \text{PlotLegends} \to \text{BarLegend}[\{\text{GrayLevel}, \{0,1\}\},$
$\text{LegendMarkerSize} \to 320], \text{PerformanceGoal} \to \text{Quality}]$
(*next line*)
$t = 100;$
$\text{ContourPlot}[\text{distribution}(x,y,t), \{x,0,a\}, \{y,0,b\}, \text{Mesh} \to \text{False},$
$\text{FrameLabel} \to \{\text{x}, \text{y}\}, \text{BaseStyle} \to \{\text{FontWeight} \to \text{Bold}, \text{FontSize} \to 12\},$
$\text{ColorFunction} \to \text{GrayLevel}, \text{PlotLegends} \to \text{BarLegend}[\{\text{GrayLevel}, \{0,1\}\},$
$\text{LegendMarkerSize} \to 320], \text{PerformanceGoal} \to \text{Quality}]$

4.11.5 Brusselator Equation One-Dimensional (Figure 4.12)

```
A = 4.5; B = 16.; d1 = 8.; d2 = 16.; T = 10; L = 6.π;
pde = {D[u[t, x], t] == d1D[u[t, x], x, x] + A − (B + 1)u[t, x] + (u[t, x])^2 v[t, x],
D[v[t, x], t] == d2D[v[t, x], x, x] + Bu[t, x] − (u[t, x])^2 v[t, x]};
bc = {u[t, −L] = 0, u[t, L] = 0, v[t, −L] = 0, v[t, L] = 0};
(*bc = {Derivative[0, 1][u][t, −L] == 0, Derivative[0, 1][u][t, L] == 0,
Derivative[0, 1][v][t, −L] == 0, Derivative[0, 1][v][t, L] == 0}; *)
ic = {u[0, x] = A + Sin[2x], v(0, x) = B/A + Sin[2x]};
eqns = {pde, bc, ic};
sol = NDSolve[eqns, {u, v}, {t, 0, T}, {x, −L, L}, MaxStepSize → 0.1]
Plot3D[u(t, x)/. sol, {t, 0, T}, {x, −L, L}, AxesLabel → {t, x, u(x,t)},
BaseStyle → {FontWeight → Bold, FontSize → 12}, PlotRange → All,
BoundaryStyle → Directive[Black, Thick], PlotStyle → None]
```

4.11.6 Brusselator Equation Two-Dimensional (Figure 4.13)

```
n = 300.; a = 4.5; b = 18; dt = 0.01; du = 8.; dv = 16.;
totaliter = 10000;
u = a + 0.3RandomReal[NormalDistribution[0, 1], {n, n}];
v = b/a + 0.3RandomReal[NormalDistribution[0, 1], {n, n}];
cf2 = With[{a = a, b = b}, Compile[{{uIn, _Real, 2}, {vIn, _Real, 2},
{aIn, _Real}, {bIn, _Real}, {duIn, _Real}, {dvIn, _Real}, {dtIn, _Real},
{iterationsIn, _Integer}}, Block[{u = uIn, v = vIn, lap, dt = dtIn, k = bIn + 1,
kern = N[{{0, 1, 1}, {1, −4, 1}, {0, 1, 0}}], du = duIn, dv = dvIn},
Do[lap = RotateLeft[u, {1, 0}] + RotateLeft[u, {0, 1}] + RotateRight[u, {1, 0}]
 + RotateRight[u, {0, 1}] − 4u;
u = dt(a + du lap − u(k − uv)) + u;
lap = RotateLeft[v, {1, 0}] + RotateLeft[v, {0, 1}] + RotateRight[v, {1, 0}]
 + RotateRight[v, {0, 1}] − 4v;
v = dt(u(b − uv) + dvlap) + v; , {iterationsIn}]; {u, v}]]];
Timing[c2 = cf2(u, v, a, b, du, dv, dt, totaliter); ]
ListDensityPlot[c2[[1]], ColorFunction → GrayLevel, AspectRatio → Automatic,
AxesLabel → Automatic, FrameLabel → Automatic, PlotLegends →
BarLegend[Automatic, LegendMarkerSize → 370, LegendMargins → 10]]
```

4.12 Matlab Codes

4.12.1 Heat Flow (Figure 4.2)

```
function(c, f, s) = pdex1pde(x, t, u, dudx)
c = 1; f = D * dudx; s = 0;
end
function u0 = pdex1ic(x)
u0 = 0.5;
end
function(pl, ql, pr, qr) = pdex1bc(xl, ul, xr, ur, t)
p1 = u1; q1 = 0; pr = ur - 1; qr = 0;
end

D = 0.475; m = 0;
x = linspace(0, 1, 20); t = linspace(0, 5, 20);
sol = pdepe(m, @pdex1pde, @pdex1ic, @pdex1bc, x, t);
u = sol(:, :, 1);
figure(1)
plot(x, u(end, :))
figure(2)
surf(x, t, u)
```

4.12.2 Brusselator Equation (Figure 4.12)

```
function(c, b, s) = eqn2(x, t, u, DuDx)
global A B
c = [1; 1];
b = [1; 1]. * DuDx;
s = [A - (B + 1) * u(1) + (u(1))^2 * u(2); B * u(1) - (u(1))^2 * u(2)];
end
function(pl, ql, pr, qr) = bc2(xl, ul, xr, ur, t)
p1 = [0; u1(2)]; q1 = [1; 0]; pr = [ur(1); 0]; qr = [0; 1]
end
```

```
function value = initial2(x)
global A B
value = [A + sin(2 * x); B/A + sin(2 * x);
end

global A B
m = 0;
x = linspace(0, 10, 10); t = linspace(0, 10, 10);
sol = pdepe(m, @eqn2, @bc2, @initial2, x, t);
u1 = sol(:, :, 1);  u2 = sol(:, :, 2);
surf(x, t, u1)
```

4.12.3 Wave Equation (Figure 4.7(a))

```
c = 1;  n = 21;
x = linspace(0, 1, n);
dx = 1/(n - 1);  dt = dx;
u(:, 1) = sin(pi * x);  u(1, 2) = 0;
for i = 2 : n - 1
u(i, 2) = 0.5 * (dt^2 * c^2 * (u(i + 1, 1) - 2 * u(i, 1) + u(i - 1, 1))/(dx)^2
 + 2 * u(i, 1));
end
u(n, 2) = 0;  error = 1;  k = 1;
while k < 100
k = k + 1;
u(1, k + 1) = 0;
for i = 2 : n - 1
u(i, k + 1) = dt^2 * c^2 * (u(i + 1, k) - 2 * u(i, k) + u(i - 1, k))/(dx)^2
 + 2 * u(i, k) - u(i, k - 1);
end
u(n, k + 1) = 0;
end
plot(x, u),  xlabel('x'),  ylabel('y')
```

4.13 Miscellaneous Examples

Problem 4.13.1 *A rod of length L, whose sides are insulated, is kept at uniform temperature u_0. Both ends of the rod are suddenly cooled at 0^oC and are kept at that temperature. If $u(x,t)$ represents the temperature function at any point x at time t,*
(i) formulate a mathematical model of the given situation using PDE, stating clearly the boundary and initial conditions.
(ii) Using the method of separation of variables, find the temperature function $u(x,t)$.

Solution: (i) The mathematical model of the given situation represents an initial boundary value problem of heat conduction and is given by

$$\frac{\partial u}{\partial t} = c^2 \frac{\partial^2 u}{\partial x^2}, \; 0 \le x \le L, \; t > 0. \tag{4.50}$$

Boundary Conditions (BCs): $u(0,t) = 0, u(L,t) = 0, \; t > 0.$

(since both ends of the rod are cooled suddenly at 0^oC).

Initial Condition (IC): $u(x,0) = u_0, \; 0 \le x \le L.$

(ii) Let $u(x,t) = X(x)T(t)$ be a solution of (4.50), which on substitution gives

$$\frac{X''}{X} = \frac{1}{c^2}\frac{T'}{T} = -\lambda^2 \text{ (separation constant)}. \tag{4.51}$$

Since the boundary conditions are periodic and homogenous in x, the periodic solution of (4.50) exists if the separation constant is negative. One can also consider the other two cases, that is, the separation constant to be positive and zero but will arrive at the same conclusion. Basically, a negative separation constant gives a physically acceptable general solution. Solving (4.51) we get,

$$X(x) = A_1 \cos(\lambda x) + A_2 \sin(\lambda x) \;\text{ and }\; T(t) = A_3 e^{-\lambda^2 c^2 t}. \tag{4.52}$$

Therefore, the complete solution of (4.50) is given by

$$u(x,t) = [C_1 \cos(\lambda x) + C_2 \sin(\lambda x)] \; e^{-\lambda^2 c^2 t} \text{ where } \; C_1 = A_1 A_2, C_2 = A_2 A_3.$$

Using the boundary conditions, $u(0,t) = 0, u(L,t) = 0$, we obtain

$$u(0,t) = 0 \Rightarrow (C_1 \cos 0 + C_2 \sin 0)e^{-\lambda^2 c^2 t} = 0 \;\Rightarrow C_1 = 0.$$
$$u(L,t) = 0 \Rightarrow [C_1 \cos(\lambda L) + C_2 \sin(\lambda L)]e^{-\lambda^2 c^2 t} = 0$$
$$\Rightarrow C_2 \sin(\lambda L) = 0 \;\Rightarrow \sin(\lambda L) = 0 \text{ (for non-trivial solution } C_2 \ne 0),$$
$$\Rightarrow \lambda = \frac{n\pi}{L}, \text{ n being an integer.}$$

Therefore, the required solution is of the form

$$u(x,t) = C_2 \sin\left(\frac{n\pi}{L}x\right) e^{\frac{-n^2\pi^2c^2}{L}t}.$$

Noting that the heat conduction equation is linear, we use the principle of superposition to obtain its most general solution as

$$u(x,t) = \sum_{n=1}^{\infty} B_n \sin\left(\frac{n\pi}{L}x\right) e^{\frac{-n^2\pi^2c^2}{L}t}.$$

Using the initial condition, we obtain

$$u(x,0) = u_0 = \sum_{n=1}^{\infty} B_n \sin\left(\frac{n\pi}{L}x\right),$$

which is a half-range Fourier series, where

$$B_n = \frac{2}{L}\int_0^L u_0 \sin\left(\frac{n\pi}{L}x\right) dx = \frac{2u_0}{n\pi}(1-\cos n\pi) = \frac{2u_0}{n\pi}\left[1-(-1)^n\right].$$

Therefore,

$$B_n = \begin{cases} \dfrac{4u_0}{(2m-1)\pi}, & n = 2m-1 \text{ (odd)},\ m = 1,2,3,... \\ 0, & n = 2m \text{ (even)},\ m = 1,2,3,... \end{cases}.$$

Hence, the temperature function is given by

$$u(x,t) = \sum_{m=1}^{\infty} \frac{4u_0}{(2m-1)\pi} \sin\left(\frac{(2m-1)\pi x}{L}\right) e^{\frac{-(2m-1)^2\pi^2c^2}{L}t}.$$

Problem 4.13.2 *A rod of length 10 cm, whose sides are insulated, is kept at temperature* $0\,°\mathrm{C}$ *and* $100\,°\mathrm{C}$ *at its ends A and B, respectively, until the steady-state condition prevails. The temperature at end A is suddenly increased to* $20\,°\mathrm{C}$*, and at end B, it is decreased to* $60\,°\mathrm{C}$*. Formulate a mathematical model of the given situation and obtain the temperature function at any time* t*.*

Solution: The mathematical model of the given situation is given by

$$\begin{aligned} &\frac{\partial u}{\partial t} = c^2\frac{\partial^2 u}{\partial x^2}, \quad 0 \le x \le 10, \\ &\text{BCs}: u(0,t) = 0,\ u(10,t) = 100. \end{aligned}$$

Before the temperature at the ends of the rod is changed, the heat flow in the rod is independent of time when steady state is reached. Therefore, at the steady state

$$\frac{\partial u}{\partial t} = 0 \Rightarrow \frac{\partial^2 u}{\partial x^2} = 0 \Rightarrow \frac{d^2 u}{dx^2} = 0 \Rightarrow u_s(x) = C_1 x + C_2.$$

Applying boundary conditions we get, $C_1 = 10, C_2 = 0$. Thus, the initial steady temperature distribution in the rod is

$$u_s(x) = 10x.$$

In a similar manner, when the temperature at the ends of the rod are changed to 20 °C and 60 °C, respectively, the final steady temperature in the rod is

$$u_s(x) = 4x + 20,$$

which will happen after a long time (several hours). To obtain the temperature distribution $u(x,t)$ in the intermediate period, we write

$$u(x,t) = u_s(x) + u_1(x,t),$$

where $u_1(x,t) \to 0$ as $t \to \infty$ and is called the transient temperature distribution, which satisfies the model. Therefore, the general solution is given by

$$u(x,t) = (4x + 20) + e^{-c^2\lambda^2 t}(A_1 \cos \lambda x + A_2 \sin \lambda x).$$

Boundary conditions:

$$u(0,t) = 20 \Rightarrow 20 + A_1 e^{-c^2\lambda^2 t} = 20 \Rightarrow \quad A_1 = 0.$$
$$u(10,t) = 60 \Rightarrow 60 + A_2 \sin(10\lambda) e^{-c^2\lambda^2 t},$$
$$\Rightarrow \sin(10\lambda) = 0 \quad (\text{since } A_2 \neq 0 \text{ for non-trivial solution}) \Rightarrow \lambda = \frac{n\pi}{10}.$$

Using the superposition principle, we obtain,

$$u(x,t) = (4x + 20) + \sum_{n=1}^{\infty} B_n \sin\left(\frac{n\pi x}{10}\right) e^{\frac{-c^2 n^2 \pi^2}{100} t}.$$

The initial condition $u(x,0) = 10x$ gives

$$4x + 20 + \sum_{n=1}^{\infty} B_n \sin\left(\frac{n\pi x}{10}\right) = 10x \Rightarrow \sum_{n=1}^{\infty} B_n \sin\left(\frac{n\pi x}{10}\right) = 6x - 20,$$

$$\text{where } B_n = \frac{2}{10}\int_0^{10} (6x - 10) \sin\left(\frac{n\pi x}{10}\right) dx = \frac{1}{5}\left[(-1)^n \frac{800}{n\pi} - \frac{200}{n\pi}\right].$$

Problem 4.13.3 *A homogenous flexible string in a guitar is stretched between two fixed points $(0,0)$ and $(2,0)$, the length of the string being 2 units. The string of the guitar is initially plucked from rest from a position $\sin^3\left(\frac{\pi x}{2}\right)$. Find the displacement $u(x,t)$ of the string of the guitar at time t.*

Solution: Assuming that the string of the guitar is pulled aside and released, mathematically, the model can be formulated as follows:

$$\frac{\partial^2 u}{\partial t^2} = c^2 \frac{\partial^2 u}{\partial x^2}, \quad 0 \le x \le L, \ t > 0.$$

Boundary conditions (BC): $u(0,t) = 0 = u(2,t)$.
Initial conditions (IC): $u(x,0) = \sin^3\left(\frac{\pi x}{2}\right), \frac{\partial u(x,0)}{\partial t} = 0$.
The physically acceptable solution is of the form

$$u(x,t) = (C_1 \cos \lambda x + C_2 \sin \lambda x)[C_3 \cos(c\lambda t) + C_4 \sin(c\lambda t)].$$

Applying boundary conditions we get $C_1 = 0$ and $C_2 \sin(2) = 0 \Rightarrow \lambda = \frac{n\pi}{2}$ $(C_2 \neq 0,)$ $n = 1, 2, \ldots$. Also, $\dfrac{\partial u(x,0)}{\partial t} = 0 \Rightarrow C_4 = 0$. Using the principle of superposition, the possible solution is given by

$$u(x,t) = \sum_{n=1}^{\infty} A_n \sin\left(\frac{n\pi x}{2}\right) \cos\left(\frac{c\pi n t}{2}\right).$$

Using the initial condition $u(x,0) = \sin^3\left(\frac{\pi x}{2}\right)$, we get,

$$\begin{aligned}
\sin^3\left(\frac{\pi x}{2}\right) &= \sum_{n=1}^{\infty} A_n \sin\left(\frac{n\pi x}{2}\right), \\
\Rightarrow \frac{3}{4}\sin\left(\frac{\pi x}{2}\right) - \frac{1}{4}\sin\left(\frac{3\pi x}{2}\right) &= A_1 \sin\left(\frac{\pi x}{2}\right) + A_2 \sin\left(\frac{2\pi x}{2}\right) \\
&+ A_3 \sin\left(\frac{3\pi x}{2}\right) + A_4 \sin\left(\frac{4\pi x}{2}\right) + \ldots
\end{aligned}$$

Comparing, we get,

$$A_1 = \frac{3}{4} \quad \text{and} \quad A_3 = -\frac{1}{4},$$

while all other $A_i^{'}s$ are zeros. Therefore, the required solution is

$$u(x,t) = \frac{3}{4}\sin\left(\frac{\pi x}{2}\right)\cos\left(\frac{\pi c t}{2}\right) - \frac{1}{4}\sin\left(\frac{3\pi x}{2}\right)\cos\left(\frac{3\pi c t}{2}\right).$$

Problem 4.13.4 *Find the traffic density $\rho(x,t)$, satisfying*

$$\frac{\partial \rho}{\partial t} + x\sin(t)\,\frac{\partial \rho}{\partial x} = 0$$

with initial condition $\rho_0(x) = 1 + \frac{1}{1+x^2}$

Solution: The characteristic base curves for this initial value problem are solutions of

$$\begin{aligned}
&\frac{dx}{dt} = x\sin(t), \quad x(0) = x_0, \\
&\Rightarrow \ln(x) = -\cos(t) + \ln(x_0), \\
&\Rightarrow x(t) = x_0 e^{1-\cos(t)}.
\end{aligned}$$

Along the characteristic base curves, the function ρ is conserved and hence we have

$$\rho(x(t), t) = \rho(x(0), 0) = \rho(x_0),$$

Now, $x_0 = x(t)e^{-1+\cos(t)}$ and from $\rho_0(x_0) = 1 + \frac{1}{1+x_0^2}$, we obtain,

$$\rho(x,t) = 1 + \frac{1}{1 + x^2 e^{-2+2\cos(t)}}.$$

Problem 4.13.5 *Find the traffic density $\rho(x,t)$, satisfying*

$$\frac{\partial \rho}{\partial t} + e^t \frac{\partial \rho}{\partial x} = 2\rho,$$

with initial condition $\rho_0(x) = 1 + \sin^2 x$.

Solution: The characteristic base curves for this initial value problem are solutions of

$$\frac{dx}{dt} = e^t, \quad x(0) = x_0,$$
$$\Rightarrow x(t) = x_0 + e^t - 1,$$

and along these curves

$$\frac{d\rho(x(t),t)}{dt} = 2\rho(x(t),t).$$

Solving, we get

$$\rho(x(t),t) = \rho(x(0),0)e^{2t} = \rho(x_0)e^{2t},$$
$$\Rightarrow \rho(x(t),t) = (1 + \sin^2 x_0)e^{2t},$$

after substituting the initial value for $\rho_0(x)$. Replacing x_0 by $x - e^t + 1$, we get the traffic density as

$$\rho(x,t) = [1 + \sin^2(x - e^t + 1)]e^{2t}.$$

Problem 4.13.6 *A rectangular plate with an insulated surface that is 8 cm wide is such that its length can be considered to be infinite compared to its width. If the temperature along the short edge $y = 0$ is given by $u(x,0) = 100 \sin(\frac{\pi x}{8})$ while the two long edges $x = 0$ and $x = 8$, as well as the other short edges, are kept at $0\,°\mathrm{C}$, find the steady-state temperature function $u(x,y)$.*

Solution: The steady state temperature is obtained by solving

$$\frac{\partial^2 u}{\partial x^2} + \frac{\partial^2 u}{\partial y^2} = 0, \text{ with boundary conditions}$$
$$u(0,y) = 0 = u(8,y), \text{ for } 0 \le y \le \infty,$$
$$u(x,y) \to 0 \text{ as } y \to \infty \text{ for } 0 \le y \le \infty,$$
$$u(x,0) = 100\sin(\frac{\pi x}{8}) \text{ for } 0 \le x \le 8.$$

We proceed as in the section(4.4.1), and obtain

$$u(x,y) = \sum_{n=1}^{\infty} A_n \sin\left(\frac{\pi x}{8}\right) e^{-\frac{n\pi y}{8}}. \text{ Now,}$$

$$u(x,0) = 100 \sin(\frac{\pi x}{8}) = \sum_{n=1}^{\infty} A_n \sin\left(\frac{n\pi x}{8}\right),$$

$$\Rightarrow u(x,0) = A_1 \sin\left(\frac{\pi x}{8}\right) + A_2 \sin\left(\frac{2\pi x}{8}\right) + ... \quad .$$

Comparing, we get $A_1 = 100$ and $A_n = 0$ for $n \geq 2$. Therefore,

$$u(x,y) = 100 \sin\left(\frac{n\pi x}{8}\right) e^{-\frac{n\pi y}{8}}.$$

Problem 4.13.7 *The spread of a population is given by*

$$\frac{\partial N(x,t)}{\partial t} = N\left[r - a(N-b)^2\right] + D\frac{\partial^2 N(x,t)}{\partial x^2}.$$

Suppose the population is confined to a region with length dimension L, that is, $0 \leq x \leq L$. This will imply the following boundary conditions:

$$\frac{\partial N(0,t)}{\partial x} = 0, \quad \frac{\partial N(L,t)}{\partial x} = 0.$$

Let the initial population size be 2 units, that is, $u(x,0) = 2$. For $r = 0.5$, $a = 0.2$, $b = 2$, $L = 10$ find the non-zero steady-state solutions. Check their stabilities for $D = 0$ and $D = 0.01$.

Solution: The non-zero steady-state solutions are given by

$$r - a(N-b)^2 = 0 \Rightarrow b + \sqrt{\frac{r}{a}}, \quad b - \sqrt{\frac{r}{a}}.$$

For the given set of parameter values, the equilibrium solutions are 3.58 and 0.42. Let $f(N) = N\left[r - a(N-b)^2\right]$, then $f'(N) = r - a(N-b)^2 - 2aN(N-b)$. When D=0, $f'(3.58) = -2.26 < 0 \Rightarrow$ the system is stable about $N^* = 3.58$ and $f'(0.42) = 0.169 > 0 \Rightarrow$ the system is unstable about $N^* = 0.42$. When $D = 0.1 (\neq 0)$, the condition for stability is $f'(N^*) - Dq^2 < 0$, where q is the wave number, whose value need to be specified (say, $q = 6.71$ in this case). Now, $f'(N^*) - Dq^2 = -2.26 - 0.01 \times (6.71)^2 = -2.71 < 0 \Rightarrow$ the system is stable about $N^* = 3.58$ and $f'(N^*) - Dq^2 = 0.26544 - 0.01 \times (6.71)^2 = -0.1848 < 0 \Rightarrow$ the system is stable about $N^* = 0.42$. This shows that diffusion maintains stability in a stable one-species system (about the equilibrium point $N^* = 3.58$). Also, it can stabilized an unstable one (about the equilibrium point $N^* = 0.42$). It can be easily verified that for $q = 4.71$, the system will remain unstable with diffusion.

Problem 4.13.8 *A reaction-diffusion model is given by*

$$\begin{aligned}\frac{\partial M_1}{\partial t} &= r(a - M_1 + M_1^2 M_2) + \frac{\partial^2 M_1}{\partial x^2},\\ \frac{\partial M_2}{\partial t} &= r(b - M_1^2 M_2) + D\frac{\partial^2 M_2}{\partial x^2},\end{aligned}$$

where M_1 and M_2 are morphogen concentrations. All the parameters r, a, b, c, D are positive constants.
(i) Explain the model and find the non-zero homogenous steady state.
(ii) Show that the condition stability (without diffusion) is $b - a - (a+b)^3 < 0$.
(iii) Linearize the system about the non-zero steady state and find the condition for diffusive instability.

Solution: A spatially homogeneous steady state (m_1^*, m_2^*) of the model is given by

$$\begin{aligned}a - m_1^* + (m_1^*)^2 m_2^* &= 0,\\ b - (m_1^*)^2 m_2^* &= 0.\end{aligned}$$

$$\text{Solving we get,} \quad (m_1^*, m_2^*) = \left(a + b, \frac{b}{(a+b)^2}\right).$$

Let

$$M_1(x,t) = m_1(x,t) - m_1^*, \tag{4.53}$$

$$M_2(x,t) = m_2(x,t) - m_2^*, \tag{4.54}$$

be small non-homogeneous perturbations of the uniform steady state. Substituting (4.53) and (4.54) in the given system and retaining the linearized term we get,

$$\frac{\partial m_1}{\partial t} = a_{11}m_1 + a_{12}m_2 + \frac{\partial^2 m_1}{\partial x^2}, \tag{4.55}$$

$$\frac{\partial m_2}{\partial t} = a_{21}m_1 + a_{22}m_2 + D\frac{\partial^2 m_2}{\partial x^2}, \tag{4.56}$$

where,

$$a_{11} = \left(\frac{\partial f_1}{\partial M_1}\right)_{(m_1^*, m_2^*)} = \frac{b-a}{b+a}; \quad a_{12} = \left(\frac{\partial f_1}{\partial M_2}\right)_{(m_1^*, m_2^*)} = (a+b)^2;$$

$$a_{21} = \left(\frac{\partial f_2}{\partial M_1}\right)_{(m_1^*, m_2^*)} = \frac{-2b}{a+b}; \quad a_{22} = \left(\frac{\partial f_2}{\partial M_2}\right)_{(m_1^*, m_2^*)} = -(a+b)^2.$$

Let $m_1 = A_1 e^{\lambda t}\cos(qx)$ and $m_2 = A_2 e^{\lambda t}\cos(qx)$ be the solution of (4.55) and (4.56), then we get

$$A_1\left(\lambda - \frac{b-a}{b+a} + q^2\right) - A_2(a+b)^2 = 0,$$

$$A_1\frac{2b}{a+b} + A_2\left(\lambda + (a+b)^2 + Dq^2\right) = 0.$$

For a non-trivial solution of A_1 and A_2, we must have

$$\begin{vmatrix} \lambda - \frac{b-a}{b+a} + q^2 & -(a+b)^2 \\ \frac{2b}{a+b} & \lambda + (a+b)^2 + Dq^2 \end{vmatrix} = 0,$$

$$\begin{aligned} \Rightarrow \lambda^2 &- \left(\frac{b-a}{b+a} - (a+b)^2 - q^2(1+D)\right)\lambda \\ &+ \left(q^2 - \frac{b-a}{b+a}\right)\left((a+b)^2 + Dq^2\right) + 2b(a+b) = 0. \end{aligned} \tag{4.57}$$

Without diffusion, (4.57) reduces to

$$(\lambda)^2 - \left\{\frac{b-a}{b+a} - (a+b)^2\right\}\lambda + (a+b)^2 = 0. \tag{4.58}$$

Therefore, the system will be stable (without diffusion) if

$$\begin{aligned} \frac{b-a}{b+a} - (a+b)^2 < 0, \\ (b-a) - (a+b)^3 < 0. \end{aligned} \tag{4.59}$$

With diffusion, the system will be stable if

$$\frac{b-a}{b+a} - (a+b)^2 - q^2(1+D) < 0, \tag{4.60}$$

$$(q^2 - \frac{b-a}{b+a})(Dq^2 + (a+b)^2) + 2b(a+b) > 0. \tag{4.61}$$

Clearly, (4.60) is always negative by virtue of (4.59). Therefore, diffusive instability will occur if (4.61) is violated, that is, if

$$\begin{aligned} (q^2 - \frac{(b-a)}{(b+a)})(Dq^2 + (a+b)^2) + 2b(b+a) < 0, \\ \Rightarrow Dq^4 + \{(a+b)^2 - D\frac{(b-a)}{(b+a)}\}q^2 + (a+b)^2 < 0. \end{aligned}$$

4.14 Exercises

1. A uniform rod of length L with diffusivity D of the material of the rod, whose sides are insulated, is kept initially at temperature $\sin\left(\frac{\pi x}{L}\right)$. Both ends of the rod are suddenly cooled at 0^oC and are kept at that temperature.
   (i) If $u(x,t)$ represents the temperature function at any point x at time t, formulate a mathematical model of the given situation, stating clearly the boundary and initial conditions.
   (ii) Using the method of separation of variables, find the temperature function $u(x,t)$. Obtain the numerical solution of the model for $L = 30$ cm, $D = 0.5$ and plot the graph.
   (iii) If the initial temperature is changed to $\sin^3\left(\frac{\pi x}{L}\right)$, find the temperature function $u(x,t)$. Also, obtain the numerical solution of the modified model for $L = 30$ cm, $D = 0.5$ and plot the graph.

2. Let $u(x,t)$ denote the temperature function in a rod of length L at any point x at any time t. The sides of the rod are insulated and kept initially at temperature $\sin\left(\frac{\pi x}{L}\right)$. The two ends of the rod are then quickly insulated such that the temperature gradient is zero at each end.
   (i) Formulate a mathematical model of the given situation using partial differential equations, stating clearly the boundary and initial conditions.
   (ii) Using the method of separation of variables, obtain an expression for the temperature $u(x,t)$. Obtain the numerical solution of the model for $L = 30, D = 0.5$ and plot the graph.
   (iii) If the initial temperature is changed to $x(L - x)$, find the temperature function $u(x,t)$. Also, obtain the numerical solution of the model for $L = 30, D = 0.5$ and plot the graph.

3. Consider a laterally insulated rod of length unity with diffusivity D of the material of the rod, whose one end is insulated at $x = 0$ and its other end is kept at temperature $0\,°\text{C}$. The initial temperature is 1, when $0 \leq x \leq 0.5$ and $2(1 - x)$, when $0.5 \leq x \leq 1$.
   (i) If $u(x,t)$ represents the temperature function at any point x at time t, formulate a mathematical model of the given situation, stating clearly the boundary and initial conditions.
   (ii) Using the method of separation of variables, find the temperature function $u(x,t)$. Obtain the numerical solution of the problem for $D = 0.475$ and plot the graph.

4. Let $u(x,t)$ denotes the temperature function in a uniform rod (with diffusivity D of the material of the rod) of length L at any point x at time t. One end of the rod is kept at zero temperature, and the other end is poorly insulated, which radiates energy into the medium at a constant rate α. The rod is kept initially at temperature $x, 0 \leq x \leq L$.
(i) Formulate a mathematical model of the given situation using partial differential equations, stating clearly the boundary and initial conditions.
(ii) Using the method of separation of variables, obtain an expression for the temperature $u(x,t)$. Obtain the numerical solution of the model for $D = 0.5$ and plot the graph.

5. Consider a homogenous flexible string that is stretched between two fixed points, $(0,0)$ and $(L,0)$. The string is released from rest from a position $\sin\left(\frac{\pi x}{L}\right)$.
(i) Formulate a mathematical model of the given situation stating clearly the boundary and initial conditions, taking c to be the speed of the propagation of the wave.
(ii) Using the method of separation of variables, find an expression for the displacement $u(x,t)$ of the string at any time t. Obtain the numerical solution of the model with $c = 0.5, L = 10$ and plot the graph.
(iii) If the mid-point of the string is pulled up to a small height h and released from rest at time $t = 0$, what will be the initial condition? Obtain the solution of the wave equation with this initial condition. Now, find the numerical solution of the model taking $c = 0.5$ and plot the graph.

6. A homogenous flexible string of length L is stretched between $(0,0)$ and $(L,0)$ and is released with a velocity $10\sin\left(\frac{3\pi x}{L}\right)\cos\left(\frac{2\pi x}{L}\right)$ parallel to the axis of y from the equilibrium position. Let $u(x,t)$ be the displacement of the string at any time t.
(i) Formulate a mathematical model of the given situation using partial differential equations, stating clearly the boundary and initial conditions, taking c to be the speed of the propagation of the wave.
(ii) Using the method of separation of variables, obtain an expression for the displacement $u(x,t)$. Obtain the numerical solution of the model with $c = 0.5, L = 20$ and plot the graph.
(iii) If the initial velocity is $\sin^3\left(\frac{\pi x}{L}\right)$, what will be the solution of the wave equation. Now, obtain the numerical solution of the model taking $c = 0.5$ and plot the graph.

7. The four edges of a thin rectangular plate of length a and breadth b are kept at zero temperature, and the faces are perfectly insulated. The

initial temperature of the plate is given by $u(x,y,0) = \cos\frac{\pi(x-y)}{a} - \cos\frac{\pi(x+y)}{a}$, where $u(x,y,t)$ gives the temperature function of the rectangular plate.
(i) Formulate a mathematical model of the given situation using partial differential equations stating clearly the boundary and initial conditions.
(ii) Use the method of separation of variables to obtain an expression for the temperature $u(x,y,t)$ in the plate.
(iii) For $a = 1$, $b = 1$, obtain the numerical solution of the model with $c = 0.5, L = 10$ and plot the graph.
(iv) If the initial temperature of the plate is $10\sin\left(\frac{\pi x}{a}\right)\sin\left(\frac{\pi y}{b}\right)$, obtain an expression for the temperature $u(x,y,t)$ in the plate. Also, obtain the numerical solution of the model for $a = 1$, $b = 1$, $c = 0.5$, $L = 10$ and plot the graph.

8. Find the traffic density $\rho(x,t)$ satisfying

$$\frac{\partial \rho}{\partial t} + \frac{\partial(\rho u)}{\partial x} = 0,$$

where u, the velocity of the car, is a function of traffic density alone and is given by

$$u(\rho) = u_{max}\left\{1 - \left(\frac{\rho}{\rho_{max}}\right)^2\right\}, \quad 0 \le \rho < \rho_{max},$$

with initial traffic density as

$$\rho(x,0) = \begin{cases} 150, & x < 0, \\ 150\left(1 - \frac{x}{2}\right), & 0 < x < 1, \\ 80, & x > 1. \end{cases}$$

9. Find the traffic density $\rho(x,t)$ satisfying the traffic equation

$$\frac{\partial \rho}{\partial t} + (1-2\rho)\frac{\partial \rho}{\partial x} = 0,$$

with initial traffic density as

$$\rho(x,0) = \begin{cases} \frac{1}{4}, & x < 0, \\ \frac{1}{4}(1-x^2)^2, & x < 1, \\ 0, & x \ge 1. \end{cases}$$

10. Find the traffic density $\rho(x,t)$ satisfying

$$\begin{aligned} (i)\ \frac{\partial \rho}{\partial t} + 2\frac{\partial \rho}{\partial x} &= 0, \quad \rho_0(x) = e^{-x^2}. \\ (ii)\ \frac{\partial \rho}{\partial t} + 2t\frac{\partial \rho}{\partial x} &= 0, \quad \rho_0(x) = e^{-x^2}. \\ (iii)\ \frac{\partial \rho}{\partial t} + 2(1-\rho)\frac{\partial \rho}{\partial x} &= 0, \end{aligned}$$

$$\rho_0(x) = \begin{cases} 1, & x \le 0, \\ 0, & x > 0. \end{cases}$$

11. The linear growth of a population is given by [30]

$$\frac{\partial P(x,t)}{\partial t} = \alpha P(x,t) + D\frac{\partial^2 P(x,t)}{\partial x^2}.$$

(i) Obtain the solution of the given model.
(ii) Let $P(x,t) = k$ (constant). Then show that the values $\frac{x}{t}$ is given by

$$\frac{x}{t} = \mp\left[4\alpha D - \frac{2D}{t}\ln(t) - \frac{2D}{t}\ln\left(\sqrt{2\pi D}\frac{k}{P_0}\right)\right]^{1/2}.$$

12. The spread of the spruce budworm [101, 30] is given by

$$\frac{\partial B}{\partial t} = \alpha B\left(1 - \frac{B}{k}\right) - \beta\frac{B^2}{h^2 + B^2} + D\frac{\partial^2 B}{\partial x^2}.$$

(i) Explain the model and find the homogeneous equilibrium points.
(ii) Perform the linear stability analysis about the equilibrium point(s) for the non-diffusive system and obtain the condition for stability. Interpret the condition in the context of the model.
(iii) Now, perform the linear stability analysis for the diffusive system and obtain the condition for stability. Interpret the condition in the context of the model with diffusion.
(iv) For $\alpha = 1$, $k = 1000$, $\beta = 12.5$, $h = 5$, solve the model numerically to obtain graphs for $D = 0$ and $D = 0.5$ and compare the results. What changes do you observe if $\alpha = 1.1$?
(v) Taking $D = 1.0, 1.9, 2.7, 5.0$, plot the graphs and compare the results.

13. Segal and Jackson [149, 30] has modeled spatially distributed predator-prey system as

$$\begin{aligned}
\text{Prey}: \frac{\partial V}{\partial t} &= (k_0 + k_1 V) - AVE + \mu_2\nabla^2 V, \\
\text{Predator}: \frac{\partial E}{\partial t} &= BVE - ME - CE^2 + \mu_2\nabla^2 E.
\end{aligned}$$

(i) Describe the model by explaining each terms in the equations.
(ii) Assuming M $= 0$ and putting $v = \frac{VB}{k_0}$ and $e = \frac{Ec}{k_0}$, write the model in non-dimensionalized form as

$$\begin{aligned}
\frac{\partial v}{\partial t} &= (1 + kv)v - aev + \delta^2\nabla^2 v, \\
\frac{\partial e}{\partial t} &= ev - e^2 + \nabla^2 e,
\end{aligned}$$

where $k = \frac{k_1}{B}, a = \frac{A}{C}$ and $\delta^2 = \frac{\mu_1}{\mu_2}$.
(iii) Find the non-trivial homogenous steady state(s).
(iv) Show that the condition for diffusive instability is $k - \delta^2 > 2\sqrt{a-k}$.

14. The spread of yeast colonies was introduced by Gray and Kirwan [52, 30] and is given by

$$\begin{aligned}\frac{\partial Y}{\partial t} &= D_1\frac{\partial^2 Y}{\partial x^2} + kY(G - G_0),\\ \frac{\partial G}{\partial t} &= D_2\frac{\partial^2 G}{\partial x^2} - ckY(G - G_0).\end{aligned}$$

Here, $Y(x,t)$ is the density of Yeast cells $G(x,t)$ is the glucose concentration in medium at time t and the location x.
(i) Explain the model and find the homogenous steady state(s).
(ii) Perform the linear stability analysis about the equilibrium point(s) for the non-diffusive system and obtain the condition for stability. Interpret the condition in context with the model.
(iii) Now, perform the linear stability analysis for diffusive system and find the condition for diffusive instability.

15. A space dependent arm race model is given by [77]

$$\begin{aligned}\frac{\partial A_1(x,t)}{\partial t} &= aA_2^2 - mA_1 + r + D_1\frac{\partial^2 A_1}{\partial x^2},\\ \frac{\partial A_2(x,t)}{\partial t} &= bA_1^2 - nA_2 + s + D_2\frac{\partial^2 A_2}{\partial x^2},\end{aligned}$$

where A_1 and A_2 are amount spent on ARMS by two countries C_1 and C_2 ; where the parameters a, b, m, n are positive.
(i) Explain the model and find the non-trivial homogenous steady state(s).
(ii) Using Dirichlet boundary condition and suitable initial condition, solve the model numerically by taking $a = 1.0$, $b = 1.2$, $m = 0.9$, $n = 0.8$, $D_1 = 0.5$, $D_2 = 0.5$, with (a) $(r,s) = (1,2)$, (b) $(r,s) = (1,-2)$, (c) $(r,s) = (-1,-2)$, (d) $(r,s) = (-1,2)$. Comment on the results from the graphs obtained.
(iii) How the result changes if Neumann's boundary condition is used?

16. The model of a reaction-diffusion system is given by

$$\begin{aligned}\frac{\partial M_1}{\partial t} &= \alpha M_1M_2 - \beta M_1^2 + \delta\frac{\partial^2 M_1}{\partial x^2},\\ \frac{\partial M_2}{\partial t} &= M_2 - M_1M_2 + M_2^2 + \frac{\partial^2 M_1}{\partial x^2},\end{aligned}$$

where M_1 and M_2 are morphogen concentrations. All the parameters α, β, δ are positive constants
(i) Explain the model and find the non-zero equilibrium point (s).
(ii) Obtain the conditions for Turing instability and evaluate the values of δ (in terms of other parameters) for which Turing instability can take place.

17. A generalized logistic growth of a population with 1D diffusion term given by

$$\frac{\partial N}{\partial t} = \frac{rN}{\alpha}\left[1 - \left(\frac{N}{K}\right)^{\alpha}\right] + D\frac{\partial^2 N}{\partial x^2}, \quad N(x,0) = N_0, \ \alpha > 0.$$

Suppose, the growth of the population is confined to a specific region with length dimension L, that is, $0 \leq x \leq L$. This will imply the following boundary conditions:

$$\frac{\partial N(0,t)}{\partial x} = 0, \quad \frac{\partial N(L,t)}{\partial x} = 0.$$

(i) Explain the parameters r, α, K and D.
(ii) Obtain the condition for linear stability of the model. For $r = 0.5$, $\alpha = 2$, $K = 10$, $D = 0.05$, $L = 10$ and initial condition $N(x,0) = 1$, numerically solve the model and draw the graph. What conclusion can you draw from the graph?
(iii) For the same set of parameters, obtain the numerical solution for 2D diffusion. Draw the graph and comment on the dynamics of the population.

18. The given equations show the spread of love (or hate) of Majnun and Layla (see Majnun-Layla model):

$$\frac{\partial M}{\partial t} = aM + bL + f + D\frac{\partial^2 M}{\partial x^2},$$
$$\frac{\partial L}{\partial t} = cM + dL + g + D\frac{\partial^2 L}{\partial x^2}.$$

Taking $M(x,0) = m_0$, $L(x,0) = l_0$ (m_0, l_0 to be chosen by the readers, for example, $m_0 = 1, l_0 = 0$) as initial conditions and, Dirichlet's boundary condition, obtain the graphs and interpret the results for the following sets of parameters:
(i) $a = 3, b = 2, c = -1, d = 2, f = 0, g = 0, D = 0.5$.
(ii) $a = -3, b = 2, c = -1, d = 2, f = 0, g = 0, D = 0.5$.
(iii) $a = -3, b = 2, c = -2, d = 1, f = 0, g = 0, D = 0.5$.
(iv) $a = -3, b = -2, c = -2, d = -1, f = 0, g = 0, D = 0.5$.
(v) $a = -3, b = -2, c = -2, d = -1, f = 10, g = 10, D = 0.5$. Change

the values of f to $15, 20, 30$ and comment on the dynamics.
(vi) $a = 3, b = 2, c = 1, d = 2, f = 0, g = 0, D = 0.5$. Put $f = -10, g = -10$ and comment on the change in dynamics.
(vii) Now, change the boundary conditions to Neumann boundary condition and repeat the process.

19. The chlorine dioxide-iodine-malonic acid reaction proposed by Lengyel et. al. [92] is converted to spatial model:

$$\frac{\partial C(x,t)}{\partial t} = a - C(x,t) - \frac{4C(x,t)I(x,t)}{1 + C(x,t)^2} + D\frac{\partial^2 C(x,t)}{\partial x^2},$$
$$\frac{\partial I(x,t)}{\partial t} = b\, C(x,t)\left(1 - \frac{I(x,t)}{1 + C(x,t)^2}\right) + D\frac{\partial^2 I(x,t)}{\partial x^2},$$

where $C(x,t)$ and $I(x,t)$ are dimensionless concentrations of chlorine dioxide and iodine, respectively.
(i) Find the steady-state solution and perform linear stability analysis of the model with diffusion.
(ii) Obtain the condition for stability of the system.
(iii) Verify the results numerically for $a = 5, b = 1$ and D $= 0.5$.
(iv) Solve the system numerically with Dirichlet boundary condition (L=20) and appropriate initial condition to obtain the graphs. Comment on the dynamics of the model from the graphs. Again, solve the system numerically with Neumann boundary condition and obtain the graphs. What difference is obtained in both the dynamics?
(v) Repeat the whole process with $a = 10, b = 5$ and $D = 0.7$.
(vi) Convert the one-dimensional spatial model to a two-dimensional spatial model and perform stability analysis and numerical solution with the same sets of parameter values.

20. The spatial model for microbial growth of in a chemostat is given by

$$\frac{\partial m}{\partial t} = k(c) - \rho m + D_1\left(\frac{\partial^2 m}{\partial x^2} + \frac{\partial^2 m}{\partial y^2}\right),$$
$$\frac{\partial c}{\partial t} = \rho(c_0 - c) - \beta k(c)m + D_2\left(\frac{\partial^2 c}{\partial x^2} + \frac{\partial^2 c}{\partial y^2}\right),$$

where $m(t)$ is the population of micro-organisms per unit volume and $c(t)$ is the concentration of rate-limiting nutrient in the medium at time t, and $k(c) = \frac{kc}{a+c}$.
(i) Find the steady-state solutions of the model and perform linear stability analysis with diffusion.
(ii) Verify the condition for stability for $a = 10$, $\beta = 0.8$, $c_0 = 3.0$, $k = 0.2$, $\rho = 0.1$, $D_1 = 0.4$, and $D_2 = 0.6$.
(iii) Obtain density plot or contour plot of m and c with Dirichlet

boundary condition and suitable initial condition. Comment on the dynamics of the model from the plots.
(iv) Repeat the same analysis with $k(c) = k(1 - e^{\frac{c \ln 2}{a}})$.
(v) Do a similar analysis as in (iii) and (iv) with Neumann boundary condition.

21. May's prey-predator model with diffusion (1D) is given by

$$\frac{dF}{dt} = rF\left(1 - \frac{F}{k}\right) - \frac{\beta FP}{F + \alpha} + D_1 \frac{\partial^2 F}{\partial x^2},$$
$$\frac{dP}{dt} = sP\left(1 - \frac{P}{\gamma F}\right) + D_2 \frac{\partial^2 P}{\partial x^2}, \quad (r, k, \alpha, \beta, s, \gamma, D_1, D_2 > 0).$$

(i) Explain the model. Find the non-zero equilibrium point (F^*, P^*) and obtain the condition for diffusive instability.
(ii) For $r = \gamma = \beta = \alpha = 1.0, s = 1.2, k = 10, D_1 = 0.2, D_2 = 0.1$, verify the result numerically.
(iii) Suppose, both the population are confined to a specific region with length dimension L, that is, $0 \leq x \leq L$. This will imply the following boundary conditions:

$$\frac{\partial F(0,t)}{\partial x} = 0, \quad \frac{\partial F(L,t)}{\partial x} = 0, \quad \frac{\partial P(0,t)}{\partial x} = 0, \quad \frac{\partial P(L,t)}{\partial x} = 0.$$

Choosing $L = 20$, $F(x,0) = F^* + \sin(2x)$, $P(x,0) = P^* + \sin(2x)$, solve the model numerically to obtain the graphs with the given set of parameters and, comment on the dynamics of the two species. How the dynamics of the model changes if $s = 0.2$?
(iv) Convert the 1D model into a 2D model with Neumann boundary condition and initial conditions $F(x,y,0) = F^* + \sin(xy)$, $P(x,y,0) = P^* + \sin(xy)$, and obtain the Turing pattern with the given set of parameters and, comment on the dynamics of the two species. How the dynamics of the model changes if $s = 0.2$?

22. F.W. Lanchester developed combat models between two armies during World War I. Let $N_1(t)$ and $N_2(t)$ be the number of troops for army A and army B, respectively; then, the spatial Conventional combat model is given by

$$\frac{\partial N_1}{\partial t} = -\alpha N_2 + D_1 \left(\frac{\partial^2 N_1}{\partial x^2} + \frac{\partial^2 N_1}{\partial y^2}\right),$$
$$\frac{\partial N_2}{\partial t} = -\beta N_1 + D_2 \left(\frac{\partial^2 N_2}{\partial x^2} + \frac{\partial^2 N_2}{\partial y^2}\right).$$

(i) For $\alpha = 0.0001, \beta = 0.00012$, and $D_1 = D_2 = 0.01$, obtain the density

plot or contour plot and comment on the spread of the armies. What changes you observe for $\alpha = 0.0001$, $\beta = 0.0001$, and $D_1 = D_2 = 0.01$? Use both Dirichlet and Neumann boundary conditions and observe the changes in the dynamics of the system.
(ii) With the same set of parameter values, do similar studies with the Guerrilla combat model

$$\frac{\partial N_1}{\partial t} = -\alpha N_1 N_2 + D_1 \left(\frac{\partial^2 N_1}{\partial x^2} + \frac{\partial^2 N_1}{\partial y^2} \right),$$
$$\frac{\partial N_2}{\partial t} = -\beta N_1 N_2 + D_2 \left(\frac{\partial^2 N_2}{\partial x^2} + \frac{\partial^2 N_2}{\partial y^2} \right),$$

and the Mixed combat model

$$\frac{\partial N_1}{\partial t} = -\alpha N_1 N_2 + D_1 \left(\frac{\partial^2 N_1}{\partial x^2} + \frac{\partial^2 N_1}{\partial y^2} \right),$$
$$\frac{\partial N_2}{\partial t} = -\beta N_1 + D_2 \left(\frac{\partial^2 N_2}{\partial x^2} + \frac{\partial^2 N_2}{\partial y^2} \right).$$

23. When a patient takes a drug orally, it dissolves and releases the medications, which diffuses into the blood, and the bloodstream takes the medications to the site where it has a therapeutic effect . Let $C(x, y, t)$ be the concentration of drug in the compartment at time t, then, a two compartmental model describing the rate of change in oral drug administration is given by

$$\frac{dC_1}{dt} = -k_1 C_1 + D_1 \left(\frac{\partial^2 C_1}{\partial x^2} + \frac{\partial^2 C_1}{\partial y^2} \right),$$
$$\frac{dC_2}{dt} = -k_1 C_1 - k_2 C_2 + D_2 \left(\frac{\partial^2 C_1}{\partial x^2} + \frac{\partial^2 C_1}{\partial y^2} \right),$$

where $C_1(x, y, t)$ is the concentration of the drug in the stomach and $C_2(x, y, t)$ is the concentration of the drug in the rest of the body.
(i) Write the complete model with Dirichlet and Newmann boundary conditions.
(ii) Numerically, solve the model for $C_1(x, y, t)$ and $C_2(x, y, t)$ with initial concentration of 500 units at different rate constants $k_1 = 0.9776/h$, $0.7448/h$, $0.3293/h$, $0.2213/h$ and $k_2 = 0.5697$. Obtain the density plot or contour plot for each case and compare the results.
(iii) Also, obtain the contour plot for variation of drug concentrations within the body at different initial concentrations of 600, 250, 200 and 100 units with $k_1 = 0.9776/h$ and $k_2 = 0.5697$.

24. Consider the following prey–predator system with diffusion:

$$\frac{dF}{dt} = F[F(1-F) - P] + D_1\frac{\partial^2 F}{\partial x^2}, \quad \frac{dP}{dt} = P(F-a) + D_2\frac{\partial^2 P}{\partial x^2},$$

where $F(x, y, t)$ and $P(x, y, t)$ are dimensionless populations of the prey and predator, respectively, and $a > 0$ is a control parameter.
(i) Explain the model. Find the non-zero equilibrium point and obtain the condition for diffusive instability.
(ii) For $a = 2$, and appropriate values of D_1 and D_2, verify the results numerically.
(iii) Suppose, both the population are confined to a specific region with length dimension L, that is, $0 \le x \le L$. This will imply the following boundary conditions:

$$\frac{\partial F(0,t)}{\partial x} = 0, \quad \frac{\partial F(L,t)}{\partial x} = 0, \quad \frac{\partial P(0,t)}{\partial x} = 0, \quad \frac{\partial P(L,t)}{\partial x} = 0.$$

Choosing appropriate initial conditions, obtain the graphs and, comment on the dynamics of the two species. How does the dynamics of the model change if $a = 1/3$?
(iv) Consider the cases $D_1 > D_2$, $D_1 < D_2$, $D_1 = D_2$ and compare the behavior of the species.
(v) Convert the one-dimensional model into a two-dimensional model. Choosing appropriate initial conditions and Neumann boundary conditions, obtain the Turing pattern with the given set of parameters and, comment on the dynamics of the two species. How does the dynamics of the model changes if $s = 1/3$?

25. Consider a SIS model with diffusion:

$$\frac{\partial S}{\partial t} = -\beta SI + \alpha I + +D_1\frac{\partial^2 S}{\partial x^2}, \quad \frac{\partial I}{\partial t} = \beta SI - \alpha I + +D_2\frac{\partial^2 I}{\partial x^2},$$

where $S(x, y, t)$ are susceptible and $I(x, y, t)$ are infected.
(i) Explain the model. Taking infection rate $\beta = 0.00005$, recovery rate $\alpha = 0.0005$, obtain the numerical solution of the model with $S(x, y, 0) = 999$, $I(x, y, 0) = 1$, considering both Dirichlet and Neumann boundary conditions and appropriate values of D_1 and D_2. Comment on the spread of infection after plotting the graphs in both the cases.
(ii) Convert the one-dimensional model into a two-dimensional model. Choosing the same initial and boundary conditions, obtain a contour plot to show the spread of the disease in the population.

4.15 Project

1. The herd behavior of the preys is related to group defense, in which the preys at the boundary of the group hurt most from the attacks of the predators. The number of preys remaining on the border of the group is proportional to the length of the perimeter of the ground region occupied by the group [17], which in turn is directly proportional to the square root of the area of that grounded region. Hence, it is reasonable and logical to use square root terms for the prey population to portray the model with herd behavior. Consider the spatial predator–prey model with herd behavior:

$$\frac{\partial U(x,t)}{\partial t} = rU(x,t)\left(1 - \frac{U(x,t)}{K}\right) - \frac{\alpha\sqrt{U(x,t)}\,V(x,t)}{1 + t_h\alpha\sqrt{U(x,t)}} + D_1\frac{\partial^2 U(x,t)}{\partial x^2},$$

$$\frac{\partial V(x,t)}{\partial t} = \frac{c\,\alpha\sqrt{U(x,t)}\,V(x,t)}{1 + t_h\alpha\sqrt{U(x,t)}} - sf(V) + D_2\frac{\partial^2 V(x,t)}{\partial x^2},$$

where $U(x,t)$ and $V(x,t)$ are the densities of prey and predator respectively. The parameters r, K, α, t_h, c, s are positive constants, $D_1(>0)$ and $D_2(>0)$ are diffusive coefficients.
(i) Explain the model by taking into account all the parameters and f(V) = V (linear mortality).
(ii) Non-dimensionalize the model by introducing proper dimensionless variables.
(iii) Find all the equilibrium points and perform linear stability analysis about each of them. Hence, find the condition of diffusive instability about its non-zero steady state.
(iv) State the suitable boundary conditions and initial condition.
(v) Obtain a set of suitable parameter values so that the stability condition is satisfied and plot the numerical solution. What can you say about the dynamics of the model from the solution?
(vi) Modify the model by adding two-dimensional diffusion terms, clearly stating the boundary and initial conditions.
(vii) Perform linear stability analysis and find the condition for diffusive instability.
(viii) Obtain a set of suitable parameter values so that the stability condition is satisfied and plot the numerical solution of the two-dimensional model. What can you say about the dynamics of the model from the graphs? (ix) Obtain another parameter set for diffusive instability and obtain the graphs showing Turing patterns. Interpret the results.
(x) Repeat the process from (i) – (ix) with $f(V) = V^2$ (quadratic mortality) and $f(V) = V + V^2$.

2. Immunotherapy is an important approach in cancer treatments. For a better understanding of tumor and immune system, the roles of predictive mathematical models are studied by experimental researchers and clinicians. Consider the following tumor-immune reaction-diffusion system:

$$\frac{\partial T}{\partial t} = rT\left(1 - \frac{T}{K}\right) - \alpha_1\ T\ I + D_1\frac{\partial^2 T}{\partial x^2},$$
$$\frac{\partial I}{\partial t} = \beta\ f(T)\ I - d_1 I - \alpha_2\ T\ I + D_2\frac{\partial^2 I}{\partial x^2},$$

where, T, I denotes the densities of tumor cells and immune components, r is the intrinsic growth rate of tumor cells, d_1 is the natural death rate of immune cells, K is the carrying capacity, α_1 is the rate at which immune cells kill the tumor cells, α_2 denotes the death rate of disappearing immune cells interacting with tumor cells, β is the activating rate of immune cells and D_1, D_2 are the diffusion rates of tumor and immune cells, respectively.

(i) Find all the equilibrium points and perform linear stability analysis about each of them. Hence, find the condition of diffusive instability about its non-zero steady state.

(ii) Suppose, both the population are confined to a specific region with length π, that is, $0 \leq x \leq \pi$. This will imply the following boundary conditions:

$$\frac{\partial T(0,t)}{\partial x} = 0, \quad \frac{\partial T(\pi,t)}{\partial x} = 0, \quad \frac{\partial I(0,t)}{\partial x} = 0, \quad \frac{\partial I(\pi,t)}{\partial x} = 0.$$

Choosing appropriate initial conditions, $T(x,0) = T_0$, $I(x,0) = I_0$, solve the model numerically for $r = 0.18$, $K = 5 \times 10^6$, $\alpha_1 = 1.101 \times 10^{-7}$, $\alpha_2 = 3.422 \times 10^{-10}$, $d = 0.0152$, $\beta = 6.2 \times 10^{-9}$, $D_1 = 0.02$ and $D_2 = 0.2$. How does the dynamics of the model change if $D_1 = 0.2$ and $D_2 = 0.02$?

(iii) Convert the one-dimensional model into a two-dimensional model. Obtain spatiotemporal pattern formation and interpret.

(iv) Choose new functional forms of $f(T)$ to form new models and repeat from (i) to (iii).

Chapter 5

Modeling with Delay Differential Equations

5.1 Introduction

What are Delay Differential Equations (DDE), is the first question that comes to mind when you begin reading this chapter. In layman's term, a DDE is a differential equation in which the derivatives of some unknown functions at the present time are dependent on the values of the functions at previous times. Let us consider a general DDE of the first order the form

$$\frac{dx(t)}{dt} = f(t, x(t), x(t-\tau_1), x(t-\tau_2), ..., x(t-\tau_n)), \tag{5.1}$$

where τ_i's are positive constants (fixed discrete delays).

When we solve an ODE (initial value problem), we only need to specify the initial values of the state variables. However, while solving a DDE, we have to look back to the earlier values of x at every time step. Assuming that we start at time $t = 0$, we, therefore, need to specify an initial function, which gives the behavior of the system prior to time $t = 0$.

Consider a DDE with a single discrete delay, that is,

$$\frac{dx(t)}{dt} = f(t, x(t), x(t-\tau)), \tag{5.2}$$

where $x(t-\tau) = \{x(\tau) : \tau \leq t\}$ gives the trajectory of the solution in the past. Here, the function f is a functional operator from $\Re \times \Re^n \times C^1$ to $\Re^n$ and x(t)$\in \Re^n$. We will not provide a detailed discussion on DDEs that fall into the class of functional differential equations. Interested readers are advised to consult the references [49, 85, 102]. For this DDE, the initial function would be a function $x(t)$ defined on the interval $[-\tau, 0]$. The solution to (5.2) is a mapping from functions on the interval $[t-\tau, t]$ into the functions on the interval $[t, t+\tau]$. Thus, the solution of (5.2) can be defined as a sequence of functions $f_0(t)$, $f_1(t)$, $f_2(t)$,... defined over a set of adjacent time intervals of length τ. The points $t = 0$, τ, 2τ,... , where the solution segments meet are called knots [138].

DOI: 10.1201/9781351022941-5

Now, what about the solutions of DDEs? It is not easy to solve a DDE analytically. To give an idea of the process, let us consider a simple DDE of the form

$$\frac{dx}{dt} = -x(t-\tau), \qquad t > 0.$$

Initial history: x(t)=1, $-\tau \leq t \leq 0$. Clearly, with $\tau = 0$, $x(t) = x(0)e^{-t}$. However, the presence of τ makes the situation a bit tricky. Hence, in the interval $0 \leq t \leq \tau$, we have

$$\frac{dx}{dt} = -x(t-\tau) = -1,$$

$$\Rightarrow x(t) = x(0) + \int_0^t (-1)ds = 1 - t, \qquad 0 \leq t \leq \tau.$$

In $\tau \leq t \leq 2\tau$, we get, $0 \leq t - \tau \leq \tau$ and so we have,

$$\frac{dx(t)}{dt} = -x(t-\tau) = -[1-(t-\tau)],$$

$$\Rightarrow x(t) = x(\tau) + \int_\tau^t [-\{1-(s-\tau)\}]ds,$$

$$\Rightarrow \ x(t) = 1 - t + \frac{(t-\tau)^2}{2}, \quad \tau \leq t \leq 2\tau,$$

and so on. In general, it can be shown (use mathematical induction) that

$$x(t) = 1 + \sum_{k=1}^{n}(-1)^k \frac{[t - \overline{k-1}\ \tau]^k}{k!}, \quad (n-1)\tau \leq t \leq n\tau, \quad n \geq 1.$$

The above method is known as a procedure of steps.

5.2 Linear Stability Analysis

An equilibrium point (steady-state solution) is a point in the state space for which $x(t) = x^*$ is a solution for all t. Therefore, for DDE of the form (5.1), equilibrium points satisfy

$$f\left(x^*, x^*, x^*, ..., x^*\right) = 0.$$

We consider the delay differential equation

$$\frac{dx}{dt} = -x(t-1), \tag{5.3}$$

and discuss its stability. Clearly, $x = 0$ is the only steady-state solution. Since the equation is linear, we try the exponential solution $x = ce^{\lambda t}$, $(c \neq 0)$ which gives

$$c\lambda e^{\lambda t} = -ce^{\lambda(t-1)} = -ce^{\lambda t}e^{-\lambda}.$$

$$\text{Let } g(\lambda) \equiv \lambda + e^{-\lambda} = 0, (\text{since } ce^{\lambda t} \neq 0)$$

This is the characteristics equation of (5.3), which is transcendental and has infinitely many solutions. It is easy to check that the characteristics equation $\lambda + e^{-\lambda} = 0$ has no real solution since g(λ) has an absolute minimum of 1 at $\lambda = 0$. Therefore, we substitute $\lambda = a + ib$ in the characteristics equation for complex solutions and get

$$\begin{aligned}
& a + ib + e^{-a-ib} = 0, \\
& \Rightarrow a + ib + e^{-a}(\cos b - i \sin b) = 0, \\
& \Rightarrow (e^{-a}\cos b + a) + i(b - e^{-a}\sin b) = 0.
\end{aligned}$$

Equating the real and imaginary parts we get,

$$e^{-a}\cos b = -a, \tag{5.4}$$

$$e^{-a}\sin b = b. \tag{5.5}$$

Without any loss of generality, we assume $b > 0$. We want to check whether (5.4) and (5.5) can have solutions with positive values of real part a. Let us assume that equations (5.4) and (5.5) have solutions with $a > 0$. Then, from (5.4) we conclude that $\cos b < 0$, which implies $b > \frac{\pi}{2}$ (since $\cos b > 0$ for acute b). Now, for $a > 0$, $e^{-a} < 1$ and $|\sin b| < 1$, implying $|b| < 1$ (from 5.5). This leads to a contradiction as b cannot simultaneously be greater than $\frac{\pi}{2}$ and numerically less than 1 in magnitude. Hence, we conclude that the real part of the characteristic root cannot be positive; that is, it is negative, and hence the equilibrium point $x = 0$ is stable.

5.2.1 Linear Stability Criteria

The stability analysis in the previous section shows the complicated nature of dealing with linear stability analysis of delay differential equations. In this section, three theorems (without proofs) are stated, which may be used directly to investigate linear stability analysis of delay differential equations.

Theorem 5.2.1.1 *Consider a linear autonomous delay differential equation of the form*

$$\frac{dx(t)}{dt} = Ax(t) + Bx(t - \tau),$$

whose equilibrium solution is $x = 0$, *where* A, B *are constants. The corresponding characteristic equation is given by*

$$\lambda = A + Be^{-\lambda\tau}.$$

Then [153],
(i) If $A + B > 0$*, then* $x = 0$ *is unstable.*
(ii) If $A + B < 0$ *and* $B \geq A$*, then* $x = 0$ *is asymptotically stable.*
(iii) If $A + B < 0$ *and* $B < A$*, then there exists* $\tau^* > 0$ *such that* $x = 0$ *is asymptotically stable for* $0 < \tau < \tau^*$ *and unstable for* $\tau > \tau^*$*. Also, there exists a pair of imaginary roots at* $\tau = \tau^* = \cos^{-1}\left(\frac{-A/B}{B^2 - A^2}\right)$*.*

Theorem 5.2.1.2 *Consider a linear autonomous delay differential equation of the form*

$$\frac{dx}{dt} + a_1 x(t - \tau_1) + a_2 x(t - \tau_2) = 0, \tag{5.6}$$

where a_1*,* a_2*,* τ_1*,* $\tau_2 \in [0, \infty)$*. The transcendental characteristic equation of (5.6) can be obtained by substituting the ansatz* $x(t) = ce^{\lambda t}$ *in (5.6), which results in*

$$\lambda + a_1 e^{-\lambda \tau_1} + a_2 e^{-\lambda \tau_2} = 0, \tag{5.7}$$

where λ *is a complex number.*
(i) Let $a_2 = 0$ *and* $a_1, \tau_1 \in (0, \infty)$*. A necessary and sufficient condition for all roots of* $\lambda + a_1 e^{-\lambda \tau_1} = 0$ *to have negative real parts is* $0 < a_1 \tau_1 < \frac{\pi}{2}$*.*
(ii) Let $a_1, a_2, \tau_1, \tau_2 \in (0, \infty)$*. A sufficient condition for all roots of (5.7) to have negative real parts is* $a_1 \tau_1 + a_2 \tau_2 < 1$ *and a necessary condition for the same is* $a_1 \tau_1 + a_2 \tau_2 < \pi/2$*.*

Theorem 5.2.1.3 *Consider the characteristic equation of the form*

$$P(\lambda) + Q(\lambda) e^{-\lambda \tau} = 0,$$

where P *and* Q *are polynomials with real coefficients and* τ (> 0) *is the discrete time delay. Let,*
(i) $P(\lambda) \neq 0$ *for all* $Re(\lambda) \geq 0$*,*
(ii) $Q(ib) < |P(ib)|$*,* $0 \leq b < \infty$*,*
(iii) $\lim_{|\lambda| \to \infty,\ Re(\lambda) \geq 0} \left|\frac{Q(\lambda)}{P(\lambda)}\right| = 0$*, then,* $Re(\lambda) < 0$ *for every root* λ *and all* $\tau \geq 0$*.*

Corollary: *Suppose* $P(\lambda)$ *has leading coefficient one and let* $Q(\lambda) = c$ *(constant). If*
(i) all roots of the polynomial $P(\lambda)$ *are real and negative and* $|P(0)| > |c|$ *or*
(ii) $P(\lambda) = \lambda^2 + a\lambda + b, a, b > 0$ *and either*
(A1) $b > |c|$ *and* $a^2 \geq 2b$ *or (A2)* $a\sqrt{4b - a^2} > 2|c|$ *and* $a^2 < 2b$*,*
then, $Re(\lambda) < 0$*, for every root* λ *and all* $\tau \geq 0$*.*

5.3 Different Models with Delay Differential Equations

5.3.1 Delayed Protein Degradation

Let $P(t)$ be the concentration of proteins at any time t in a system, then the production of proteins at any time is given by [14]

$$\frac{dP(t)}{dt} = \alpha - \beta P(t) - \gamma P(t-\tau), \tag{5.8}$$

where α is the constant rate of protein production, β is the rate of non-delayed protein degradation, and γ is the rate of delayed protein degradation. The discrete time delay τ is due to the fact that the protein degradation machine degrades the protein after a time τ after initiation. The equilibrium point of (5.8) is given by

$$\alpha - \beta P^* - \gamma P^* = 0 \Rightarrow P^* = \frac{\alpha}{\beta+\gamma}.$$

Using the transformation $P = P' + P^*$ in (5.8), we obtain

$$\frac{dP'(t)}{dt} = -\beta P'(t) - \gamma P'(t-\tau). \tag{5.9}$$

Putting $P' = A_1 e^{\lambda t}$ in (5.9), we get the characteristic equation as $\lambda = -\beta - \gamma e^{-\lambda\tau}$. Comparing with (5.2.1.1), we get $A+B = -\beta-\gamma < 0$. Since, $-\gamma > -\beta$ (assumed), the system (5.8) is asymptotically stable about $P^* = \frac{\alpha}{\beta+\gamma}$. Fig. 5.1 shows the degradation of protein for various parameter values, obtained from [14].

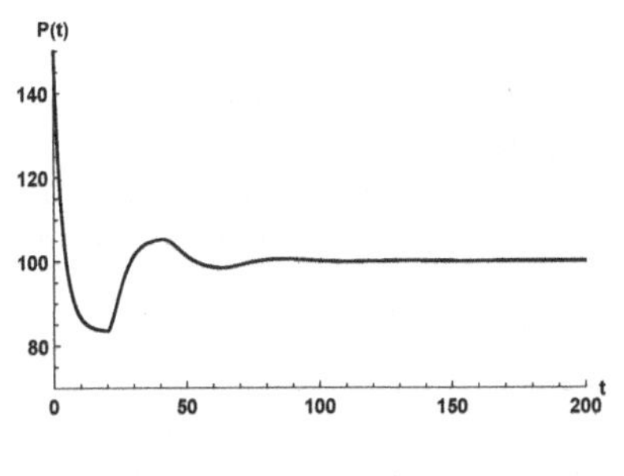

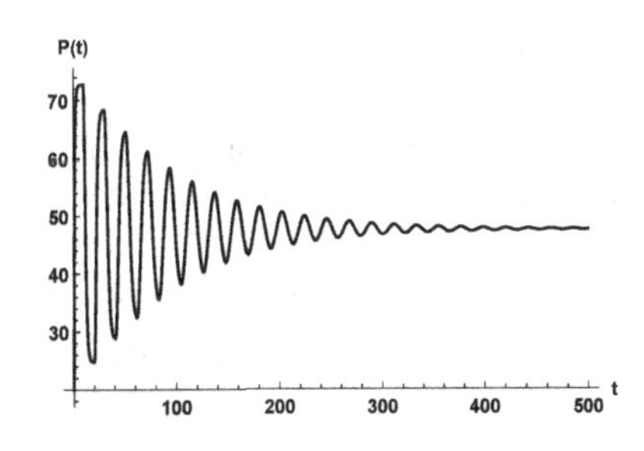

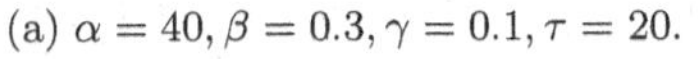
(a) $\alpha = 40, \beta = 0.3, \gamma = 0.1, \tau = 20$.

(b) $\alpha = 100, \beta = 1.1, \gamma = 1, \tau = 10$.

FIGURE 5.1: *The figures show delay-induced protein degradation. (a) Protein concentration delay with initial history* 150. *(b) Oscillatory behavior of protein degradation with initial history* 20.

5.3.2 Football Team Performance Model

R.B. Banks [12] proposed a delay-induced mathematical model to analyze the performance of a National Football League (NFL) football team during the last 40 years. The proposed model is

$$\frac{dU}{dt} = b\left[\frac{1}{2} - U(t-\tau)\right], \tag{5.10}$$

where $U(t)$ is the fraction of games won by an NFL team during one season and it lies between 0 and 1, and b is the growth rate. The computational formula for $U(t)$ is given by

$$U(t) = \frac{1 \times \text{no. of games won} + \frac{1}{2} \times \text{no. of games tied} + 0 \times \text{no. of games lost}}{\text{Total no. of games}}.$$

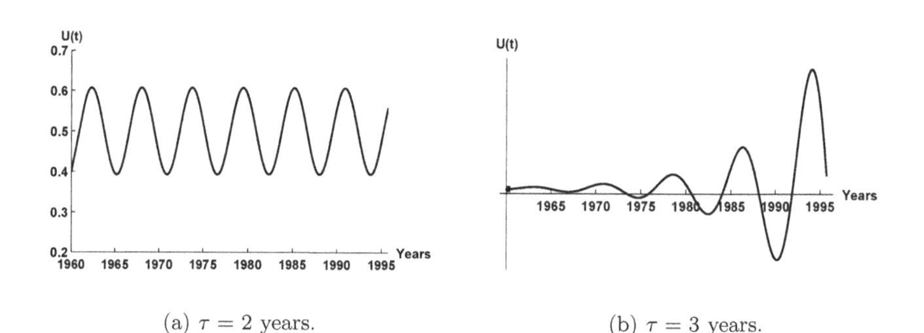

(a) $\tau = 2$ years. (b) $\tau = 3$ years.

FIGURE 5.2: *The performance of an NFL team from 1960 to 1992, with parameter values* $b = 0.785, \tau = 2$ *years and initial history* 0.4. *The performance of the team shows (a) a periodic solution and (ii) an unstable situation, which matches with data [12].*

Basically, the proposed model says that at the present time, the rate of change of U is proportional to the difference between $U = \frac{1}{2}$ (average values) and the values of U at some previous time $t - \tau$. The equilibrium point of (5.10) is given by $b\left[\frac{1}{2} - U^*\right] = 0 \Rightarrow U^* = \frac{1}{2}$. Using the transformation $U = U' + U^*$ in (5.10), we obtain

$$\frac{dU'(t)}{dt} = -b\, U'(t-\tau). \tag{5.11}$$

Putting $U' = A_1 e^{\lambda t}$ in (5.11), we get the characteristic equation as $\lambda = -be^{-\lambda\tau}$. Comparing with (5.2.1.1), we get $A+B = -b < 0$ and $B < A$. Hence, the system (5.10) is asymptotically stable about $U^* = \frac{1}{2}$ for $0 < \tau < \tau^*$ and unstable for $\tau > \tau^*$. Banks [12] concluded from his model that the time delay τ plays an important role in the ups and downs of the football team. It experiences a simple periodicity (fig. 5.2(a)) and an unstable situation (fig. 5.2(b)).

5.3.3 Shower Problem

People enjoy showering, especially when they are able to control the water temperature. The dynamics of human behavior while taking a shower when the water temperature is not comfortable is quite interesting. A simple DDE model is proposed to capture such dynamics. We assume that the speed of water is constant (uniform flow) from the faucet to the shower head, which takes the time τ seconds (say).

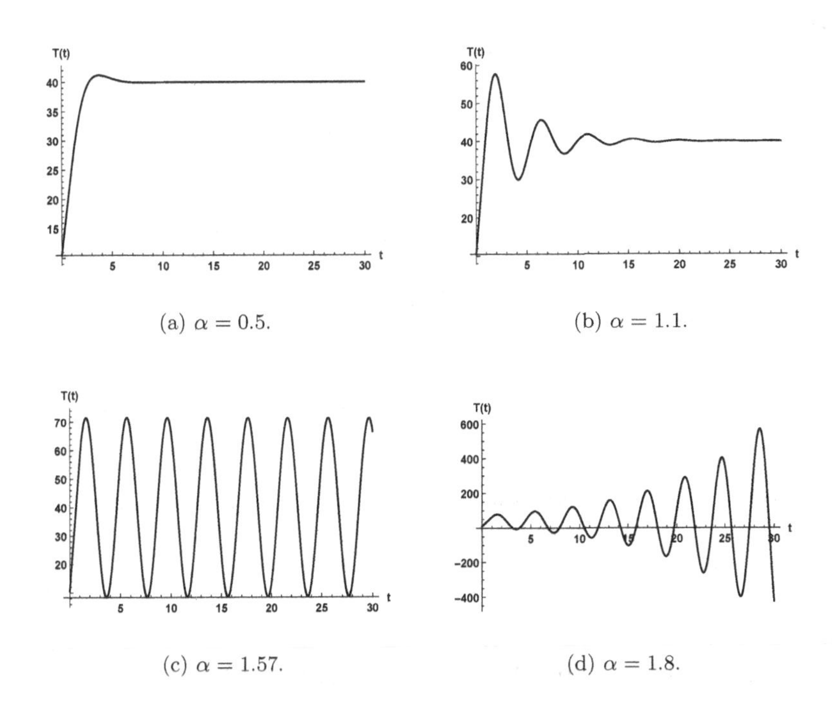

(a) $\alpha = 0.5$.

(b) $\alpha = 1.1$.

(c) $\alpha = 1.57$.

(d) $\alpha = 1.8$.

FIGURE 5.3: *Varied temperatures of water for different values of* α*. The parameter values are* $T_d = 40\,^\circ\mathrm{C}$, $\tau = 1$ *and initial history* $= 0.5$.

Let $T(t)$ denote the temperature of water at the faucet at time t, then the temperature evolution is given by

$$\frac{dT}{dt} = -\alpha[T(t-\tau) - T_d], \tag{5.12}$$

where T_d is the desired temperature and α gives the measure of a person's reaction due to the wrong water temperature. One type of person might prefer a low value of α, whereas another type of person would choose a higher value. The equilibrium point of (5.12) is given by $T^* - T_d = 0 \Rightarrow\ T^* = T_d$. Using the transformation $T = T' + T^*$ in (5.12), we obtain

$$\frac{dT'(t)}{dt} = -\alpha\, T'(t-\tau). \tag{5.13}$$

Putting $T' = A_1 e^{\lambda t}$ in (5.13), we get the characteristic equation as $\lambda = -\alpha\, e^{-\lambda\tau}$. Comparing with (5.2.1.1), we get $A + B = 0 + (-\alpha) = -\alpha < 0$ and $B < A$. Hence, the system (5.12) is asymptotically stable about $T^* = T_d$.

For $\alpha = 0.5$, the temperature of the water goes to 40 °C, which is comfortable to the body and the person remains calm (fig. 5.3(a)). For $\alpha = 1.1$, after initial fluctuation, the temperature of the water goes to 40 °C. A person may show initial discomfort with the start of the shower (fig. 5.3(b)). One person may prefer the value of $\alpha = 1.57$ while showering (a bathroom singer?), which shows cyclic behavior of water temperature (fig. 5.3(c)). For $\alpha = 1.8$, the temperature of the water is erratic and unpleasant while taking shower (fig. 5.3(d)).

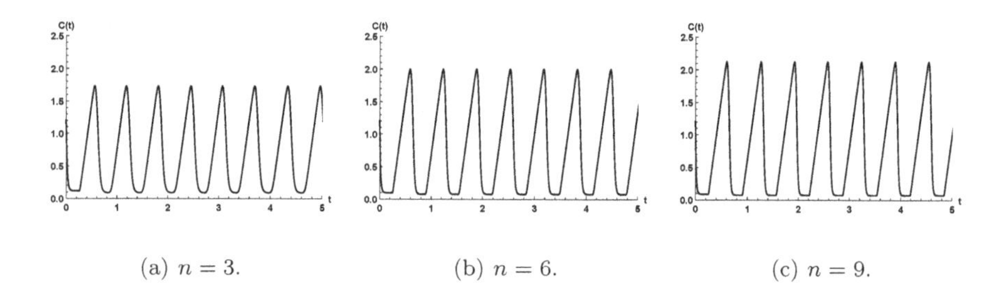

(a) $n = 3$. (b) $n = 6$. (c) $n = 9$.

FIGURE 5.4: *The oscillatory behavior of carbon dioxide content for (a) n=3, (b) n=6, (c) n=9. The parameter values, obtained from [78], are* $\lambda = 6, \alpha = 1.0, V_{max} = 80, \theta = 1,\ \tau = 0.25$ *and initial history=* 1.2.

5.3.4 Breathing Model

The arterial carbon dioxide level controls our rate of breathing. A mathematical model was first developed by Mackey and Glass [46], where they assumed that carbon dioxide is produced at a constant rate λ due to metabolic activity and its removal from the bloodstream is proportional to both the current carbon dioxide concentration and to ventilation. Ventilation, which is the volume of gas exchanged by the lungs per unit of time, is controlled by the carbon dioxide level in the blood. The process is complex and involves the detection of carbon dioxide levels by receptors in the brain stem. This carbon dioxide detection and its subsequent adjustment to ventilation is not an instantaneous process; there is a time lag due to the fact that the blood transport from the lungs to the heart and then back to the brain requires time [78]. Thus, if C is the concentration of the carbon dioxide, then the rate of change of concentration of carbon dioxide due to breathing is given by

$$\frac{dC(t)}{dt} = \lambda - \alpha V_{max} C(t)\dot{V}(t-\tau),$$

where $\dot{V}(t)$ is the rate of ventilation and is assumed to follow the Hill function, that is $\dot{V}(t) = \frac{(C(t))^n}{\theta^n+(C(t))^n}$; $V_{\max}$, θ, n, and α are constants. Thus, the rate of change of concentration of carbon dioxide is given by

$$\frac{dC(t)}{dt} = \lambda - \alpha V_{max} C(t) \frac{(C(t-\tau))^n}{\theta^n + (C(t-\tau))^n}.$$

Fig. 5.4 shows the oscillatory solutions of the Mackey-Glass equation, representing the carbon dioxide content for $n = 3, 6, 9$. As n increases, the amplitude of the oscillation also increases gradually.

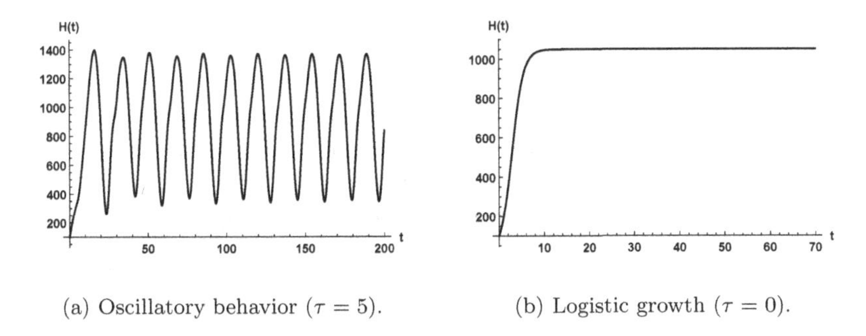

(a) Oscillatory behavior ($\tau = 5$).

(b) Logistic growth ($\tau = 0$).

FIGURE 5.5: *(a) The oscillatory behavior of the adult houseflies. Parameter values [142, 160]:* $d_1 = 0.147, \beta = 1.81, k = 0.5107, M = 0.000226$, *initial history* 100. *(b) The adult houseflies follow a logistic growth for* $\tau = 0$.

5.3.5 Housefly Model

Taylor and Sokal [160] proposed a model to describe the behavior of the adult housefly *Musca* domestic in laboratory conditions. To capture the dynamics of the housefly, the model is represented using as

$$\frac{dH}{dt} = -d_1 H(t) + \beta H(t-\tau)[k - \beta M H(t-\tau)].$$

Here, $H(t)$ represents the number of adult houseflies at any time t, d_1 is the natural death of the houseflies, τ (> 0) is the discrete time delay, which is the time from laying eggs until their emerging from the pupal case (oviposition and eclosion of adults), β is the number of eggs laid per adult, and assuming the number of eggs laid is proportional to the number of adults, the number of new eggs at time $t - \tau$ would be $\beta H(t-\tau)$. The term $k - \beta M H(t-\tau)$ gives the egg-to-adult survival rate, k and M being the maximum egg-adult survival rate and reduction in survival for each egg, respectively. Fig. 5.5(a) shows periodic solution as observed in the behavior of adult houseflies in laboratory conditions [160]. Please note that for $\tau = 0$, the housefly population follows a logistic growth (fig. 5.5(b)).

5.3.6 Two-Neuron System

A two-neuron system of self-existing neurons is given by [9, 164]

$$\frac{du_1}{dt} = -u_1(t) + a_1 \tanh[u_2(t - \tau_{21})], \quad \frac{du_2}{dt} = -u_2(t) + a_2 \tanh[u_1(t - \tau_{12})],$$

where $u_1(t)$ and $u_2(t)$ are the activities of the first and second neurons respectively, τ_{21} is the delay in signal transmission between the second neuron and the first neuron (τ_{12} can be explained in a similar manner) and a_1, a_2 are the weights of the connection between the neurons. By taking $a_1 = 2, a_2 = -1.5, \tau_{21} = 0.2, \tau_{12} = 0.5$ such that $\tau_{21} + \tau_{12} < 0.8$, numerically it has been shown that the model is asymptotically stable about the origin (fig. 5.6(a), fig. 5.6(b)). For $\tau_{21} = 0.4, \tau_{12} = 0.6$ such that $\tau_{21} + \tau_{12} > 0.8$, a periodic solution bifurcates from the origin; that bifurcation is supercritical and the bifurcating periodic solution is orbitally asymptotically stable (fig. 5.6(c), fig. 5.6(d)). Interested readers may look into Ruan et al.[164], to learn more about the analytical calculations and restrictions on τ_{21} and τ_{12}.

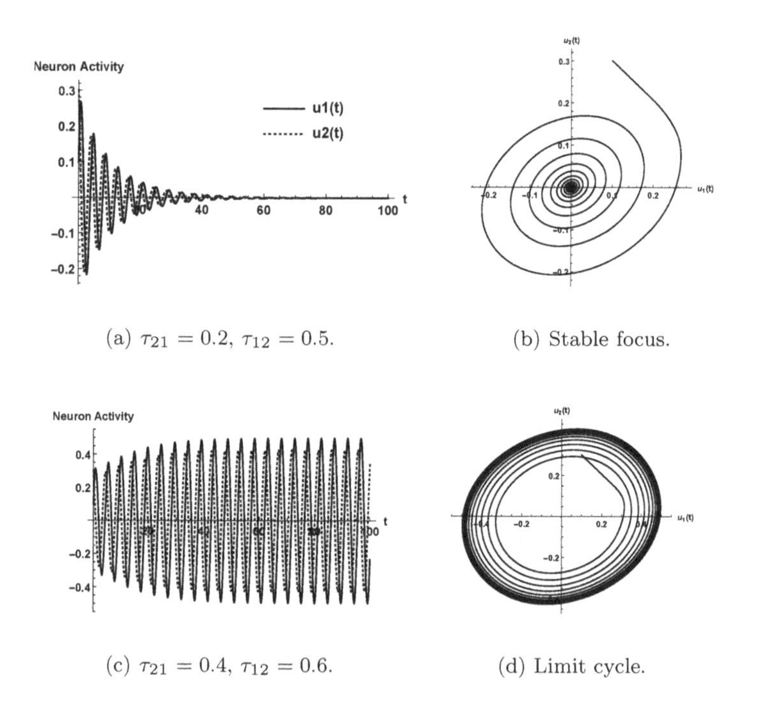

(a) $\tau_{21} = 0.2, \tau_{12} = 0.5$. (b) Stable focus.

(c) $\tau_{21} = 0.4, \tau_{12} = 0.6$. (d) Limit cycle.

FIGURE 5.6: *The figures show the activities of two self-exiting neurons. Parameter values [142]: $a_1 = 2, a_2 = -1.5$ with initial history (0.1,0.3). (a) Asymptotically stable, (b) stable focus, (c) periodic solution, (d) limit cycle.*

5.4 Immunotherapy with Interleukin-2, a Study Based on Mathematical Modeling [11] (a Research Problem)

5.4.1 Background of the Problem

The mechanism of establishment and destruction of cancer, one of the greatest killers in the world, is still a puzzle. Modern treatment involves surgery, chemotherapy, and radiotherapy, yet relapses occur. Hence, the need for more successful treatment is clear. Developing schemes for immunotherapy or its combination with other therapy methods are the major focus at present and aim at reducing the tumor mass, heightening tumor immunogenicity, and removal of immunosuppression induced in an organism in the process of tumor growth. Recent progress has been achieved through immunotherapy, which refers to the use of cytokines (protein hormones that mediate both natural and specific immunities) usually together with adoptive cellular immunotherapy (ACI) [40, 79, 133, 134, 148].

The main cytokine responsible for lymphocyte activation, growth, and differentiation is interleukin-2 (IL-2), which is mainly produced by T helper cells (CD4+ T-cells) and in relatively small quantities by cytotoxic T-lymphocytes (CD8+ T-cells). CD4 lymphocytes differentiate into T-Helper 1 and T-Helper 2 functional subjects due to the immune response. IL-2 acts in an autocrine manner on T-Helper 1 and also induces the growth of T-Helper 2 and CD8 lymphocytes in a paracrine manner. The T-lymphocytes themselves are stimulated by the tumor to induce further growth. Thus the complete biological assumption of adoptive cellular immunotherapy is that the immune system is expanded in number artificially (*ex vivo*) in cell cultures by means of human recombinant interleukin-2. This can be done in two ways, either by (i) lymphokine-activated killer cell therapy, where the cells are obtained from the *in vitro* culturing of peripheral blood leukocytes removed from patients with a high concentration of IL-2, or (ii) tumor-infiltrating lymphocyte therapy (TIL), where the cells are obtained from lymphocytes recovered from the patient tumors, which are then incubated with high concentrations of IL-2 *in vitro* and are comprised of activated natural killer (NK) cells and cytotoxic T-lymphocyte (CTL) cells. The TIL is then returned into the bloodstream, along with IL-2, where they can bind to and destroy the tumor cells. It has been established clinically that immunotherapy with IL-2 has enhanced CTL activity at different stages of the tumor [133, 134, 148]. Also, there is evidence of the restoration of the defective NK cell activity as well as enhancement of polyclonal expansion of CD4+ and CD8+ T cells [135, 159].

Kirschner and Panetta have studied the role of IL-2 in tumor dynamics, particularly long-term tumor recurrence and short-term oscillations, from mathematical perspective [83]. The model proposed there deals with three populations, namely, the activated immune-system cells (commonly called effector cells), such as cytotoxic T cells, macrophages, and NK cells that are cytotoxic to the tumor cells, the tumor cells and the concentration of IL-2. The important parameters in their study are antigenicity of tumor (c), a treatment term that represents the external source of effector cells (s_1), and a treatment term that represents an external input of IL-2 into the system (s_2). Their results can be summarized as follows: (i) For non-treatment cases ($s_1 = 0, s_2 = 0$), the immune system has not been able to clear the tumor for low-antigenic tumors, while for highly antigenic tumors, reduction to a small dormant tumor is the best-case scenario. (ii) The effect of adoptive cellular immunotherapy (ACI) ($s_1 > 0, s_2 = 0$) alone can yield a tumor-free state for tumors of almost any antigenicity, provided the treatment concentration is above a given critical level. However, for tumors with small antigenicity, early treatment is needed while the tumor is small, so that the tumor can be controlled. (iii) Treatment with IL-2 alone ($s_1 = 0, s_2 > 0$) shows that if IL-2 administration is low, there is no tumor-free state. However, if IL-2 input is high, the tumor can be cleared, but the immune system grows without bounds, causing problems such as capillary leak syndrome. (iv) Finally, the combined treatment with ACI and IL-2 ($s_1 > 0, s_2 > 0$) gives the combined effects obtained from the monotherapy regime. For any antigenicity, there is a region of tumor clearance. These results indicate that treatment with ACI may be a better option either as a monotherapy or in conjunction with IL-2. Here I have proposed a modification of the model studied by Kirschner and Panetta by adding a discrete time delay which exists when activated T-cells produce IL-2.

5.4.2 The Model

The proposed model is an extension of the Kirschner-Panetta ordinary differential equation model [83]

$$\begin{aligned}
\frac{dE}{dt_1} &= cT + \frac{p_1 E I_L}{g_1 + I_L} - \mu_2 E + s_1, \\
\frac{dT}{dt_1} &= r_2(1 - bT)T - \frac{aET}{g_2 + T}, \\
\frac{dI_L}{dt_1} &= \frac{p_2 ET}{g_3 + T} - \mu_3 I_L + s_2,
\end{aligned}$$

to a DDE model with proper biological justifications and is given by

$$\begin{aligned}
\frac{dE}{dt_1} &= cT + \frac{p_1 E(t_1 - \tau) I_L(t_1 - \tau)}{g_1 + I_L(t_1 - \tau)} - \mu_2 E + s_1, \\
\frac{dT}{dt_1} &= r_2(1 - bT)T - \frac{aET}{g_2 + T}, \\
\frac{dI_L}{dt_1} &= \frac{p_2 ET}{g_3 + T} - \mu_3 I_L + s_2.
\end{aligned}$$

Using the following scaling [83]
$x = \frac{E}{E_0}$, $y = \frac{T}{T_0}$, $z = \frac{I_L}{I_{L_0}}$, $t = t_s t_1$; $\bar{c} = \frac{cT_0}{t_s E_0}$, $\bar{p_1} = \frac{p_1}{t_s}$,
$\bar{g_1} = \frac{g_1}{I_{L_0}}$, $\bar{\mu_2} = \frac{\mu_2}{t_s}$, $\bar{g_2} = \frac{g_2}{T_0}$, $\bar{b} = bT_0$, $, \bar{r_2} = \frac{r_2}{t_s}$, $\bar{a} = \frac{aE_0}{t_s T_0}$,
$\bar{\mu_3} = \frac{\mu_3}{t_s}$, $\bar{p_2} = \frac{p_2 E_0}{t_s I_{L_0}}$, $\bar{g_3} = \frac{g_3}{T_0}$, $\bar{s_1} = \frac{s_1}{t_s E_0}$, $\bar{s_2} = \frac{s_2}{t_s I_{L_0}}$,
the given system is non-dimensionalized, given by (after dropping the overbar notation for convenience)

$$\begin{aligned} \frac{dx}{dt} &= cy + \frac{p_1 x(t-\tau)z(t-\tau)}{g_1 + z(t-\tau)} - \mu_2 x + s_1 \\ \frac{dy}{dt} &= r_2(1-by)y - \frac{axy}{g_2+y} \\ \frac{dz}{dt} &= \frac{p_2 xy}{g_3+y} - \mu_3 z + s_2 \end{aligned} \tag{5.14}$$

subject to the following initial conditions

$$\begin{aligned} x(\theta) = \psi_1(\theta), y(\theta) = \psi_2(\theta), z(\theta) = \psi_3(\theta) \\ \psi_1(\theta) \geq 0, \psi_2(\theta) \geq 0, \psi_3(\theta) \geq 0; \theta \in [-\tau, 0] \\ \psi_1(0) > 0, \psi_2(0) > 0, \psi_3(0) > 0; \end{aligned} \tag{5.15}$$

where $C_+ = (\psi_1(\theta), \psi_2(\theta), \psi_3(\theta)) \in C([-\tau, 0], R^3_{+0})$, the Banach space of continuous functions mapping the interval $[-\tau, 0]$ into R^3_{+0}, where R^3_{+0} is defined as

$$\begin{aligned} R^3_{+0} &= ((x,y,z) : x,y,z \geq 0) \text{ and } R^3_+, \text{ the interior of } R^3_{+0} \text{ as} \\ R^3_+ &= ((x,y,z) : x,y,z > 0) \end{aligned}$$

In the system described by (5.14), $x(t), y(t)$, and $z(t)$, respectively, represent the effector cells, the tumor cells, and the concentration of IL-2 in the single site compartment. The first equation of the system (5.14) describes the rate of change for the effector cell population. The effector cells grow due to the direct presence of the tumor, given by the term cy, where c is the antigenicity of the tumor. It is also stimulated by IL-2 that is produced by effector cells in an autocrine and paracrine manner (the term $\frac{p_1 xz}{g_1+z}$, where p_1 is the rate at which effector cells grow, and g_1 is the half-saturation constant).

Clinical trials show that there are immune stimulation effects from treatment with IL-2 [135, 159], and there is a time lag between the production of IL-2 by activated T-cells and the effector cell stimulation from treatment with IL-2. Hence, a discrete time delay is being added to the second term of the first equation of the system (5.14), which modifies to $\frac{p_1 x(t-\tau)z(t-\tau)}{g_1+z(t-\tau)}$, where $\mu_2 x$ gives the natural decay of the effector cells and s_1 is the treatment term that represents the external source of the effector cells such as ACI. A similar type of term was introduced by Galach [39] in his model equation, where he

Parameters	Values	Scales Values
c (antigenicity of tumor)	$0 \leq c \leq 0.05$	$0 \leq c \leq 0.278$
p_1 (growth rate of effector cells)	0.1245	0.69167
g_1 (half saturation constant)	2×10^7	0.02
μ_2 (natural decay rate of effector cells)	0.03	0.1667
r_2 (growth rate of tumor cells)	0.18	1
b (1/carrying capacity of tumor cells)	1.0×10^{-9}	1
a (decay rate of tumor)	1	5.5556
g_2 (half saturation constant)	1×10^5	0.0001
μ_3 (natural decay rate of IL-2)	10	55.556
p_2 (growth rate of IL-2)	5	27.778
g_3 (half saturation constant)	1×10^3	0.000001

TABLE 5.1: **Parameter Values Used for Numerical Results.**

assumed that the source of the effector cells is the term $x(t-\tau)y(t-\tau)$, as the immune system needs some time to develop a suitable response.

The second equation of the system (5.14) shows the rate of change of the tumor cells, which follows logistic growth (a type of limiting growth). Due to tumor-effector cell interaction, there is a loss of tumor cells at the rate a and which is modeled by Michaelis Menten kinetics to indicate the limited immune response to the tumor (the term $\frac{axy}{g_2+y}$, where g_2 is a half-saturation constant). The third equation of the system (5.14) gives the rate of change for the concentration of IL-2. Its source is the effector cells that are stimulated by interaction with the tumor and also has Michaelis Menten kinetics to account for the self-limiting production of IL-2 (the term $\frac{p_2xy}{g_3+y}$, where p_2 is the rate of production of IL-2 and g_3 is a half-saturation constant), $\mu_3 z$ is the natural decay of the IL-2 concentration and s_2 is a treatment term that represents an external input of IL-2 into the system.

Proper scaling is needed as the system is numerically stiff, and numerical routines used to solve these equations will fail without scaling or inappropriate scaling (in this case, a proper choice of scaling is $E_0 = T_0 = I_{L_0} = 1/b$ and $t_s = r_2$ [83]). The parameter values have been obtained from [83], which is put in tabular form (Table 5.1). The units of the parameters are in day^{-1}, except for g_1, g_2, g_3, and b, which are in volumes.

The aim of this problem is to study this modified model and to explore any changes in the dynamics of the system that may occur when a discrete time delay has been added to the system and to compare with the results obtained by Kirschner and Panetta in [83].

5.4.3 Positivity of the Solution

The system of equations is now put in a vector form by setting

$$X = col(M, N, Z) \in R^3_{+0},$$

$$F(X) = \begin{pmatrix} F_1(X) \\ F_2(X) \\ F_3(X) \end{pmatrix} = \begin{pmatrix} cy + \frac{p_1 x(t-\tau) z(t-\tau)}{g_1 + z(t-\tau)} - \mu_2 x + s_1 \\ r_2(1-by)y - \frac{axy}{g_2+y} \\ \frac{p_2 xy}{g_3+y} - \mu_3 z + s_2 \end{pmatrix},$$

where $F : C_+ \to R^3_{+0}$ and $F \in C^\infty(R^3_{+0})$. Then system (5.14) becomes

$$\dot{X} = F(X_t), \tag{5.16}$$

where $\cdot \equiv d/dt$ and with $X_t(\theta) = X(t+\theta), \theta \in [-\tau, 0]$ [55]. It is easy to check in equation (5.16) that whenever we choose $X(\theta) \in C_+$ such that $X_i = 0$, then we obtain $F_i(X)|_{X_i(t)=0, X_t \in C_+} \geq 0, i = 1, 2, 3$. Due to the lemma in [166], any solution of equation (5.16) with $X(\theta) \in C_+$, say, $X(t) = X(t, X(0))$, is such that $X(t) \in R^3_{+0}$ for all $t > 0$.

5.4.4 Linear Stability Analysis with Delay

The equilibria for the system (scaled) are as follows: (i) The $x - z$ planar equilibrium is $(\frac{s_1(g_1\mu_3+s_2)}{\mu_2(g_1\mu_3+s_2)-p_1 s_2}, 0, \frac{s_2}{\mu_3})$ and exists if $\mu_2 > \frac{p_1 s_2}{g_1\mu_3+s_2}$.
(ii) The interior equilibrium is $E_*(x^*, y^*, z^*)$ where $x^* = \frac{r_2}{a}(1-by^*)(g_2+y^*)$, $z^* = (\frac{p_2 r_2(1-by^*)(g_2+y^*)}{a\mu_3(g_3+y^*)} + \frac{s_2}{\mu_3})$ and y^* is given by the equation $cy^* - \mu_2 x^* + \frac{p_1 x^* z^*}{g_1+z^*} + s_1 = 0$.

In the case of positive delay, the characteristic equation for the linearized equation around the point (x^*, y^*, z^*) is given by

$$P(\lambda) + Q(\lambda)e^{-\lambda\tau} = 0, \tag{5.17}$$

where $P(\lambda) = \lambda^3 + a_1\lambda^2 + a_2\lambda + a_3, \quad Q(\lambda) = b_1\lambda^2 + b_2\lambda + b_3,$

$$a_1 = \mu_2 + \mu_3 + br_2 y^* - \frac{ax^* y^*}{(g_2+y*)^2},$$

$$a_2 = br_2 y^*(\mu_3 - \frac{p_1 z^*}{g_1+z^*}) + \frac{ap_1 x^* y^* z^*}{(g_2+y^*)^2(g_1+z^*)} - \frac{\mu_3 p_1 z^*}{g_1+z^*}$$
$$+ \frac{acy^*}{(g_2+y^*)} + \mu_2\{\mu_3 + br_2 y^* - \frac{ax^* y^*}{(g_2+y^*)^2}\},$$

$$a_3 = b\mu_2\mu_3 r_2 y^* - \frac{a\mu_2\mu_3 x^* y^*}{(g_2+y*)^2} + \frac{ac\mu_3 y^*}{(g_2+y^*)} - \frac{bg_1 p_1 p_2 r_2 x^*(y^*)^2}{(g_3+y^*)(g_1+z^*)^2}$$
$$+ \frac{ag_1 g_3 p_1 p_2 (x^*)^2 y^*}{(g_2+y^*)(g_3+y^*)^2(g_1+z^*)^2} + \frac{ag_1 p_1 p_2 (x^*)^2(y^*)^2}{(g_2+y^*)^2(g_3+y^*)(g_1+z^*)^2}$$
$$+ \frac{a\mu_3 p_1 x^* y^* z^*}{(g_2+y^*)^2(g_1+z^*)} - \frac{b\mu_3 p_1 r_2 y^* z^*}{g_1+z^*} + \frac{bg_1 p_1 p_2 r_2 x^*(y^*)^2}{(g_3+y^*)(g_1+z^*)^2},$$

$$b_1 = -\frac{p_1 z^*}{g_1 + z^*}, \quad b_2 = -\frac{g_1 p_1 x^* y^*}{(g_3 + y^*)(g_1 + z^*)^2} < 0,$$
$$b_3 = \frac{a g_1 p_1 p_2 \{g_2 g_3 + 2 g_3 y^* + (y*)^2\} x^* y^*}{(g_2 + y^*)^2 (g_3 + y^*)^2 (g_1 + z^*)^2} - \frac{b r_2 g_1 p_1 p_2 x^* (y^*)^2}{(g_3 + y^*)(g_1 + z^*)^2}.$$

The steady state is stable in the absence of delay ($\tau = 0$) if the roots of

$$P(\lambda) + Q(\lambda) = 0 \Rightarrow \lambda^3 + (a_1 + b_1)\lambda^2 + (a_2 + b_2)\lambda + a_3 + b_3 = 0$$

have negative real parts. This occurs if and only if $a_1 + b_1 > 0$, $a_3 + b_3 > 0$ and $(a_1 + b_1)(a_2 + b_2) - (a_3 + b_3) > 0$ (by Routh Hurwitz's criteria). This implies

$$\begin{aligned}
\mu_2 + \mu_3 + b r_2 y^* - \frac{a x^* y^*}{(g_2 + y*)^2} - \frac{p_1 z^*}{g_1 + z^*} &> 0 \\
p_1 \left\{ \frac{g_1 p_2 x^* y^*}{\mu_3 (g_3 + y^*)(g_1 + z^*)^2} + \frac{z^*}{g_1 + z^*} \right\} &< \mu_2 < p_1 \{ \frac{z^*}{g_1 + z^*} \\
+ \frac{g_1 p_2 x^* (g_2 g_3 + 2 g_3 y^* + (y*)^2)}{\mu_3 (g_3 + y^*)^2 (g_1 + z^*)^2} \}. &
\end{aligned}$$

(The above criteria is satisfied with the set of parameters shown in Section 5.4.2, provided $0 \leq c \leq 0.278$, $s_2 < \frac{\mu_2 \mu_3 g_1}{p_1 - \mu_2}$.) Now substituting $\lambda = i\omega$ (where ω is positive) in equation (5.17) and separating the real and imaginary parts, we obtain the system of transcendental equations

$$a_1 \omega^2 - a_3 = (b_3 - b_1 \omega^2) cos(\omega\tau) + b_2 \omega sin(\omega\tau) \tag{5.18}$$
$$\omega^3 - a_2 \omega = b_2 \omega cos(\omega\tau) - (b_3 - b_1 \omega^2) sin(\omega\tau) \tag{5.19}$$

Squaring and adding (5.18) and (5.19) we get,

$$\begin{aligned}
(b_3 - b_1 \omega^2)^2 + b_2^2 \omega^2 &= (a_1 \omega^2 - a_3)^2 + (\omega^3 - a_2 \omega)^2 \\
\Rightarrow \rho^3 + A_1 \rho^2 + A_2 \rho + A_3 &= 0, \text{ where } \rho = \omega^2 \qquad (5.20)
\end{aligned}$$

where

$$\begin{aligned}
A_1 &= a_1^2 - 2a_2 - b_1^2 \\
&= \mu_2^2 + \mu_3^2 + b^2 r_2^2 (y^*)^2 + \frac{a^2 (x^*)^2 (y^*)^2}{(g_2 + y*)^4} - \frac{2 a y^* \{b r_2 x^* y^* + c(g_2 + y^*)\}}{(g_2 + y^*)^2} \\
&+ \frac{p_1^2 (z^*)^2}{(g_1 + z^*)^2} - \frac{2 \mu_2 p_1 z^*}{g_1 + z^*},
\end{aligned}$$

$$
\begin{aligned}
A_2 &= a_2^2 - b_2^2 - 2a_1a_3 + 2b_1b_3 \\
&= \frac{-g_1^2p_1^2p_2^2(x^*)^2(y^*)^2}{(g_3+y*)^2(g_1+z^*)^4} + \{\frac{-a\mu_3x^*y^*}{(g_2+y*)^2} - \frac{\mu_3p_1z^*}{g_1+z*} + \frac{acy^*}{g_2+y*} + \mu_2(\mu_3 \\
&+ br_2y^* - \frac{-ax^*y^*}{(g_2+y*)^2}) + \frac{ap_1x^*y^*z^*}{(g_2+y*)^2(g_1+z^*)} + br_2y^*(\mu_3 - \frac{p_1z^*}{g_1+z}))\}^2 \\
&- 2\mu_3y^*[\mu_2 + \mu_3 + br_2y^* - \frac{ax^*y^*}{(g_2+y*)^2} - \frac{p_1z^*}{g_1+z^*}]\ [br_2(\mu_2 - \frac{p_1z^*}{g_1+z^*} \\
&+ \frac{a\{-\mu_2x + c(g_2+y) + \frac{\mu_3p_1z^*}{g_1+z*}\}}{(g_2+y^*)^2}], \\
A_3 &= a_3^2 - b_3^2 = (a_3+b_3)(a_3-b_3) \\
&= y^*[br_2\{\mu_2\mu_3 - \frac{g_1p_1p_2x^*y^*}{(g_3+y*)(g_1+z^*)^2} - \frac{\mu_3p_1z^*}{g_1+z^*}\} + \frac{a}{(g_2+y^*)^2}\{c\mu_3 \\
&(g_2+y^*) + x^*(-\mu_2\mu_3 + \frac{p_1p_2g_1x^*(g_2g_3+2g_3y^*+(y^*)^2)}{(g_3+y*)^2(g_1+z*)^2} \\
&+ \frac{\mu_3p_1z^*}{g_1+z^*})\ \}\ ] \times y^*[br_2\{\mu_2\mu_3 + \frac{g_1p_1p_2x^*y^*}{(g_3+y*)(g_1+z^*)^2} - \frac{\mu_3p_1z^*}{g_1+z^*}\} \\
&+ \frac{a}{(g_2+y^*)^2}\{c\mu_3(g_2+y^*) + \frac{\mu_3p_1z^*}{g_1+z^*}) \\
&+ x^*(-\mu_2\mu_3 + \frac{-p_1p_2g_1x^*(g_2g_3+2g_3y^*+(y^*)^2)}{(g_3+y*)^2(g_1+z*)^2}\ \}\ ].
\end{aligned}
$$

Assuming A_1 to be positive (this is satisfied with the parameter values from Table 5.1), the simplest assumption that (5.20) will have a positive root is $A_3 = a_3^2 - b_3^2 < 0$. Since $(a_3 + b_3)$ is positive (from the non-delay case), we must have $(a_3 - b_3) < 0$ and this gives

$$
\begin{aligned}
y^*[br_2\{\mu_2\mu_3 + \frac{p_1(\frac{g_1p_2x^*y^*}{g_3+y^*} - \mu_3z^*(g_1+z^*))}{(g_1+z^*)^2}\}br_2y^* &- \frac{ax^*y^*}{(g_2+y*)^2} \\
&- \frac{p_1z^*}{g_1+z^*} > 0
\end{aligned}
$$

$$
\begin{aligned}
p_1\{\frac{g_1p_2x^*y^*}{\mu_3(g_3+y^*)(g_1+z^*)^2} &+ \frac{z^*}{g_1+z^*}\} < \mu_2 \\
&< p_1\{\frac{g_1p_2x^*(g_2g_3+2g_3y^*+(y*)^2)}{\mu_3(g_3+y^*)^2(g_1+z^*)^2} + \frac{z^*}{g_1+z^*}\}
\end{aligned}
$$

Hence, we can say that there is a positive ω_0 satisfying (5.20); that is, the characteristic equation (5.17) has a pair of purely imaginary roots of the form $\pm\ i\omega_0$. Eliminating $\sin(\tau\omega)$ from (5.18) and (5.19), we get,

$$
\cos(\omega\tau) = \frac{(a_1\omega^2 - a_3)(b_3) + (\omega^3 - a_2\omega)(b_2\omega)}{(b_3)^2 + (b_2\omega)^2}.
$$

Then τ_n^* corresponding to ω_0 is given by

$$\tau_n^* = \frac{1}{\omega_0}\arccos\left[\frac{(a_1\omega_0^2 - a_3)(b_3) + (\omega_0^3 - a_2\omega_0)(b_2\omega_0)}{(b_3)^2 + (b_2\omega_0)^2}\right] + \frac{2n\pi}{\omega_0} \quad (5.21)$$

For $\tau = 0$, E_* is stable. Hence, E_* will remain stable for $\tau < \tau_0^*$ $(n = 0)$ [38].

5.4.5 Delay Length Estimation to Preserve Stability

The linearized form of the system (5.14) is

$$\begin{aligned}
\frac{dx}{dt} &= \left(\frac{p_1 z^*}{g_1 + z^*} - \mu_2\right)x + \frac{p_1 z^*}{g_1 + z^*}x(t-\tau) + cy + \frac{p_1 g_1 x^*}{(g_1 + z^*)^2}z(t-\tau)\\
\frac{dy}{dt} &= -\frac{ay^*}{g_2 + y^*}x + \left(\frac{ax^*y^*}{(g_2 + y^*)^2} - r_2 b y^*\right)y\\
\frac{dz}{dt} &= -\frac{p_2 y^*}{g_3 + y^*}x + \frac{p_2 g_3 x^*}{(g_3 + y^*)^2}y - \mu_3 z
\end{aligned}$$

Taking the Laplace transform of the above linearized system we get,

$$\begin{aligned}
\left(s + \mu_2 - \frac{p_1 z^*}{g_1 + z^*}\right)\bar{x}(s) &= \frac{p_1 z^*}{g_1 + z^*}e^{-s\tau}\bar{x}(s) + \frac{p_1 z^*}{g_1 + z^*}e^{-s\tau}K_1(s)\\
&+ c\bar{y}(s) + \frac{p_1 g_1 x^*}{(g_1 + z^*)^2}e^{-s\tau}\bar{z}(s)\\
&+ \frac{p_1 g_1 x^*}{(g_1 + z^*)^2}e^{-s\tau}K_2(s) + x(0),\\
\left(s + r_2 b y^* - \frac{ax^*y^*}{(g_2 + y^*)^2}\right)\bar{y}(s) &= -\frac{ay^*}{g_2 + y^*}\bar{x}(s) + y(0) +\\
(s + \mu_3 z) &= -\frac{p_2 y^*}{g_3 + y^*}\bar{x}(s) + \frac{p_2 g_3 x^*}{(g_3 + y^*)^2}\bar{y}(s) + z(0),
\end{aligned}$$

where

$$K_1(s) = \int_{-\tau}^{0} e^{-st}x(t)dt, \quad K_2(s) = \int_{-\tau}^{0} e^{-st}z(t)dt,$$

and $\bar{x}(s)$, $\bar{y}(s)$ and $\bar{z}(s)$ are the Laplace transforms of $x(t), y(t)$ and $z(t)$, respectively.

Following the lines of [37] and using the Nyquist criterion, it can be shown that the conditions for the local asymptotic stability of $E_*(x^*, y^*, z^*)$ are given by

$$Im\ \ H(i\ \eta_0)\ \ >\ \ 0, \tag{5.22}$$
$$Re\ \ H(i\ \eta_0)\ \ =\ \ 0, \tag{5.23}$$

where, $H(s) = s^3 + a_1 s^2 + a_2 s + a_3 + e^{-s\tau}(b_1 s^2 + b_2 s + b_3)$ and η_0 is the smallest positive root of (5.23). In this case, (5.22) and (5.23) gives

$$a_2\eta_0 - \eta_0^3 \ > \ -b_2\eta_0\cos(\eta_0\tau) + b_3\sin(\eta_0\tau) - b_1\eta_0^2\sin(\eta_0\tau), \tag{5.24}$$
$$a_3 - a_1\eta_0^2 \ = \ b_1\eta_0^2\cos(\eta_0\tau) - b_3\cos(\eta_0\tau) - b_2\eta_0\sin(\eta_0\tau). \tag{5.25}$$

Now, equations (5.24) and (5.25), if satisfied simultaneously, are sufficient conditions to guarantee stability, which are now used to get an estimate on the length of time delay. The aim is to find an upper bound η_+ on η_0, independent of τ and then to estimate τ so that (5.24) holds true for all values of η, $0 \le \eta \le \eta_+$ and hence in particular at $\eta = \eta_0$.
(5.25) is rewritten as

$$a_1\eta_0^2 \ = \ a_3 + b_3\cos(\eta_0\tau) - b_1\eta_0^2\cos(\eta_0\tau) + b_2\eta_0\sin(\eta_0\tau). \tag{5.26}$$

Maximizing $a_3 + b_3\cos(\eta_0\tau) - b_1\eta_0^2\cos(\eta_0\tau) + b_2\eta_0\sin(\eta_0\tau)$, subject to $|sin(\eta_0\tau)| \le 1$, $|cos(\eta_0\tau)| \le 1$, we obtain,

$$|a_1|\eta_0^2 \ \le \ |a_3| + |b_3| + |b_1|\eta_0^2 + |b_2|\eta_0. \tag{5.27}$$

Hence, if

$$\eta_+ \ = \ \frac{1}{2(|a_1| - |b_1|)}[\ |b_2| + \sqrt{b_2^2 + 4(|a_1| - |b_1|)(|a_3| + |b_3|)}\ ], \tag{5.28}$$

then clearly from (5.27), we have $\eta_0 \le \eta_+$.

From the inequality (5.24), we obtain

$$\eta_0^2 \ < \ a_2 + b_2\cos(\eta_0\tau) + b_1\eta_0\sin(\eta_0\tau) - \frac{b_3\sin(\eta_0\tau)}{\eta_0}. \tag{5.29}$$

As $E_*(x^*, y^*, z^*)$ is locally asymptotically stable for $\tau = 0$, therefore, for sufficiently small $\tau > 0$, inequality (5.29) will continue to hold. Substituting (5.26) in (5.29) and rearranging we get,

$$(b_3 - b_1\eta_0^2 - a_1b_2)[\cos(\eta_0\tau) - 1] + \{(b_2 - a_1b_1)\eta_0 + \frac{a_1b_3}{\eta_0}\}\sin(\eta_0\tau)$$
$$< a_1a_2 - a_3 - b_3 + b_1\eta_0^2 + a_1b_2. \tag{5.30}$$

Using the bounds,

$$\begin{aligned}(b_3 - b_1\eta_0^2 - a_1b_2)[\cos(\eta_0\tau) - 1] &= (b_1\eta_0^2 + a_1b_2 - b_3)2\sin^2(\frac{\eta_0\tau}{2}) \\ &\leqslant \frac{1}{2}|(b_1\eta_+^2 + a_1b_2 - b_3)|\eta_+^2\tau^2,\end{aligned}$$

and

$$\{(b_2 - a_1b_1)\eta_0 + \frac{a_1b_3}{\eta_0}\}\sin(\eta_0\tau) \leqslant \{|(b_2 - a_1b_1)|\ \eta_+^2 + |a_1||b_3|\}\ \tau,$$

we obtain from (5.30),

$$L_1\tau^2 + L_2\tau < L_3, \tag{5.31}$$

where,

$$\begin{aligned}L_1 &= \frac{1}{2}|(b_1\eta_+^2 + a_1b_2 - b_3)|\eta_+^2, \\ L_2 &= \{|(b_2 - a_1b_1)|\ \eta_+^2 + |a_1||b_3|\}, \\ L_3 &= a_1a_2 - a_3 - b_3 + b_1^2\eta_+ + a_1b_2\end{aligned}$$

Hence, if

$$\tau_+ = \frac{1}{2L_1}(-L_2 + \sqrt{L_2^2 + 4L_1L_3}), \tag{5.32}$$

then for $0 \leq \tau < \tau_+$, the Nyquist criterion holds true and τ_+ estimates the maximum length of delay preserving the stability.

5.4.6 Numerical Results

The model is now studied numerically to see the effect of discrete time delay on the system. The scaled parameter values have been used for numerical calculations using MATLAB.

Case 1 ($s_1 > 0$, $s_2 = 0$): In the model, the time delay has no qualitative effect on adoptive cellular immunotherapy (ACI). Therefore, the results will be the same as obtained in [83]. So is the case $s_1 = 0$, $s_2 = 0$. Hence, these two cases are not discussed thoroughly.

Case 2 ($s_1 = 0$, $s_2 > 0$): Fig. 5.7 explores the input of concentration IL-2 into the system, if the input of concentration of IL-2 is administered and the effector cells are stimulated after 0.7227 days = 17.346 hours and 0.529 days = 12.7 hours, respectively (obtained by using (5.21) and scaled parameter values). For a low antigenic tumor and low input of concentration of IL-2

($c = 0.0056$, $s_2 = 0.05$), the tumor cell regresses and the concentration of IL-2 decreases alarmingly to almost zero (fig. 5.7(A)). For a higher concentration of IL-2 ($c = 0.0056$, $s_2 = 0.2$), the same scenario happens, only in this case, the concentration of IL-2 does not reduce to zero (fig. 5.7(B)).

For tumors with high antigenicity ($c = 0.222$, $s_2 = 0.05$), the volume of the tumor increases in the beginning, and when there is an input of IL-2 concentration after 12.7 hours, the tumor volume reduces and ultimately is cleared off (fig. 5.7(C)); at the same time the concentration of IL-2 decreases alarmingly. But with high input of IL-2 on tumors with high antigenicity (c=0.222,s_2=0.2), the tumor regresses as well as both the immune system and the concentration of IL-2 stabilizes (fig. 5.7(D)). This is a new and interesting positive result. According to [83], large amounts of administrated IL-2 together with any degree of antigenicity shows that the tumor is cleared, but the immune system grows unbounded as the IL-2 concentration reaches a steady-state value (fig. 5.8). This uncontrolled growth of the immune system represents a situation that is detrimental to the host. However, in our case, due to the time delay effect, the situation is under control. The tumor is cleared off, and the immune system stabilizes.

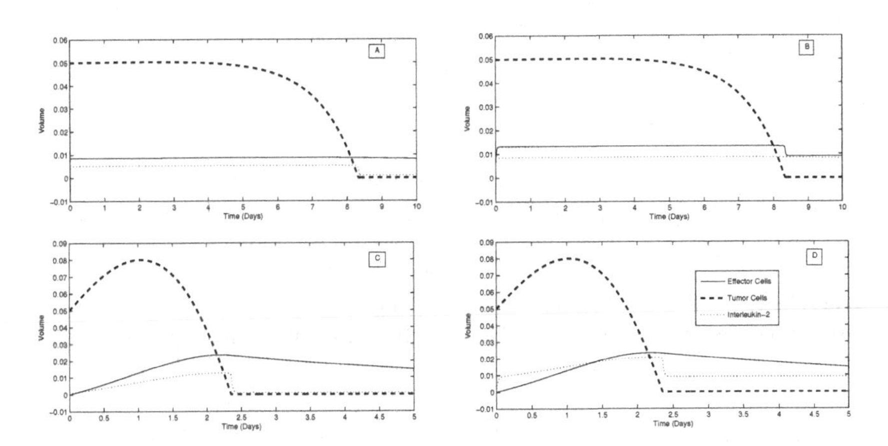

FIGURE 5.7: *Effector cells, tumor cells and IL-2 vs. time. All the parameter values have been scaled accordingly. A: c=0.0056, $s_1 = 0$, $s_2 = 0.05$; B: c=0.0056, $s_1 = 0$, $s_2 = 0.2$; C: c=0.222, $s_1 = 0$, $s_2 = 0.05$; D: c=0.222, $s_1 = 0$, $s_2 = 0.2$; $\tau = \tau_0^*$=0.723 days =17.346 hours for cases A and B; $\tau = \tau_0^*$=0.529 days =12.7 hours for cases C and D.*

Case 3 ($s_1 > 0$, $s_2 > 0$)**:** Fig. 5.9 shows the effect of immunotherapy with both ACI and IL-2, if the input of both ACI and the concentration of IL-2 is administered and the stimulation of effector cells by IL-2 takes place after 0.7228 days = 17.348 hours and 0.528 days = 12.67 hours, respectively. Irrespective of the antigenicity of the tumor, the dynamics of all the figures (figs. 5.9 (A), (B), (C), (D)) are the same, that is, the volume of the tumor

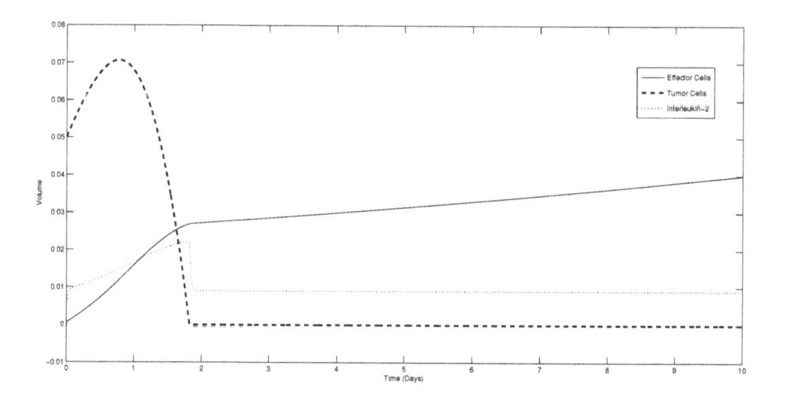

FIGURE 5.8: *Effector cells, tumor cells, and IL-2 vs. time in the case of non-delay (that is, $\tau = \tau_0^* = 0$). All the parameter values have been scaled accordingly. Here c=0.222, $s_1 = 0$, $s_2 = 0.5$.*

decreases significantly when both ACI and IL-2 are administered in various concentrations.

5.4.7 Conclusion

The aim of this chapter is to see the effect of time delay during immunotherapy with interleukin-2 (IL-2). The effect of immunotherapy with IL-2 on the modified model has been explored, and under what circumstances the tumor can be eliminated is described. The model represented by a set of delay differential equations contains treatment terms s_1 and s_2, that represent the external source of the effector cells by adoptive cellular immunotherapy (ACI) and external input of IL-2 into the system, respectively. However, the effects of IL-2 on tumor immune dynamics with time delay is the main focus. It is shown that treatment with IL-2 alone can offer a satisfactory outcome.

When there is an external input of concentration of IL-2, and the effector cells are being stimulated after 96.38 hours, during which IL-2 production reaches its peak value to generate more effector cells, tumors with medium to high antigenicity show regression, and the concentration of IL-2 stabilizes. Unlike in [83], the immune system also stabilizes, indicating that side effects such as capillary leak syndrome do not arise here. In other words, a patient need not endure many side effects before IL-2 therapy will successfully clear the tumor. It is found in a study by S.A. Rosenberg et al. [134] on the effectiveness of high-dose bolus treatment with IL-2, that many patients are in complete remission for 7 to 91 months. Hence, this model predicts that it is indeed possible to render a patient cancer free with immunotherapy with IL-2 alone.

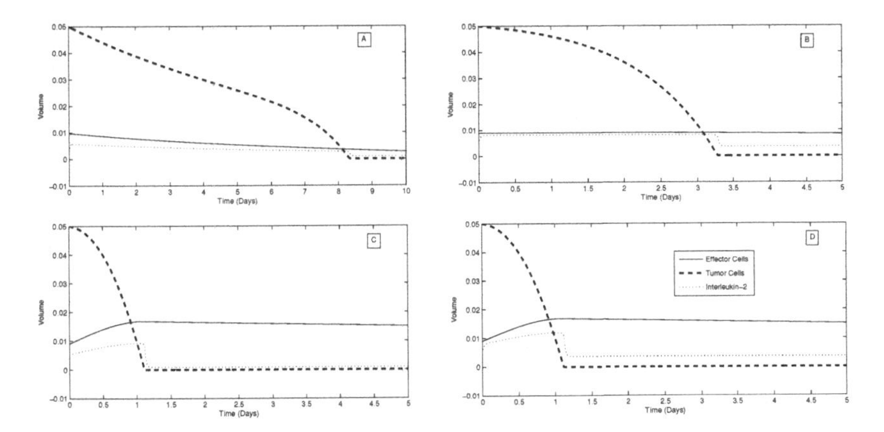

FIGURE 5.9: *Effector cells, tumor cells, and IL-2 vs. time. All the parameter values have been scaled accordingly. A: c=0.0056,* $s_1 = 0.00000246446$, $s_2 = 0.05$; *B: c=0.0056,* $s_1 = 0.0000010144$, $s_2 = 0.2$; *C: c=0.222,* $s_1 = 0.00000246446$, $s_2 = 0.05$; *D: c=0.222,* $s_1 = 0.0000010144$, $s_2 = 0.2$; $\tau = \tau_0^* = 0.7228$ *days =17.348 hours for cases A and B;* $\tau = \tau_0^* = 0.528$ *days =12.67 hours for cases C and D.*

Finally, the above findings shed some light on immunotherapy with IL-2 and can be helpful to medical practitioners, experimental scientists, and others to control this killer disease cancer. Extension along this line of work will be to examine the effects of other cytokines such as IL-10, IL-12, and Interferon-γ, which are involved in the cellular dynamics of the immune system response to tumor invasion, and to study how these cytokines affect the dynamics of the system.

5.5 Miscellaneous Examples

Problem 5.5.1 ***Mackey-Glass Model:*** *Mackey and Glass [46] proposed a delayed model for the growth of density of blood cells*

$$\frac{dB}{dt} = \frac{\lambda \alpha^m B(t-\tau)}{\alpha^m + B^m(t-\tau)} - \beta B(t),$$

where λ, α, m, β *and* τ *are positive constants.*
(i) Explain the model in terms of the system parameters.
(ii) Solve the system numerically by taking $\lambda = 0.2, \alpha = 0.1, \beta = 0.1, m = 10, \tau = 4$ *and initial history* 0.1 *and represent it graphically. Interpret the*

result. What changes do you observe in the dynamics of the system for $\tau = 6$ and $\tau = 20$?

Solution: (i) $B(t)$ is the density of blood cells in the circulating blood, λ is the rate at which it is produced, α is the dissociation constant, and β is the rate of natural death of the cells. There is a release of mature cells from the bone marrow into the blood due to the reduction in cells in the bloodstream, but there is a delay of approximately 6 days. Here, τ is the delay between the production of blood cells in the bone marrow and its release into the bloodstream. It is also assumed that the density of blood cells that enter the bloodstream depends on cell density at an earlier time $\mathrm{B}(t-\tau)$.

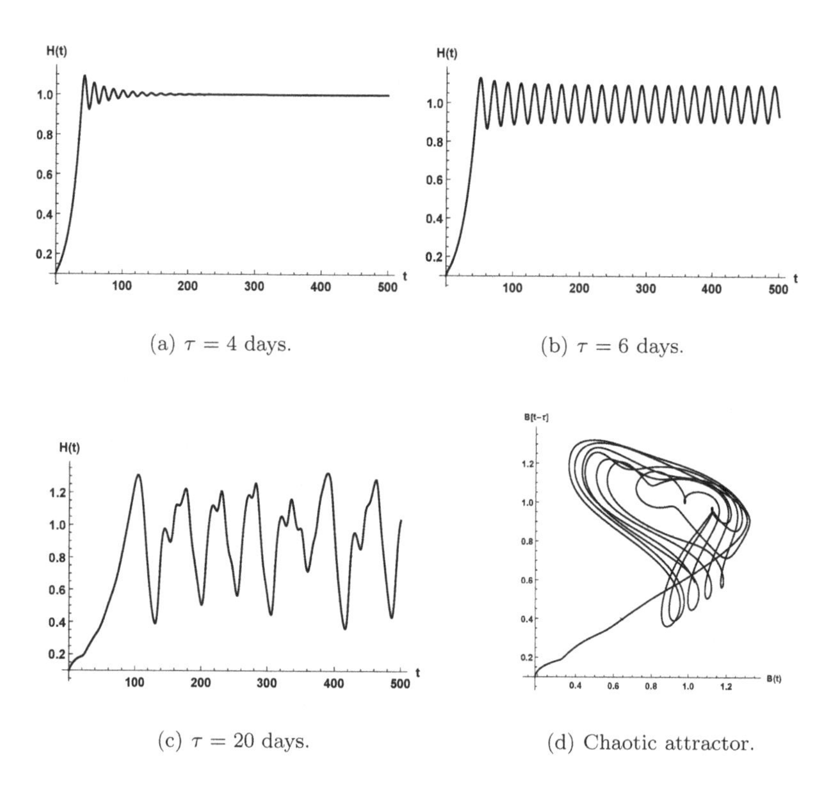

(a) $\tau = 4$ days.

(b) $\tau = 6$ days.

(c) $\tau = 20$ days.

(d) Chaotic attractor.

FIGURE 5.10: *Dynamics of the density of blood cells with parameter values $\lambda = 0.2, \alpha = 1.0, \beta = 0.1, m = 10$ and initial history 0.1, obtained from [142]. (a) Damped oscillation, (b) periodic pattern, (c) chaotic pattern, (d) chaotic attractor.*

(ii) The numerical solution of the model shows that the density of blood cells follows a damped oscillation and tends toward equilibrium for $\tau = 4$ days (fig. 5.10(a)). As the value of τ (=6 days) is increased, the system exhibits a periodic pattern, which can be seen as modeling a healthy variation of blood cell density (fig. 5.10(b)). The pattern becomes chaotic for $\tau = 20$ days, which can be seen as pathological blood cell density variation (fig. 5.10(c)). Fig. 5.10(d) shows the Mackey-Glass attractor for $\tau = 20$ days.

Problem 5.5.2 *A second-order delayed feedback system is given by [153]*

$$\frac{d^2x}{dt^2} + \frac{dx}{dt} + 3x = -\sqrt{5}\ \tanh[x(t-\tau)].$$

Simulate the model numerically for $\tau = 0.4, 0.5555,$ *and* 0.7 *and comment on the dynamics of the system.*

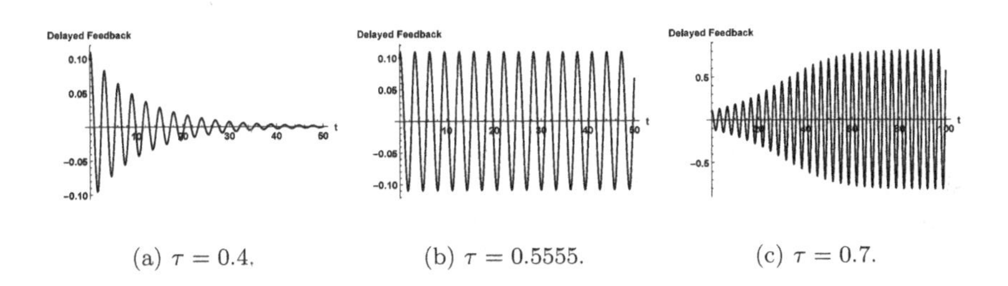

(a) $\tau = 0.4$. (b) $\tau = 0.5555$. (c) $\tau = 0.7$.

FIGURE 5.11: *The figures show the effect of delays on the feedback system with initial history (0.1,0.1). (a) Asymptotically stable, (b) periodic solution (c) periodic solution with high amplitude.*

Solution: The second-order delayed feedback system can be expressed as a system of first-order delayed differential equations as

$$\frac{dx}{dt} = y_1, \quad \frac{dy_1}{dt} = -y_1 - 3x - \sqrt{5}\ \tanh[x(t-\tau)],$$

The system is solved numerically for different values of τ. For $\tau = 0.4$, the system is asymptotically stable (fig. 5.11(a)). For $\tau = 0.5555$, the system bifurcates to periodic solution (fig. 5.11(b)) and for $\tau = 0.7$, the system is in oscillatory mode with high amplitude of oscillation (fig. 5.11(c)).

Problem 5.5.3 *Cooke proposed an epidemic model [96, 102] given by*

$$\frac{dy}{dt} = by(t-7)[1-y(t)] - cy(t),$$

where $y(t)$ denotes the fraction of a population infected at time t.
(i) Find the equilibrium point(s) of the model and comment on their existence.
(ii) Investigate the stability of the system about the equilibrium point(s).
(iii) Simulate the model numerically for $b = 2$, $c = 1$, *initial history* $= 0.8$ *and comment on the dynamics of the system.*

Solution: (i) The equilibrium points are $y^* = 0$ and $y^* = \frac{b-c}{b}$. For the existence of non-zero equilibrium points, we must have $b > c$.

(ii) Putting $y = Y + y^*$ in the model equation and retaining the linear terms only we get,

$$\frac{dy}{dt} = b(1-y^*)y(t-7) - (by^* + c)y(t). \text{ For } y^* = 0, \frac{dy}{dt} = -cy(t) + by(t-7).$$

Using Theorem 5.2.1.1, we get $A+B = b-c > 0$ (or else non-zero equilibrium points do not exist). Hence, the system is unstable about the equilibrium point $y^* = 0$.

Note: One can also conclude if $b > c$, the system is unstable and if $b < c$, the system is asymptotically stable about $y^* = 0$. However, the system will have only one positive equilibrium point in that case.

$$\text{For } y^* = \frac{b-c}{b}, \frac{dy}{dt} = -by(t) + cy(t-7).$$

Using Theorem 5.2.1.1, we get $A + B = -(b-c) < 0$. The system is asymptotically stable about the equilibrium point $y^* = \frac{b-c}{b}$ if $b > c$. However, if $b < c$, then there exists $\tau^* > 0$ such that $x = 0$ is asymptotically stable for $0 < \tau < \tau^*$ and unstable for $\tau > \tau^*$.

(iii) Fig. 5.12(a) shows the simulation of the model for $b = 2$, $c = 1$ and initial history $= 0.8$. The numerical solution agrees with the theory, and the graph shows that the system is stable and reaches the equilibrium solution $y^* = \frac{b-c}{b} = 0.5$.

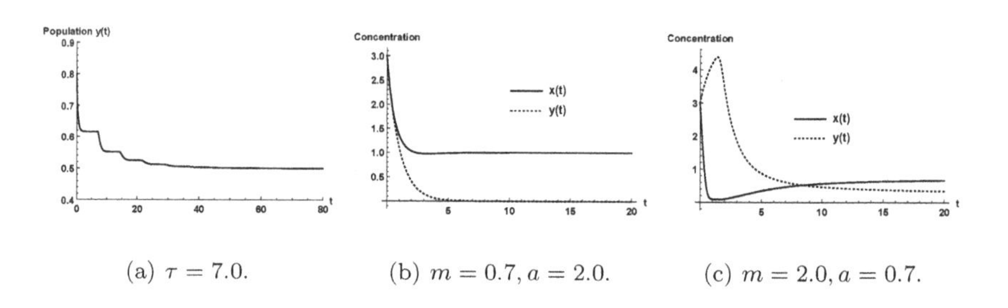

(a) $\tau = 7.0$. (b) $m = 0.7, a = 2.0$. (c) $m = 2.0, a = 0.7$.

FIGURE 5.12: *(a) The behavior of the population with time for* $b = 2, c = 1$, *and initial history* 0.8. *(b) Washout state. (c) Survival state.*

Problem 5.5.4 ***Microbial Growth Model with Delay:*** *A delayed bacterial growth model in a chemostat is given by (after scaling time and the dependent variables) [33, 153]*

$$\frac{dS(t)}{dt} = 1 - S(t) - \frac{mS(t)B(t)}{a+S(t)}, \quad \frac{dB(t)}{dt} = \frac{mS(t-\tau)}{a+S(t-\tau)}B(t-\tau) - B(t),$$

where $S(t)$ is the substrate concentration (food for bacteria) and $B(t)$ is the biomass concentration of bacteria, m and a are positive constants, and τ is the time of cellular absorption resulting in the increase of bacterial biomass.
(i) Find the equilibrium point(s) of the model and comment on their existence.
(ii) Investigate the stability of the system about the equilibrium point(s).
(iii) Simulate the model numerically for $m = 0.7$, $a = 2$, $\tau = 1.1$, initial history $= [3, 3]$ and comment on the dynamics of the system. What changes do you observe for $m = 2.0$, $a = 0.7$?

Solution: (i) The equilibrium solutions (S^*, B^*) are obtained by solving

$$1 - S^* - \frac{mS^*B^*}{a + S^*} = 0, \quad \frac{mS^*}{a + S^*}B^* - B^* = 0,$$

where (S^*, B^*) are (i) $(1, 0)$, washout state and (ii) $\left(\frac{a}{m-1}, \frac{m-1-a}{m-1}\right)$, survival state. The washout state always exists, and the survival state exists if $m > 1 + a$.
(ii) For linear stability analysis, the characteristic equation is

$$\begin{vmatrix} -1 + \frac{amB^*}{(a+S^*)^2} - \lambda & \frac{-mS^*}{(a+S^*)} \\ \frac{amB^*}{(a+S^*)^2}e^{-\lambda\tau} & -1 + \frac{mS^*}{a+S^*}e^{-\lambda\tau} - \lambda \end{vmatrix} = 0.$$

For the washout state $(1, 0)$, the characteristic equation is given by

$$\begin{vmatrix} -1 - \lambda & \frac{-m}{1+a} \\ 0 & -1 + \frac{m}{a+1}e^{-\lambda\tau} - \lambda \end{vmatrix} = 0.$$

This is a transcendental equation, which has infinitely many eigenvalues. One of the eigenvalues is $\lambda = -1$. Rest of them are given by $\lambda = -1 + \frac{m}{a+1}e^{-\tau\lambda}$. Using Theorem 5.2.1.1, we get

$$A = -1, \; B = \frac{m}{a+1} \Rightarrow A + B = -1 + \frac{m}{a+1} = \frac{m - a - 1}{a + 1}.$$

Therefore, the washout state is unstable if $m > a + 1$ $(A + B > 0)$ and stable if $m < a + 1$ (please note that the survival state exists if the washout state is unstable).

In a similar manner, for the survival state, the characteristic equation is of the form

$$P(\lambda) + Q(\lambda)e^{-\lambda\tau} = 0, \text{ where}$$
$$P(\lambda) = 4\lambda^2 + \left(\frac{2m}{a} + \frac{2}{am} + \frac{2}{m} - \frac{4}{a} + 2\right)\lambda + \left(\frac{m}{a} + \frac{1}{am} + \frac{1}{m} - \frac{2}{a}\right),$$
$$Q(\lambda) = -(2\lambda + 1).$$

Using Theorem 5.2.1.3 and parameter values $m = 2$, $a = 0.7$, $\tau = 0.1$ and

initial history $[0.3, 0.3]$, it can be easily shown that the survival state (S^*, B^*) is always stable (absolute stability), as stability holds for every value of the delay.
(iii) The model is solved numerically, and graphs are obtained. Fig. 5.12(b) shows the stability of the washout state and fig. 5.12(c) shows the stability of the survival state.

Problem 5.5.5 *The delayed Lotka-Volterra competition system is given by*

$$\frac{dx}{dt} = x(t)\left[2 - \alpha x(t) - \beta y(t-\tau)\right], \quad \frac{dy}{dt} = y(t)\left[2 - \gamma x(t-\tau) - \delta y(t)\right].$$

(i) Obtain the steady-state solution(s).
(ii) Investigate the stability of the non-zero steady state(s) for (a) $\alpha = \delta = 2$, $\beta = \gamma = 1$; (b) $\alpha = \delta = 1$, $\beta = \gamma = 2$.
(iii) Simulate the model for (a) $\alpha = \delta = 2$, $\beta = \gamma = 1$; (b) $\alpha = \delta = 1$, $\beta = \gamma = 2$ and comment on the dynamics of the system.

Solution:(i) The steady-state solution(s) (x^*, y^*) are the solutions of

$$x^*(2 - \alpha x^* - \beta y^*) = 0, \; y^*(2 - \gamma x^* - \delta y^*) = 0,$$

which give the following: $(0,0)$; $\left(\frac{2}{\alpha}, 0\right)$; $\left(0, \frac{2}{\delta}\right)$; $\left(\frac{2(\delta-\beta)}{\alpha\delta-\beta\gamma}, \frac{2(\alpha-\gamma)}{\alpha\delta-\beta\gamma}\right)$. For both the species to exist, we must have $\alpha > \gamma$ and $\delta > \beta$. For linear stability analysis, the characteristic equation is

$$\begin{vmatrix} 2 - 2\alpha x^* - \beta y^* - \lambda & -\beta x^* e^{-\lambda\tau} \\ -\gamma y^* e^{-\lambda\tau} & 2 - \gamma x^* - 2\delta y^* - \lambda \end{vmatrix} = 0.$$

(iia) For $\alpha = \delta = 2, \beta = \gamma = 1$, the non-zero steady-state solution is $\left(\frac{2}{3}, \frac{2}{3}\right)$ and the corresponding characteristic equation is of the form

$$P(\lambda) + Q(\lambda)e^{-\lambda\tau} = 0, \text{ where } P(\lambda) = \left(\lambda + \frac{4}{3}\right)\left(\lambda + \frac{4}{3}\right) \text{ and } Q(\lambda) = -\frac{4}{3}.$$

Clearly, the leading coefficient of $P(\lambda)$ is one and $Q(\lambda) = -\frac{4}{3}$ is a constant. Moreover, all the roots of $P(\lambda)$ are real and negative and $|P(0)| = \frac{16}{3} > |Q(\lambda)| = \frac{4}{3}$. Therefore, by the corollary of Theorem 5.2.1.3, we conclude that $Re(\lambda) < 0$ for every root λ and all $\tau \geq 0$. Therefore, the given system is asymptotically stable.

(iib) For $\alpha = \delta = 1, \beta = \gamma = 2$, the non-zero steady-state solution is $\left(\frac{2}{3}, \frac{2}{3}\right)$ and the corresponding characteristic equation is of the form

$$P(\lambda) + Q(\lambda)e^{-\lambda\tau} = 0, \text{ where } P(\lambda) = \left(\lambda + \frac{2}{3}\right)\left(\lambda + \frac{2}{3}\right) \text{ and } Q(\lambda) = -\frac{16}{3}.$$

Clearly, the leading coefficient of $P(\lambda)$ is one and $Q(\lambda) = -\frac{16}{3}$ is a constant.

Moreover, all the roots of $P(\lambda)$ are real and negative but $|P(0)| = \frac{4}{3} \not> |Q(\lambda)| = \frac{16}{3}$. Therefore, one of the criteria of the corollary of Theorem 5.2.1.3 is violated, and we conclude that the given system is unstable.

(iii) The model is solved numerically and plotted. Fig. 5.13(a) shows the dynamics of the competition model where both the species coexists ($\alpha = 2$, $\beta = 1$, $\gamma = 1$, $\delta = 2$) for $\tau = 0.7$ units and the first species goes to extinction ($\alpha = 1$, $\beta = 2$, $\gamma = 2$, $\delta = 1$) for $\tau = 3.0$ units (fig. 5.13(b)). Initial history is $(0.3, 1.0)$ units for both the cases.

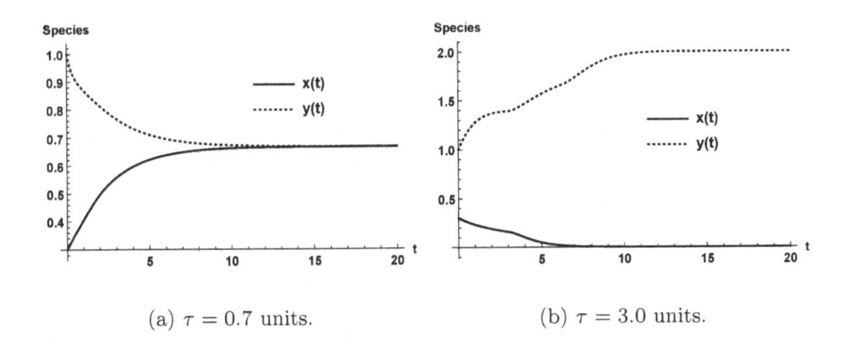

(a) $\tau = 0.7$ units.

(b) $\tau = 3.0$ units.

FIGURE 5.13: *The figures show the dynamics of the delayed Lotka-Volterra competition system. (a) Co-existence, (b) Extinction of one species.*

Problem 5.5.6 ***Delayed Gene Regulatory System:*** *A negative feedback gene regulatory system is described by a system of DDEs as [153]*

$$\frac{dR(t)}{dt} = \frac{g_m}{1 + \left(\frac{P(t-\tau)}{k}\right)^n} - \alpha_1 R(t), \quad \frac{dP(t)}{dt} = R(t) - \alpha_2 P(t),$$

where $R(t)$ and $P(t)$ represent the intracellular mRNA and the protein product of the gene, respectively. The growth of mRNA follows a Hill function; α_1 and α_2 are the rates of degradation of mRNA and the protein product of the gene, respectively. The discrete time delay τ is the time taken by mRNA to leave the nucleus, undergo protein synthesis in the ribosome, whereupon the protein re-enters the nucleus and suppresses its own mRNA production. Solve the system numerically by taking $g_m = 1$, $k = 0.5$, $n = 3$, $\alpha_1 = 1$, $\alpha_2 = 1$, $\tau = 2.0$ and represent it graphically. What changes do you observe in the dynamics of the system for $\tau = 3.5, 10.5$?

Solution: (i) The model equation is solved numerically, and the graphical representation is shown in fig. 5.5. Both the state variables $R(t)$ and $P(t)$ converge to equilibrium solution for $\tau = 2.0$ (fig. 5.14(a)). For $\tau = 3.5$, we observe sustained oscillations (fig. 5.14(b)). With a further increase of τ (= 10.5), the amplitude of the oscillation increases (fig. 5.14(c)).

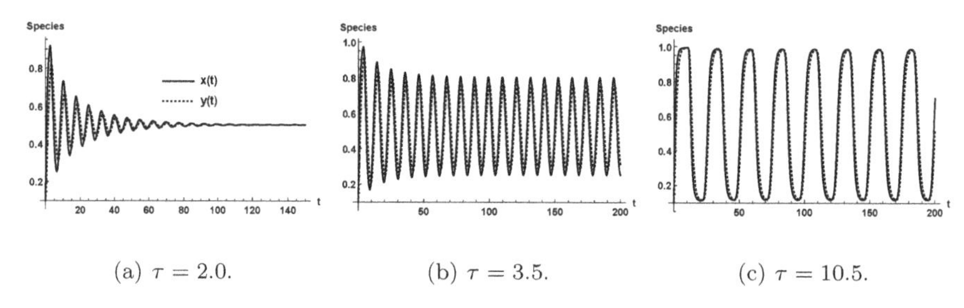

(a) $\tau = 2.0$. (b) $\tau = 3.5$. (c) $\tau = 10.5$.

FIGURE 5.14: *The figures show (a) asymptotically stable system, (b) sustained oscillation, (c) sustained oscillation with high amplitude. Parameter values:* $g_m = 1$, $k = 0.5$, $n = 3$, $\alpha_1 = 1$, $\alpha_2 = 1$ *and initial history* $[0.1, 0.1]$.

Problem 5.5.7 ***Cargo Pendulation Reduction Problem:*** *Bridge cranes are used in shipyards and warehouses to lift several hundred tons of containers and move them to another place. A delay-induced mathematical model for safe control of the crane pendulation is given by [57, 107]*

$$\frac{d^2\theta}{dt^2} + \epsilon\frac{d\theta}{dt} + \sin\theta = -k\cos\theta\ [\theta(t-\tau) - \theta],$$

where θ *represents the angle due to pendulation. The aim is to stabilize the system by reducing* θ *significantly at the end of the motion for safe operation.*
(i) Show that for small θ *and small* τ*, the system reduces to*

$$\frac{d^2\theta}{dt^2} + (\epsilon - k\ \tau)\frac{d\theta}{dt} + \theta = 0.$$

(ii) Solve the system numerically for $\tau = 12$, $\epsilon = 0.1$, $k = -0.15$, *initial history* $\theta'(0) = 0$, $\theta(0) = 1$ *for* $-\tau < t \leq 0$ *and comment on the result.*
(iii) How do the dynamics change if $\theta'(0) = 0$ *and* $\theta = 1.5$ *for* $-\tau < t \leq 0$*?*

Solution: (i) Clearly, $\theta = 0$ is an equilibrium point. Near the steady-state solution $\theta = 0, \sin(\theta) \approx \theta, \cos(\theta) \approx 1$ and the model becomes

$$\frac{d^2\theta}{dt^2} + \epsilon\frac{d\theta}{dt} + \theta = -k[\theta(t-\tau) - \theta)]. \tag{5.33}$$

Now, for small τ, $\theta(t-\tau) \approx \theta - \tau\frac{d\theta}{dt}$, and (5.33) reduces to

$$\frac{d^2\theta}{dt^2} + \epsilon\frac{d\theta}{dt} + \theta = -k[\theta - \tau\frac{d\theta}{dt} - \theta)] \Rightarrow \frac{d^2\theta}{dt^2} + (\epsilon - k\ \tau)\frac{d\theta}{dt} + \theta = 0.$$

(ii) Graphical representation of the numerical solution for the given parameters shows that the oscillations are damped (fig. 5.15(a)) for small amplitude perturbation about the steady-state solution $\theta = 0$.

(iii) When the value of $\theta(0)$ is changed from 1.0 to 1.5, the graphical representation of the numerical solution shows that the perturbation is large enough to make the oscillation of the crane pendulation sustained (fig. 5.15(b)).

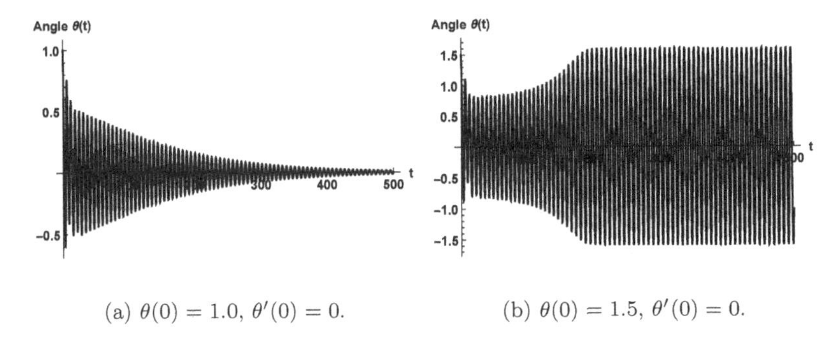

(a) $\theta(0) = 1.0$, $\theta'(0) = 0$.   (b) $\theta(0) = 1.5$, $\theta'(0) = 0$.

FIGURE 5.15: *The figures show (a) damped oscillation, (b) sustained oscillation. Parameter values:* $\epsilon = 0.1$, $k = -0.15$ *and* $\tau = 12$.

5.6 Mathematica Codes

Mathematica codes of selected figures.

5.6.1 Delayed Protein Degradation (Figure 5.1)

$\alpha = 40;\ \beta = 0.3;\ \gamma = 0.1;\ \tau = 20.0;\ t0 = 150;$
$\text{sol1} = \text{NDSolve}[\{x'[t] = \alpha - \beta x[t] - \gamma x[t-\tau], x[t/;t \leq 0] = \text{t0}\}, x, \{t, 0, 200\}]$
$\text{Plot}[\text{Evaluate}[x[t]/.\ \text{First}[\text{sol1}]], \{t, 0, 200\}, \text{PlotRange} \rightarrow \{0, 200\}, \{70, 150\}\}]$

5.6.2 Two Neuron System (Figure 5.6)

$a_1 = 2.0; a_2 = -1.5; \tau_1 = 0.2; \tau_2 = 0.5; t1 = 0.1; t2 = 0.3;$
$\text{sol2} = \text{First}[\text{NDSolve}[\{x'[t] = a_1 \tanh[y[t-\tau_1]] - x[t], y'[t] = a_2 \tanh[x[t-\tau_2]]$
$- y[t], x[t/;t \leq 0] = \text{t1}, y[t/;t \leq 0] = \text{t2}\}, \{x, y\}, \{t, 0, 100\}]]$
$\text{Plot}[\text{Evaluate}[\{x(t), y(t)\}/.\ \text{sol2}], \{t, 0, 100\}, \text{PlotRange} \rightarrow \text{All},$
$\text{AxesLabel} \rightarrow \{\text{“t”}, \text{“Neuron Activity”}\}, \text{PlotStyle} \rightarrow \{\text{Black}, \{\text{Black}, \text{Dotted}\}\},$
$\text{BaseStyle} \rightarrow \{\text{FontWeight} \rightarrow \text{Bold}, \text{FontSize} \rightarrow 12\}]$

Note: Other figures can be obtained by modifying the codes in section 5.6.1 and 5.6.2.

5.7 Matlab Codes

Matlab codes of selected figures.

5.7.1 Delayed Protein Degradation (Figure 5.1)

```
function sol = hepat1
sol = dde23(@hepat1f, 20., 150, [0, 200]);
plot(sol.x, sol.y)
xlabel('t'); ylabel('P(t)');
end
function dy = hepat1f(t, y, Z)
alpha = 40; beta = 0.3; gamma = 0.1;
ylag1 = Z(:, 1);
dy = alpha - beta * y(1) - gamma * ylag1(1);
end
```

5.7.2 Two Neuron System (Figure 5.6)

```
function sol = neuron
sol = dde23(@neuronf, [0.2 0.5], [0.1 0.3], [0, 100]);
figure(1)
plot(sol.x, sol.y)
xlabel('t'); ylabel('NeuronActivity');
figure(2)
plot(sol.y(1, :), sol.y(2, :))
xlabel('u1(t)'); ylabel(''u2(t)');
end
function dy = neuronf(t, y, Z)
al = 2; a2 = -1.5;
ylag1 = Z(:, 1); ylag2 = Z(:, 2);
dy = [-y(1) + a1 * tanh(ylag2(2));
    - y(2) + a2 * tanh(ylag1(1))];
end
```

5.8 Exercises

1. **Delayed Logistic Equation or Hutchinson's Equation:** The delayed logistic equation was first proposed by Hutchinson in 1948. Hutchinson was studying the growth of the species *Daphnia*, a small planktonic crustacean, commonly known as the water flea. Assuming that the process of reproduction is not instantaneous, he modeled their growth using the logistic equation as

$$\frac{dD}{dt} = rD(t)\left[1 - \frac{D(t-\tau)}{k}\right],$$

where τ (> 0) is the discrete time delay because of the time taken for egg formation before hatching, r is the intrinsic growth rate, and k is the carrying capacity.
(i) Find the steady state(s) of the model and perform linear stability analysis.
(ii) Solve the DDE numerically, taking $r = 0.15$, $k = 1.00$, $\tau = 8$ and initial history to be 0.5.
(iii) Comment on the dynamics of the system for $\tau = 8$ and $\tau = 11$. Compare them with the result in (ii).

2. **Nicolson's Blowflies Model:** Let B(t) denote the population of blowflies who are sexually mature, then the entire population dynamics is given by

$$\frac{dB}{dt} = rB(t-\tau)e^{-\frac{B(t-\tau)}{B_0}} - d_1 B(t),$$

where r is the maximum per capita daily egg production rate, d_1 is the per capita daily death rate, and B_0 is the size at which the blowfly population rate is at maximum.
(i) Find the positive equilibrium point.
(ii) Solve the DDE numerically by taking $r = 8, B_0 = 4, d_1 = 0.175$ and $\tau = 15$ and represent it graphically.
(iii) Redraw the graph by taking $r = 8, B_0 = 4, d_1 = 0.475$ and $\tau = 15$.
(iv) What conclusions do you draw from the above graphs?

3. **Ikeda Delay Differential Equation:** Ikeda, in 1979, modeled a nonlinear absorbing medium with two-level atoms placed in a ring cavity [66, 67]. The system is subjected to a constant input of light. Using Maxwell-Block equations, the DDE formulated by Ikeda (in dimensionless form) is given by

$$\frac{dx}{dt} = -x + \mu \sin[x(t-\tau) - x_0].$$

(i) Show that the equation admits a chaotic solution with $\mu = 20$, $x_0 =$

$\frac{\pi}{4}$, and $\tau = 5$.
(ii) Consider the simplest form of the equation as $\frac{dx}{dt} = \sin[x(t-\tau)]$. Comment on the dynamics of the system for $\tau = 0.3679$, $\frac{\pi}{2}$, 3.894, 4.828, 4.991, 5.535 and 9.28.

4. **Recruitment Model:** A general single-species population model with a discrete time delay was proposed by Blythe et al. [16] as

$$\frac{dP}{dt} = f[P(t-\tau)] - DP(t),$$

where $P(t)$ is the population size at any time t, the first term is the recruitment term, the second term is the death term, and τ is the maturation period.
(i) Taking the recruitment term as

$$f[P(t-\tau)] = \frac{bP^2(t-\tau)}{P(t-\tau)+P_0}\left[1-\frac{P(t-\tau)}{k}\right],$$

solve the equation numerically for some suitable parameter values of b, P_0, k, D, and τ. (ii) How do the dynamics of the system change if you take the recruitment term as

$$f[P(t-\tau)] = bP(t-\tau)e^{\frac{-P(t-\tau)}{P_0}},$$

keeping the parameters b, P_0, and τ same?

5. **Allee Effect:** It is a general notion from the classical view of population dynamics that the growth rate of the population increases when population size is small and decreases when population size is large (due to intra-specific competition. that is, competition for resources). However, Warder Clyde Allee showed that the reverse is also true in some cases, which he demonstrated for the growth of goldfish in a tank. The "Allee effect" introduced the phenomenon that the growth rate of individuals increases when the population size falls below a certain critical level. A single species delayed population model with the Allee effect was proposed by Gopalswamy and Ladas [51] as

$$\frac{dx}{dt} = x(t)\left[a + bx(t-\tau) - cx^2(t-\tau)\right],$$

where $x(t)$ is the population density at any time t, the per capita growth rate given by the term $a + bx(t-\tau) - cx^2(t-\tau)$ is quadratic and is subjected to time delay $\tau(\geq 0)$, $a > 0, c > 0$ and b is the real constant.
(i) Find the steady-state solution(s) of the model and perform linear stability analysis.
(ii) Solve the DDE numerically for $a = 1, b = 1, c = 0.5$ and $\tau = 0.2$ and comment on its behavior.
(iii) Show that, if the delay is sufficiently large, the solution of the model oscillates about the positive equilibrium.

6. **Delayed Food-Limited Model:** The delayed food-limited model was introduced by Gopalswamy et al. [50] as

$$\frac{dx}{dt} = rx(t)\left[\frac{k - x(t-\tau)}{k + rcx(t-\tau)}\right].$$

(i) Explain the model with respect to the parameters.
(ii) Taking $r = 0.15$, $k = 100$, $c = 1$ and $\tau = 8$, obtain the numerical solution, represent it graphically and comment on its dynamics. What happens when $\tau = 12.8$?

7. **Wazewska-Czyzewsha and Lasuta Model:** Wazewska-Czyzewsha and Lasuta [163] proposed a model for the growth of blood cells, which takes the form

$$\frac{dB(t)}{dt} = pe^{-rB(t-\tau)} - \mu B(t),$$

where $B(t)$ gives the number of red blood cells at any time t, μ is the natural death of the red blood cells, p and r are positive constants related to the recruitment term for the red blood cells, and τ is the time required for producing red blood cells.
(i) Solve the model equation numerically for $p = 2$, $r = 0.1$, $\mu = 0.5$ and $\tau = 5$ and comment on its behavior.
(ii) Change the values of $p, r \in (0, \infty)$ and $\mu \in (0, 1)$ to observe the changes in the dynamics of the system (if any).

8. **Vector Disease Model:** Cooke [19] proposed a delayed vector disease model given by

$$\frac{dy}{dt} = by(t-\tau)[1 - y(t)] - ay(t),$$

where $y(t)$ is the infected population, b is the contact rate, and a is the cure. The discrete time delay τ gives the incubation period before the disease agent can infect a host. Cooke assumed that the total population is constant and scaled so that $x(t) + y(t) = 1$, $x(t)$ being the uninfected population. He also assumed that an infected population is not subject to death, immunity, or isolation.
(i) Find the steady-state solution(s) of the model.
(ii) Find the condition(s) for linear stability (if any) about the steady-state solution(s).
(iii) Obtain the numerical solution for (a) $a = 5.8$, $b = 4.8$ $(a > b)$, $\tau = 5$ and (b) $a = 38$, $b = 4.8$ $(a < b)$, $\tau = 5$.
(iv) Compare and comment on the graphs.

9. **Currency Exchange Rate Model:** Pavol Brunovsky [18] proposed a mathematical model on the short time fluctuation of an asset, namely,

the price of a foreign currency in a domestic reference, that is, the foreign exchange rate. The model is given by the delay differential equation

$$\frac{dx}{dt} = a[x(t) - x(t-1)] - |x(t)|x(t),$$

where $x(t)$ is the deviation of the value of a foreign currency and $a > 0$ is the parameter which measures the sensitivity to the changes in the exchange rate.
(i) Show that the model has a single steady-state solution $x^* = 0$ for $a > 0$ and the equilibrium solution $x^* = 0$ is asymptotically stable for $a < 1$.
(ii) Also, solve the DDE numerically and show that for $a > 1$, the system shows a periodic behavior.

10. A single species growth model by Arino et al. [69] is given by

$$\frac{dN}{dt} = \frac{\gamma\mu N(t-\tau)}{\mu e^{\mu\tau} + \alpha(e^{\mu\tau} - 1)N(t-\tau)} - \mu N - \alpha N^2,$$

where γ, μ, α, $\tau > 0$.
(i) Show that the system admits a unique positive equilibrium $\overline{N} = \frac{\sqrt{\mu^2+4\gamma\mu k} - \mu(1+2k)}{2\alpha k}$, where, $k = e^{\pi\tau} - 1$ (called delayed carrying capacity).
(ii) For $\gamma = 1.0, \mu = 0.5$ and $\alpha = 0.005$, show that the positive equilibrium $\overline{N}$ decreases as the time lag τ increases from 0 to 1.4 (increment by steps of 0.1).

11. A second-order delayed feedback system is given by [153]

$$\frac{d^2y}{dt^2} + \frac{dy}{dt} - \sin(y) = -py(t-\tau),$$

Simulate the model numerically for
(i) $p = 1.1$, $\tau = 1.0$, $t = [0, 300]$ and initial history $(0.05, 0)$.
(ii) $p = 0.99$, $\tau = 0.8$, $t = [0, 300]$ and initial history $(0.05, 0)$.
(iii) Compare the two results and comment on the dynamics of the system.

12. Nazarenko [109] proposed a model for control of a single population of cells, given by

$$\frac{dc}{dt} + c - \frac{qc}{1 + c^p(t-\tau)} = 0,$$

where $q > 1$, $p > 0$ and τ is the discrete time delay.
(i) Obtain the equilibrium points.
(ii) Determine the characteristic equation(s) after linearization about the non-zero equilibrium point(s) and perform linear stability analysis.
(iii) Solve the DDE numerically for $q = 3$, $p = 2$ and $\tau = 0.5$, 1.0, 1.7, 2.8, and compare them.

13. The delayed prey–predator model is given by

$$\frac{dx}{dt} = x(t)[m - x(t) - y(t)], \quad \frac{dy}{dt} = y(t)[-1 + ax(t - \tau)],$$

where m, a, $\tau > 0$.
(i) Obtain the positive steady-state solution(s) for $m = 2$, $a = 1$.
(ii) Investigate the stability of the positive steady state(s) for case (i).
(iii) Numerically show how the dynamics of the prey–predator model changes by taking different values of τ.

14. Israelsson and Johnsson [68] proposed a model to describe the geotrophic circumnutations of *Helianthus annuus*, commonly known as sunflower. The proposed equation, also known as the sunflower equation, is given by

$$\frac{d^2y}{dt^2} + \frac{a}{\tau}\frac{dy}{dt} + \frac{b}{\tau}\sin[y(t - \tau)] = 0,$$

where y denotes the angle. Show that for $a = 4.8$ and $b = 0.186$, there is a periodic solution for τ between 35 and 80 minutes.

15. The two delay logistic equations in a single species is given by [142]

$$\frac{dx}{dt} = rx(t)[1 - a_1x(t - \tau_1) - a_2x(t - \tau_2)$$

where $r_1, a_1, a_2, \tau_1, \tau_2 > 0$.
(i) Obtain the positive equilibrium points for $r = 0.15, a_1 = 0.25$, and $a_2 = 0.75$.
(ii) Solve the model equation numerically and show that the system is stable about the positive equilibrium for $\tau_1 = 15$ and $\tau_2 = 5$.
(iii) What happens when $\tau_1 = 15$ and $\tau_2 = 10$?

16. Monk [108] proposed a model of Hes1 protein and the transcription of messenger RNAs (mRNAs), given by

$$\frac{dM}{dt} = \frac{\alpha_m}{1 + \left(\frac{P(t-\tau)}{P_0}\right)^n} - \mu_m M, \quad \frac{dP}{dt} - \alpha_p M - \mu_p P,$$

where $M(t)$ and $P(t)$ are the concentrations of Hes1 mRNA and Hes1 protein, μ_m and μ_p are the degradation rates of mRNA and Hes1 protein, respectively, α_m is the transcript initiation rate in the absence of Hes1 protein, and α_p is the production rate of Hes1 protein from Hes1 mRNA, the discrete time delay τ results from the processes of translation and transcription, P_0 is the repression threshold, and n is a Hill coefficient.
(i) Introducing the following dimensionless variables $m \equiv \mu_m \frac{M}{\alpha_m}, p \equiv \frac{P}{P_0}$ and $s \equiv \mu_m t$, obtain the following dimensionless equations

$$\frac{dm}{ds} = \frac{1}{1 + (p(s - \theta))^n} - m, \quad a\frac{dp}{ds} = b\, m - p,$$

where a, b and θ need to be determined.
(ii) Find the steady states of the dimensionless equation.
(iii) Taking $n = 5$, $\theta = 0.56$, $a = 1$ and $b = 11.11$, obtain the solution numerically and comment on the dynamics of the system. What happens when $\theta = 0.90$?

17. A system of DDEs is given by [110]

$$\begin{aligned}\frac{dy_1}{dt} &= y_5(t-1) + y_3(t-1), \quad \frac{dy_2}{dt} = y_1(t-1) + y_2(t-0.5),\\ \frac{dy_3}{dt} &= y_3(t-1) + y_1(t-0.5), \quad \frac{dy_4}{dt} = y_5(t-1) + y_4(t-1),\\ \frac{dy_5}{dt} &= y_1(t-1).\end{aligned}$$

Solve the system numerically on $[0, 1]$ with initial history $y_1(t) = e^{t+1}$, $y_2(t) = e^{t+0.5}$, $y_3(t) = \sin(t+1)$, $y_4(t) = e^{t+1}$, $y_5(t) = e^{t+1}$ for $t \leq 0$.

18. Pinney [124] used Minorsky's equation to discuss the problem of sound generated by a speaker. The non-linear delay differential equation is given by

$$y''(t) + ay'(t) + y(t) = -b\, y'(t-\tau) + \epsilon\, c\, (y'(t-\tau))^3,$$

where $\varepsilon = a << 1, b$ and c are positive. Numerically solve the DDE by taking $a = 0.1, c = 1, \varepsilon = 0.1$ and $\tau = 3\pi$ and comment on the dynamics of the system.

19. Wheldon proposed a model of chronic granulocytic leukemia, given by [34]

$$\frac{dx}{dt} = \frac{\alpha}{1 + \beta x^{\gamma}(t-\tau)} - \frac{\lambda x(t)}{1 + \mu y^{\delta}(t)}, \quad \frac{dy}{dt} = \frac{\lambda x(t)}{1 + \mu y^{\delta}(t)} - wy(t).$$

Solve the model numerically for the parameter values $\alpha = 1.1 \times 10^{10}, \beta = 10^{-12}, \gamma = 1.25, \lambda = 10, \mu = 4 \times 10^{-8}, \delta = 0.5, w = 2.43$, initial history: $x(t) = 100, y(t) = 100$ for $t \leq 0$ and $\tau = 0, 7, 20$. Compare the graphs.

20. Consider a suitcase with two wheels. As the suitcase is pulled by a person, there is the possibility that it may begin to rock from side to side. The person pulling it then applies restoring moment to the handle to balance the suitcase and makes it vertical. Suherman et al. [155] modeled this scenario using DDE as

$$\frac{d^2\theta}{dt^2} + sign(\theta(t))\gamma\cos(\theta(t)) - sin(\theta(t)) + \beta\theta(t-\tau) = A\sin(\Omega t + \eta).$$

Solve the DDE numerically for parameter values $\gamma = 2.48, \beta = 1, \tau = 0.1, A = 0.75, \Omega = 1.37, \eta = \sin^{-1}\left(\frac{\gamma}{A}\right)$ and initial history $\theta(t) = 0$ for $t \leq 0$.

5.9 Project

1. In 1932, the first zombie film (*white zombie*) was released, which told the story of a woman being transformed into a zombie by a voodoo practitioner when she was visiting Haiti (North America). Later, in 1968, in the movie *Night of the living dead*, a different kinds of zombies were shown, feeding on human flesh. However, in the 2002 release of *28 days later*, a new generation of zombies were created, who were not truly dead but infected with a virus and it turns the victim into a violent creature very fast. The common mode of this virus transmission is via bite, as shown in zombie fiction.

   Let normal live humans are considered as susceptible (S), Z are zombies, and R are dead (not permanently). The encounters between zombies and humans can result in humans becoming zombies or the zombie becoming dead. We assume that the conversion of humans to zombies is not an instantaneous process but followed by a discrete time lag τ. Also, sometimes the dead becomes zombies spontaneously. The differential equations modeling this scenario is given by

$$\frac{dS}{dt} = -\beta S(t-\tau)Z(t-\tau),$$
$$\frac{dZ}{dt} = \beta S(t-\tau)Z(t-\tau) - \alpha SZ + \gamma R,$$
$$\frac{dR}{dt} = \alpha SZ - \gamma R.$$

   (i) Explain the model and find its equilibrium solution.
   (ii) Perform linear stability analysis and obtain the critical value τ^* ($0 < \tau < \tau^*$) for which the system is stable.
   (iii) Estimate the length of delay to preserve stability.
   (iv) Taking $\beta = 1.0$, $\alpha = 4.0$, $\gamma = 0.2$, $S(0) = 0.99$, $Z(0) = 0.01$, $R(0) = 0$ and $\tau = 1$, obtain the numerical solution of the model and obtain the graph. Comment on the dynamics of the model. You can change the parameter values and try to obtain different dynamics.
   (v) Now assume that conversion of zombies to dead is not an instantaneous process but followed by a discrete time lag τ_1. Modify the model and repeat the process.

2. (a) Consider the love affair model (Majnun-Layla model) described by the delay differential equation

$$\frac{dM}{dt} = aM(t-\tau) + bL + f, \quad \frac{dL}{dt} = cM + dL + g.$$

What does τ signifies? For the given values of a, b, c, d, g, f, characterize the romantic styles of Majnun and Layla, classify the fixed point and its implication for the affair, sketch $M(t)$ and $L(t)$ as functions of t (time-series graph), assuming $M(0) = m_0$, $L(0) = l_0$ (m_0, l_0 and τ to be chosen by the readers, for example, $m_0 = 1, l_0 = 0$). Also, draw the phase portraits.
(i) $a = 3, b = 2, c = -1, d = 2, f = 0, g = 0$.
(ii) $a = -3, b = 2, c = -1, d = 2, f = 0, g = 0$.
(iii) $a = -3, b = 2, c = -2, d = 1, f = 0, g = 0$.
(iv) $a = -3, b = -2, c = -2, d = -1, f = 0, g = 0$.
(v) $a = -3, b = -2, c = -2, d = -1, f = 10, g = 10$. Change the values of f to $15, 20, 30$ and comment on the dynamics.
(vi) $a = 3, b = 2, c = 1, d = 2, f = 0, g = 0$. Put $f = -10, g = -10$ and comment on the change in dynamics.

(b) Repeat the process for the models

(i) $\frac{dM}{dt} = aM + bL(t - \tau) + f, \ \ \frac{dL}{dt} = cM + dL + g$.
(ii) $\frac{dM}{dt} = aM + bL + f, \ \ \frac{dL}{dt} = cM(t - \tau) + dL + g$.
(iii) $\frac{dM}{dt} = aM + bL + f, \ \ \frac{dL}{dt} = cM + dL(t - \tau) + g$.

Chapter 6

Modeling with Stochastic Differential Equations

6.1 Introduction

Considering the title of this chapter, I must admit that a whole book can be written on this topic. However, I shall here discuss how stochasticity affects the dynamics of models, mostly by solving them numerically. To begin with, we present a few terminologies and definitions.

6.1.1 Random Experiment

Whenever we perform an experiment under nearly identical conditions, we expect to obtain results that are essentially the same. However, there are experiments in which the results will not be essentially the same, even though the conditions may be nearly identical. For example, if we throw two coins simultaneously, the results are TT, TH, HT or HH. We form the set of all possible outcomes as {TT, HT, TH, HH}. Each time we perform this experiment, the outcome is uncertain, although it will be one of the elements of the set {TT, HT, TH, HH}. Such an experiment is called a random experiment, where the result depends on chance.

6.1.2 Outcome

The results of the random experiment are known as the outcome. For example, in the random experiment of throwing two coins simultaneously, there are four possible outcomes, namely, TT, TH, HT or HH.

6.1.3 Event

Any phenomenon that occurs in a random experiment is called an event. An event can be elementary or composite. An elementary event corresponds to a single possible outcome, whereas a composite event corresponds to more than a single possible outcome. For example, when a dice is thrown, the event

DOI: 10.1201/9781351022941-6

"multiple of 2" is composite because it can be decomposed into elementary events 2, 4, 6.

6.1.4 Sample Space

A sample space is a collection of all possible outcomes of a random experiment. In the random experiment of throwing two coins simultaneously, the sample space S = {TT, HT, TH, HH}.

6.1.5 Event Space

An event space (Σ) contains all possible events for a given random experiment. Sometimes, event space is confused with sample space. Consider "Toss" of a coin. Two possible outcomes are either head or tail; hence the sample space is S={H,T}. However, event space is a little different. In a 'toss' of a coin, the possible events are
(i) $\{H\} \to$ flipping the coin and getting head
(ii) $\{T\} \to$ flipping the coin and getting tail
(iii) $\{H, T\} \to$ flipping the coin and getting either head or tail.
Then, event space $\Sigma = \{\{H\}, \{T\}, \{H, T\}\}$.

6.1.6 Axiomatic definition of Probability

Let E be a random experiment described by the event space S and A be any event connected with E. Then the probability of event A, denoted by $P(A)$, is a real number that satisfies the following axioms (Kolmogorov's axioms):
(a) P(A) ≥ 0
(b) P(S) = 1 (probability of a certain event is 1)
(c) If $A_1, A_2, ...$ be a finite or infinite sequence of pairwise mutually exclusive events (that is, $A_i \cap A_j = \emptyset, i \neq j, i, j = 1, 2, ...$), then,
$P(A_1 \cup A_2 \cup ....) = P(A_1) + P(A_2) + ...$.

6.1.7 Probability Function

A probability function is a mapping, $P : \Sigma \to [0, 1]$, that assigns probabilities to the events in Σ and satisfies Kolmogorov's axioms.

6.1.8 Probability Space

A three-tuple (S, Σ, P) whose components are sample space (S), event space (Σ) and probability function (P) is called a probability space.

6.1.9 Random Variable

Let S be a sample space of the random experiment E. A random variable (or a variate) is a function or mapping $X : S \to \mathbf{R}$ (set of real numbers), which assigns to each element $\omega \in S$, one and only one number $X(\omega) = a$. The set of all values which X takes, that is, the range of the function X, is called the spectrum of the random variable X, which is a set of real numbers $B = \{a : a = X(\omega), \omega \in S\}$. A random variable is called a discrete random variable if it takes on a finite or countably infinite number of values and it is called continuous random variable if it takes on a noncountably infinite number of values.

Consider two "tosses" of an unbiased coin. Then, the sample space is

$$S = \{\omega_1 = (HH), \omega_2 = (HT), \omega_3 = (TH), \omega_4 = (TT)\}.$$

We now define a function or a mapping $X : S \to \mathbf{R}$ such that $X(\omega_i) = \lambda_i$, where λ_i is the number of heads, $i = 1, 2, 3, 4, ...$. Then, $X(\omega_1) = 2, X(\omega_2) = 1, X(\omega_3) = 1, X(\omega_4) = 0$. Here, X is a random variable defined in the domain S. The spectrum of X is $\{0, 1, 2\}$. Since, the spectrum is countable, X is a discrete random variable. Consider the height of a group of high school students which lies between 155 cm and 185 cm. Here, $S = \{\omega : 155 < \omega < 185\}$. Since, X takes any positive real values between 155 and 185, X is a continuous random variable.

6.1.10 Sigma Algebra

A collection Σ of subsets of S is called σ-algebra if
(i) $\phi \in \Sigma$, (ii) $A \in \Sigma \Rightarrow A^c \in \Sigma$
(iii) If $A_1, A_2, A_3, ...$ is a countable collection of subsets in Σ, then $\bigcup_{i=1}^{\infty} A_i \in \Sigma$.
$\sigma -$ algebra is closed under countable union and countable intersection. Then 2-tuple (S, Σ) is called a measurable space.

6.1.11 Measure

Let (S, Σ) be a measurable space. A measure on (S, Σ) is a function $\mu : \Sigma \to [0, \infty)$ such that (i) $\mu(\phi) = 0$ (ii) If $\{A_i, i \geq 1\}$ is a sequence of disjoint sets in Σ, then the measure of the union (of countably infinite disjoint sets) is equal to the sum of measures of individual sets, that is, $\mu(\bigcup_{i=1}^{\infty} A_i) = \Sigma_{i=1}^{\infty} \mu(A_i)$. The triplet (S, Σ, μ) is called a measure space. μ is said to be a finite measure if $\mu(S) < \infty$, otherwise μ is called infinite.

6.1.12 Probability Measure

A probability measure is a function $P : \Sigma \to [0, 1]$ such that (i)$P(\phi) = 0$ (ii) $P(S) = 1$ (iii) if $\{A_i, i \geq 1\}$ is a sequence of disjoint sets in Σ, then $P(\bigcup_{i=1}^{\infty} A_i) = \Sigma_{i=1}^{\infty} P(A_i)$. The triplet (S, Σ, P) is called a probability space.

6.1.13 Mean and Variance

The mean of the random variable X is defined as

$$\mu = E(X) = \sum_{-\infty}^{\infty} x_i f_i, \text{ for a discrete distribution,}$$

$$\mu = E(X) = \int_{-\infty}^{\infty} x f_X(x)\, dx, \text{ for a continuous distribution.}$$

The mean gives a rough position of the bulk of the distribution, and hence called the measure of location.

The variance of the random variable X is defined as

$$\sigma^2 = Var(X) = E[(X - \mu)^2] = E[X^2 + \mu^2 - 2\mu X] = E(X^2) - [E(X)]^2.$$

The variance is a characteristic which describes how widely the probability masses are spread about the mean.

6.1.14 Independent Random Variables

The cumulative distribution function (CDF) of a random variable X is defined as

$$F_X(x) = P(-\infty < X \leq x) \text{ for all } x \in \mathbf{R}.$$

The joint distribution function of two random variables X and Y is defined by

$$F_{XY}(x, y) = P(-\infty < X \leq x, -\infty < Y \leq y), \text{ for all } x, y \in \mathbf{R},$$

where the event $(-\infty < X \leq x, -\infty < Y \leq y)$ means the joint occurrence of the two events $-\infty < X \leq x$ and $-\infty < Y \leq y$. If the events $(-\infty < X \leq x)$ and $(-\infty < Y \leq y)$ are the independent for all x, y, then

$$P(-\infty < X \leq x, -\infty < Y \leq y) = P(-\infty < X \leq x)\; P(-\infty < Y \leq y)$$
$$\Rightarrow F_{XY}(x, y) = F_X(x)\; F_Y(y).$$

Thus, the necessary and sufficient condition for the independence of the random variable X and Y is that their joint distribution function $F_{XY}(x, y)$ can be written as the product of the marginal distribution functions.

6.1.15 Gaussian Distribution (Normal Distribution)

A random variable X is said to be normally distributed with mean μ and variance σ^2, if its probability density function (pdf) is

$$f_X(x) = \frac{1}{\sqrt{2\pi}\sigma} e^{-\frac{(x-\mu)^2}{2\sigma^2}}, \quad -\infty < x < \infty.$$

Usually, the Gaussian or Normal distribution of X is represented by

$$X \sim N(\mu, \sigma^2).$$

The probability density function (pdf) of the Gaussian or Normal distribution is a bell-shaped curve, symmetric about the mean μ (attains its maximum value $\frac{1}{\sqrt{2\pi}\sigma}$ there) and is completely characterized by the two parameters, namely, μ (mean) and σ^2 (variance). The mean μ is the centroid of the pdf, and in this case, it is also the point at which the pdf is maximum. The variance σ^2 gives the measure of the dispersion of the random variable around the mean. If $\mu = 0, \sigma = 1$, then the random variable X is said to follow Standard Normal Distribution.

6.1.16 Characteristic Function

The characteristic function of the random variable X is a complex-valued function of a real variable t and is defined as

$$\phi_X(t) = E(e^{itx}) = E[\cos(tX) + i\sin(tX)],$$

where $i = \sqrt{-1}$ and t is a real number. The characteristic function satisfies the following properties:
(i) $\phi_X(0) = 1$ and $|\phi_X(t)| \leq 1, \forall t \in \mathbf{R}$.
(ii) If $Y = aX + b$, then $\phi_Y(t) = e^{ibt}\, \phi_X(at)$.
(iii) If X and Y are independent random variables and $Z = X + Y$, then, $\phi_Z(t) = \phi_X(t)\, \phi_Y(t)$.

6.1.17 Characteristic Function of Gaussian Distribution

Let $X \sim N(\mu, \sigma^2)$, then probability density function of X is

$f_X(x) = \frac{1}{\sqrt{2\pi}\sigma} e^{-\frac{(x-\mu)^2}{2\sigma^2}}, -\infty < x < \infty$. Now,

$$\phi_X(t) = E(e^{itX}) = \frac{1}{\sqrt{2\pi}\sigma}\int_{-\infty}^{\infty} e^{itx}\, f_X(x)\, dx = \frac{1}{\sqrt{2\pi}\sigma}\int_{-\infty}^{\infty} e^{-\frac{(x-\mu)^2}{2\sigma^2}+itx}\, dx,$$

$$= e^{i\mu t - \frac{\sigma^2 t^2}{2}} \times \frac{1}{\sqrt{2\pi}\sigma}\int_{-\infty}^{\infty} e^{-\frac{(x-\mu-it)^2}{2\sigma^2}}\, dx = e^{i\mu t - \frac{\sigma^2 t^2}{2}} \times 1,$$

$$\Rightarrow \phi_X(t) = e^{i\mu t - \frac{\sigma^2 t^2}{2}}.$$

6.1.18 Inversion Theorem

Let X be a continuous random variable, having probability density function $f_X(x)$, then the corresponding characteristic function is given by

$$\psi_X(t) = \int_{-\infty}^{\infty} e^{itx}\ f_X(x)\ dx.$$

Also, the probability density function $f_X(x)$ can be obtained from the characteristic function as

$$f_X(x) = \frac{1}{2\pi} \lim_{T\to\infty} \int_{-T}^{+T} e^{-itx}\ \psi_X(t)\ dt,$$

at every point where $f_X(x)$ is differentiable. Now,

$$f_X(x) = \frac{1}{2\pi} \lim_{T\to\infty} \int_{-T}^{T} e^{-itx} e^{i\mu t - \frac{\sigma^2 t^2}{2}}\, dt$$

$$= \frac{1}{2\pi} e^{-\frac{(x-\mu)^2}{2\sigma^2}} \lim_{T\to\infty} \int_{-T}^{+T} e^{-\frac{\sigma^2}{2}(t+i\frac{x-\mu}{\sigma^2})^2}\ dt = \frac{1}{\sqrt{2\pi}\ \sigma} e^{-\frac{(x-\mu)^2}{2\sigma^2}}.$$

6.1.19 Convergence of Random Variables and Limit Theorems

Convergence in Probability: A sequence of random variables $X_1, X_2, ..., X_n, ...$ converges to a random variable X in probability, denoted by $X_n \xrightarrow{\text{P}} X$, if for any $\epsilon > 0$,

$$\lim_{n\to\infty} P\{|X_n - X| < \epsilon\} = 1 \text{ or } \lim_{n\to\infty} P\{|X_n - X| \geq \epsilon\} = 0.$$

Almost Sure Convergence : Consider a sequence of random variables $X_1, X_2, ..., X_n, ...$, all defined on the same sample space S. For every $\omega \in S$, we obtain sample sequence $X_1(\omega), X_2(\omega), ..., X_n(\omega), ...$. A sequence of random variables $X_1, X_2, ..., X_n, ...$ converges to a random variable X almost surely (also known as with probability 1), denoted by $X_n \xrightarrow{\text{a.s.}} X$, if

$$P\{\omega : \lim_{n\to\infty} X_n(\omega) = X(\omega)\} = 1,$$

or equivalently, if for every $\epsilon > 0$,

$$\lim_{n\to\infty} P\{|X_n - X| < \varepsilon \text{ for every } n \geq m\} = 1.$$

Convergence in Mean Square: A sequence of random variables $X_1, X_2, ..., X_n, ...$ converges to a random variable X in mean square (m.s.), denoted by $X_n \xrightarrow{\text{m.s.}} X$, if

$$\lim_{n\to\infty} E[(X_n - X)^2] = 0.$$

Converge in Distribution: A sequence of random variables X_1, X_2, ..., X_n, ... converges in distribution to a random variable X, denoted by $X_n \xrightarrow{d} X$, if

$$\lim_{n\to\infty} F_{X_n}(x) = F_X(x),$$

for every x at which $F_X(x)$ is continuous.

Notes:
(i) Convergence in probability implies convergence in distribution but the converse is not true. However, convergence in distribution implies convergence in probability when the limiting random variable X is a constant.
(ii) Almost surely convergence implies convergence in probability and hence implies convergence in distribution (fig. 6.1).
(iii) Convergence in mean square implies convergence in probability and hence implies convergence in distribution (fig. 6.1).
(iv) Convergence in probability does not necessarily implies almost surely convergence or mean square convergence. Convergence in probability is weaker than both almost surely convergence and convergence in mean square.
(v) Almost sure convergence does not imply convergence in mean square. Also, convergence in mean square does not imply almost sure convergence.

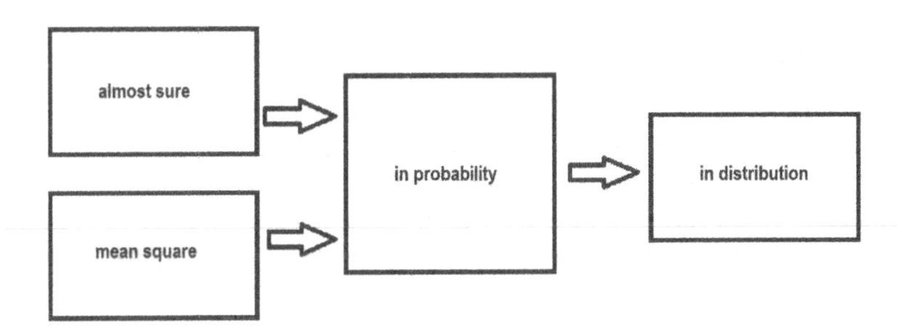

FIGURE 6.1: *The figure shows the relationship between different types of convergences.*

Weak Law of Large Numbers (WLLN): Let $X_1, X_2, ..., X_n, ...$ be independent and identically distributed random variables (each r.v. has the same probability distribution as the others and all are mutually independent) with finite variance. Then, for any $\epsilon > 0$,

$$\lim_{n\to\infty} P(|\overline{X} - \mu| \geq \varepsilon) = 0, \text{ where } \overline{X} = \frac{1}{n}\sum_{1}^{n} X_i.$$

Thus, the weak law of large numbers says that the probability of the difference between the sample mean and the true mean by a fixed number ϵ (> 0)

becomes smaller and smaller, and converges to zero as n goes to infinity.

Central Limit Theorem: Let $X_1, X_2, ..., X_n, ...$ be independent and identically distributed random variables with $E(X_i) = \mu$ (finite) and $Var(X_i) = \sigma^2$ (finite). Then, the random variable

$$Z_n = \frac{\bar{X} - \mu}{\sigma/\sqrt{n}} = \frac{(X_1 + X_2 + ... + X_n) - n\mu}{\sigma\sqrt{n}}$$

converges in distribution to the standard normal variable as n goes to infinity, that is,

$$\lim_{n \to \infty} P(Z_n \leq x) = \Phi(x), \text{ for all } x \in \mathbb{R},$$

where $\Phi(x)$ is the standard normal distribution.

Statistical methods like testing of hypothesis and construction of confidence intervals are used in data analysis and these methods assume that the population is normally distributed. According to central limit theorem, we can assume the sampling distribution of an unknown or non-normal distribution as normal (it does not matter what the distribution is). This is the significance of central limit theorem.

6.1.20 Stochastic Process

A stochastic process is a collection of random variables $\{X_t, t \in T\}$, defined on some probability space (S,Σ,P). We call the values of X_t as state space denoted by Ω. The index set T from where t takes its value is called a parameter set or a time set. A stochastic process may be discrete or continuous according to whether the index set T is discrete or continuous. The range (possible values) of the random variables in a stochastic process is called the state space of the process.

Example 6.1.1 *A stochastic process $\{X_n : n = 0, 1, 2, 3, \cdots\}$ with discrete index set $\{0, 1, 2, 3, \cdots\}$ is a discrete time stochastic process.*

Example 6.1.2 *A stochastic process $\{X_t : t \geq 0\}$ with continuous index set $\{t : t \geq 0\}$ is a continuous time stochastic process.*

Example 6.1.3 *$\{X_n : n = 0, 1, 2, 3, \cdots\}$, where the state space of X_n is $\{1, 2\}$, which represents whether an electronic component is acceptable or defective, and time n corresponds to the number of components produced.*

Example 6.1.4 *$\{X_t : t \geq 0\}$, where the state space of X_t is $\{0, 1, 2, \cdots\}$, which represents the number of cars parked in the parking 1 to t in front of a movie theater and t corresponds to hours.*

A **filtration** $\{\Sigma_t\}_{t\geq 0}$ is a family of sub-sigma algebras of some sigma-algebra Σ with the property that $s < t$, then $\Sigma_s \subset \Sigma_t$. Thus, a stochastic process $\{X_t : 0 \leq t < \infty\}$ is **adapted** to $\{\Sigma_t\}_{t\geq 0}$ means that for any t, X_t is Σ_t is measurable.

6.1.21 Markov Process

A Markov process is a stochastic process with the following properties:
(i) It has a finite number of possible outcomes or states.
(ii) The outcome at any stage depends only on the outcome of the previous stage.
(iii) Over time, the probabilities are constant.

Mathematically, a Markov Process is a sequence of random variables $X_1, X_2, X_3, ...$ such that

$$P\left(X_n = x_n / X_{n-1} = x_{n-1}, ..., X_1 = x_1\right) = P\left(X_n = x_n / X_{n-1} = x_{n-1}\right).$$

Let the initial state of the system be denoted by x_0. Then, there is a matrix A, which gives the state of the system after one iteration by the vector Ax_0. Thus, we obtain a chain of state vectors, namely, x_0, Ax_0, A^2x_0,..., where the state of the system after n iterations is given by $A^n x_0$. Such a chain is called a Markov Chain and the matrix A is called a transition matrix.

Example 6.1.5 *A metro ride in a city was studied. After analyzing several years of data, it was found that* 25% *of the people who regularly ride on the metro in a given year do not prefer the metro rides in the next year. It was also observed* 31% *of the people who did not ride on the metro regularly in that year began to ride the metro regularly the next year.*

In a given year, 8000 *people ride the metro and* 9000 *do not ride the metro. Of the persons who currently ride the metro,* 75% *of them will continue to do so and of the persons who do not ride the metro,* 31% *will start doing so.*

In order to find the distribution of metro riders/metro non-riders in the next year, we first obtain the number of people who will ride the metro next year. Therefore, the number of persons who will ride the metro next year = $b_1 = 0.75 \times 8000 + 0.31 \times 9000 = 8790$. *Similarly, the number of persons who will not ride the metro next year* = $b_2 = 0.25 \times 8000 + 0.69 \times 9000 = 8210$.

This can be expressed in matrix notation as $Ax = b$ *where*

$$A = \begin{bmatrix} 0.75 & 0.31 \\ 0.25 & 0.69 \end{bmatrix}, \; x = \begin{bmatrix} 8000 \\ 9000 \end{bmatrix}, \; \textit{and } b = \begin{bmatrix} b_1 \\ b_2 \end{bmatrix} = \begin{bmatrix} 8790 \\ 8210 \end{bmatrix}.$$

After two years, we use the same matrix A*, but* x *is replaced by* b *and the* *distribution becomes* $Ab = A^2x$. *Thus,*

$$A^2x = \begin{bmatrix} 0.75 & 0.31 \\ 0.25 & 0.69 \end{bmatrix}^2 \begin{bmatrix} 8000 \\ 9000 \end{bmatrix} = \begin{bmatrix} 0.64 & 0.4464 \\ 0.36 & 0.5536 \end{bmatrix} \begin{bmatrix} 8000 \\ 9000 \end{bmatrix} = \begin{bmatrix} 9137.6 \\ 7862.4 \end{bmatrix}.$$

After 2 years, the number of persons who will ride the metro is 9138 *and the number of persons who will not ride the metro is* 7862. *In general, the distribution is* $A^n x$ *after* n *years.*

6.1.22 Gaussian (Normal) Process

A Gaussian Process is a stochastic process such that the joint distribution of every finite subset of random variables is a multivariate normal distribution. Thus, a random process $\{X(t), t \in T\}$ is said to be a Gaussian (Normal) process if, for all $t_1, t_2, ..., t_n \in T$, the random variables $X(t_1)$, $X(t_2)$,...,$X(t_n)$ are jointly normal.

6.1.23 Wiener Process (Brownian Motion)

A Wiener process or a Brownian motion is a zero-mean continuous process with independent Gaussian increments (by independent increments we mean a process X_t, where for every sequence $t_0 < t_1 < \cdots < t_n$, the random variables $X_{t_1} - X_{t_0}, X_{t_2} - X_{t_1}, \cdots, X_{t_n} - X_{t_{n-1}}$ are independent).

Mathematically, we can say that a one-dimensional standard Wiener process or Brownian motion $B(t) : \mathbb{R}_+ \longrightarrow \mathbb{R}$ is a real-valued stochastic process on some probability space (S, Σ, P) adapted to $\{\Sigma_t\}$ with the following properties:
(i) $B(0) = 0$ (with probability 1).
(ii) $B(t)$ is continuous for all t (with probability 1).
(iii) $B(t)$ has independent increment.
(iv) $B(t) - B(s)$ has a Gaussian or Normal distribution with mean zero and variance $t - s$ for every $t > s \geq 0$. The density function of the random variable is given by

$$f(x; t, s) = \frac{1}{\sqrt{2\pi(t-s)}} e^{-\frac{x^2}{2(t-s)}}.$$

The consequences of this definition are
(i) The process $\{W(t)\}$ is Gaussian.
(ii) $E[W(t)] = 0$ and $E[W(s)W(t)] = min(s, t)$, for all $s, t \geq 0$ and in particular, $E[W^2(s)] = s$.

6.1.24 Gaussian White Noise

We consider a stochastic process $\{X_t, t \in T\}$ such that the random variables X_t are independent and

(i) $E\{X_t\} = 0$.
(ii) $E\{X_t - X_s\} = \delta(t-s)$, where $\delta(t-s)$ is the dirac-delta function.
This process is known as Gaussian white noise.

Notes: (i) $\delta(x)$ is not really a function. It is a mathematical entity called a distribution and is defined as $\delta(x) = 0, x \neq 0$ and becomes infinity at $x = 0$, with $\int_{-\infty}^{\infty} \delta(x)dx = 1$. (ii) The generalized derivative of the Wiener process is called Gaussian white noise, though the trajectories of the Wiener process are not differentiable.

6.1.25 Ito Integral:

Let Σ_t be the filtration generated by Brownian motion up to time t and $X(t) \in \Sigma_t$ be an adapted stochastic process. Corresponding to the Riemann sum approximation to the Riemann integral, we define the following approximations to the Ito integral

$$Y_{\Delta t}(t) = \sum_{t_k<t} X(t_k)\Delta W_k, \tag{6.1}$$

where $t_k = k\Delta t$ and $\Delta W_k = W(t_{k+1}) - W(t_k)$. Then, the Ito integral is

$$Y(t) = \lim_{\Delta\to 0} Y_{\Delta t}(t), \text{provided the limit exists.}$$

We consider the integral

$$Y(T) = \int_0^T W(t)\ dW(t), \text{ where } W(t) \text{ is a Wiener process.}$$

Suppose $W(t)$ is differentiable with derivative $\dot{W}(t)$ and the limit of (6.1) can be calculated as

$$\int_0^T W(t)\ \dot{W}(t)\ dt = \frac{1}{2}\int_0^T \frac{\partial}{\partial t}[W(t)]^2 dt = \frac{1}{2}[W(t)]^2.$$

However, this is not correct. If we use the definition (6.1) with actual rough path Brownian motion, we get,

$$\begin{aligned}
Y_{\Delta t} &= \sum_{k<n} W(t_k)\left[W(t_{k+1}) - W(t_k)\right] \\
&= \sum_{k<n}\left(\frac{1}{2}\left[W(t_{k+1}) + W(t_k)\right] - \frac{1}{2}\left[W(t_{k+1}) - W(t_k)\right]\right) \times \left[W(t_{k+1}) - W(t_k)\right] \\
&= \sum_{k<n}\left(\frac{1}{2}\left[W^2(t_{k+1}) - W^2(t_k)\right] - \frac{1}{2}\left[W(t_{k+1}) - W(t_k)\right]^2\right).
\end{aligned}$$

Now,

$$\sum_{k<n} \left[W^2(t_{k+1}) - W^2(t_k)\right]$$
$$= W^2(t_1) - W^2(t_0) + W^2(t_2) - W^2(t_1) + W^2(t_3)) - W^2(t_2) + ...$$
$$+ W^2(t_{n-1}) - W^2(t_{n-2}) + W^2(t_n) - W^2(t_{n-1}) = W^2(t_n) - W^2(t_0)$$
$$= W^2(t_n) \; [W(t_0) = W(0) = 0, \text{ as } W(t) \text{ is a Wiener process.}]$$

The second term

$$\sum_{k<n} [W(t_{k+1}) - W(t_k)]^2$$

is a sum of n independent random variables. Each term has expected value $\frac{\Delta t}{2}$ and variance $\frac{(\Delta t)^2}{2}$. Therefore, the sum is a random variable with mean $\frac{n\Delta t}{2}$ and variance $\frac{n(\Delta t)^2}{2}$. This implies

$$\sum_{t_k<T} [W(t_{k+1}) - W(t_k)]^2 \to \frac{T}{2} \text{ as } \Delta t \to 0.$$

Let $0 = t_0 < t_1 < ... < t_n = T$, where $t_i = \frac{iT}{n}$ be a partition of the interval $[0, T]$ into n equal parts and $\Delta_i W = W(t_{i+1}) - W(t_i)$. Then,

$$E[(\Delta_i W)^2] = E[(W(t_{i+1}) - W(t_i))^2]$$
$$= E[(W(t_{i+1}) - W(t_i))^2] - E[W(t_{i+1}) - W(t_i)]$$
$$= Var[W(t_{i+1}) - W(t_i)] = t_{i+1} - t_i = \frac{T}{n}.$$

For Gaussian distribution, all odd-order moments about mean vanish and even order moments about mean are given by

$$E[X^{2n}] = 1.3.5... \;.(2n-1)\; Var(X)]^{2n}. \text{ Therefore,}$$
$$E[(\Delta_i W)^4] = E[(W(t_{i+1}) - W(t_i))^4]$$
$$= 1.3 \;\left(Var\left[W(t_{i+1}) - W(t_i)\right]\right)^2 = 3\left(\frac{T}{n}\right)^2.$$

$$E\left[\sum_{k<n} (W(t_{k+1}) - W(t_k))^2 - T\right]^2 = \sum_{k=0}^{n-1} E\left[(W(t_{k+1}) - W(t_k))^2 - \frac{T}{n}\right]^2$$
$$= \sum_{k=0}^{n-1}\left[E\left[W(t_{k+1}) - W(t_k)\right]^4 + \left(\frac{T}{n}\right)^2 - 2\frac{T}{n}E\left[W(t_{k+1}) - W(t_k)\right]^2\right]$$
$$= \sum_{k=0}^{n-1}\left[3\left(\frac{T}{n}\right)^2 + \left(\frac{T}{n}\right)^2 - 2\frac{T}{n}\left(\frac{T}{n}\right)\right] = \sum_{k=0}^{n-1}\left[2\;\frac{T^2}{n^2}\right]$$
$$= 2\;\frac{T^2}{n} \;\to 0 \text{ as } n \to \infty.$$

Hence, from the definition of mean square convergence, we say

$$\lim_{n\to\infty} \sum_{k<n} [W(t_{k+1}) - W(t_k)]^2 = T \text{ in } L^2 \text{ norm. Hence,}$$

$$\int_0^T W(t)\, dW(t) = \frac{1}{2}\left[W^2(t) - T\right], \text{ which is the correct answer.}$$

6.1.26 Ito's Formula in One-Dimension

Suppose $X(t)$ is an Ito process $dX = Adt + BdW$, where A and B are the drift and the diffusion coefficients. Let $f(x,t) : R^+ \times R \to R$ be a function with continuous partial derivative $\frac{\partial f}{\partial t}$, and continuous second partial space derivative $\frac{\partial^2 f}{\partial x^2}$. Then $F(t) = f(X(t),t)$ is an Ito process, and

$$dF = \frac{\partial f(X(t),t)}{\partial t}dt + \frac{\partial f(X(t),t)}{\partial x}dX + \frac{1}{2}B^2(t)\frac{\partial^2 f(X(t),t)}{\partial x^2}dt,$$

or equivalently,

$$dF = \left(\frac{\partial f}{\partial t} + \frac{1}{2}B^2\frac{\partial^2 f}{\partial x^2}\right)dt + \frac{\partial f}{\partial t}dX.$$

We again calculate $\int_0^t W(t)dW(t)$.

In the definition of Ito process $dX(t) = A(t)dt + B(t)dW(t)$, putting $A = 0$, $B = 1$, we get $X(t) = W(t)$. Let $F(t) = f(X(t),t) = \frac{X^2}{2}$. Applying Ito formula,

$$dF = \frac{\partial f}{\partial t}\, dt + \frac{\partial f}{\partial x}dW + \frac{1}{2}\frac{\partial^2 f}{\partial x^2}\, dt$$

$$\Rightarrow \frac{1}{2}dX^2 = 0 + XdW + \frac{1}{2}dt \Rightarrow \frac{1}{2}\, dW^2 = WdW + \frac{1}{2}\, dt$$

Integrating we obtain, $\int_0^t W(t)dW(t) = \frac{1}{2}\, W^2(t) - \frac{t}{2}.$

6.1.27 Stochastic Differential Equation (SDE)

A stochastic differential equation is of the form

$$dX(t) = f(X(t),t)dt + g(X(t),t)dW(t),\ X_0 = x_0,\ 0 \le t \le T, \qquad (6.2)$$

where $X(t)$ is a stochastic process, $W(t)$ is a m-dimensional Brownian motion, $f : R^d \times [0,T] \to R^d$ and $g : R^d \times [0,T] \to R^{d\times m}$. The solution of (6.2) is a path or a function $X(t)$ (an R^d - valued stochastic process) satisfying

$$X_t = \int_0^T f(X(t),t)dt + \int_0^T g(X(t),t)\, dW(t).$$

One can refer to Bernt Oksendal [116] for more information on stochastic differential equations.

Existence and Uniqueness
If the coefficients of the stochastic differential equation

$$dX(t) = f(X(t),t)dt + g(X(t),t)dW(t), \ X_0 = x_0, \ 0 \leq t \leq T,$$

satisfy a space-variable Lipschitz condition

$$|f(x,t) - f(y,t)|^2 + |g(x,t) - g(y,t)|^2 \leq k_1|x-y|^2$$

and the spatial growth condition

$$|f(x,t)|^2 + |g(x,t)|^2 \leq k_2\left(1+|x|^2\right),$$

then there exists a continuous adapted solution $X(t)$ of (6.2), that is, uniformly bounded in

$$L^2 : E\left[\sup_{0\leq t\leq T} X^2(t)\right] < \infty.$$

Moreover, if $X(t)$ and $Y(t)$ are both continuous L^2 bounded solutions of equation (6.2), then

$$P\left(X(t) = Y(t) \text{ for all t } \in [0,T]\right) = 1.$$

6.1.28 Stochastic Stability

We consider the d-dimensional stochastic differential equation of the form

$$dX(t) = f(X(t),t)dt + g(X(t),t)dW(t), \tag{6.3}$$

for all $t \geq t_0$ with $X(t_0) = x_0 \in R^d$.

We also assume that $f(0,t) \equiv g(0,t) \equiv 0$, so that the trivial solution $X(t) = 0$ holds for $x_0 = 0$.

We will discuss three different types of stochastic stability, namely, stable in probability, almost sure exponential stability and moment exponential stability.

6.1.28.1 Stable in Probability

The trivial solution of (6.3) is said to be *stable in probability*, if every pair of $\epsilon \in (0,1)$ and $r > 0$, there exists a $\delta(\epsilon, r, t_0) > 0$ such that [105]

$$P\left[\ |X(t;t_0,x_0)| < r \text{ for all } t \geq t_0\right] \geq 1-\epsilon, \text{ whenever } |x_0| < \delta.$$

Otherwise, it is said to be stochastically unstable.

6.1.28.2 Almost sure Exponential Stability

The trivial solution of (6.3) is said to be *almost surely (a.s.) exponentially stable* if [105]

$$\lim_{t\to\infty} \sup \frac{1}{t} \ln |X(t; t_0, x_0)| < 0 \text{ a.s. for all } x_0 \in R^d.$$

6.1.28.3 Moment Exponential Stability

The trivial solution of (6.3) is said to be p^{th} *moment exponentially stable* if there is a pair of positive constants λ and C such that [105]

$$E[\ |X(t; t_0, x_0)|\, p] \le c|x_0|^p\ \bar{e}^{\lambda(t-t_0)} \text{ on } t \ge t_0 \text{ for all } x_0 \in R^d.$$

When $p = 2$, the trivial solution of (6.3) is said to be *exponentially stable in mean square.*

Note: One can refer to Xuerong Mao [105] for more information on stochastic differential equations and their stability analysis.

6.2 Stochastic Models

In stochastic modeling, we take into account a certain degree of randomness or unpredictability. The million-dollar question is when to use deterministic models and when we really need to use stochastic ones. People argue that stochasticity put realism in models, and hence it should be added to make the model more realistic. However, I prefer that a stochastic model should be built when it is absolutely necessary and then stochasticity should be put in those parts of the model that are absolutely necessary to be stochastic, and then control the rest to improve the understanding of the model.

6.2.1 Stochastic Logistic Growth

The famous logistic growth model for a single species is given by (in the deterministic case)

$$\frac{dx}{dt} = rx\left(1 - \frac{x}{k}\right), \ x(0) = x_0, \tag{6.4}$$

where r is the intrinsic growth rate and k is the carrying capacity. Clearly, $x = 0$ and $x = k$ are the two points of equilibria. It can be easily shown that the solution of (6.4) is

$$x(t) = \frac{e^{rt}x_0}{(e^{rt}-1)\frac{x_0}{k}+1}.$$

Suppose the logistic growth model for a single species is now subjected to the environment stochasticity or randomness $\eta(t)$, which is a Gaussian white noise with a time-varying intensity $\sigma^2(t)$. Then $\eta(t)dt = \sigma\, dW$, where $W(t)$ is a Wiener process and σ is the intensity of the noise. The stochastic version of the model is given by

$$dX(t) = rX(t)\left(1 - \frac{X(t)}{k}\right)dt + \sigma dW.$$

It can be shown that the logistic model is stochastically stable if $\sigma^2 < \frac{2r}{k}$, $t \geq 0$ [48]. Fig. 6.2(a) numerically confirms the result and shows that the equilibrium point $x^* = k$ is stochastically stable.

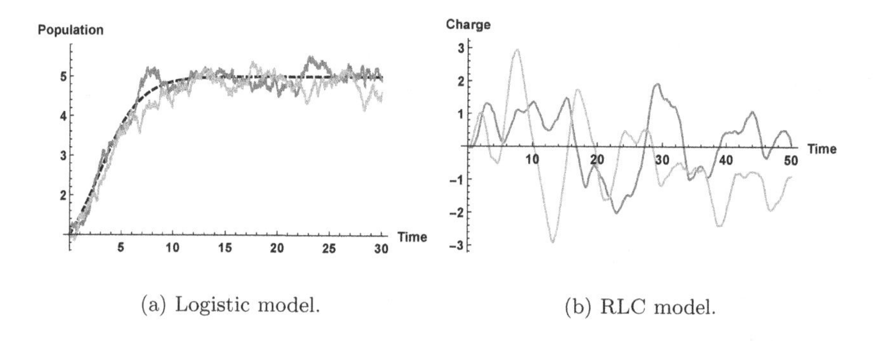

(a) Logistic model. (b) RLC model.

FIGURE 6.2: *(a) Effect of stochasticity on the logistic growth model. For $r = 0.5, k = 5.0, \sigma = 0.3$ and initial condition $x(0) = 1.0$, the steady-state solution $x^* = 5$ is stochastically stable. (b) The voltage source influenced by some randomness for $L = 2, C = 2, R = 0.75$, $\omega = 1.5, \sigma = 1.25$.*

6.2.2 RLC Electric Circuit with Randomness

We consider an Resistor-Inductor-Capacitor (RLC) electric circuit with a coil of inductance L, a resistor of resistance R, a capacitor of capacitance C, and a voltage source $V(t)$ arranged in series. Let $Q(t)$ be the charge on the capacitor and the current flowing in the circuit $I(t)$, then the respective voltage across R, L and C are RI, $L\frac{dI}{dt}$ and $\frac{Q}{C}$. Using Kirchhoff's law (the voltage between any two points is independent of the path), we get,

$$L\frac{dI}{dt} + RI(t) + \frac{Q(t)}{C} = V(t).$$

Since $I(t) = \frac{dQ(t)}{dt}$, the differential equation satisfying the charge is given by

$$L\frac{d^2Q(t)}{dt^2} + R\frac{dQ(t)}{dt} + \frac{Q(t)}{C} = V(t),\ Q(0) = Q_0 \text{ , and } Q'(0) = I_0.$$

with initial condition $Q(0) = Q_0$ and $Q'(0) = I_0$.

The second order differential equation can be written as a system of first-order ordinary differential equations(ODEs) as

$$\frac{dQ(t)}{dt} = I(t),\ \frac{dI}{dt} = -\frac{R}{L}I(t) - \frac{1}{LC}Q(t) + \frac{V(t)}{L}.$$

Let the voltage source be influenced by some randomness, which is mathematically described as noise (to be more specific, a stochastic process called the Gaussian white noise process) and is denoted by $\eta(t)$. The stochastic version of the model is given by

$$\begin{aligned} dQ(t) &= I(t)dt, \\ dI(t) &= -\frac{R}{L}I(t)dt - \frac{1}{LC}Q(t)dt + \frac{V(t)}{L} + \frac{\sigma}{L}dW(t), \end{aligned}$$

where $\eta(t)\ dt = \sigma\ dW(t)$, as the white noise $\eta(t)$ is the time derivative of the Wiener process $W(t)$ and σ is the intensity of the process. The system is solved numerically by taking $V(t) = \sin(\omega t)$, and fig. 6.2(b) shows the voltage source due to randomness.

6.2.3 Heston Model

In 1993, Steven Heston developed a stochastic volatility model for analyzing bond and currency options. The model is given by [167]

$$\begin{aligned} dS(t) &= \mu S(t)\ dt + \sqrt{V(t)}\ S(t)\ dW_1(t), \\ dV(t) &= k\ (\theta - V(t))\ dt + \sigma\ \sqrt{V(t)}\ dW_2(t),\quad dW_1(t)\ dW_2(t) = \rho\ dt, \end{aligned}$$

where $S(t)$ and $V(t)$ are the stock price and volatility (its return variance) process respectively, $W_1(t)$ and $W_2(t)$ are correlated Wiener processes with correlation coefficient ρ. Here, k is the mean reverting speed, θ is the long run mean, σ is the intensity of volatility. Fig. 6.3(a) shows the behavior of the price of the asset and fig. 6.3(b), its corresponding volatility due to a fluctuating market.

6.2.4 Two Species Stochastic Competition Model

Lotka and Volterra proposed a two species competition model, which was later studied by Gause empirically. The proposed model is given by [30]

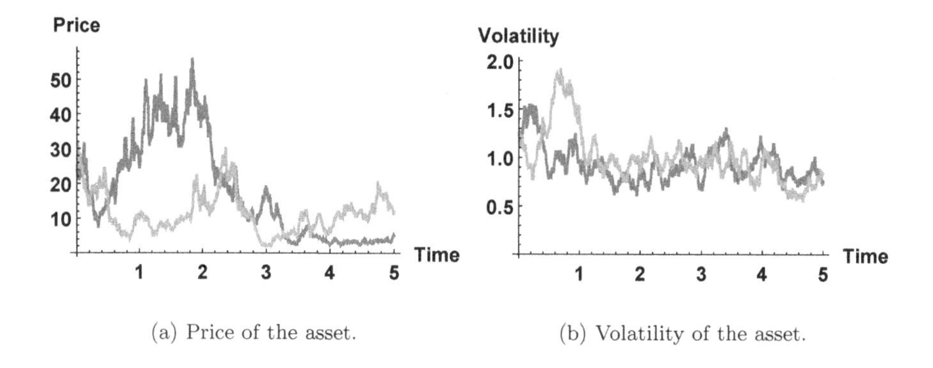

(a) Price of the asset.

(b) Volatility of the asset.

FIGURE 6.3: *Heston model showing the price of the asset and its corresponding volatility. The parameter values are* $\mu = 0.2,\ k = 2,\ \theta = 1,\ \sigma = 0.5,\ \rho = -0.35$ *with initial condition:* $(25, 1.25)$.

$$\begin{aligned}
\frac{dN_1(t)}{dt} &= r_1\ N_1(t)\ \left(\frac{K_1 - N_1(t) - \alpha_{12}\ N_2(t)}{K_1}\right), \\
\frac{dN_2(t)}{dt} &= r_2\ N_2(t)\ \left(\frac{K_2 - N_2(t) - \alpha_{21}\ N_1(t)}{K_2}\right),
\end{aligned}$$

where $N_1(t)$ and $N_2(t)$ are densities of species 1 and species 2, respectively, at any time t. The meaning and interpretation of the positive parameters r_1, K_1, β_{12}, r_2, K_2, β_{21} are left for the readers.

The model is now subjected to external noises and we obtain the stochastic two species competition model as

$$\begin{aligned}
dN_1(t) &= r_1\ N_1(t)\ \left(\frac{K_1 - N_1(t) - \alpha_{12}\ N_2(t)}{K_1}\right)\ dt + \sigma_1\ dW_1(t), \\
dN_2(t) &= r_2\ N_2(t)\ \left(\frac{K_2 - N_2(t) - \alpha_{21}\ N_1(t)}{K_2}\right)\ dt + \sigma_2\ dW_2(t),
\end{aligned}$$

with initial conditions $N_1(0) = N_2(0) = 50$, where $W_1(t)$, $W_2(t)$ are two independent Wiener processes and σ_1, σ_2 are the intensities of the noise.

The model is solved numerically by taking $r_1 = 0.22, r_2 = 0.06, K_1 = 13, K_2 = 5.8, \alpha_{12} = 3.15, \alpha_{21} = 0.44$ and different values of the intensities σ_1, σ_2. Fig. 6.4(a) shows the deterministic model ($\sigma_1 = 0,\ \sigma_2 = 0$), where, species 1 wins and species 2 dies off. With $\sigma_1 = 0.7,\ \sigma_2 = 0.7$, the stochastic model shows similar dynamics, that is, due to competition, species 2 goes to extinction (fig. 6.4(b)). However, it is observed that by manipulating the intensities of the noise, the dynamics of the model can be changed. Taking

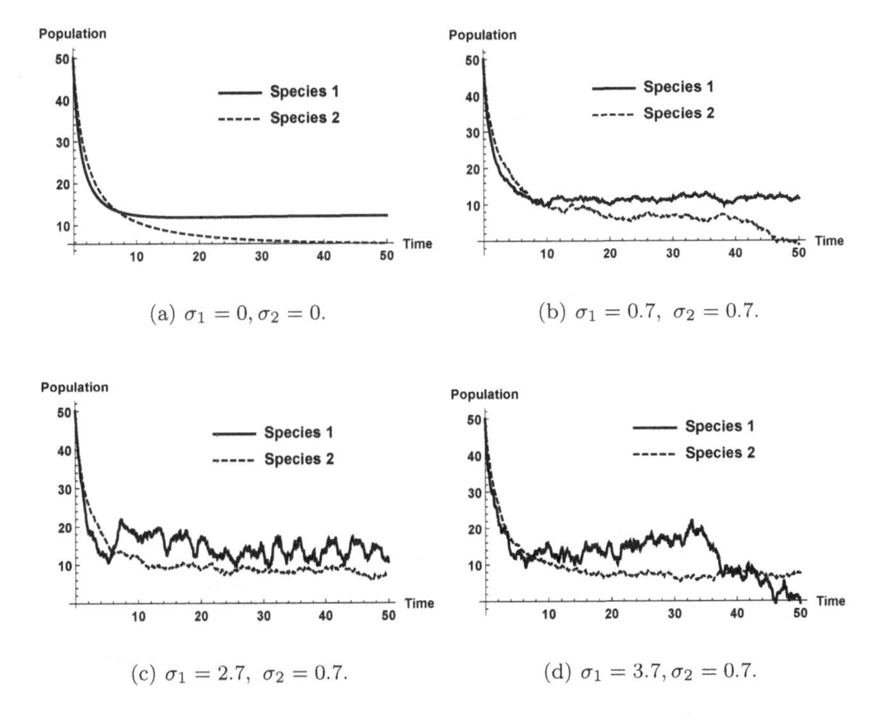

(a) $\sigma_1 = 0, \sigma_2 = 0$.

(b) $\sigma_1 = 0.7, \ \sigma_2 = 0.7$.

(c) $\sigma_1 = 2.7, \ \sigma_2 = 0.7$.

(d) $\sigma_1 = 3.7, \sigma_2 = 0.7$.

FIGURE 6.4: *The figures show the dynamics of two species stochastic competing model with different values of the intensities of the noise. (a) Deterministic model: Species 1 wins, (b) stochastic model: Species 1 wins, (c) stochastic model: Coexistence of both the species, (d) stochastic model: Species 2 wins.*

$\sigma_1 = 2.7, \ \sigma_2 = 0.7$, we observe that neither species can contain the other and it is the case of stable coexistence (fig. 6.4(c)) of both the species. But, with $\sigma_1 = 3.7, \ \sigma_2 = 0.7$, species 2 dominates over species 1 and forces it to go to extinction (fig. 6.4(d)).

Therefore, from the behavior of the model, it is concluded that with proper intensities of the noise (which needs to be interpreted in terms of biology/ecology), different scenarios are obtained to give rich dynamics of the two species stochastic competition model.

Note: Please note that the stability analysis of the stochastic models discussed is actually research problems; hence they are left for the readers to research.

6.3 Research Problem: Cancer Self-Remission and Tumor Stability–a Stochastic Approach [145]

6.3.1 Background of the Problem

A great deal of human and economical resources are devoted, with successful results as well as failures, to cancer research, with particular emphasis on experimental and theoretical immunology. Cancer is one of the greatest killers in the world. Patients with advanced cancer beyond the possibility of cure are often sent home to die, only to show up again five or seven years later free of disease. No one knows the real reason behind this. This spectacular phenomenon of spontaneous cancer remission persists in the medical annals, totally inexplicable but real. A Medline search (1966–1992) yields 11,231 references to the terms "Spontaneous Regression" or "Spontaneous Remission" [35, 119]. High prolonged temperature or hyperthermic condition may be one reason behind this spontaneous regression, say, 105 °F fever for over a week [132]. Majumder and Roy [104], in their Frank H. George award-winning paper, developed a theoretical foundation of a tumor's self-control, homeostasis, and regression induced by thermal radiation or oxygenation fluctuations (using the Prigogine-Glansdorff Langevin stability theory and biocybernetic principles).

On the biomedical front, reasonable levels of progress have been and are still being made in the fight against disseminated cancers and precancerous disorders. In certain instances, an appreciable increase has been recorded in remission. Despite the advances, however, challenges remain in detection, treatment, and management of these diseases that engender multidisciplinary approaches in many circumstances. Several authors have also suggested different mathematical models of the disseminated cancers, which are used to capture some essential characteristics of cancer cell kinetics [26, 27]. However, the therapeutic applicability in some of the tumors which are difficult to treat conventionally, and estimation of system parameters under the effect of stochastic fluctuations in such cases (for example, radiation, chemotherapy, hemodynamic perfusion of the tumor) requires special attention.

In this work, we have developed a model for spontaneous tumor regression and progression, which is an interaction between the anticancer agent or immune cell, namely, cytotoxic T-lymphocytes (CTLs) and macrophages, which are natural killer cells that destroy the malignant (tumor) cells, that is, a predator–prey-like relationship. CTLs have cytoplasmic granules that contain the proteins perforin and granzymes. When the CTL binds to its target, the contents of the granules are discharged by exocytosis. Perforin molecules insert themselves into the plasma membrane of target cells and this enables the

granzymes to enter the cell. Granzymes are serine proteases that once inside the cell proceed to cleave the precursors of caspases, thus activating them to cause the cell to self-destruct by apoptosis. We can translate this interaction in terms of standard Parallel Distributing Processing (PDP) models of receiving, storing, processing, and sending information. Here, the immune cells (CTLs) have the following informational modes:
(i) Receiving the informational specifics of a malignant cell by immunologically "hunting" for it.
(ii) Storing the information when the immune cell engulfs or attaches to the cancer cell; is a "resting" state of the immune cell.
(iii) Processing the cytolytic information, that is, digesting or destroying the cancer cell.
(iv) Sending information to the immunological network on the lysis of the malignant cell.

Local stability analysis about the equilibrium point(s) is performed on the proposed mathematical model and the corresponding biological implications have been stated. Next, we contend that spontaneous cancer regression can be taken as fluctuation regression and the deterministic system is extended to a stochastic one by allowing random fluctuations about the positive interior equilibrium. Conditions for stochastic stability about the positive equilibrium have been obtained and the numerical results are justified with proper biological explanations. At the end, the control of tumor progression has been proposed under a stochastic situation.

6.3.2 The Deterministic Model

In this section, we construct the spontaneous tumor regression and progression system as a prey–predator-like system. The two following cellular species are clear in case of tumor. The predator is T-lymphocytes and cytotoxic macrophages/natural killer cells of the immune system, which attack, destroy or ingest the tumor cells. The prey is the tumor cells, which are attacked and destroyed by the immune cells. The predator has two states, hunting and resting, and destroys the prey (tumor cells). The tumor cells are caught by macrophages which can be found in all tissues in the body and circulate in the blood system. Macrophages absorb tumor cells, eat them and release series of cytokines (fast diffusing substance) which activate the resting T-lymphocytes (predator cells) that coordinate the counterattack. The resting predator cells can also be directly stimulated to interact with antigens. These resting cells cannot kill tumor cells directly, but they are converted to a special type of T-lymphocyte cells called natural killer or hunting cells and begin to multiply and release other cytokines that further stimulate more resting cells. This stimulation or conversion between hunting and resting cells results in a degradation of resting cells undergoing natural growth and activation of hunting cells.

Keeping in mind the above biological scenario, we consider that both tumor and hunting cells follow logistic growth and that the tumor cells are being destroyed at a rate proportional to the densities of tumor cells and hunting predator cells according to the law of mass action. Next, we assume that the resting predator cells are converted to the hunting cells either by direct contact with them or by contact with a fast diffusing substance (cytokines) produced by the hunting cells. We also consider that once a cell has been converted, it will never return to the resting stage and active cells die at a constant probability per unit of time. We assume that all resting predator cells and tumor cells are nutrient-rich and undergoing mitosis. We suppose that the tumor cells have a proliferative advantage over the normal cells [20]. Hence, we consider two different carrying or packing capacities for tumor cells and resting predator cells, respectively, where the carrying capacity of tumor cells is greater than that of the normal cells. This results in the following tumor-immune interaction model:

$$\frac{dM}{dt} = q + rM(1 - \frac{M}{k_1}) - \alpha MN, \tag{6.5}$$

$$\frac{dN}{dt} = \beta NZ - d_1 N, \tag{6.6}$$

$$\frac{dZ}{dt} = sZ(1 - \frac{Z}{k_2}) - \beta NZ - d_2 Z, \tag{6.7}$$

where M is the density of the tumor cells, N is the density of the hunting predator cells, Z is the density of the resting predator cells, r is the growth rate of tumor cells, q is the conversion of normal cells to malignant ones (fixed input), k_1 is the carrying capacity of tumor cells, k_2 is the carrying capacity of resting cells (also, $k_1 > k_2$), α is the rate of predation/destruction of tumor cells by the hunting cells, β is the rate of conversion of resting cell to hunting cell, d_1 is the natural death of hunting cells, s is the growth rate of resting predator cells and d_2 is the natural death rate of resting cells. System (6.5)-(6.7) must be analyzed with the following initial conditions: $M(0) > 0$, $N(0) > 0$, $Z(0) > 0$. All parameters of the model are positive.

6.3.3 Equilibria and Local Stability Analysis

We now find all biologically feasible equilibria admitted by the system (6.5)–(6.7) and study the dynamics of the system around each equilibrium. The equilibria for the system (6.5)–(6.7) are as follows:
(i) There exists an equilibrium on the boundary of the first octant, namely,

$$E_1 = \left[\frac{k_1}{2}\left(1 + \sqrt{1 + \frac{4q}{rk_1}}\right), 0, 0\right].$$

(ii) The M-Z planer equilibrium

$$E_2 = \left[\frac{k_1}{2}\left(1 + \sqrt{1 + \frac{4q}{rk_1}}\right), 0, k_2\left(1 - \frac{d_2}{s}\right)\right], \quad \text{which exists if } s > d_2.$$

(iii) The interior equilibrium

$$E_3 = \left[M^*, N^* = \frac{s}{\beta}(1 - \frac{d_1}{\beta k_2}) - \frac{d_2}{\beta}, Z^* = \frac{d_1}{\beta}\right],$$

which exists if $\beta > \frac{sd_1}{k_2(s-d_2)}$ and M^* is the solution of

$$\frac{r}{k_1}(M^*)^2 + [\frac{\alpha s}{\beta}(1 - \frac{d_1}{\beta k_2}) - \frac{\alpha d_2}{\beta} - r]M^* - q = 0, \quad \text{that is,}$$

$$M^* = \frac{-[\frac{\alpha s}{\beta}(1 - \frac{d_1}{\beta k_2}) - \frac{\alpha d_2}{\beta} - r] + \sqrt{[\frac{\alpha s}{\beta}(1 - \frac{d_1}{\beta k_2}) - \frac{\alpha d_2}{\beta} - r]^2 + \frac{4rq}{k_1}}}{2\frac{r}{k_1}}$$

(the negative sign is not admissible for the existence of a positive interior equilibrium).

The variational matrix of the system (6.5)–(6.7) at E_1 is

$$V_1 = \begin{bmatrix} -r\sqrt{1+\frac{4q}{rk_1}} & -\alpha\frac{k_1}{2}(1+\sqrt{1+\frac{4q}{rk_1}}) & 0 \\ 0 & -d_1 & 0 \\ 0 & 0 & (s-d_2) \end{bmatrix}.$$

The eigenvalues of the variational matrix V_1 are $\lambda_1 = -r\sqrt{1+\frac{4q}{rk_1}}(< 0)$, $\lambda_2 = -d_1(< 0)$ and $\lambda_3 = s - d_2$ (> 0 from the existence condition of E_2). Clearly, this steady state is unstable if the planer equilibrium point E_2 exists. The variational matrix of the system (6.5)–(6.7) at E_2 is

$$V_2 = \begin{bmatrix} -r\sqrt{1+\frac{4q}{rk_1}} & -\alpha\frac{k_1}{2}(1+\sqrt{1+\frac{4q}{rk_1}}) & 0 \\ 0 & \frac{\beta k_2(s-d_2)}{s} - d_1 & 0 \\ 0 & -\frac{\beta k_2(s-d_2)}{s} & -(s-d_2) \end{bmatrix}.$$

The eigenvalues of the variational matrix V_2 are $(\lambda_1)' = -r\sqrt{1+\frac{4q}{rk_1}}(< 0)$, $(\lambda_2)' = \frac{\beta k_2(s-d_2)}{s} - d_1 = \frac{k_2}{s}(s - d_2)\{\beta - \frac{sd_1}{k_2(s-d_2)}\}$($>0$ from the existence condition of E_3) and $(\lambda_3)' = -(s - d_2)$ (<0). Therefore, this steady state is also unstable (saddle point) if the positive interior equilibrium point E_3 exists.

The variational matrix of the system (6.5)–(6.7) at E_3 is

$$V_3 = \begin{bmatrix} -\sqrt{[\frac{\alpha s}{\beta}(1-\frac{d_1}{\beta k_2}) - \frac{\alpha d_2}{\beta} - r]^2 + \frac{4rq}{k_1}} & -\alpha M^* & 0 \\ 0 & 0 & s(1-\frac{d_1}{\beta k_2}) - d_2 \\ 0 & -d_1 & -\frac{sd_1}{\beta k_2} \end{bmatrix}.$$

The eigenvalues of the variational matrix V_3 are
$\lambda_1^{''} = -\sqrt{[\frac{\alpha s}{\beta}(1-\frac{d_1}{\beta k_2}) - \frac{\alpha d_2}{\beta} - r]^2 + \frac{4rq}{k_1}}(<0)$,
$\lambda_2^{''} = \frac{-p+\sqrt{p^2-4m}}{2}$, $\lambda_3^{''} = \frac{-p-\sqrt{p^2-4m}}{2}$,
where $p = \frac{sd_1}{\beta k_2}(>0)$ and $m = \{s(1-\frac{d_1}{\beta k_2}) - d_2\}d_1$ (>0, from the existence condition). Since $\lambda_1^{''} < 0$ and the roots $\lambda_2^{''}$ and $\lambda_3^{''}$ have negative real part (since p>0), system (6.5)-(6.7) is asymptotically stable around E_3.

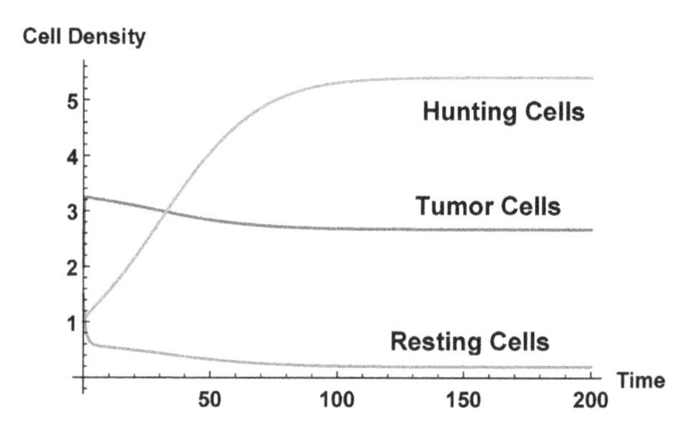

FIGURE 6.5: *Solution of the deterministic system (6.5)–(6.7), showing that E^* is asymptotically stable. The parameter values are given in Table 6.1. The initial condition is $M(0) = 2.0,\ N(0) = 1.5,\ Z(0) = 0.5$.*

6.3.4 Biological Implications

We mainly concentrate on the case (iii) as the first two cases are clearly unstable. Case (iii) shows that the system is asymptotically stable around the positive interior equilibrium E_3. Comparing the densities of malignant tumor cells of E_3 with E_2, we observe that if

$$\alpha < \frac{2r\beta}{s(1-\frac{d_1}{\beta k_2}) - d_2},$$

then the density of malignant tumor cells of E_3 decreases. We have observed analytically that the existence of E_3 implies the instability of E_2. We may

Deterministic Parameters	Values	Stochastic Parameters	Values
q	$10\ cells/day$	ω_3	0.05
r	$0.9/day$	ω_4	0.08
k_1	$0.8/cell$	σ_1	4.0 (unstable situation)
k_2	$0.7/cell$	σ_1	3.4 (stable situation)
α	0.3/cell/day	---	---
β	$0.1/cell/day$	σ_2	0.005 (for both cases)
s	$0.8/day$	σ_3	0.005 (for both cases)
d_1	$0.02/day$	---	---
d_2	$0.03/day$	---	---

TABLE 6.1: Parameter Values for numerical results.

think that if no mechanism works to convert resting cells to hunting ones (existence of E_2), then malignant tumor growth increases and there is no way to control such a situation. On the other hand, if the conversion of resting cells to hunting ones occurs (existence of E_3), then naturally, the question arises, what will be the range of system parameters so that we can control the growth of tumor cells. This justifies our approach to compare the tumor cell densities between two states E_2 and E_3. To support our analytical study, we have performed numerical simulations and consider a hypothetical set of parameters given in Table 6.1.

In our numerical study, we observe that the system is asymptotically stable for $\beta = 0.1$ (fig. 6.5). Moreover, the tumor cell density at E_2 is 3.41 and 2.64 at E_3 and, in this case, $\alpha < 0.32$. This clearly confirms our analytical observations. Further, we may conclude that if we can activate the hunting predator cells, keeping in mind the above condition, then we can control the growth of malignant tumor cells. Due to the lack of clinical results/experimental observations that match our model equations and corresponding parameter values, we have considered the hypothetical values and have tried to give a rough qualitative idea about the dynamics of the model. We have mentioned earlier that the oncological applications of fluctuation theory and energy dissipation play an important role in the behavior of tumor progression and regression. Furthermore, increasing fluctuations can destabilize a system as stated in the Glansdorff-Prigogine Langevin stability theory. In our subsequent analysis, we extend our deterministic model allowing stochastic perturbations in the form of white noise processes, following the approach of Carletti [21].

6.3.5 The Stochastic Model

In model (6.5)–(6.7), we assume that stochastic perturbations of the variables around their values at E_3 are of the white noise type, which are proportional to the distances of M, N, Z from values M^*, N^*, Z^* [15]. Thus, system (6.5–(6.7) results in

$$\begin{aligned} dM &= [q + rM(1 - \frac{M}{k_1}) - \alpha MN]dt + \sigma_1(M - M^*)d\xi_t^1, \quad (6.8) \\ dN &= [\beta NZ - d_1 N]dt + \sigma_2(N - N^*)d\xi_t^2, \quad (6.9) \\ dZ &= [sZ(1 - \frac{Z}{k_2}) - \beta NZ - d_2 Z]dt + \sigma_3(Z - Z^*)d\xi_t^3, \quad (6.10) \end{aligned}$$

where σ_i (i = 1, 2, 3) are real constants and can be defined as the intensities of stochasticity and $\xi_t = (\xi_t^1, \xi_t^2, \xi_t^3)$ is a three-dimensional white noise process [43, 44, 45]. We wonder whether the dynamical behavior of model (6.5)–(6.7) is robust with respect to such stochastic perturbations by investigating the asymptotic stochastic stability behavior of equilibrium E_3 for (6.8)–(6.10) and comparing the results with those obtained for (6.5)–(6.7).

Equations (6.5)–(6.7) can be represented as an Ito Stochastic differential system of the type

$$dX_t = f(t, X_t)dt + g(t, X_t)d\xi_t, \quad X_{t_0} = X_0, \ t \in [t_0, t_f], \quad (6.11)$$

where the solution $\{X_t, t \in [t_0, t_f](t > 0)\}$ is an Ito process, f is the slowly varying continuous component or drift coefficient, g is the rapidly varying continuous random component or diffusion coefficient [84] and ξ_t is a multidimensional stochastic process having scalar Wiener process components with increments $\Delta\xi_t^j = \xi_{t+\Delta t}^j - \xi_t^j = \xi_j(t + \Delta t) - \xi_j^t$, which are independent Gaussian random variables N(0, Δt)-distributed.

Comparing (6.8)–(6.10), we have

$$X_t = (M, N, Z)^T, \xi_t = (\xi_t^1, \xi_t^2, \xi_t^3)^T,$$

$$f = \begin{bmatrix} q + rM(1 - \frac{M}{k_1}) - \alpha MN \\ \beta NZ - d_1 N \\ sZ(1 - \frac{Z}{k_2}) - \beta NZ - d_2 Z \end{bmatrix},$$

$$g = \begin{bmatrix} \sigma_1(M - M^*) & 0 & 0 \\ 0 & \sigma_2(N - N^*) & 0 \\ 0 & 0 & \sigma_3(Z - Z^*) \end{bmatrix}. \quad (6.12)$$

Since the diffusion matrix (6.12) depends on the solution $X_t = (M, N, Z)^T$, system (6.8)–(6.10) is said to have multiplicative noise. Furthermore, from the diagonal form of the diffusion matrix (6.11), the system (6.8)–(6.10) is said to have (multiplicative) diagonal noise.

6.3.6 Stochastic Stability of the Positive Equilibrium

Introducing the variables $u_1 = M - M^*$, $u_2 = N - N^*$, $u_3 = Z_Z^*$, the stochastic differential system (6.8)–(6.10) can be centered at its positive equilibrium

$$E_3 = \left(\frac{-(\alpha N^* - r) + \sqrt{(\alpha N^* - r)^2 + \frac{4rq}{k_1}}}{2\frac{r}{k_1}}, \frac{s}{\beta}(1 - \frac{d_1}{\beta k_2}) - \frac{d_2}{\beta}, \frac{d_1}{\beta} \right),$$

which exists provided that $\beta > \frac{sd_1}{k_2(s-d_2)}$.

We now have to show that system (6.8)–(6.10) is asymptotically stable in the mean square sense (or in probability 1). Linearizing the vector function f in (6.11) around the positive equilibrium E_3, we obtain a set of stochastic differential equations (SDEs).

From the Jacobian matrix of E_3, the linearized SDEs around E_3 take the form

$$du(t) = f(u(t))dt + g(u(t))d\xi(t), \tag{6.13}$$

$u(t) = \mathrm{col}(u_1(t), u_2(t), u_3(t))$,

$$f(u(t)) = \begin{bmatrix} \delta_1 u_1 & -\delta_2 u_2 & 0 \\ 0 & 0 & \frac{\beta}{\alpha}\delta_2 u_3 \\ 0 & -d_1 u_1 & -\frac{sd_1}{\beta k_2} u_3 \end{bmatrix}, \text{ and } g(u(t)) = \begin{bmatrix} \sigma_1 u_1 & 0 & 0 \\ 0 & \sigma_2 u_2 & 0 \\ 0 & 0 & \sigma_3 u_3 \end{bmatrix},$$

where $\delta_1 = -\sqrt{[\frac{\alpha s}{\beta}(1 - \frac{d_1}{\beta k_2}) - \frac{\alpha d_2}{\beta} - r]^2 + \frac{4rq}{k_1}}$ (<0) and $\delta_2 = \frac{\alpha s}{\beta}(1 - \frac{d_1}{\beta k_2}) - \frac{\alpha d_2}{\beta} > 0$ (from the existence condition of E_3). Obviously in (6.13), the positive equilibrium E_3 corresponds to the trivial solution $(u_1, u_2, u_3) = (0, 0, 0)$.

We consider the set $\Psi = \{(t \geq t_0) \times R^3, t_0 \in R^+\}$. If $V \in C_2(\Psi)$ is twice continuously differentiable function with respect to u and a continuous function with respect to t, then we can state the following theorem by Afanasev [2].

Theorem 6.3.6.1 *Suppose there exists a function $V(u,t) \in C_2(\Psi)$ satisfying the inequalities*

$$K_1|u|^p \leq V(u,t) \leq K_2|u|^p, \tag{6.14}$$

$$LV(u,t) \leq -K_3|u|^p, K_i > 0, p > 0 (i = 1, 2, 3). \tag{6.15}$$

Then, the trivial solution of (6.13) is exponentially p-stable, for $t \geq 0$.

Note that in (6.14) and (6.15), if $p = 2$, then the trivial solution of (6.13) is said to be exponentially mean square stable. Furthermore, the trivial solution of (6.13) is globally asymptotically stable in probability. With reference to (6.13), $LV(u,t)$ is defined as follows:

$$\begin{aligned} LV(u,t) &= \frac{\partial V(u(t),t)}{\partial t} + f^T(u(t))\frac{\partial V(u,t)}{\partial u} \\ &+ \frac{1}{2}Tr[g^T(u(t))\frac{\partial^2 V(u,t)}{\partial u^2}g(u(t))], \\ \text{where, } \frac{\partial V}{\partial t} &= col\left(\frac{\partial V}{\partial u_1}, \frac{\partial V}{\partial u_2}, \frac{\partial V}{\partial u_3}\right), \\ \frac{\partial^2 V(u,t)}{\partial u^2} &= \left(\frac{\partial^2 V}{\partial u_j \partial u_i}\right)_{i,j=1,2,3} \end{aligned} \tag{6.16}$$

and the superscript T means transposition.
We now state our main result in the form of the following theorem:

Theorem 6.3.6.2 *Assume that for all positive real values ω_3, ω_4 the following inequality holds true:*

$$(2\delta_1' - \sigma_1^2)(2d_1\omega_4 - (\omega_2^* + \omega_4)\sigma_2^2) > \delta_2^2. \tag{6.17}$$

Then, if

$$\sigma_1^2 < 2\delta_1', \sigma_2^2 < \frac{2d_1\omega_4}{(\omega_2^* + \omega_4)}, \sigma_3^2 < \frac{2\frac{sd_1}{\beta k_2} - \omega_3}{(\omega_3 + \omega_4)} \tag{6.18}$$

where $\omega_2^* = \frac{\alpha}{\beta\delta_2}[(d_1 - \frac{\beta\delta_2}{\alpha} + \frac{sd_1}{\beta k_2})\omega_4 + d_1\omega_3]$ and $\delta_1' = -\delta_1 (> 0)$ (since $\delta_1 < 0$), $0 < \delta_2 < \frac{sd_1}{\alpha k_2}$, the zero solution of the system (6.8)–(6.10) is asymptotically mean square stable.

Proof: We consider the Lyapunov function

$$V(u(t),t) = \frac{1}{2}[u_1^2 + \omega_2 u_2^2 + \omega_3 u_3^2 + \omega_4(u_2 + u_3)^2],$$

where ω_i $(i = 2,3,4)$ are real positive constants to be chosen later.

Applying (6.16) to the system (6.8)–(6.10), we get

$$\begin{aligned} LV(u(t)) &= (\delta_1 u_1 - \delta_2 u_2)u_1 + \frac{\beta}{\alpha}\delta_2\omega_2 u_2 u_3 + (-d_1 u_2 - \frac{sd_1}{\beta k_2}u_3)\omega_3 u_3 \\ &+ (-d_1 u_2 + (\frac{\beta}{\alpha}\delta_2 - \frac{sd_1}{\beta k_2})u_3)\omega_4(u_2 + u_3) \\ &+ \frac{1}{2}Tr[g^T(u(t))\frac{\partial^2 V}{\partial u^2}g(u(t))] \\ &= \delta_1 u_1^2 - \delta_2 u_2 u_1 + [\frac{\beta}{\alpha}\delta_2\omega_2 - d_1\omega_4 + (\frac{\beta}{\alpha}\delta_2 - \frac{sd_1}{\beta k_2})\omega_4 - d_1\omega_3]u_2 u_3 \\ &- d_1\omega_4 u_2^2 - [\frac{sd_1}{\beta k_2}\omega_3 + \frac{sd_1}{\beta k_2}\omega_4 - \frac{\beta}{\alpha}\delta_2\omega_4]u_3^2 \\ &+ \frac{1}{2}Tr[g^T(u(t))\frac{\partial^2 V}{\partial u^2}g(u(t))]. \end{aligned}$$

Also,

$$\frac{\partial^2 V}{\partial u^2} = \begin{bmatrix} 1 & 0 & 0 \\ 0 & \omega_2+\omega_4 & \omega_4 \\ 0 & \omega_4 & \omega_3+\omega_4 \end{bmatrix},$$

which implies

$$g^T(u(t))\frac{\partial^2 V}{\partial u^2}g(u(t)) = \begin{bmatrix} \sigma_1^2 u_1^2 & 0 & 0 \\ 0 & (\omega_2+\omega_4)\sigma_2^2 u_2^2 & \omega_4\sigma_2\sigma_3 u_2 u_3 \\ 0 & \omega_4\sigma_2\sigma_3 u_2 u_3 & (\omega_3+\omega_4)\sigma_3^2 u_3^2 \end{bmatrix}$$

and

$$\frac{1}{2}Tr[g^T(u(t))\frac{\partial^2 V}{\partial u^2}g(u(t))] = \frac{1}{2}[\sigma_1^2 u_1^2 + (\omega_2+\omega_4)\sigma_2^2 u_2^2 + (\omega_3+\omega_4)\sigma_3^2 u_3^2].$$

Therefore,

$$\begin{aligned} LV(u(t)) &= \delta_1 u_1^2 - \delta_2 u_2 u_1 + [\frac{\beta}{\alpha}\delta_2\omega_2 - d_1\omega_4 + (\frac{\beta}{\alpha}\delta_2 - \frac{sd_1}{\beta k_2})\omega_4 - d_1\omega_3]u_2 u_3 \\ &- d_1\omega_4 u_2^2 - [\frac{sd_1}{\beta k_2}\omega_3 + \frac{sd_1}{\beta k_2}\omega_4 - \frac{\beta}{\alpha}\delta_2\omega_4]u_3^2 \\ &+ \frac{1}{2}[\sigma_1^2 u_1^2 + (\omega_2+\omega_4)\sigma_2^2 u_2^2 + (\omega_3+\omega_4)\sigma_3^2 u_3^2]. \end{aligned}$$

If we choose ω_2 such that

$$\frac{\beta}{\alpha}\delta_2\omega_2 - d_1\omega_4 + (\frac{\beta}{\alpha}\delta_2 - \frac{sd_1}{\beta k_2})\omega_4 - d_1\omega_3 = 0, \quad \text{that is,}$$

$$\omega_2^* = \frac{\alpha}{\beta\delta_2}[(d_1 - \frac{\beta}{\alpha}\delta_2 + \frac{sd_1}{\beta k_2})\omega_4 + d_1\omega_3] \text{ and } \delta_1^{'} = -\delta_1(>0), \quad (\text{since } \delta_1 < 0),$$

then,

$$\begin{aligned} LV(u,t) &= -\delta_1^{'} u_1^2 - \delta_2 u_2 u_1 - d_1\omega_4 u_2^2 - \frac{sd_1}{\beta k_2}\omega_3 u_3^2 - (\frac{sd_1}{\beta k_2} - \frac{\alpha}{\beta}\delta_2)\omega_4 u_3^2 \\ &+ \frac{1}{2}[\sigma_1^2 u_1^2 + (\omega_2^*+\omega_4)\sigma_2^2 u_2^2 + (\omega_3+\omega_4)\sigma_3^2 u_3^2] = -u^T Q u \quad (6.19) \end{aligned}$$

for all ω_3, $\omega_4 > 0$, where

$$Q = \begin{bmatrix} q_{11} & q_{12} & 0 \\ q_{21} & q_{22} & 0 \\ 0 & 0 & q_{33} \end{bmatrix}$$

and $q_{11} = \delta_1^{'} - \frac{\sigma_1^2}{2}$, $q_{12} = \frac{\delta_2}{2}$, $q_{21} = \frac{\delta_2}{2}$, $q_{22} = d_1\omega_4 - \frac{1}{2}(\omega_2^*+\omega_4)\sigma_2^2$, $q_{33} = \frac{sd_1}{\beta k_2}\omega_3 - \frac{1}{2}(\omega_3+\omega_4)\sigma_2^2$.

Thus, Q is a real symmetric positive definite matrix and hence its eigenvalues λ_1, λ_2, λ_3 will be positive real quantities if the following conditions hold:

$$(2\delta_1^{'} - \sigma_1^2)(2d_1\omega_4 - (\omega_2^* + \omega_4)\sigma_2^2) > \delta_2^2,$$

$$\sigma_1^2 < 2\delta_1^{'},\ \sigma_2^2 < \frac{2d_1\omega_4}{(\omega_2^* + \omega_4)},\ \sigma_3^2 < \frac{\frac{2sd_1}{\beta k_2} - \omega_3}{\omega_3 + \omega_4},$$

where $\omega_2^* = \frac{\alpha}{\beta\delta_2}[(d_1 - \frac{\beta}{\alpha}\delta_2 + \frac{sd_1}{\beta k_2})\omega_4 + d_1\omega_3]$ and $0 < \delta_2 < \frac{sd_1}{\alpha k_2}$.

If λ_m denotes the minimum of the three positive eigenvalues λ_1, λ_2, λ_3 then from (6.19), we get,

$$LV(u,t) \leq -\lambda_m |u(t)|^2,$$

and we conclude that the zero solution of system (6.8)–(6.10) is asymptotically mean square stable.

6.3.7 Numerical Results and Biological Interpretations

We numerically simulate the strong solution of the SDEs (6.8)–(6.10) along one path with an initial condition (M(0), N(0), Z(0)) = (2.0,1.5,0.5). The approximate strong solution of the Ito system of SDEs (6.8)–(6.10) was computed by the Euler-Maruyama method, which is of strong order 0.5. In this case, we observe that the stochastic threshold depends on the parameters σ_i(i = 1,2,3). In a deterministic case, we observe that the system is stable and the tumor cells can be controlled if $\alpha < 0.32$ (fig. 6.5). It is interesting to note that for $\alpha < 0.32$ and $\sigma_1 = 4.0$, $\sigma_2 = 0.005$, $\sigma_3 = 0.005$, the system is stochastically unstable (fig. 6.6(a)). Hence, in the realistic situation, that is, under the effect of stochastic fluctuations, the deterministically stable system may become unstable due to particular intensities of white noises. We also observe that if we set $\sigma_1 < 3.68$, $\sigma_2 < 0.14$, $\sigma_3 < 1.64$, then the system becomes stochastically stable (fig. 6.6(b)). Such parameter values clearly satisfy the inequalities (6.17) and (6.18) and provide thresholds for the intensity of the white noises, which may be useful to control the system under stochastic fluctuations. This may be helpful to experimental scientists who conduct clinical experiments of this type.

Cancer is one of the most prolific killers in the world and the control of tumor growth requires special attention. In this section, we have developed a deterministic predator–prey-like model considering the fact that spontaneous tumor regression and progression is an interaction between an anticancer agent and an immune cell (T-lymphocytes and cytotoxic macrophages), which destroys the malignant (tumor) cells. Here, the immune cell also possesses the resting stage and may be activated to the hunting stage for destroying the

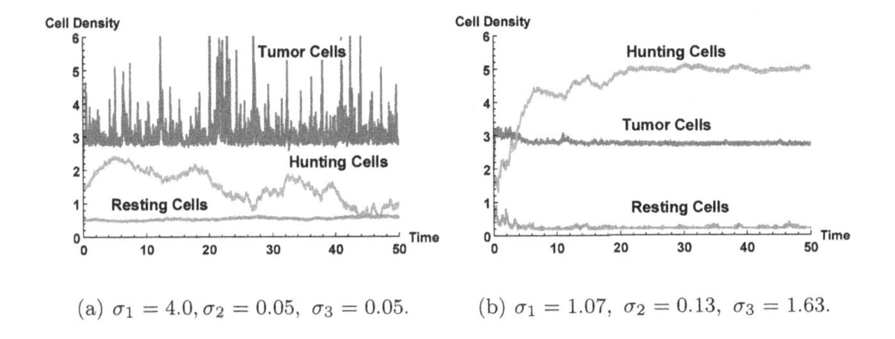

(a) $\sigma_1 = 4.0, \sigma_2 = 0.05,\ \sigma_3 = 0.05$.

(b) $\sigma_1 = 1.07,\ \sigma_2 = 0.13,\ \sigma_3 = 1.63$.

FIGURE 6.6: *The figures show the solutions of the Ito system of stochastic differential equations (6.8)–(6.10) for different sets of values of the intensities of the noise. The parameter values are given in Table 6.1. The initial condition for both the cases is* $M(0) = 2.0$, $N(0) = 1.5$, $Z(0) = 0.5$. *(a)* E^* *is stochastically unstable, (b)* E^* *is stochastically stable.*

tumor cells. Our analytical, as well as numerical studies, reveal that the system has stability properties around the positive interior equilibrium. Consequently, by comparing the densities of malignant cells for hunting cell free equilibrium (E_2) and positive interior equilibrium (E_3), we have obtained a threshold for the rate of destruction of tumor cells by the hunting cells (α) in terms of the rate of conversion of resting cells to hunting cells (β) as well as the other system parameters. Such thresholds clearly give an idea about the control of the malignant tumor growth in deterministic situations.

We contend that spontaneous cancer regression can be taken as fluctuation regression (mentioned earlier). Hence, the deterministic system is extended to a stochastic one by allowing random fluctuations about the positive interior equilibrium. In the stochastic case, we investigate the dynamical behavior of the model (6.8)–(6.10) by computing the strong solution along one sample path. Our simulations confirm that stochastic mean square stability is achieved under particular conditions for the intensities of the fluctuations according to conditions (6.17) and (6.18). If the conditions (6.17) and (6.18) are satisfied, then system (6.8)–(6.10) is asymptotically stable (in mean square sense).

Biologically, this means that, if the intensities of the stochastic fluctuations remain below some threshold values, the density of the malignant tumor cells decreases to a very low value; that is, there occurs a phase transition from macro cancer focus to micro cancer focus. This corresponds to the regression and elimination of malignancy. However, the internal stochasticity of the system cannot be estimated, but by conditions (6.17) and (6.18), we get an idea of it, provided we know the other system parameters. On the other hand, if we consider external stochastic fluctuations, for example, (i) radiation flux, (ii) cytotoxic chemical flux, (iii) immune cell concentration, (iv) tumor temperature, (v) glucose level of the blood impinging on the tumor, (vi) oxygen partial pressure, that is, oxygenation level in tumor matrix, (vii) hemodynamic perfusion of the tumor, and so on, then our procedure may be applied as well to get an estimation of the system parameters like α (rate of predation of tumor cells by hunting cells) and β (rate of conversion of resting cells to hunting cells) for controlling the growth of the malignant tumor cells.

The model we developed is a general one. However, we placed special emphasis on the therapeutic applicability in some of the tumors which are really difficult to treat conventionally: (i) radio-resistant Ewing's bone tumor (temperature variation therapy) [141], (ii) lung carcinoma oxygenation by endostatin therapy [140, 150] and (iii) neurogranuloma [139, 141]. Finally, we can say that the above thresholds will also be helpful to medical practitioners, experimental scientists, and others to control this killer disease.

6.4 Mathematica Codes

Mathematica codes of selected figures.

6.4.1 Stochastic Logistic Growth (Figure 6.2)

```
k = 5.; r = 0.5; x0 = 1;
aa1 = NDSolve[{x'[t] == rx[t](1 - x[t]/k), x[0] == x0}, x, {t, 0, 30}];
pp1 = Plot[x[t]/. aa1, {t, 0, 30}, PlotStyle → Directive[Black, Thick, Dashed],
PlotRange → All, PlotLegends → Automatic]
sto = ItoProcess[dX[t] = rX[t](1 - X[t]/k)dt + σdW(t), X[t],
{X, X0}, t, W ≈ WienerProcess[]]
sto1 = sto/. {X0 → 1, k → 5, σ → 0.3, r → 0.5};
sample1 = RandomFunction[sto1, {0, 30, 0.01}, 2, Method → EulerMaruyama];
pp2 = ListLinePlot[sample1]
Show[pp1, pp2, AxesLabel → {"Time", "Population"},
BaseStyle → {FontWeight → "Bold", FontSize → 12}]
```

6.4.2 Stochastic Competition Model (Figure 6.4)

```
r1 = 0.22; r2 = 0.06; K1 = 13; K2 = 5.8;
α12 = 0.15; α21 = 0.05; σ1 = 3.7; σ2 = 0.7
CModel = ItoProcess[{dN1[t] == r1N1[t](K1 − N1[t] − α12N2[t])/K1 dt
+ σ1dW1[t], dN2[t] = r2N2[t](K2 − N2[t] − α21N1[t])/K2 dt + σ2dW2[t]},
{N1[t], N2[t]}, {{N1, N2}, {50, 50}}, t,
{W1 ≈ WienerProcess[], w2 ≈ WienerProcess[]}];
sol1 = RandomFunction[CModel, {0., 50, 0.1}, 1,
Method → "StochasticRungeKutta"];
p1 = ListLinePlot[sol1, PlotRange → All,
PlotStyle → {{Black, Thick}, {Black, Dashed}}
AxesLabel → {"Time", "Population"},
BaseStyle → {FontWeight → "Bold", FontSize → 12},
PlotLegends → Placed[{"Species 1", "Species 2"}, {Scaled[{0, 0.5}], {0, 0.5}}]]
```

6.5 Matlab Codes

6.5.1 Stochastic Logistic Growth (Figure 6.2)

```
T = 30; k = 200; r = 0.5; b = 0.03; K = 5;
deltat = T/k; deltat2 = sqrt(deltat); nh = 3;
for j = 1 : nh
if j = nh
b = 0.;
end
X(1) = 1.;
for i = 1 : k
X(i + 1) = X(i) + r * X(i) * (1. − X(i)/K) * deltat + b * X(i) * deltat2 * randn;
end
plot ([0 : deltat : T], X,'LineWidth', 1) ;
hold on
end
```

6.5.2 Stochastic Competition Model (Figure 6.4(c))

```
clear; clf;
T = 50;
k = 200;
r1 = 0.22;
r2 = 0.06;
K1 = 13;
K2 = 5.8;
alpha12 = 0.15;
alpha21 = 0.44;
b1 = 0.35;
b2 = 0.15;
dt = T/k;
nh = 1;
for j = 1 : nh
X(1) = 50.;
Y(1) = 50;
for i = 1 : k
X(i + 1) = X(i) + r1 * X(i) * (1. - X(i)/K1 - alpha12 * Y(i)) * dt
+ b1 * X(i) * sqrt(dt) * randn;
Y(i + 1) = Y(i) + r2 * Y(i) * (1. - Y(i)/K2 - alpha21 * X(i)) * dt
+ b2 * Y(i) * sqrt(dt) * randn;
end
plot ([0 : dt : T], X, [0 : dt : T], Y,'LineWidth', 1) ;
legend ('Species 1', 'Species 2')
end
xlabel ('Time') ; ylabel ('Population') ;
```

6.6 Exercises

1. A Schlögl model is described as an artificial chemical system used to describe bistable behavior in the state variable for certain parameter values. The model equation is given by [6]

   $$dx = (C_1x^2 + C_4 - C_2x^3 - C_3x)dt + (C_1x^2 + C_4 + C_2x^3 + C_3x)\ dW.$$

   Solve the system numerically by taking $C_1 = C_4 = 6$, $C_2 = 1$, $C_3 = 11$ and comment on the graph.

2. A simple stochastic model for the rainfall at a certain location over a period of decades is given by [6]

   $$d\gamma(t) = \mu\ dt + \sigma\ dW(t),$$

   where $\gamma(t)$ denotes the total rainfall at time t where $\gamma(0) = 0$, μ is the drift term, σ is the diffusion term, and $W(t)$ is the Wiener process. Taking $\mu = 18.57$ and $\sigma = 6.11$, solve the model numerically to obtain an annual rainfall from 1920 to 2000.

3. The Vasicek interest rate model is given by [6, 13]

   $$d\gamma(t) = \alpha[\gamma_e - \gamma(t)]\ dt + \sigma\ dW(t),$$

   and the Cox-Ingersoll-Ross (CIR) interest rate model is given by [6, 13]

   $$d\gamma(t) = \alpha[\gamma_e - \gamma(t)]\ dt + \sigma\sqrt{\gamma(t)}\ dW(t)$$

   where $\gamma(t)$ is the instantaneous interest rate. By taking $\alpha = 0.7$, $\gamma_e = 3.5$, $\sigma = 0.5$ and $\gamma(0) = 5$, solve both models numerically and compare the graphical results.

4. The composite index of Budapest is modeled using a stochastic differential equation as [88]

   $$dv(t) = -\alpha\ v(t)\ dt + \beta\ v(t)\ dW_1 + \gamma\ dW_2,$$

   where $v(t)$ is the velocity (relative) index changes for sufficiently short time, α, β and γ are positive constants. W_1 and W_2 are Wiener processes. The two separate Wiener terms can be interpreted as an internal noise characterizing the trading dynamics ($\beta\ v(t)\ dW_1$) and an external driving noise ($\gamma\ dW_2$) representing information and market news.
   (i) Taking $\alpha = 1.1$, $\beta = 1.0$, $\gamma = 0.0006$ and $v(0) = 6000$, simulate the model numerically between $[0, 2 \times 10^6]$.
   (ii) Introducing an additional term $(-qv^3dt)$, solve the model again numerically and compare the graphs (take $q = 0.7$).

5. A new stochastic differential equation based on pharmacokinetics is given by [6]

$$dC(t) = -K\ C(t)\ dt + \gamma\ C(t)\ dW,$$

where $C(t)$ is the concentration of the drug in the compartment (the human body may be represented as a set of compartments) at time t.
(i) Solve the model numerically by taking $K = 4$, $\gamma = 0$, $C(0) = 1$ and plot the graph. Does stochasticity have any effect on this graph?
(ii) Now, putting $\gamma = 2$, solve the model again and compare the graph with (i).
(iii) A more realistic assumption for this model is when K is randomly perturbed. Then, the system becomes

$$\begin{aligned} dK(t) &= [K^* - K(t)]\ dt + \gamma\ \sqrt{K(t)}\ dW, \\ dC(t) &= -K(t)\ C(t)\ dt. \end{aligned}$$

Taking $K^* = 4$, $\gamma = 2$, $C(0) = 1$, $K(0) = 3.5$, obtain the numerical solution, plot them and compare them with (ii).

6. The stochastic system of SDE that describes the dynamics of the spring-mass system has the form [80, 87]

$$\begin{aligned} dx(t) &= v(t)dt, \\ m\ dv(t) &= [-kx(t) - b\ v(t)]\ dt + \sqrt{2\gamma^2\lambda}\ dW(t), \end{aligned}$$

where $x(t)$ is the displacement of the mass from equilibrium, $v(t)$ is the velocity, m is the mass, $M(t) = mv(t)$ is the momentum and $W(t)$ is the Wiener process. Assuming $k = 1$, $b = 0.5$, $\gamma^2 = 0.25$, $\lambda = 0.4$, $m = 20$, $x(0) = 7$ and $v(0) = 2$, obtain the numerical solution of the system, plot them and comment on the dynamics of the system.

7. The time dependent behavior of a nuclear reaction is governed by a system of SDE, given by [6, 58]

$$\begin{aligned} dn(t) &= [\lambda C(t) + ((1-\beta)\nu - 1)n(t)\sigma_f v - n(t)\sigma_c v + Q]dt \\ &+ \sqrt{v_{11}}\ dW_1 + \sqrt{v_{12}}\ dW_2, \\ dC(t) &= (\beta\nu n(t)\sigma_f v - \lambda_c)dt + \sqrt{v_{21}}\ dW_1 + \sqrt{v_{22}}\ dW_2. \end{aligned}$$

Here, $n(t)$ is the neutron population, $C(t)$ is the number of atoms at time t of a radioactive isotope which spontaneously decays by neutron emission, Q is an extraneous neutron source, λ is the rate of decay of fission product C(t), σ_f is the probability per unit distance for a neutron to cause a fission, σ_c is the probability per unit distance of a neutron loss by capture in an atom, ν is the total number of neutrons per fission,

v is the neutron speed, $\beta\nu$ is the number of atoms of fission product C(t) produced per fission, W_1 and W_2 are two Wiener processes and

$$
\begin{aligned}
V &= \begin{bmatrix} v_{11} & v_{12} \\ v_{21} & v_{22} \end{bmatrix} \text{ where} \\
v_{11} &= \lambda C(t) + [(1-\beta)\nu - 1]^2 n(t)\sigma_f v + n(t)\sigma_c v + Q, \\
v_{12} &= v_{21} = -\lambda C(t) + \beta\nu[(1-\beta)\nu - 1]n(t)\sigma_f v, \\
v_{22} &= \lambda C(t) + \beta^2\nu^2 n(t)\sigma_f v.
\end{aligned}
$$

Taking $\lambda = 0.79/\text{sec}$, $\beta = 0.0079$, $\nu = 2.432$, $\sigma_f v = 4111.84/\text{sec}$, $\sigma_c v = 5858.16/\text{sec}$ and $Q = 10000/\text{sec}$, solve the system of SDE numerically and plot the neutron population from $t = 0$ to $t = 0.1$ seconds.

8. A stochastic susceptible-infective-susceptible (SIS) epidemic model is considered, which consists of susceptible $S(t)$ and infected $I(t)$ populations. The susceptible becomes infected, recovers, and becomes susceptible again. The stochastic version of the model is given by [6, 80]

$$
\begin{aligned}
dS(t) &= \left(-\alpha\frac{S(t)I(t)}{N} + \beta I(t)\right) dt \\
&+ \frac{1}{\sqrt{2}}\sqrt{\alpha\frac{S(t)I(t)}{N} + \beta I(t)}\ (dW_1 - dW_2), \\
dI(t) &= \left(\alpha\frac{S(t)I(t)}{N} - \beta I(t)\right) dt \\
&+ \frac{1}{\sqrt{2}}\sqrt{\alpha\frac{S(t)I(t)}{N} + \beta I(t)}\ (-dW_1 + dW_2),
\end{aligned}
$$

where $S(0) + I(0) = S(t) + I(t) = N$ (Total population, constant), α is the rate at which susceptible becomes infected, β is the rate at which infected individuals after recovery become susceptible again and W_1 and W_2 are two Wiener processes. Taking $\alpha = 0.04$, $\beta = 0.01$, $S(0) = 950$, $I(0) = 50$ and time period $= [0, 100]$, solve the system numerically and graphically represent them. Compare the graphs with the deterministic solution of the model and comment on the results.

9. The susceptible-infective-removed (SIR) epidemic model, where the population is divided into susceptible $S(t)$, infected $I(t)$, and recovered $R(t)$, in deterministic form is given by

$$
\begin{aligned}
\frac{dS(t)}{dt} &= -\frac{\alpha I(t)S(t)}{N}, \\
\frac{dI(t)}{dt} &= \alpha\frac{I(t)S(t)}{N} - \beta I(t), \\
\frac{dR(t)}{dt} &= \beta I(t),
\end{aligned}
$$

where $N = S(t) + I(t) + R(t) = S(0) + I(0) + R(0)$. The corresponding stochastic model is given by [6, 7, 80]

$$\begin{bmatrix} dS \\ dI \end{bmatrix} = \begin{bmatrix} -\alpha\frac{IS}{N} \\ \alpha\frac{IS}{N} - \beta I \end{bmatrix} dt + \frac{\sqrt{\frac{\alpha SI}{N}}}{\sqrt{2 + \frac{\beta N}{\alpha S} + 2\sqrt{\frac{\beta N}{\alpha S}}}} \begin{bmatrix} 1 + \frac{\beta N}{\alpha S} & -1 \\ -1 & 1 + \frac{\beta N}{\alpha S} + \sqrt{\frac{\beta N}{\alpha S}} \end{bmatrix} \begin{bmatrix} dW_1 \\ dW_2 \end{bmatrix},$$

where W_1 and W_2 are two Wiener processes.

(i) Taking $\alpha = 0.04$, $\beta = 0.01$, $S(0) = 950$, $I(0) = 50$ and time period $[0, 100]$, obtain the numerical solution of the system and graphically represent them.
(ii) Compare the graphs with the deterministic solution of the model and comment on the results.

10. A single species population dynamics in a random environment is given by [6]

$$\begin{aligned} dx &= (Bx - Dx)dt + \sqrt{Bx - Dx}\, dW_1, \\ dB &= 2\alpha_1\beta_1(\overline{B} - B)dt + \alpha_1\sqrt{2q_1}\, dW_2, \\ dD &= 2\alpha_2\beta_2(\overline{D} - D)dt + \alpha_2\sqrt{2q_2}\, dW_3, \end{aligned}$$

where $x(t)$ is the population size of a single species and B and D are per capita birth and death rates, respectively. There is an effect on the birth and death rates due to environmental noise. W_1, W_2 and W_3 are three independent Wiener processes. The terms $\beta_1(\overline{B} - B)$ and $\beta_2(\overline{D} - D)$ correspond to the drift $\overline{B}$, $\overline{D}$ being the averages, q_1, q_2 are with the diffusion process.

(i) Taking $2\alpha_1\beta_1 = 2\alpha_1\beta_1 = 1$, $\alpha_1\sqrt{2q_1} = \alpha_2\sqrt{2q_2} = 0.5$, $\overline{B} = 1$, $\overline{D} = 1.4$, $x(0) = 30$, $B(0) = 1$ and $D(0) = 1.4$, solve the system numerically and plot them.
(ii) Compare the graphs with the one, when there is no environmental noise.

Chapter 7

Hints and Solutions

Solutions to Chapter 1

1. $a = 1/2, b = 1/2, c = 0,\ \ u = k\sqrt{\lambda g}$.

3. $h(x) = \sqrt{x^2 + (x-4)^2}$. $h(x)$ is minimum at $x = 2$ ft.

5. (i) The carrying capacity is growing faster than the population, and hence it can sustain the population (continued growth).
(ii) The population levels off smoothly below the carrying capacity (sigmoid growth).
(iii) Overshoot and oscillation: The population exceeds the carrying capacity without inflicting permanent damage and then oscillates around the limit before leveling off.
(iv) The population exceeds the carrying capacity with severe damage to the resource base and is forced to decline rapidly to achieve a new balance with a reduced carrying capacity (overshoot and collapse).

7. Total cost of the plastic sheet, $C = 0.2\left(\frac{0.002}{r} + 4\pi r^2\right)$. The minimum cost per plastic contained is \$0.01395 = 1.395 cents.

9. $\frac{1}{x}\frac{dx}{dt} = r\ln\left(\frac{k}{x}\right) \ \Rightarrow dt \ \Rightarrow \int \frac{d(\ln x)}{\ln x - \ln k} = -\int r dt \Rightarrow \ln\left(\frac{x}{k}\right) = Ae^{-rt}$.
At t=0, $x = x_0 \ \Rightarrow A = \ln\left(\frac{x_0}{k}\right) \ \Rightarrow \ x = k\ \exp\left[\ln\left(\frac{x_0}{k}\right)e^{-rt}\right]$.

11. (i) See fig.7.1.

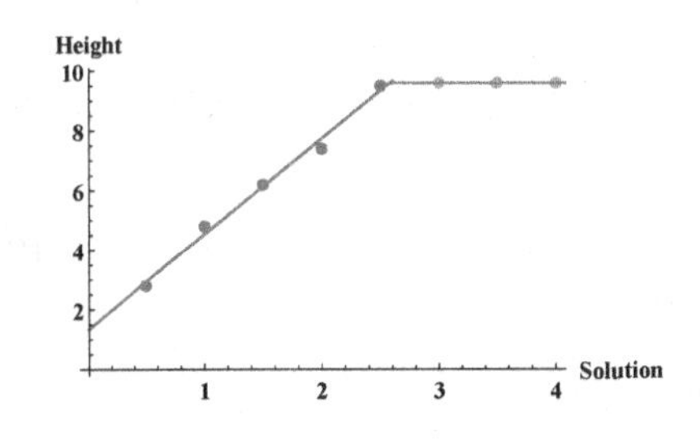

FIGURE 7.1: *Piecewise linear function, which fits the given data.*

(ii) $f(x) = 3.2x + 1.34,\ x < 2.6$, and $= 9.6,\ x \geq 2.6$.

DOI: 10.1201/9781351022941-7

(iii) The increasing portion of the graph represents the period when the reaction is going on, and the horizontal section represents the period when the potassium iodide is used up.

13. Let x be the number of boxes, then total cost is $f(x) = (6x) + 250 + (3x) = 9x + 250$.

15. Let x be the number of boxes of chocolates, then

$$\begin{aligned} f(x) &= 4.50, \ x \leq 15, \\ &= 3.75, \ 15 < x \leq 60, \\ &= 2.75, \ x > 60. \end{aligned}$$

Draw the figure of $f(x)$.

Solutions to Chapter 2

1. (i) $A_n = (1+r)^n A_0 + \frac{[(1+r)^n - 1]}{r} d$. (ii) $A_{36} = \$57,384.57$. (iii) $\$342,629.17$.

3. (i) $P_n = (1+R)^n \left[P_0 + \frac{k}{R}\right] - \frac{k}{R}$. (ii) $P_4 = 5956$. (iii) $k = -190$ (migration), $P_4 = 4269$.

5. (i) Let θ_n be the greatest angle when the pendulum crosses the vertical axis n times, $\theta_0 =$ initial angle and $\alpha =$ fraction. Then, $\theta_{n+1} = -\alpha\theta_n$, $\Rightarrow \theta_1 = -\alpha\theta_0, \theta_2 = -\alpha\theta_1 = (-\alpha)^2\theta_0, \theta_3 = (-\alpha)^3\theta_0, ..., \theta_n = (-\alpha)^n\theta_0$. Note that in each step, the sign does change because the pendulum moves to the opposite side.
(ii) $\theta_n = (-0.90)^n\theta_0$. We want to find the smallest n such that $|\theta_n| \leq 1$, because we are interested in the magnitude of the angle, not its direction. Now,
$|\theta_n| \leq 1 \Rightarrow |(-0.90)^n \times 20| \leq 1 \Rightarrow 20\ (0.90)^n \leq 1 \Rightarrow n\ln(0.9) \leq \ln(1/20) \Rightarrow (-0.1057)n \leq (-2.996) \Rightarrow n \geq 28.34 \Rightarrow n \approx 29$.
(iii) Although the swing distance decreases, a swing always takes 4 seconds. So, the answer is $29 \times 4 = 116$ seconds.

7. (i) $T_n = (1-r)^n T_0 + \frac{(P-H)[1-(1-r)^n]}{r} = (0.97)^n T_0 + 10^5[1 - (0.97)^n]$, (ii) 100,000.

9. (i) $I_n = (1-R)^n I_0 + \frac{N[1-(1-R)^n]}{R}$, $I_{14} = 4314$. (ii) 5000.

11. (i) $I_{n+1} = I_n - rI_n + \alpha I_n(N - I_n)$.
(ii) $I_{n+1} = 0.2I_n + 1.3013I_n - 1.3013 \times 10^{-6} I_n^2$.
(iii) Same as (ii), (iv) Same as (ii), (v) Do yourself.

13. (i) $w_{n+1} = w_n + rw_n$.
(ii) $w_{15} \approx 546$. (iii) Modify the code in (2.8.1) to draw the graph.
(iv) $n = 23.45$ years.
(v) As $n \to \infty, w_n \to \infty$. The population size will grow unboundedly. In reality, it will not happen because the resources necessary for growth are limited. (vi) Same as (iv).

15. (i) Newton's Law of cooling: $T_{n+1} - T_n = k(T_n - S)$, where S is the surrounding temperature and T_n = Body temperature at time.
Suppose at time $n = 0$, the person is murdered. Then, $T_0 = 98.6\,°\text{F}$ and $S = 72\,°\text{F}$.

$$T_1 = (k+1)T_0 - kS,\ T_2 = (k+1)T_1 - kS = (k+1)^2T_0 - kS[1 + (1+k)],$$
$$T_3 = (k+1)T_2 - kT_s = (k+1)^3T_0 - kS[1 + (1+k) + (k+1)^2],$$

$$T_n = (k+1)T_{n-1} - kS$$
$$= (k+1)^nT_0 - kS[1 + (1+k) + (k+1)^2 + ... + (k+1)^{n-1}],$$
$$T_n = (k+1)^nT_0 - kS\frac{[(k+1)^n - 1]}{1+k-1} = (k+1)^n(T_0 - S) + S.$$

Let us assume that after the murder, the forensics arrive after P minutes.

$$T_P = (1+k)^p(26.6) + 72 = 87.5\ , T_{P+60} = (1+k)^{p+60}(26.6) + 72 = 80.4,$$
$$\Rightarrow\ (1+k)^P = 0.5827 \text{ and } (1+k)^{P+60} = 0.3158$$
$$\Rightarrow\ (1+k)^{60} = 0.542\ \Rightarrow\ k = -0.0101. \text{ Now, } (0.9898)^P = 0.5827.$$
$$\Rightarrow\ P = 53.2 \approx 53 \text{ minutes}\ \Rightarrow \text{The victim was murdered at 4:07 am.}$$

17. (i) $w_n = w_{n-1} + k,\ w_n = w_0 + nk$. (ii) $k = 10$.

19. (i) $p_n = p_{n-1} - 0.45p_{n-1} - 0.35p_{n-1} - 0.15p_{n-1} + 0.02l_{n-1} + 0.05b_{n-1} + 2,\ l_n = l_{n-1} + 0.35p_{n-1} - 0.02l_{n-1},\ b_n = b_{n-1} + 0.15p_{n-1} - 0.05b_{n-1} - 0.1b_{n-1} + 1$.
(ii) $(4.24, 74.24, 10.91)$. The system is stable.
(iii) 0.75 mg of chemical B; 8mg (Chemical B); 52.5 mg (vitamin A).

21. (i) $X_{n+1} = r_1X_n - s_1Y_n,\ Y_{n+1} = r_2Y_n - s_2X_n$.
(ii) $X_{n+1} = r_1X_n - s_1Y_n \pm k_1,\ Y_{n+1} = r_2Y_n - s_2X_n \pm k_2$; we consider $(+k_1)$ for immigration and $(-k_1)$ for migration.
(iii) $X_{n+1} = 1.5X_n - 0.4Y_n,\ Y_{n+1} = 1.25Y_n - 0.4X_n$; system is unstable.
(iv) $X_{n+1} = 1.5X_n - 0.4Y_n + 2500,\ Y_{n+1} = 1.25Y_n - 0.4X_n - 1200$; equilibrium point: $(4142.86, 114.28.6)$; system is unstable.

23. (i) $A_n = A_0 - n\,r$. (ii) 18; 180 km.

25. $A_n = A_0 + n\,r$; $n = 35$; $A_{35} = 1000.42$ m.

27. (i) Let a be the amount of water wasted per person while shaving, $W_n=$ water wasted until n people have shaved. $W_n = W_{n-1} + a,\ W_1 = W_0 + a,\ W_0 = 0$ (As no water wasted when none of them shaved), $W_1 = a,\ W_2 = W_1 + a = 2a,\ W_3 = W_2 + a = 3a,\ W_n = n\ a.$
(ii) Water wasted during 1 shave = a $\Rightarrow$ water wasted in 5 shaves $= 5a$. Water wasted per day in shaving $=\frac{5a}{7}$.

29. Let C_n = calorie consumed during n^{th} week, W_n = weight of person after n^{th} week, then,

$$C_(n) = C_{n-1} - 200, \quad W_n = W_{n-1} + \frac{1}{3600}[C_n - 130W_{n-1}],$$

$$\Rightarrow C_n = 21000 - 200\ n, \quad W_{n-1} + \frac{1}{3600}[(21000 - 200\ n) - 130W_{n-1}],$$

$$\Rightarrow W_n = \frac{347}{360}W_{n-1} + \frac{21000 - 200\ n}{3600}. \text{ Now } W_1 = \frac{347}{360}W_0 + \frac{21000 - 200}{3600}$$

$$W_2 = \frac{347}{360}W_1 + \frac{21000 - 200 \times 2}{3600}$$

$$= \left(\frac{347}{360}\right)^2 W_0 + \left(\frac{347}{360}\right)\frac{21000 - 200 \times 1}{3600} + \frac{21000 - 200 \times 2}{3600},$$

$$\Rightarrow W_n = \left(\frac{347}{360}\right)^n W_0 + \sum_{r=1}^{n}\left(\frac{347}{360}\right)^{n-r}\frac{21000 - 200\ r}{3600}.$$

31. The fixed point $x^* = a - 1$, exists for all $a > 1$, is stable for $1 < a < 5$ and unstable for $a > 5$.

33. (i) $I^* = 0,\ I^* = 10^6\left(1 \pm \sqrt{\frac{r}{s}}\right)$. Since we are interested in disease-free equilibrium (disease will eventually die out on its own for any I_0 between 0 and 10^6), we now need to check the stability of $I^* = 0$.

$$f'(I) = 1 - r + s - \frac{sI}{250000} + \frac{3sI^2}{10^{12}} \Rightarrow f'(I^* = 0) = 1 - r + s$$

$$|f'(0)| < 1 \Rightarrow |1 - r + s| < 1 \Rightarrow -1 < 1 - r + s < 1$$

$$\Rightarrow -1 + r - 1 < 1 - r + s + r - 1 < 1 + r - 1 \Rightarrow r - 2 < s < r.$$

This condition holds if $r > 2$. If $r < 2$, the condition will be $0 < s < r$ as $s > 0$.
For $r = 0.25, |1 - 0.25 + s| < 1 \Rightarrow |s + 0.75| < 1 \Rightarrow -1 < s + 0.75 < 1 \Rightarrow -1.75 < s < 0.25$. Since $s > 0$, disease-free equilibrium will be stable for $0 < s < 0.25$. (ii) $\frac{1}{r} < s < \frac{(r+1)^2}{r}$.

35. $0 < A < 2$, Largest value of A is 2.

37. (i) Equilibrium points: $x = -1 - \sqrt{3}, -1 + \sqrt{3}$.
(ii)$(-1 - \sqrt{3})$ and $(-1 + \sqrt{3})$ constitute a 2-cycle, which is stable.

39. Let $f(x) = \frac{rx}{1+x^2}$. The fixed points are obtained by solving

$$\frac{rx^*}{1+x^{*2}} = x^* \Rightarrow x^*\left(1-r+x^{*2}\right) = 0 \Rightarrow x^* = 0,\ x_r{}^* = \pm\sqrt{(r-1)}.$$

The system has three fixed points for $r > 1$ and one fixed point for $r < 1$. Now,

$$f'(x) = \frac{(1+x^2)r - rx.2x}{(1+x^2)^2} = \frac{r(1-x^2)}{(1+x^2)^2}$$
$$f'(0) = r, f'(\pm\sqrt{(r-1)}) = \frac{r(1-r+1)}{(1+r-1)^2} = \frac{2}{r} - 1$$

If $r < 1$, $|f'(0)| < 1$ and $|f'(\pm\sqrt{(r-1)})| > 1$, implying $x^* = 0$ is stable and $x_r{}^* = \pm\sqrt{(r-1)}$ are unstable.
If $r > 1$, $|f'(0)| > 1$ and $|f'(\pm\sqrt{(r-1)})| < 1$, implying $x^* = 0$ is unstable and $x^* = \pm\sqrt{(r-1)}$ are stable.
There is a change in stability from stable to unstable at $r = 1$ and this occurs simultaneously with $x_r{}^*$ coming into existence at the same location, since $x_r{}^* = 0$ when $r = 1$. Again,

$$f'(x)|_{(x^*,r^*)=(0,1)} = \frac{r(1-x^{*2})}{(1+x^{*2})^2}|_{(x^*,r^*)=(0,1)} = 1,$$

implying there is a neutral fixed point at $x = 0$ for $r = 1$. Also,

$$f(x^*, r^*) = f(0,1) = 0,\ \frac{\partial f(0,1)}{\partial x} = 1,$$
$$\frac{\partial f(x^*, r^*)}{\partial r} = \frac{x}{1+x^2}|_{(0,1)} = 0,$$
$$\frac{\partial^2 f(0,1)}{\partial x \partial r} = \frac{(1+x^2) - x.2x}{(1+x^2)^2} = \frac{1-x^2}{(1+x^2)^2} = 1 \neq 0,$$
$$\frac{\partial f^2(0,1)}{\partial x^2} = \frac{(1+x^2)^2(-2xr) - r(1-x^2)2(1+x^2).2x}{(1+x^2)^4}|_{(0,1)} = 0,$$
$$\frac{\partial^3 f(0,1)}{\partial x^3} = -\frac{6(1-6x^2+x^4)}{(1+x^2)^4}|_{(0,1)} = -6 \neq 0.$$

Hence, the system undergoes a pitchfork bifurcation at $(x^*, r^*) = (0,1)$ (modify the code in (2.8.3) to draw the graph).

41. Clearly, $(0,0)$ satisfies the system

$$x_{n+1} = -rx_n - (1+r){x_n}^2 + (2rx_n + 2r + r^2)y_n$$
$$y_{n+1} = -rx_n - (1+r){x_n}^2 + (2rx_n + 2r + r^2)y_n - x_n$$

and hence it is a fixed point. The Jacobian matrix at $(0,0)$

is given by $J(r) = \begin{bmatrix} -r & 2r+r^2 \\ -r-1 & 2r+r^2 \end{bmatrix}$, whose eigenvalues are $\lambda_{1,2} = \frac{r+r^2 \pm i\sqrt{(4(2r+r^2)-(r+r^2)^2)}}{2}$. Clearly, $\lambda_1(r) = \overline{\lambda_2(r)}$ and $\left|\lambda_1\left(r^* = \frac{\sqrt{5}-2}{2}\right)\right| = 1$, $\frac{d}{dr}(|\lambda_1(r^* = \frac{\sqrt{5}-2}{2})|) \neq 0$. Therefore, the system exhibits a Neimark-Sacker bifurcation at the origin when $r = \frac{\sqrt{5}-2}{2}$.

43. Fixed points: $x^* = 0, \ \pm\sqrt{(r-1)}$. The system has three fixed points for $r > 1$ and one fixed point for $r < 1$. Now, $f'(x) = r - 3x^2$, $f'(0) = r$, $f'(\pm\sqrt{(r-1)}) = r - 3(r-1) = 3 - 2r$. If $r < 1, |f'(0)| < 1$ and $f'(\pm\sqrt{(r-1)}) > 1$, implying $x^* = 0$ is stable and $x_r^* = \pm\sqrt{(r-1)}$ are unstable. If $r > 1$, $|f'(0)| > 1 \ \Rightarrow x^* = 0$ is unstable. For stability of $x_r^* = \pm\sqrt{(r-1)}$, we must have

$$f'(\pm\sqrt{(r-1)}) < 1 \ \Rightarrow |3-2r| < 1 \ \Rightarrow -1 < 3-2r < 1 \ \Rightarrow 1 < r < 2.$$

The system undergoes a pitchfork bifurcation at (x^*, r^*)=(0,1).
2-cycle is stable if $1 < r < 2$.

45. Differentiating $f(x)$ with respect to x, we get

$$\begin{aligned} f'(x) &= r, x < 0 \\ &= -r, x > 0 \end{aligned}$$

Therefore, $|f'(x)| = r$. The Lyapunov exponent of the map under consideration is given by,

$$\lambda = \lim_{n\to\infty} \frac{1}{n} \sum_{i=0}^{n-1} \ln|f'(x_i)| = \ln(r) \lim_{n\to\infty} \frac{1}{n} \times n = \ln(r).$$

For $r \leq 0$, $\lambda = \ln(r) < 0$, the given map is non-chaotic and for $1 < r \leq 2$, $\lambda = \ln(r) > 0$, implying that the given map is chaotic (it can easily be checked that all solutions remain bounded for those values of r).

47. Same as Example (2.7.16). Modify the code given in subsection 2.8.2 to obtain the numerical results.

49. Same as Example (2.7.18). Modify the code given in subsection 2.8.4 to obtain numerical results.

Solutions to Chapter 3

1. $t = 0, \ \pm\sqrt{\frac{k}{\alpha}}$. System is stable at $t = 0$ and unstable at $t = \pm\sqrt{\frac{k}{\alpha}}$.

3. Same as (1).

5. (i) $\frac{b/a}{1-\frac{\left(N_0-\frac{b}{a}\right)}{N_0}e^{bt}}$ (ii) Equilibrium points: 0, b/a. The system is stable for $N^*=0$ and unstable for $N^*=b/a$. (iii) As $t\to\infty, N(t)\to 0$ for $N_0>\frac{b}{a}$ or $N_0<\frac{b}{a}$. For $N_0=\frac{b}{a}, N(t)\to\frac{b}{a}$.

7. $\lambda=2\pm\sqrt{9-k}$. $k=9$ (unstable node), $5<k<9$ (unstable node) , $k>9$ (unstable spiral).

9. $\frac{dc}{dt}=-ke^{-c}\Rightarrow e^{-c(t)}=kt+C_1\Rightarrow c(t)=-\ln(kt+e^{-y_0})$. Therefore, $c(T)=y_0-\ln(kT+e^{-y_0})$. This is now used at the new initial condition to calculate the constant C_1. Now, $e^{-c(T)}=kT+C_1\Rightarrow e^{-y_0+\ln(kT+e^{-y_0})}=kT+C_1\Rightarrow C_1=e^{-y_0+\ln(kT+e^{-y_0})}-kT$. Therefore, $e^{-c(t)}=kt+e^{-y_0+\ln(kT+e^{-y_0})}-kT\Rightarrow c(2T)=-\ln(1+e^{-y_0})kT+e^{-2y_0}$.

11. Same as 3.5.3.

13. (i) Substituting $i=\frac{dQ}{dt}$ in $V-\frac{Q}{C}=iR$, we get, $\frac{dQ}{Q-VC}=-\frac{1}{CR}dt$. Integrate to get the result.
(ii) The circuit contains no source of e.m.f, implies, $iR+\frac{Q}{C}=0\Rightarrow R\frac{dQ}{dt}=-\frac{Q}{C}$.

15. Solve $\frac{dv}{dt}=-kv^2$ and $v\frac{dv}{dx}=-kv^2$.

17. Since, F is the fractional resistance and P is the pull, the differential equation modeling the motion is given by $M\dfrac{dv}{dt}=P-F=\frac{R}{v}-F$ ($Pv=R$). At the maximum velocity, $\frac{dv}{dt}=0\Rightarrow F=\frac{R}{W}$. Then we have,
$M\frac{dv}{dt}=\frac{R}{vW}(W-v)\Rightarrow Mv^2\frac{dv}{dx}=\frac{R}{W}(W-v)\Rightarrow x=\frac{MW}{R}\int_0^V\{-v-W+\frac{W^2}{W-v}\}dv$
$\Rightarrow x=\frac{MW}{R}[-\frac{v^2}{2}-vW-W^2\log(W-v)]_0^V=\frac{MW^3}{R}[\log\frac{W}{W-V}-\frac{V}{W}-\frac{1}{2}\frac{V^2}{W^2}]$.

19. (i) $\frac{d^2x}{dt^2}=-\mu[x+\frac{a^4}{x^3}],\Rightarrow\frac{dx}{dt}=-\frac{\sqrt{x}\sqrt{a^4-x^4}}{x}$, $t_1=\frac{1}{\sqrt{\mu}}\int_0^a\frac{x}{\sqrt{a^4-x^4}}dx=\frac{1}{2\sqrt{\mu}}\int_0^{\frac{\pi}{2}}d\theta=\frac{\pi}{4\sqrt{\mu}}$ $x^2=a^2sin\theta$). For **(ii)** and **(iii)**, follow **(i)**.

21. Equation of motion is given by $\frac{d^2x}{dt^2}=-\frac{\mu_1}{x^2}=-\frac{ga^2}{x^2}$ ($\frac{\mu_1}{a^2}=g$, on the surface of the Earth)$\Rightarrow$ $\left(\frac{dx}{dt}\right)^2=2ga^2\left(\frac{1}{x}-\frac{1}{b}\right)$, (when $x=b, \frac{dx}{dt}=0$). If v_1 be the velocity on reaching the surface, $\frac{dx}{dt}=v_1$, when x = a $\Rightarrow v_1^2=2ag\left(1-\frac{a}{b}\right)$.
Equation of motion (inside the earth):
$\frac{d^2x}{dt^2}=-\mu_2x=-\frac{gx}{a}$, since, $\mu_2a=g$, on the surface of the Earth.
$\Rightarrow\left(\frac{dx}{dt}\right)^2=-\frac{g}{a}x^2+$ Constant. At $x=a, \frac{dx}{dt}=\sqrt{2ag\left(1-\frac{a}{b}\right)}\Rightarrow$ Constant $=ag\left(3-\frac{2a}{b}\right)$. Therefore, $\left(\frac{dx}{dt}\right)^2=-\frac{g}{a}x^2+ag\left(3-\frac{2a}{b}\right)$. On reaching the center of the Earth $x=0$ and we obtain v_2^2.

23. (i) $m\frac{d^2x}{dt^2} = mg - T = mg - 2mg\frac{\frac{a}{x}+x}{a} = -\frac{2mg}{a}x$, (ii) $\frac{(\sqrt{5}+1)a}{2}$,
(iii) $t_1 = \sqrt{\frac{a}{2g}}\int_{-\frac{a}{2}}^{\frac{a\sqrt{5}}{2}} \frac{dx}{\sqrt{\frac{5}{4}a^2-x^2}} = \sqrt{\frac{a}{2g}}\left[\frac{\pi}{2} + \sin^{-1}\frac{1}{\sqrt{5}}\right]$, $t_2 = \frac{2a}{g}$. Required time=$2(t_1 + t_2)$.

25. (a) $\frac{dT}{dt} = k(T_S - T) \Rightarrow T_S + (T_0 - T_S)e^{-kt}$, T_0 is the initial temperature of the body; $k = -0.0123$, $T(t) = 35 + 130e^{-0.0123t}$; 106.68 °F.
(b) 11:12 pm. (c) $\frac{dT}{dt} = k(T_{S0} + \beta t - T)$, $\tau = \frac{1}{k}\ln\left[\frac{k}{\beta}(T_0 - T_{S0}) + 1)\right]$.

27. (a)Rate of increase of salt content due to inflow = 150 × 5 gms/min=750 gms/min. Let k gms be the concentration of salt at time t in water. Therefore, the rate of decrease of salt due to outflow = $k \times 5$ gms/min. The initial volume of water = 1000 liters. The volume of liquid at time t is $(1000 + t)$ litres containing q grams of salt, that is,

$$k = \frac{q}{1000+t} \Rightarrow \text{Rate of decrease of salt} = \frac{5q}{1000+t} \text{ gms/min.}$$

The differential equation modeling this situation is $\frac{dq}{dt} = 750 - \frac{5q}{1000+t}$

$\Rightarrow \frac{dq}{dt} + \frac{5q}{1000+t} = 750$, $q(0) = 0$. The solution of this first order linear differential equation is $q(t) = \frac{125}{(1000+t)^5}(t^6 + 6000t^5 + 15 \times 10^6 t^4 + 20 \times 10^9 t^3 + 15 \times 10^{12} t^2 + 6 \times 10^{15} t)$.

(b) Proceed as (a).

29. $M(t) = M_0 e^{\frac{r}{\alpha}(1-e^{-\alpha t})}$.

31. (i) It is assumed that individuals emigrate from occupied patches and have a certain probability to settle in another patch. The number of migrants that are traveling between the patches was therefore assumed to be proportional to the number of occupied patches, p. Only if the migrants colonize an empty patch, which occurs with probability $(1-p)$, the migration event should be counted as an increase in the fraction of occupied patches. Here, c is the colonization parameter, which represents the number of migrants leaving an occupied path and the rate at which land in another patch, m is the extinction parameter for mortality, which occurs when an occupied patch looses the species of interest.
$p^* = 1 - \frac{m}{c}$, which simply says that the degree at which patches are occupied depends on the ratio of the extinction and colonization parameters. The fraction of empty patches at steady state, that is, $\frac{m}{c}$, reflects the ratio of the extinction and the colonization rate. The colonization parameter will be large if the habitats are located close to each other, and the extinction parameter will be small when the habitats are large.

(ii) The colonization rate was estimated from the average distance between suitable habitats, and was written as $c = c_0 e^{-\alpha D}$, where D is the average distance between the habitats. The extinction rate was estimated from the average size of the habitat, and was written as $m = m_0 e^{-\beta A}$, where A is the average area of the preferred habitats. The parameters α and β describe how strongly colonization and extinction depend on the distance and the area size.
$p^* = 1 - \frac{m_0}{c_0} e^{\alpha D - \beta A}$, which decreases with the distance and increases with the area size. Habitat destruction by removing suitable patches, which increases the average distance D, or by decreasing the size of the habitats, A, are therefore expected to have a similar detrimental effect on the distribution of the butterflies over Finland.

33. $N(t) = \frac{15000}{25e^{-t} - 10}$. As $t \to \infty$, $N(t) \to -1500$, however, in reality, tiger population cannot be negative. Therefore, the tiger population will become extinct with time.

35. (i) When $b = 0$, the system reduces to the classic Lotka-Volterra prey–predator model.
(ii) Equilibrium points are $(0,0)$, $(\frac{k}{\lambda}, \frac{a}{c})$ and $(\frac{k}{\lambda}, \frac{a}{c} - \frac{bk}{c\lambda})$.
(iii) Condition for stability: $a\lambda > bk$. Biologically, it means that the product of the growth rates of the predator and the prey must be greater than the product of their death rates.
(iv) For the given parameters, the eigenvalues are $-0.0125 \pm i0.06$, implying that the system is a stable spiral. The time series solution of the system with $P(0) = 20$ and $S(0) = 15$ is also obtained, which shows that both the sea-turtle coexists.

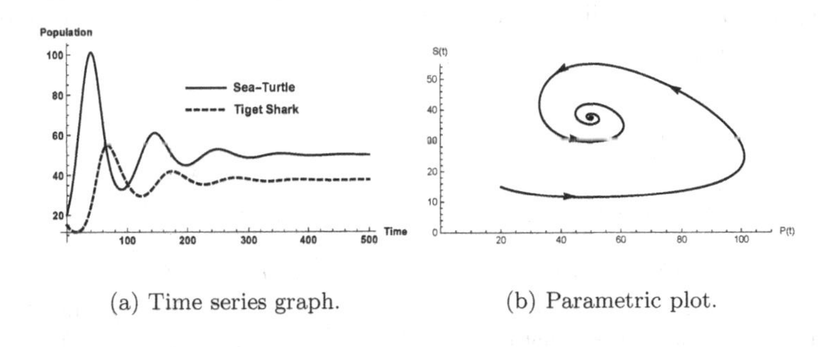

(a) Time series graph. (b) Parametric plot.

FIGURE 7.2: *The figures show coexistence of both the species and stable spiral.*

When $c = 0$, the equilibrium points are $(0,0)$ and $(\frac{a}{b} = 200, 0)$. The system is a saddle (unstable) about both the equilibria.
When $c = 0$ as well as $\lambda = 0$, we get the equilibrium points are $(0,0)$

and ($\frac{a}{b} = 200, 0$). But, now the system is a stable node about $(200, 0)$ and saddle about $(0, 0)$.
(v) The only difference with the previous model is that now there is competition among the predators (intra-specific competition) through the term $(-\sigma S^2)$. The rest of the problem is similar to (i)–(iii).
(vi) Similar to (iv).

37. $\frac{dC}{dt} = -\beta C$ and $\frac{dS}{dt} = \alpha SC$, $C(t) = C_0 e^{-\beta t}$, $S(t) = S_0 e^{(-\frac{\alpha C_0}{\beta} e^{-\beta t})}$.

39. Proceed as in (35).

41. Proceed as 3.5.9.4.

43. (a)(i) This is an example of a food chain, where y eats x and z eats y, but z does not eats x, that is, y is the predator of x and z is the predator of y. ax is the growth term of the prey x, $-ky$, and $-ez$ are death rates of predators due to starvation. c and m are the rates at which y and z eat their respective preys; fxy, syz are the growth term of the predators after they have consumed their respective preys.
(ii) The equilibrium points are (0,0,0), $(0, \frac{e}{s}, \frac{-k}{m})$ and $(\frac{k}{f}, \frac{a}{c}, 0)$. Use Routh-Hurwitz condition, to check asymptotic stability. (iii) Unstable node. (iv) Proceed as (iii). Unstable spiral to stable spiral. Numerical solution left to the reader. (b) Same as (a).

45. (i) Do yourself. (ii) Modify the Mathematica code in 3.10.1 and 3.10.2 for time-series graph and phase portrait.

47. (i) $(0, 0)$ and $\left(\frac{1-\alpha\beta}{\alpha\beta}, \frac{1-\alpha\beta}{\alpha}\right)$. Unstable node and saddle. Modify the Mathematica code in 3.10.1 and 3.10.2 for time-series graph and phase portrait. (ii)(a) Three equilibrium points, (b) Two equilibrium points, (c) $(0, 0)$ is the only equilibrium point.

49. (a) $\frac{d^2y}{dt^2} = -\frac{\mu}{y^2}$, $\frac{d^2x}{dt^2} = 0$. At $t = 0, x = 0, y = 2a$, $\frac{dx}{dt} = \sqrt{\frac{\mu}{a}}$, $\frac{dy}{dx} = -\sqrt{\frac{2a-y}{y}}$. Put $y = 2a\cos^2\theta = 2a(1 + \cos 2\theta)$ to get $x = (2\theta + \sin 2\theta)$, which represents a cycloid. (b) Proceed as 49(a).

51. We consider a function $V(x, y, z) = a_1x^2 + a_2y^2 + a_3z^2$, where a_1, a_2 and a_3 are some positive constants to be determined. Clearly, the function $V(x, y, z)$ is positive definite as $V(x, y, z) > 0 \; \forall \; (x, y, z) \neq (0, 0, 0)$ and $V(0, 0, 0) = 0$.

$$\text{Also, } \dot{V}(x, y, z) = \frac{dV}{dt} = \frac{\partial V}{\partial x}\dot{x} + \frac{\partial V}{\partial y}\dot{y} + \frac{\partial V}{\partial z}\dot{z}$$
$$= [2a_1x\sigma(y - x) + 2a_2y(rx - y - xz) + 2a_3z(xy - bz)]$$
$$= -2a_1\sigma x - 2a_2y^2 - 2a_3bz^2 + xy(2a_1\sigma + 2ra_2) + (2a_3 - 2a_2)xyz.$$

If we choose $a_1 = \frac{1}{\sigma}$, $a_2 = 1$ and $a_3 = 1$, then

$$\dot{V}(x, y, z) = -2\left[x^2 + y^2 + bz^2 - (1+r)xy\right]$$
$$= -2\left[\left(x - \frac{r+1}{2}y\right)^2 + \left(1 - \left(\frac{r+1}{2}\right)^2\right)y^2 + bz^2\right]$$

Since $0 < r < 1, 0 < \frac{r+1}{2} < 1$, implying $\dot{V}(x,y,z) < 0, \forall\ (x,y,z) \neq (0,0,0)$, and $= 0,\ \forall\ (x,y,z) = (0,0,0)$. Therefore, $\dot{V}(x,y,z)$ is negative definite on the entire space R^3 and the system is globally asymptotically stable about the origin.

53. $ML\ddot{\theta} = -Mg\sin(\theta)$. Let $x = \theta$ and y $= \dot{\theta}$, then $\dot{x} = \dot{\theta} = y$, $\dot{y} = \ddot{\theta} = -\frac{g}{L}\sin(\theta)$. Let $V(x,y) = \frac{1}{2}M(L\dot{\theta})^2 + MgL(1-\cos(\theta)) = \frac{1}{2}ML^2y^2 + MgL(1-\cos(x))$, $V(x,y) > 0\ \forall\ (x,y) \neq (0,0)$ and $V(0,0) = 0$, $\dot{V}(x,y) = \frac{dV}{dt} = \frac{\partial V}{\partial x}\dot{x} + \frac{\partial V}{\partial y}\dot{y} = MgL\sin(x)y + ML^2y\left(-\frac{g}{L}\sin(x)\right) = MgLy\sin(x) - MgLy\sin(x) = 0$, $\Rightarrow V(x,y)$ is negative semi-definite, the system is stable (not asymptotically) about the equilibrium point $(0,0)$.

55. (a) $\mu - x = e^{-x}$ and $\frac{d}{dx}(\mu - x) = \frac{d}{dx}(e^{-x})$, implies $-1 = -e^{-x} \Rightarrow x = 0$ and $\mu = 1$. Therefore, the bifurcation occurs at the point $x = 0$ for $\mu = 1$. Saddle-node bifurcation at $\mu = 1$, which occurs at $x = 0$. Modify the Mathematica code in 3.10.3 for bifurcation diagram. (b)$\mu = 2$ is a saddle-node bifurcation point at $x = -1$.

57. (a) The fixed points of the given system are x=0 and $x = \frac{\mu-1}{\mu}$. Bifurcation point for the ODE $\frac{dx}{dt} = x - \mu x(1-x) = f(x,\mu)$ must satisfy $f(x^*) = 0$ and $f'(x^*) = 0$. The system exhibits a transcritical bifurcation at $\mu = 1$, which occurs at $x = 0$. Modify the Mathematica code in 3.10.3 for bifurcation diagram. (b) Proceed as in 57(a).

59. (a) Putting $x = r\ \cos\theta, y = r\ \sin\theta$ and using the relations $x\frac{dx}{dt} + y\frac{dy}{dt} = r\frac{dr}{dt}$ and $x\frac{dy}{dt} - y\frac{dx}{dt} = r^2\frac{d\theta}{dt}$, we obtain,

$$\frac{dr}{dt} = 1 - r^2,\ \frac{d\theta}{dt} = -1 \Rightarrow \frac{1}{2}\ln\frac{1+r}{1-r} + \frac{1}{2}\ln k = t \text{ and } \theta = -t + t_0.$$
$$\Rightarrow \left(\frac{1+r}{1-r}\right)k = e^{2t} \Rightarrow r = \frac{1-ke^{-2t}}{1+ke^{-2t}}.$$

Taking $t_0 = 0$ (since it is arbitrary), we get the solution of the given systems as $x = \frac{1-ke^{-2t}}{1+ke^{-2t}}\cos t, y = -\frac{1-ke^{-2t}}{1+ke^{-2t}}\sin t$ when k=0, we obtain the circle $x^2 + y^2 = 1$. When $k \neq 0$, we obtain non-closed paths with spiral behavior. If $k > 0$, we obtain spirals which lie inside the circle $x^2 + y^2 = 1$ and (i) approaches the circle as $t \to +\infty$ (ii) approaches the fixed point $(0,0)$ of the system as $t \to -\infty$ .If $k < 0$, we again get spirals which lie outside the circle $x^2 + y^2 = 1$ and approaches the circle as $t \to +\infty$; while as $t \to ln\sqrt{(-k)}$, both $|x|$ and $|y|$ become infinite. Thus as $t \to \infty$, the closed path $x^2 + y^2 = 1$ approaches spirally both from inside and outside by non-closed paths. Hence, $x^2 + y^2 = 1$ is a limit cycle of the system. (b) Proceed as 59(a).

61. (a),(b) Proceed as in example 3.6.7. Modify the code of fig. 3.24 to generate figures.

63. (a),(b) Proceed as in example 3.6.7. Modify the code of fig. 3.24 to generate bifurcation figures.

65. (a) Proceed as in example 3.7.3. (b) Proceed as in example 3.7.4.

67. Proceed as in miscellaneous example 3.9.9. Modify the Mathematica code in 3.10.1 and 3.10.2 for time-series graph and phase portrait.

69. Proceed as in example 3.7.5. Initial guess for (iv) is $r = 0.5$, $k = 30$, $\beta = 0.5$, $\gamma = 0.9$.

71. Proceed as in example 3.7.6. Initial Guess: $a_1 = 0.2, b_1 = 0.01, c_1 = 0.04, a_2 = 0.05, b_2 = 0.007, c_2 = 0.005$.

73. Proceed as in example 3.6.7. Modify the code of fig. 3.24 to generate bifurcation figures.

75. Proceed as in miscellaneous example 3.9.9. Modify the Mathematica code in 3.10.1 and 3.10.2 for time-series graph and phase portrait.

77. In the rumor model, I are the ignorant, who have not heard the rumor, S are the spreaders of rumor and R are the repressor, who knows the rumor but no longer spreading it and $N = I+S+R$, the total population of the network remains constant . When an ignorant meets a spreader, it is turned into a new spreader at a rate β (k represents the average number of contacts of each individual). The spreading of rumor decays either due to the forgetting process or because the spreaders who already know the rumor has lost its value. The decaying process occurs when a spreader meets another spreader or a repressor at a rate α.
Equilibrium point: S=0 and under this condition, all I and R such that $I + R = 1$. The system is asymptotically stable if $0 < I < \frac{\alpha}{\alpha+\beta}$ and unstable if $\frac{\alpha}{\alpha+\beta} < I < 1$. Proceed as in miscellaneous example 3.9.9. Modify the Mathematica code in 3.10.1 and 3.10.2 for time-series graph and phase portrait.

79. Proceed as in exercise (48).

81. Proceed as in exercise (79).

83. Proceed as (82).

85. Proceed as (81).

87. (i) Proceed as in miscellaneous example 3.9.9. Modify the Mathematica code in 3.10.1 and 3.10.2 for time-series graph and phase portrait.
(ii) Proceed as in example 3.6.7. Modify the code of fig. 3.24 to generate bifurcation figures.

89. $\frac{dZ}{dt} = (\beta - \alpha)SZ + rR \Rightarrow \frac{dZ}{dt} = (\beta - \alpha)(-\frac{1}{\beta}\frac{dS}{dt}) + rR,$
$\Rightarrow (1 - \frac{\alpha}{\beta})\frac{dS}{dt} = r(n+1) - rZ - rS - \frac{dZ}{dt},$
$\Rightarrow (1 - \frac{\alpha}{\beta})\frac{dS}{dt} + rS = r(n+1) - rZ - \frac{dZ}{dt} = k$ (say),
$\Rightarrow S(t) = \frac{k+(nr-k)e^{\frac{t\beta r}{\alpha-\beta}t}}{r} = \frac{k+(nr-k)e^{-\frac{t\beta r}{\beta-\alpha}}}{r}, \ Z(t) = \frac{e^{-rt}(k-nr)+(n+1)r-k}{r}.$
As $t \to \infty, S(t) \to \frac{k}{r}, Z(t) \to (n+1) - \frac{k}{r},\ R(t) \to, 0,$where
$R(t) = \frac{e^{-rt}}{\beta r^3}[r^2 n(\beta - \alpha)\{r^2 n(\beta - \alpha) + \beta + n\alpha e^{-\frac{\beta r t}{\beta-\alpha}} - (n+1)\beta e^{-\frac{\alpha r t}{\beta-\alpha}}\}$
$+k^2\{\alpha^2 - \alpha\beta + \beta^2 + \alpha\beta r t - \alpha(\alpha - \beta)e^{-\frac{\beta r t}{\beta-\alpha}} - \alpha\beta e^{rt} + (\alpha\beta - \beta^2)e^{-\frac{r\alpha}{\beta-\alpha}}\}$
$+ kr\{3n\alpha\beta - 2n\alpha^2 - \beta^2 - n\alpha\beta r t + n\alpha\beta e^{rt} - 2n\beta^2\} +$
$\{(2n\alpha^2 - 2n\alpha\beta)e^{-\frac{r\beta t}{\beta-\alpha}} + (2n+1)\beta(\beta - \alpha)e^{-\frac{r\alpha t}{\beta-\alpha}}\}].$
Modify the Mathematica code in 3.10.1 for time-series graph.

Solutions to Chapter 4

1. Proceed as in example 4.2.1 and obtain

$$u(x,t) = \sum_{n=1}^{\infty} B_n \sin\left(\frac{n\pi x}{L}\right) e^{\frac{-n^2\pi^2 D}{L^2}t}.$$

(i) Using the initial condition $u(x,0) = \sin\left(\frac{\pi x}{L}\right)$, we obtain

$$\sin\left(\frac{\pi x}{L}\right) = \sum_{n=1}^{\infty} B_n \sin\left(\frac{n\pi x}{L}\right) \Rightarrow B_1 = 1, B_2 = B_3 = ... = 0.$$

(ii) Proceed as (i). For numerical result, modify the code of 4.11.1.

3. The mathematical model of the given situation represents an initial boundary value problem of heat conduction and is given by

$$\frac{\partial u}{\partial t} = D\frac{\partial^2 u}{\partial x^2};\ 0 \le x \le 1, t > 0.$$

Boundary Condition (BCs): $\dfrac{\partial u(0,t)}{\partial x} = 0, u(1,t) = 0;\ t > 0$ (one end is insulated and the other is kept at zero).

Initial Condition (ICs): $u(x,0) = \begin{cases} 1.0, & 0 \le x \le 0.5, \\ 2(1-x), & 0.5 \le x \le 1. \end{cases}$

$$u(x,t) = \sum_{n=1}^{\infty} A_n \cos\left(\frac{(2n-1)\pi x}{2}\right) e^{\frac{-(2n-1)^2}{4}\pi^2 Dt}, \text{ where}$$
$$A_n = \frac{16}{(2n-1)^2\pi^2} \cos\left(\frac{(2n-1)\pi}{4}\right).$$

For numerical result, modify the code of 4.11.2.

5. Mathematically, we can formulate the model as follows:

$$\frac{\partial^2 u}{\partial t^2} = c^2 \frac{\partial^2 u}{\partial x^2}, \; 0 \le x \le L, \; t > 0.$$

Boundary conditions (BCs): $u(0,t) = 0 = u(L,t)$, for all $t \ge 0$.
(a) Initial conditions (ICs): $u(x,0) = \sin\left(\frac{\pi x}{L}\right)$, $\frac{\partial u(x,0)}{\partial t} = 0$.
Solution: $u(x,t) = \sin\left(\frac{\pi x}{L}\right)\cos\left(\frac{\pi ct}{L}\right)$.
(b) Initial condition (ICs): $u(x,0) = \begin{cases} \frac{2hx}{L}, & 0 \le x \le L/2, \\ \frac{2hx}{L-x}, & L/2 \le x \le L. \end{cases}$
Solution: $u(x,t) = \sum_{n=1}^{\infty} A_n \sin\left(\frac{n\pi x}{L}\right)\cos\left(\frac{n\pi ct}{L}\right)$, where $A_n = \frac{8h}{n^2\pi^2}\sin\left(\frac{n\pi x}{L}\right)$.
If $n = 2m-1$ (odd) with $m = 1, 2, 3, ..., A_{2m-1} = \frac{8h(-1)^{m+1}}{(2m-1)^2\pi^2}$.
If $n = 2m$ (even) with $m = 1, 2, 3, ..., A_{2m} = 0$.
For numerical result, modify the code of 4.11.3.

7 (i),(ii) Proceed as in example 4.6.1

$$A_{mn} = \frac{4}{\pi^2}\int_{x=0}^{\pi}\int_{y=0}^{\pi} [\cos(x-y) - \cos(x+y)] \; \sin(mx) \; \sin(my) \; dy \; dx$$
$$\Rightarrow A_{mn} = \frac{4}{\pi^2}\int_{x=0}^{\pi}\int_{y=0}^{\pi} 2\sin(x) \; \sin(y) \; \sin(mx) \; \sin(my) \; dy \; dx$$
$$\Rightarrow A_{mn} = \frac{2}{\pi^2}\int_{x=0}^{\pi} 2\sin(x) \; \sin(mx) \; dx \; \times \int_{y=0}^{\pi} 2\sin(y) \; \sin(my) \; dy$$
$$\Rightarrow A_{mn} = \frac{2}{\pi^2}\left[\frac{2\sin(m\pi)}{1-m^2} \times \frac{2\sin(n\pi)}{1-n^2}\right] = 0, \; m \ne 1, n \ne 1$$
$$\Rightarrow A_{mn} = \frac{2}{\pi^2} \times \pi \times \pi = 2, \; m = 1, n = 1$$
$$\lambda_{mn}{}^2 = \lambda_{11}{}^2 = \pi^2 \; k^2 \left(\frac{1^2}{\pi^2} + \frac{1^2}{\pi^2}\right) = 2k.$$

Therefore, the required solution is $u(x,y,t) = 2\sin(x)\sin(y)e^{-2kt}$.

(iv) $u(x,y,t) = 10\sin\left(\frac{\pi x}{a}\right)\sin\left(\frac{\pi y}{b}\right)e^{-\pi^2 k\left(\frac{1}{a^2}+\frac{1}{b^2}\right)t}$.
For numerical result, modify the code of 4.11.3.

9. Same as corollary 1 of Section 4.8.

11. (i) $P(x,t) = \frac{P_0}{2\sqrt{\pi Dt}}\ exp\left(\alpha t - \frac{x^2}{4Dt}\right)$, where $P(0,0) = P_0$.
    (ii) $\frac{k}{P_0} = \frac{1}{2\sqrt{\pi Dt}}\ exp\left(\alpha t - \frac{x^2}{4Dt}\right) \Rightarrow \ln\left(\sqrt{2\pi D}\frac{k}{P_0}\right) = \ln\left(\frac{1}{\sqrt{2t}}\right) + \alpha t - \frac{x^2}{4Dt}$
    $\Rightarrow \frac{x^2}{t^2} = 4D\alpha - \frac{2D}{t}\ln(t) - \frac{4D}{t}\ln\left(\sqrt{2\pi Dt}\frac{k}{P_0}\right)$.

13. Proceed as in problem 4.12.8.

15. Proceed as in example 4.10.2. Modify the code 4.11.5 to obtain the numerical results.

17. Proceed as in example 4.10.1. Modify the code 4.11.5 to obtain the numerical results.

19. Proceed as in corollary of section 4.10.2. Modify the code 4.11.5 to obtain the numerical results for both 1D and 2D.

21. Proceed as in example 4.10.2. Modify the code 4.11.6 to obtain the numerical results.

23. Modify the code 4.11.5 to obtain the density plots or contour plots.

25. Modify the code 4.11.5 to obtain the numerical results.

Solutions to Chapter 5

1. Proceed as in Problem 5.5.3. Modify the Mathematica code 5.6.1 to get numerical results.

3. Modify the Mathematica code 5.6.1 to get numerical graphs.

5. Proceed as in Problem 5.5.3. Modify the Mathematica code 5.6.1 to get numerical results.

7. Proceed as in Problem 5.5.3. Modify the Mathematica code 5.6.1 to get numerical results.

9. Proceed as in Problem 5.3.5. Modify the Mathematica code 5.6.1 to get numerical results.

11. Proceed as in Problem 5.5.7. Modify the Mathematica code 5.6.2 to get numerical results.

13. Proceed as in Problem 5.5.5. Modify the Mathematica code 5.6.2 to get numerical results.

15. Proceed as in Problem 5.3.5. Modify the Mathematica code 5.6.1 to get numerical results.

17. Mathematical Code is as follows:

$$
\begin{aligned}
&\text{sol1} = \text{First}[\text{NDSolve}[\{\text{y1}'(t) = \text{y3}(t-1) + \text{y5}(t-1),\\
&\text{y2}'(t) = \text{y1}(t-1) + \text{y2}(t-0.5),\\
&\text{y3}'(t) = \text{y1}(t-0.5) + \text{y3}(t-1),\\
&\text{y4}'(t) = \text{y4}(t-1) + \text{y5}(t-1),\\
&\text{y5}'(t) = \text{y1}(t-1),\\
&\text{y1}(t/;t \leq 0) = \exp(t+1), \text{y2}(t/;t \leq 0) = \exp(t+0.5),\\
&\text{y3}(t/;t \leq 0) = \sin(t+1), \text{y4}(t/;t \leq 0) = \exp(t+1),\\
&\text{y5}(t/;t \leq 0) = \exp(t+1)\}, \{\text{y1}, \text{y2}, \text{y3}, \text{y4}, \text{y5}\}, \{t, 0, 5\}]]\\
&\text{Plot}[\text{Evaluate}[\{\text{y1}(t), \text{y2}(t), \text{y3}(t), \text{y4}(t), \text{y5}(t)\}/.\,\text{sol1}], \{t, 0, 5\},\\
&\text{PlotRange} \to \text{All}, \text{AxesLabel} \to \{\text{t}, \text{y}\}, \text{PlotLegends} \to\\
&\text{Placed}[\{\text{y1}, \text{y2}, \text{y3}, \text{y4}, \text{y5}\}, \{\text{Scaled}[\{0, 0.5\}], \{0, 0.5\}\}],\\
&\text{BaseStyle} \to \{\text{FontWeight} \to \text{Bold}, \text{FontSize} \to 12\}]
\end{aligned}
$$

19. Proceed as in Problem 5.5.6. Modify the Mathematica code 5.6.2 to get numerical results.

Solutions to Chapter 6

1. Modify the Mathematica code 6.4.1 to get numerical results.

3. Modify the Mathematica code 6.4.1 to get numerical results.

5. Modify the Mathematica code 6.4.2 to get numerical results.

7. Modify the Mathematica code 6.4.2 to get numerical results.

9. Modify the Mathematica code 6.4.2 to get numerical results.

Bibliography

[1] J. Abramson. *https://courses.lumenlearning.com/ivytech-collegealgebra /chapter/build-a-logistic-model-from-data/.* 2013.

[2] V.N. Afanasev, V.B. Kolmanowskii, and V.R. Nosov. *Mathematical Theory of Control Systems Design.* Kluwer, Dordrecht, 1996.

[3] E. Ahmed, A. El-Misiery, and H.N. Agiza. On controlling chaos in an inflation–unemployment dynamical system. *Chaos Solitons Fractals*, 10(9):1567–1570, 1999.

[4] Brian Albright and W.P. Fox. *Mathematical Modeling with Excel.* CRC Press, Taylor & Francis Group, New York, 2019.

[5] E. Alec Johnson. Traffic flow: Deriving a partial differential equation from a global conservation law, http://www.danlj.org/eaj/math/ summaries/trafficflow/trafficPDE.pdf, 2010.

[6] E. Allen. *Modeling with Itô Stochastic Differential Equations.* Springer, Dordrecht, The Netherlands, 2007.

[7] L.J. Allen. *An Introduction to Stochastic Processes with Applications to Biology.* Chapman and Hall/CRC, Boca Raton, FL, 2011.

[8] Isaac Amidror and D. Roger Hersch. Mathematical moire' models and their limitations. *Journal of Modern Optics*, 57(1):23–36, 2010.

[9] K.L. Babcock and R.M. Westervelt. Dynamics of a simple electronic neural networks. *Physica D: Non-linear Phenomena*, 28(3):305–316, 1987.

[10] M. Bando. Dynamical model of traffic congestion and numerical simulation. *Physical Review E*, 51:1035–1042, 1995.

[11] Sandip Banerjee. Immunotherapy with Interleukin-2: A study based on mathematical modeling. *International Journal of Applied Mathematics and Computer Science*, 18(3):1–10, 2008.

[12] R.B. Banks and T. Icebergs. *Falling Dominoes and Other Adventures in Applied Mathematics.* Princeton University Press, Princeton, NJ, 1998.

[13] J.H. Barrett and J.S. Bradley. *Ordinary Differential Equations.* International Text Book Company, Scranton, PA, 1972.

[14] M. Barrio, K. Burrage, A. Laie, and T. Tian. Oscillatory regulation of Hes1: Discrete stochastic delay modelling and simulation. *PLoS Computational Biology*, 2(9):1017–1030, 2006.

[15] E. Beretta, V. Kolmanowskii, and L. Shaikhet. Stability of epidemic model with time delays influenced by stochastic perturbations. *Mathematical Computation and Simulation*, 45:269–277, 1998.

[16] S.P. Blythe, R.M. Nisbet, and W.S.C. Gurney. Instability and complex dynamics behavior in population models with long time delays. *Theoretical Population Biology*, 22:147–176, 1982.

[17] P. A. Braza. Predator prey dynamics with square root functional responses. *Nonlinear Analysis: Real World Applications*, 13:1837–1843, 2012.

[18] P. Brunovsky, A. Erdelyi, and H.O. Walther. On a model of the currency exchange rate-local stability and periodic solutions. *Journal of Dynamics and Differential Equations*, 16(2):393–432, 2004.

[19] S. Busenberg and K.L. Cooke. Periodic solutions of a periodic nonlinear delay differential equation. *SIAM Journal of Applied Mathematics*, 35:704–721, 1978.

[20] H.M. Byrne, S.M. Cox, and C.E. Kelly. Macrophage-tumor interactions: In vivo dynamics. *Discrete and Continuous Dynamical System: Series B*, 4:81–89, 2004.

[21] M. Carletti. On the stability properties of a stochastic model for phage-bacteria interaction in open marine environment. *Mathematical Biosciences*, 175:117–129, 2002.

[22] R.F. Costantito, R.A. Deshrnals, J.M. Cushing, and B. Dennis. Chaotic dynamics in an insect population. *Science*, 275:389–391, 1995.

[23] E. Cummins. *https://study.com/academy/lesson/chaos-theory-definition -history-examples.html.* 2003.

[24] J.H.P. Dawes and M.O. Souza. A derivation of Holling's type I,II,III functional responses in predator-prey systems. *Journal of Theoretical Biology*, 327:11–22, 2013.

[25] R.H. Day. Irregular growth cycles. *American Economic Review*, 72:406–414, 1982.

[26] R.J. De Boer and P. Hogeweg. Interactions between macrophages and T-lymphocytes: Tumor sneaking through intrinsic to helper T cell dynamics. *Journal of Theoretical Biology*, 120(3):331–344, 1985.

[27] R.J. De Boer, P. Hogeweg, H.F. Dullens, R.A. De Weger, and W. Den Otter. Macrophage T-lymphocyte interactions in the anti-tumor immune response: A mathematical model. *Journal of Immunology*, 134(4):2748–2759, 1985.

[28] R.L. Devaney. *An Introduction to Chaotic Dynamical Systems.* CRC Press, Taylor & Francis Group, FL, 2018.

[29] K.I. Diamantaras and S.Y. Kung. *Principal Component Neural Networks: Theory and Applications.* John Wiley & Sons, New York, 1996.

[30] Leah Edelstein-Keshet. *Mathematical Models in Biology.* SIAM: Society for Industrial and Applied Mathematics, PHL, 1988.

[31] Saber Elaydi. *An Introduction to Difference Equations.* Springer, NY, 2005.

[32] I. Elishakoff. Differential equations of love and love of differential equations. *Journal of Humanistic Mathematics*, 9(2):226–246, 2019.

[33] S.F. Ellermeyer, J. Hendrix, and N. Glasochen. A theoretical and empirical investigation of delayed growth response in the continuous culture of bacteria. *Journal of Theoretical Biology*, 222:485–494, 2003.

[34] T. Erneux. *Applied Delay Differential Equations.* Springer, New York, 2009.

[35] T. Everson and W. Cole. *Spontaneous Regression of Cancer.* Saunders, PHL, 1966.

[36] Marotto F.R. *Introduction to Mathematical Modeling Using Discrete Dynamical Systems.* Thomson Brooks/Cole, CA, 2006.

[37] H.I. Freedman, L.H. Erbe, and V.S.H. Rao. Three species food chain models with mutual interference and time delays. *Mathematical Biosciences*, 80:57–80, 1986.

[38] H.I. Freedman and V. Sree Hari Rao. The trade-off between mutual interference and time lags in predator-prey systems. *Bulletin of Mathematical Biology*, 45:109–121, 1983.

[39] M. Galach. Dynamics of the tumor immune system competition –the effect of time delay. *International Journal of Applied Mathematics and Computer Science*, 13(3):395–406, 2003.

[40] B.L. Gause, M. Sznol, W.C. Kopp, J.E. Janik, J.W. Smith, R.G. Steis, W.J. Urba, W. Sharfman, R.G. Fenton, S.P. Creekmore, J. Holmlund, K.C. Conlon, L.A. VanderMolen, and D.L. Longo. Phase I study of subcutaneously administered interleukin-2 in combination with

interferon alfa-2a in patients with advanced cancer. *Journal of Clinical Oncology*, 14(8):2234–2241, 1996.

[41] G.F. Gause. Experimental studies on the struggle for existence. I. Mixed populations of two species of yeast. *Journal of Experimental Biology*, 9:389–402, 1998.

[42] Michael Mesterton Gibbons. *A Concrete Approach to Mathematical Modelling*. Wiley-Interscience, NY, 2007.

[43] I.I. Gikhman and A.V. Skorokhod. *The Theory of Stochastic Processes I*. Springer, Berlin, 1974.

[44] I.I. Gikhman and A.V. Skorokhod. *The Theory of Stochastic Processes II*. Springer, Berlin, 1975.

[45] I.I. Gikhman and A.V. Skorokhod. *The Theory of Stochastic Processes III*. Springer, Berlin, 1979.

[46] L. Glass and M.C. Mackey. *From Clocks to Chaos*. Princeton University Press, Princeton, PA, 1988.

[47] Samuel Goldberg. *Introduction to Difference Equations*. Dover Publications, NY, 1986.

[48] J. Golec and S. Sathananthan. Stability analysis of a stochastic logistic model. *Mathematical and Computer Modeling*, 38:585–593, 2003.

[49] K. Gopalswamy. *Stability and Oscillation in Delay Differential Equations of Population Dynamics*. Kluwer Academic, Dordrecht, 1992.

[50] K. Gopalswamy, M.R.S. Kulenović, and G. Ladas. Time lags in a food-limited population model. *Applied Analysis*, 31:225–237, 1988.

[51] K. Gopalswamy and G. Ladas. On the oscillation and asymptotic behavior of $\dot{N}(t) = N(t)[a + bN(t-\tau) - CN^2(t-\tau)]$. *Quarterly of Applied Mathematics*, 48:433–440, 1990.

[52] B.F. Gray and N.A. Kirwan. Growth rates of yeast colonies on solid media. *Biophysical Chemistry*, 1:204–213, 1974.

[53] C. Guanrong, J. Fang, Y. Hong, and H. Qin. Controlling Hopf Bifurcations: Discrete-Time Systems. *Discrete Dynamics in Nature and Society*, 5(1):29–33, 2000.

[54] Richard Haberman. *Mathematical Models: Mechanical Vibrations, Populations Dynamics and Traffic Flow*. SIAM: Society for Industrial and Applied Mathematics, PHL, 1996.

[55] J. Hale and S.V. Lunel. *Introduction to Functional Differential Equations*. Springer-Verlag, New York, 1993.

[56] I. Hanski. Single-species metapopulation dynamics: Concepts, models and observations. *Biological Journal of the Linnean Society*, 42:17–38, 1991.

[57] R.J. Henry, Z.N. Masoud, A.H. Nayfeh, and D.T. Mook. Cargo pendulation reduction on ship-mounted cranes via boom-luff angle actulation. *Journal of Vibration and Control*, 7:1253–1264, 2001.

[58] D.L. Hetrick. *Dynamics of Nuclear Reactors*. The University of Chicago Press, Chicago, 1971.

[59] C.S. Holling. Some characteristics of simple types of predation and parasitism. *The Canadian Entomologist*, 91:385–398, 1959.

[60] P. Howard. *https://www.math.tamu.edu/ phoward/m442/modbasics.pdf.* 2009.

[61] A. Hunt and A. Sykes. *Chemistry Pupil's Book.* Longman Group Ltd., FL, 1984.

[62] I.D. Huntley and D.J.G. James. *Mathematical Modelling, A Source Book of Case Studies.* Oxford University Press, NY, 1990.

[63] K.M. Ibrahim, R.K. Jamal, and F.H. Ali. Chaotic behaviour of the Rossler model and its analysis by using bifurcations of limit cycles and chaotic attractors. *IOP Conf. Series: Journal of Physics: doi :10.1088/1742-6596/1003/1/012099*, 2018.

[64] ifi. *https://www.ifi.uzh.ch/dam/jcr:00000000-2826-155d-0000-00002be3 a972/i-chapter7.pdf.* 2003.

[65] Vakalis Ignatiosaf. Pharmacokinetics: Mathematical Analysis of Drug Distribution in Living Organisms, http://www.capital.edu /uploadedFiles/Capital/Academics/ Schools and Departments/Natural Sciences, Nursing and Health/ Computational Studies/Educational Materials/Mathematics/Pharmacokinetics.pdf.

[66] K. Ikeda. Multiple-valued stationary state and its instability of the transmitted light by a ring cavity system. *Optics Communications*, 30(2):257–261, 1979.

[67] K. Ikeda, H. Daido, and O Akimoto. Optical turbulence: Chaotic behavior of transmitted light from a ring cavity. *Physical Review Letters*, 45(9):709–712, 1980.

[68] D. Israelsson and A. Johnsson. A theory for circumnutations in *Helianthus annuus*. *Plant Physiology*, 20:957–976, 1967.

[69] J. J. Arino, L. Wang, and G. Wolkowicz. An alternative formulation for a delayed logistic equation. *Journal of Theoretical Biology*, 241:109–119, 2006.

[70] S.D. Johnson and K. Bowers. The burglary as clue to the future: The beginnings of prospective hot-spotting. *European Journal of Criminology*, 1:237–255, 2004.

[71] S.D. Johnson and K. Bowers. Domestic burglary repeats and space-time clusters: The dimensions of risk. *European Journal of Criminology*, 2:67–92, 2005.

[72] S.D. Johnson, K. Bowers, and A. Hirschfield. New insights into the spatial and temporal distribution of repeat victimisation. *British Journal of Criminology*, 37:224–244, 1997.

[73] J.N. Kapur. *Mathematical Modelling.* New Age International Pvt. Ltd. Publishers, India, 1988.

[74] T. Kardi. *https://people.revoledu.com/ kardi/ tutorial/ DifferenceEquation /WhatIsDifferenceEquation.htm.* 2015.

[75] P. Kareiva and G. Odell. Swarma of predators exhibit preytaxis if individual predators use area-restricted search. *The Americal Naturalist*, 130:233–270, 1987.

[76] D. Kartofelev. *https://www.ioc.ee/ dima/YFX1520/LectureNotes-7.pdf.* 2020.

[77] Therese Keane. Combat modelling with partial differential equations. *Applied Mathematical Modelling*, 35:2723–2735, 2011.

[78] J. Keener and J. Sneyd. *Mathematical Physiology II: Systems Physiology.* Springer, New York, 2009.

[79] U. Keilholz, C. Scheibenbogen, E. Stoelben, H.D. Saeger, and W. Hunstein. Immunotherapy of metastatic melanoma with interferon-alpha and interleukin-2: Pattern of progression in responders and patients with stable disease with or without resection of residual lesions. *European Journal of Cancer*, 30A(7):955–958, 1994.

[80] Andre A. Keller. Population biology models with time-delay in a noisy environment. *WSEAS Transactions on Biology and Biomedicine*, 8(4):113–134, 2011.

[81] W.O. Kermack and A.G. McKendrick. A contribution to the mathematical theory of epidemics. *Proceedings of Royal Society London*, 115:700–721, 1927.

[82] M.A. Khanday, A. Rafiq, and K. Nazir. Mathematical models for drug diffusion through the compartments of blood and tissue medium. *Alexandria Journal of Medicine*, 53:245–249, 2017.

[83] Denise Kirschner and John Carl Panetta. Modeling immunotherapy of the tumor immune interaction. *Journal of Mathematical Biology*, 37:235–252, 1998.

[84] P.E. Kloeden and E. Platen. *Numerical Solution of Stochastic Differential Equations.* Springer, Berlin, 1995.

[85] Y. Kuang. *Delay Differential Equations with Applications in Population Dynamics.* Academic Press, NY, 1993.

[86] F.W. Lanchester. Mathematics in Warfare. *In the World of Mathematics*, 4:2138–2160, 1956.

[87] H.P. Langtangen. Numerical solution of first passage problems in random vibration. *SIAM Journal of Scientific Computing*, 15:977–996, 1994.

[88] H.P. Langtangen. Modeling the BUX index by a novel stochastic differential equation. *Physica A*, 299:273–278, 2001.

[89] A. Lasota. Ergodic problems in biology. *Asterisque*, 50:239–250, 1977.

[90] G.C. Layek. *An Introduction to Dynamical Systems and Chaos.* Springer, NY, 2015.

[91] B. Lehmann, J. McEwen, B. Lane, and S. Rai. http://staff.ulsu.ru/ semoushin/ index/ pilocus/ gist/ docs/ mycourseware/ 1-basmod/ 5-assignments/group projects/Group-project-assignments/modifying the richardson arms race model.pdf, 2014.

[92] I. Lengyel, G. Ribai, and I.R. Epstein. Experimental and modeling study of oscillations in the chlorite-iodide-malonic-acid reaction. *Journal of American Chemical Society*, 112:9104–9110, 2010.

[93] R. Levin. Some demographic and genetic consequences of environmental heterogeneity for biological control. *Bulletin of the Entomological Society of America*, 15:237–240, 1969.

[94] R. Levin and D. Culver. Regional coexistence of species and competition between rare species. *PNAS*, 68:1246–1248, 1971.

[95] E.R. Lewis. *Network Models in Population Biology.* Springer-Verlag, NY, 1977.

[96] L.F. Shampine. Solving Delay Differential Equations with dde23, http://www.radford.edu/ thompson/webddes/tutorial.pdf.

[97] W.F. Libby. Radiocarbon dating. *American Scientist*, 44:98–112, 1956.

[98] E.T. Lofgren, K.M. Collins, T.C. Smith, and R.A. Cartwright. Equations of the end: Teaching mathematical modeling using the zombie apocalypse. *Journal of Microbiology and Biology Education*, 17(1):137–142, 2016.

[99] E.N. Lorenz. Deterministic nonperiodic flow. *Journal of the Atmospheric Sciences*, 20:130–141, 1963.

[100] D. Ludwid, D.D. Jones, and C.S. Holling. Qualitative analysis of insects outbreak systems: The spruce budworm and forest. *Journal of Animal ecology*, 47:315–332, 1978.

[101] D. Ludwig, D.G. Aronson, and H.F. Weinberger. Spatial patterning of the Spruce Budworm. *Journal of Mathematical Biology*, 8:217–258, 1979.

[102] N. MacDonald. *Time Lags in Biological Models.* Springer-Verlag, Heidelberg, 1978.

[103] J.M. Mahaffy. *https://jmahaffy.sdsu.edu/courses/f11/math636/ beamer lectures/dyn sys-04.pdf.* 2011.

[104] D.D. Majumder and P. Roy. Cancer self remission and tumor instability–a cybernetic analysis: Towards a fresh paradigm for cancer treatment. *Cybernetics*, 29:896–905, 2000.

[105] Xuerong Mao. *Stochastic Differential Equations and Applications.* Woodhead Publishing, Cambridge, 2007.

[106] Frederick R. Marotto. *Introduction to Mathematical Modeling using Discrete Dynamical Systems.* Thomson Brooks/Cole, 2006.

[107] Z.N. Masoud, A.H. Nayfeh, and A. Al-Mousa. Delayed position feedback controller for the reduction of payload pendulations of rotary cranes. *Journal of Vibration Control*, 9:257–277, 2003.

[108] N.A. Monk. Oscillatory expression of Hesl, ps3 and NF-kappaB driven by transcriptional time delays. *Current Biology*, 13:1409–1413, 2003.

[109] V.G. Nazarenko. Influence of delay on auto-oscillations in cell populations. *Biofisika*, 21:352–356, 1976.

[110] K.W. Neves. Automatic integration of functional differential equations: An approach. *ACM Transactions on Mathematical Software*, 1:357–368, 1975.

[111] J.W. Nevile. The mathematical formulation of Harrod's Growth Model. *The Economic Journal*, 72(286):367–370, 1962.

[112] NIST. *https://doi.org/10.18434/M32189.* 2013.

[113] D.Q. Nykamp. *http://mathinsight.org/definition/chaos.* 2020.

[114] H. Obanawa and Y. Matsukura. Mathematical modeling of talus development. *Computers and Geosciences*, 32(9):1461–1478, 2006.

[115] J.R. Ockendon. *Mathematical Modelling in Steel Industry.* Oxford Centre for Industrial and Applied Mathematics, University of Oxford, Oxford, 1996.

[116] B. Oksendal. *Stochastic Differential Equations: An Introduction with Applications.* Springer, NY, 2003.

[117] M. Olinick. *Mathematical Modeling in the Social and Life Sciences.* Wiley, NY.

[118] Omstavan. *https://steemit.com/steemstem/@omstavan/love-affairs-a-mathematical-model-using-differential-equations.* 2017.

[119] B. O'Regan and C. Hirschberg. Spontaneous Remission, Institute of Noetic Sciences, Sausalito, CA, 1992.

[120] A.B. Özer and E Akin. Tool for Detecting Chaos. *SAÜ Fen Bilimleri Enstitüsü*, 9(1):60–66, 2005.

[121] F. Pasemann and N. Stollenwerk. Attractor switching by neural control of chaotic neurodynamics. *Network: Computation in Neural Systems*, 9(4):549–561, 1998.

[122] B. Pena and C. Perez-Garcia. Stability of Turing patterns in the Brusselator model. *Physical Review E*, 64:1–9, 2001.

[123] L. Perko. *Differential Equations and Dynamical Systems.* Springer, NY, 2006.

[124] E. Pinney. *Ordinary Difference-Differential Equations.* University of California Press, Berkeley, 1958.

[125] J.R.C Piqueira. Rumor propagation model: An equilibrium study. *Mathematical Problems in Engineering*, 2010:1–7, 2010.

[126] A. Presa, A. Hayes, and C Lin. https://brilliant.org/ wiki/ linear recurrence relations, 2020.

[127] Anatol Rapoport. Lewis F. Richardson's mathematical theory of war. *Journal of Conflict Resolution*, I:249–299, 1957.

[128] A. Rapp and R.W. Fairbridge. Geomorphology: Talus fan or cone; Scree and cliff debris. 1997.

[129] B.C. Richardson. Limitations on the use of Mathematical Models in Transportation Policy Analysis (Technical Report), 1979.

[130] L.F. Richardson. Generalized foreign politics. *British Journal of Psychology (Monogram Supplement)*, 23:98–112, 1939.

[131] S. Rinaldi. Love dynamics: The case of linear couples. *Applied Mathematics and Computation*, 95(2-3):181–192, 1998.

[132] G. Rohdenburg. Fluctuations in malignant tumors with spontaneous recession. *Journal of Cancer Research*, 3:193–201, 1981.

[133] S.A. Rosenberg and M.T. Lotze. Cancer immunotherapy using interleukin-2 and interleukin-2-activated lymphocytes. *Annual Review of Immunology*, 4:681–709, 1986.

[134] S.A. Rosenberg, J.C. Yang, S.L. Topalian, D.J. Schwartzentruber, J.S. Weber, D.R. Parkinson, C.A. Seipp, J.H. Einhorn, and D.E. White. Treatment of 283 consecutive patients with metastatic melanoma or renal cell cancer using high-dose bolus interleukin 2. *Journal of the American Medical Association*, 12:907–913, 1994.

[135] M. Rosenstein, S.E. Ettinghousen, and S.A. Rosenberg. Extravasation of intravascular fluid mediated by the systemic administration of recombinant interleukin 2. *Journal of Immunology*, 137(5):1735–1742, 1986.

[136] S.L. Ross. *Differential Equations.* Wiley, NY, 2007.

[137] O. E. Rössler. An equation for continuous chaos. *Physics Letters A*, 57(5):397–398, 1976.

[138] M.R. Roussel. Morgan & Claypool Publishers, VT.

[139] P. Roy and J. Biswas. Biological parallelism in spontaneous remission of tumor and neurogranuloma. *Tumor Biology*, 44(3):333–340, 1996.

[140] P. Roy, D.D. Majumder, and J. Biswas. Spontaneous cancer regression: Implication for fluctuation. *Indian Journal of Physics*, 73B(5):777–785, 1999.

[141] P. Roy and P.K. Sen. A dynamical analysis of spontaneous cancer regression. *Journal of Investigative Medicine*, 44(3):333–346, 1996.

[142] S. Ruan. Delay differential equations in single species dynamics (pp. 477–517), in Arino et al. eds. *Delay Differential Equations and Applications.* Springer, Netherlands, 2006.

[143] J. Sandefur. *Elementary Mathematical Modeling.* Thomson Brooks/Cole, 2003.

[144] J. Sandefur. *Elementary Mathematical Modeling, A Dynamic Approach.* Brooks Cole, MO, 2003.

[145] R.R. Sarkar and S. Banerjee. Cancer self remission and tumor stability–a stochastic approach. *Mathematical Biosciences*, 196:65–81, 2005.

[146] K. Sasaki. *https://www.lewuathe.com/covid 19 dynamics with sir model.html.* 2020.

[147] Hermann Schichl. Models and History of Modeling, www.mat.univie.ac.at/herman/papers/modtheoc.pdf.

[148] Douglas J. Schwartzentruber. In vitro predictors of clinical response in patients receiving interleukin-2-based immunotherapy. *Current Opinion in Oncology*, 5:1055–1058, 1993.

[149] L.A. Segel and J.L. Jackson. Dissipative structure: An explanation and an ecological example. *Journal of Theoretical Biology*, 37:545–559, 1972.

[150] V. Shapot. *Biochemical Aspects of Tumor Growth* (English translation). Mir Publishers, Moscow, 1990.

[151] M.B. Short, M.R. Dorsogna, V.B. Pasour, G.E. Tita, P.J. Brantingham, A.L. Bertozzi, and L.B. Chayes. A statistical model of criminal behavior. *Mathematical Models and Methods in Applied Sciences*, 18:1249–1267, 2008.

[152] K.N. Singh. Critical decisions in new product introduction and development – a mathematical modeling approach. *Journal of Academy of Business and Economics*, 4:10–16, 2004.

[153] H. Smith. *An Introduction to Delay Differential Equations with Applications to the Life Sciences.* Springer, New York, 2010.

[154] S.H. Strogatz. *Nonlinear Dynamics and Chaos.* Westview Press, CO, 2014.

[155] S. Suherman, R.H. Plaut, L.T. Watson, and S. Thompson. Effect of human response time on rocking instability of a two wheeled suitcase. *Journal of Sound and Vibration*, 207:617–625, 1997.

[156] E.R. Sullivan. *Introduction to Mathematical Modeling.* Creative Commons Attribution, 2018.

[157] M.S.W. Sunaryo, Z. Salleh, and M. Mamat. Mathematical model of three species food chain with holling type III functional response. *International Journal of Pure and Applied Mathematics*, 89(5):647–657, 2013.

[158] M. Tabor. *Chaos and Integrability in Nonlinear Dynamics: An Introduction.* Wiley, New York, 1989.

[159] E. Tartour, J.Y. Blay, T. Dorval, B. Escudier, V. Mosseri, J.Y. Douillard, L. Deneux, I. Gorin, S. Negrier, C. Mathiot, P. Pouillart, and W.H. Fridman. Predictors of clinical response to interleukin-2-based immunotherapy in melanoma patients: A French multi-institutional study. *Journal of Clinical Oncology*, 14(5):1697–1703, 1996.

[160] C.E. Taylor and R.R. Sokal. Oscillations in housefly population sizes due to time lags. *Ecology*, 57:1060–1067, 1976.

[161] W.F. Walsh. Compstat: An analysis of an emerging police managerial paradigm. *Policing*, 24:347–362, 2001.

[162] J. Wauer, D. Schwarzer, G.Q. Cai, and Y.K. Lin. Dynamical models of love with time-varying fluctuations. *Applied Mathematics and Computation*, 188:1535–1548, 2007.

[163] M. Wazewska-Czyzewsha and A. Lasuta. Mathematical problems of the dynamics of the red blood cells system. *Annales Polish Mathematical Society III Applied Mathematics*, 31:23–40, 1976.

[164] D. Wei and S. Ruan. Stability and bifurcation in a neural network model with two delays. *Physica D: Non-linear Phenomena*, 130:255–272, 1999.

[165] J.Q. Wilson and G.L. Kelling. Broken windows and police and neighborhood safety. *Atlantic Monthly*, 249:29–38, 1982.

[166] X. Yang, L. Chen, and J. Chen. Permanence and positive periodic solution for the single-species nonautonomous delay diffusive model. *Computer and Mathematics with Application*, 32:109–121, 1996.

[167] J.E. Zhan and J. Shu. Pricing S and P 500 index options with Heston's model. *IEEE*, 299:85–92, 2003.

Index